modern
carpentry

building construction details
in easy-to-understand form

by

WILLIS H. WAGNER

Professor Emeritus, Industrial Technology
University of Northern Iowa, Cedar Falls

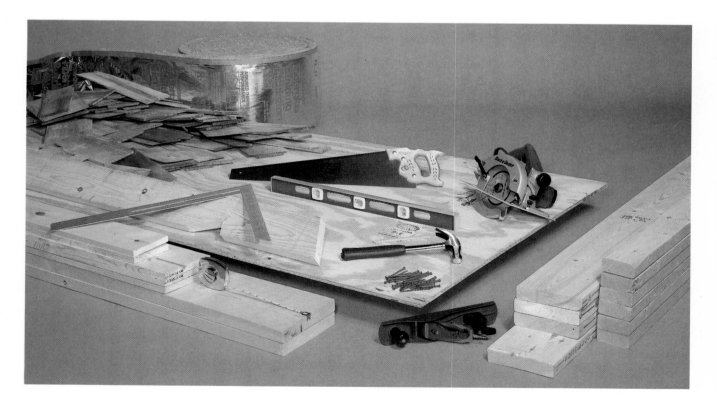

South Holland, Illinois
THE GOODHEART-WILLCOX COMPANY, INC.
Publishers

Library of Congress Cataloging in Publication Data

Wagner, Willis H.
  Modern carpentry.

  Includes index.
  1. Carpentry.   I. Title.
TH5604.W34 1987 694 87-410
ISBN 0-87006-648-X

INTRODUCTION

MODERN CARPENTRY is a colorful, easy-to-understand cyclopedia of authoritative and up-to-date information on building materials and construction methods. It provides detailed coverage of all aspects of light frame construction. Included are site clearing, site layout, foundations, framing, sheathing, roofing, windows and doors, exterior finish, and interior wall, floor, and ceiling finish. Special emphasis is placed on the use of modern tools, materials, and prefabricated components in the application of interior trim, and the construction of stairs and cabinetwork.

Also included is basic information covering post-and-beam construction, chimney and fireplaces, and prefabricated structures. Units are arranged in a logical sequence—similar to the order in which the various phases of building construction are performed.

Special units cover passive solar construction and remodeling. The unit on insulation pays special attention to energy efficiency in residential construction.

Illustrations consist of more than 1600 carefully selected photos and drawings. Many photos are in full color. Illustrations are accurately coordinated with written instructions and descriptions that are easy to read. Special photographs of wood samples show, in full color, the beauty and grain characteristics of 35 native and foreign species of wood that are commonly used in carpentry and cabinetmaking.

Information about building materials includes size and grade descriptions and also basic technical information that covers physical properties and other important characteristics. Scientific and technological discoveries have led to the development of many new materials. The proper use and application of these materials depend on a craftsperson who has considerable knowledge of the material and how it will function in the completed structure.

MODERN CARPENTRY is designed to provide basic instruction for students in high schools, vocational-technical schools, college classes, and apprentice training programs. It can also serve as a valuable reference for students in architectural drafting classes, and for journeymen carpenters and construction supervisors. MODERN CARPENTRY will enable do-it-yourselfers to handle many construction jobs that they would otherwise be reluctant to undertake.

Willis H. Wagner

CONTENTS

Modern wood construction. This completed residence is a test of skills and materials used in carpentry. Note the different roof styles and use of glass. If you will refer to the facing page you will learn many of the names for the building components you can see.   (Manville Building Materials Corp.)

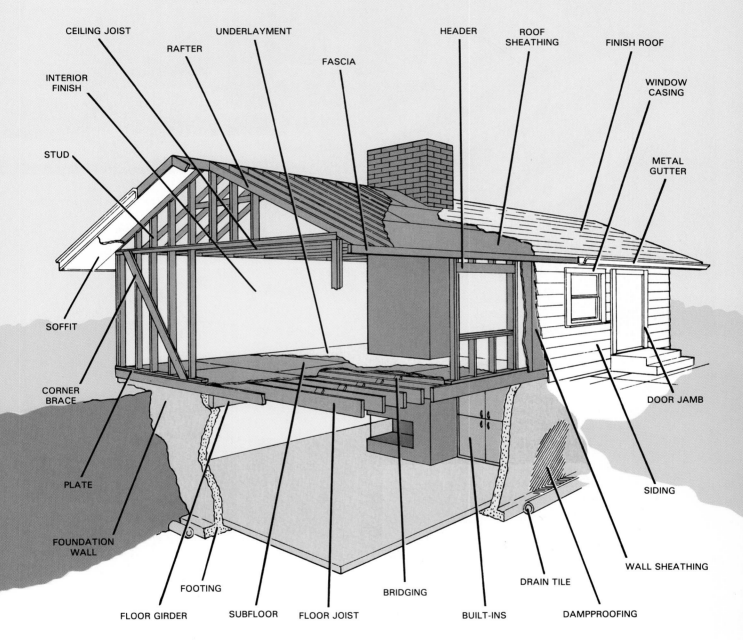

CEILING JOIST

UNDERLAYMENT

HEADER

ROOF SHEATHING

FINISH ROOF

RAFTER

FASCIA

WINDOW CASING

INTERIOR FINISH

METAL GUTTER

STUD

SOFFIT

CORNER BRACE

DOOR JAMB

PLATE

SIDING

FOUNDATION WALL

WALL SHEATHING

FOOTING

DRAIN TILE

BRIDGING

DAMPPROOFING

FLOOR GIRDER

SUBFLOOR

FLOOR JOIST

BUILT-INS

Names of basic parts of a house. Names of detailed parts are provided throughout the book.

Softwood lumber is used in various sizes for building construction. Sizes shown include 1 x 4, 1 x 6, 1 x 10, 2 x 4, 2 x 10, 2 x 12, and 4 x 4.

Many structural wood panels are manufactured for construction uses. Products shown (top to bottom) include: waferboard, structural particle board, composite plywood, oriented strand board, and plywood. Oriented strand board has layers of compressed strandlike particles arranged at right angles to each other. (Georgia Pacific)

Shingles, either of wood or composite materials, protect most buildings from the weather. Left. Composition shingles are made up of asphalt or fiberglass and mineral coatings (Certainteed) Right. Cedar is a popular and durable wood for shingles. (Shakertown Corp.)

Unit 1

BUILDING MATERIALS

Many different types of materials go into the construction of a modern residence, Fig. 1-1. A carpenter should be aware of all of them since each has special properties which makes it suited to certain applications.

All constructional materials are processed in some way so they are suitable for building. Some are composites of different materials which are designed to do a certain job as well or better than a natural material. Construction materials include:
1. Sawed lumber.
2. Plywood.
3. Particle board, hardboard, and waferboard.
4. Wood and nonwood materials for shingles and flooring.
5. Steel and aluminum.
6. Concrete.
7. Adhesives and sealers.
8. Gypsum board and fibrous manufactured ceiling tiles.

LUMBER

Wood is one of our greatest natural resources. When cut into pieces uniform in thickness, width, and length, it becomes lumber. This material has always been widely used for residential construction.

Lumber is the name given to products of the sawmill. Lumber includes:
1. Boards used for flooring, sheathing, paneling, and trim.
2. Dimension lumber used for sills, plates, studs, rafters, and other framing members.
3. Timbers used for posts, beams, and heavy stringers.
4. Numerous specialty items.

Carpenters must have a good working knowledge of lumber. They must be familiar with kinds, grades, sizes, and other details that apply to lumber selection and use. For a better understanding of how wood should be handled, carpenters should also know something about its growth, structure, and characteristics.

WOOD STRUCTURE AND GROWTH

Wood is made up of long narrow tubes or cells called fibers or TRACHEIDS. The cells are no larger around than human hair. Their length varies from about 1/25 in. in hardwoods to approximately 1/8 in. in softwoods. Tiny strands of cellulose make up the cell walls. The cells are held together with a natural cement called LIGNIN. This cellular structure makes it possible to drive nails and screws into the wood. It also accounts for the light weight, low heat transmission, and sound absorption qualities of wood.

The growing parts of a tree are:
1. The tips of the roots.
2. The leaves.

Fig. 1-1. About 80 percent of our nation's homes have a structural framework built of wood. Other products used include asphalt, aluminum, particle board or other manufactured board, concrete, and steel. (American Plywood Assoc.)

Fig. 1-2. Leaf silhouettes of several species of hardwoods and softwoods. (Gamble Brothers, Inc.)

SUGAR (HARD MAPLE)
ACER SACCHARUM

SWEET GUM
LIQUIDAMBAR STYRACIFLUA

BLACK WALNUT
JUGLANS NIGRA

RED OAK
QUERCUS BOREALIS

SILVER (SOFT) MAPLE
ACAR SACCHARINUM

DOUGLAS FIR
PSEUDOTSUGA TAXIFOLIA

WHITE OAK
QUERCUS ALBA

RED SPRUCE
PICEA RUBENS

WESTERN WHITE
(IDAHO) PINE
PINUS MONTICOLA

HEMLOCK
TSUGA CANADENSIS

AMERICAN ELM
ULMUS AMERICANA

SITKA SPRUCE
PICEA SITCHENSIS

season. These layers are called annular rings, Fig. 1-4. Each ring is composed of two layers: springwood and summerwood.

In the spring, trees grow rapidly and the cells produced are large and thin walled. As growth slows down during summer months, the cells produced are smaller, thicker walled, and appear darker in color. These annual growth rings are largely responsible for the grain patterns that are seen in the surface of boards cut from a log.

Sapwood is located inside the cambium layer. It contains living cells and may be several inches or more in thickness. Sapwood carries sap to the

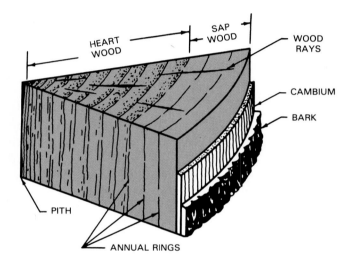

HEART WOOD

SAP WOOD

WOOD RAYS

CAMBIUM

BARK

PITH

ANNUAL RINGS

Fig. 1-3. Wood is made up of cells and a log has many basic parts.

3. A layer of cells just inside the bark called the CAMBIUM.

Water absorbed by the roots travels through the sapwood to the leaves, Fig. 1-2. Here it is combined with carbon dioxide from the air. Through the miracle of photosynthesis, sunlight changes these elements to a food known as carbohydrates. The sap carries this food back to the various parts of the tree.

New cells are formed in the cambium layer, Fig. 1-3. The inside area of the layer is called XYLEM. It develops new wood cells while the outside area, known as PHLOEM, develops cells that form the bark.

ANNULAR RINGS

Growth in the cambium layer takes place in the spring and summer. Separate layers form each

Fig. 1-4. Annular rings are formed each year and indicate the age of the tree. Drought, disease, or insects can interrupt growth causing an extra or false ring to form. (Forest Products Lab.)

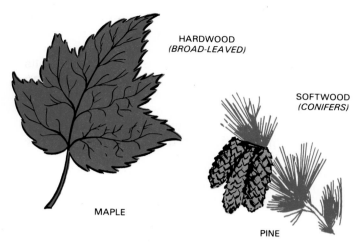

Fig. 1-5. Woods are classified as either hardwoods or softwoods. Hardwoods have broad leaves; softwoods have needles.

leaves. The heartwood of the tree is formed as the sapwood becomes inactive. Usually it turns darker in color because of the presence of gums and resins. In some woods such as hemlock, spruce, and basswood, there is little or no difference in appearance. Sapwood is as strong and heavy as heartwood but is not as durable when exposed to weather.

KINDS OF WOOD

Lumber is either softwood or hardwood. Softwoods come from the evergreen or needle bearing trees. These are called "conifers" because many of them bear cones. See Fig. 1-5. Hardwoods come from broadleaf (deciduous) trees that shed their leaves at the end of the growing season.

This classification is somewhat confusing because many of the hardwood trees produce a

SOFTWOODS	HARDWOODS
Douglas Fir Southern Pine Western Larch	Basswood Willow American Elm
Hemlock White Fir Spruce	*Mahogany Sweet Gum *White Ash
Ponderosa Pine Western Red Cedar Redwood	Beech Birch Cherry
Cypress White Pine Sugar Pine	Maple *Oak *Walnut

*Open grained wood

Fig. 1-6. List of popular woods for residential use. Usually, softwoods are used as framing lumber.

softer wood than some of the so-called softwood trees. Several of the more common kinds of commercial softwoods and hardwoods are listed in Fig. 1-6.

A number of hardwoods have large pores in the cellular structure and are called OPEN GRAIN WOODS. They require special or additional operations during finishing. Different kinds of wood will vary also in weight, strength, workability, color, texture, grain pattern, and odor.

Check the full-color wood samples, following page 29, as a first step in wood identification. To further develop ability to identify woods, study actual specimens. Several of the softwoods used in construction work are similar in appearance. Considerable experience is required to make accurate identification.

Most of the samples shown in this text were cut from plain-sawed or flat-grain stock. Edge-grain views would look different.

Availability of different species (kinds) of lumber varies from one part of the country to another. This is especially true of framing lumber which is expensive to transport long distances. It is usually more economical to select building materials found in the area.

CUTTING METHODS

Most lumber is cut so that the annular rings form an angle of less than 45 deg. with the surface of the board. This produces lumber called FLAT-GRAINED if it is softwood, or PLAIN-SAWED if it is hardwood. This method produces the least waste. Also, more desirable grain patterns are possible.

Lumber can also be cut so the annular rings form an angle of more than 45 deg. with the surface of the board, Fig. 1-7. This method produces lumber called EDGE-GRAIN if it is softwood, and QUARTERSAWED if it is hardwood. It is more difficult and expensive to use this method. However, it

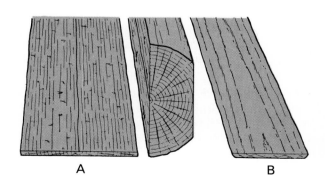

A   B

Fig. 1-7. Lumber may be sawed in different ways. Board at left is edge-grain or quartersawed. Saw cut was made roughly parallel to a line running through center of log. Right. Flat-grained or plain-sawed board.

produces lumber that swells and shrinks less across its width and is not so likely to warp.

MOISTURE CONTENT AND SHRINKAGE

Before wood can be used commercially, a large part of the moisture (sap) must be removed. When a living tree is cut, more than half of its weight may be moisture.

Lumber used for framing and outside finish should be dried to a moisture content of about 15 percent. Most cabinet and furniture woods are dried to a moisture content of 7 to 10 percent.

The amount of moisture or moisture content (M.C.) in wood is given as a percent of the oven-dry weight. To determine the moisture content, a sample is first weighed. It is then placed in an oven and dried at a temperature of about 212 °F. The drying is continued until the wood no longer loses weight. The sample is weighed again and this oven-dry weight is subtracted from the initial weight. The difference is then divided by the oven-dry weight, Fig. 1-8.

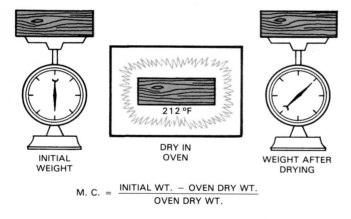

$$M.\,C. = \frac{\text{INITIAL WT.} - \text{OVEN DRY WT.}}{\text{OVEN DRY WT.}}$$

Fig. 1-8. Moisture content of wood can be determined by weighing it before and after drying.

Moisture contained in the cell cavities is called free water. That in the cell walls is called bound water. As the wood is dried, moisture first leaves the cell cavities. When the cells are empty but the cell walls are still full of moisture, the wood has reached a condition called the FIBER SATURATION POINT. For most woods this is about 30 percent, Fig. 1-9.

The fiber saturation point is important because wood does not start to shrink until this point is reached. As the M.C. drops below 30 percent, moisture is removed from the cell walls and they shrink. Fig. 1-10 shows the actual shrinkage in a 2 x 10 joist.

Wood shrinks most along the direction of the annual rings (tangentially) and about one-half as much across these rings. There is little shrinkage in the length. How this shrinkage affects lumber cut from different parts of a log is shown in Fig. 1-11. As wood takes on moisture, it swells in the same proportion as the shrinkage that took place.

EQUILIBRIUM MOISTURE CONTENT

A piece of wood will give off or take on moisture from the air around it until the moisture in the wood is balanced with that in the air. At this point the wood is said to be at equilibrium moisture content (E.M.C.). Since wood is exposed to daily and seasonal changes in the relative humidity of the air, its moisture content is always changing.

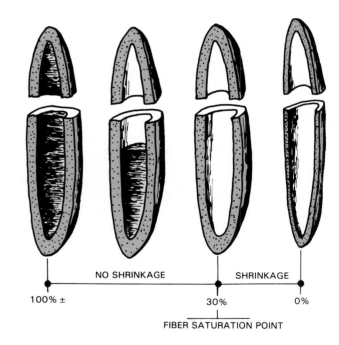

Fig. 1-9. How a wood cell dries. First the free water in the cell cavity is removed. Then the cell wall dries and shrinks.

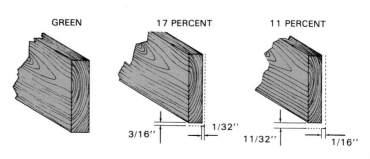

Fig. 1-10. A 2 x 10 joist may shrink 1/16 in. across its shortest dimension.

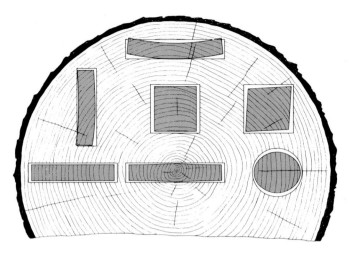

Fig. 1-11. Shrinkage and distortion of flat, square, and round pieces is affected by the direction of the annual rings.

Therefore, its dimensions are also changing. This is the reason doors and drawers often stick during humid weather.

Ideally, a wood structure should be framed with lumber at an M.C. equal to that which it will have in service. This is not practical. Lumber with such a low moisture content is seldom available and would likely gain moisture during construction. Standard practice is to use lumber with a moisture content in the range of 15 to 19 percent. In heated structures, it will eventually reach a level of about 8 percent. However, this will vary in different geographical areas, Fig. 1-12.

Carpenters understand that some shrinkage is inevitable. They make allowances where it will affect the structure. The first, and by far the greatest, change in moisture content occurs during the first year after construction, particularly during the first heating season.

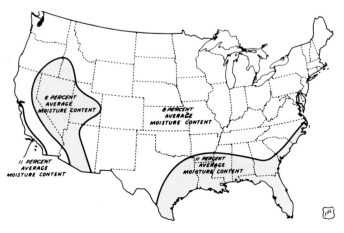

Fig. 1-12. Map shows average moisture content of interior woodwork for various regions of the United States.

When "green" lumber (more than 20 percent M.C.) is used, shrinkage will be excessive. Warping, plaster cracks, nail pops, squeaky floors, and other difficulties will be almost impossible to prevent.

SEASONING LUMBER

Seasoning is reducing the moisture content of lumber to the required level specified for its grade and use. In air-drying, the lumber is simply exposed to the outside air. It is carefully stacked with stickers (wood strips) between layers so air can circulate through the pile. Boards are also spaced within the layers so air can reach edges. Air-drying is a slower process than kiln drying. It often creates additional defects in the wood.

Lumber is kiln-dried by placing it in huge ovens where the temperature and humidity can be carefully controlled. When the green lumber is first placed in the kiln, steam is used to keep the humidity high. The temperature, meanwhile, is kept at a low level. Gradually the temperature is raised while the humidity is reduced. Fans keep the air in constant circulation around the wood. See Fig. 1-13.

Fig. 1-13. Huge kilns are used to season lumber at a modern sawmill. (Forest Products Lab.)

Bundles of lumber may carry a stamp to indicate that they have been kiln-dried. The letters "k.d." mean "kiln dried," while "p.k.d." stands for "partly kiln dried."

MOISTURE METERS

The moisture content of wood can be determined by:
1. Oven drying a sample, as previously described.
2. Using an electric moisture meter.

Although the oven drying method is the most accurate, meters are often used because readings can be secured rapidly and conveniently. The meters are usually calibrated to cover a range from 7 to 25 percent. Accuracy is plus or minus 1 percent of the moisture content.

There are two basic types of moisture meters. One determines the moisture content by measuring the electrical resistance between two pin-type electrodes that are driven into the wood. The other type measures the capacity of a condenser in a high-frequency circuit where the wood serves as the dielectric (nonconducting) material of the condenser. See Fig. 1-14.

LUMBER DEFECTS

A defect is an irregularity occurring in or on wood that reduces its strength, durability, or usefulness. It may or may not detract from appearance. For example, knots, commonly considered a defect, may add to the appearance of pine paneling. An imperfection that impairs only the appearance of wood is called a blemish. Some of the common defects include:

1. Knots — caused by an embedded branch or limb, Fig. 1-15. They are generally considered to be strength reducing. To what extent depends upon their type, size, and location. See Fig. 1-16.
2. Splits and checks — separations of the wood fibers which run along the grain and across the annular growth rings. They usually occur at the ends of lumber that has been unevenly seasoned.
3. Shakes — separations along the grain and between the annular growth rings. Shakes are likely to occur only in species with abrupt change from spring to summer growth.
4. Pitch pockets — cavities that contain or have contained pitch in solid or liquid form.
5. Honeycombing — separation of the wood fibers inside the tree. It may not be visible on the board's surface.
6. Wane — the presence of bark or the absence of wood along the edge of a board. It forms a bevel and/or reduces width.
7. Blue stain — discoloration caused by a mold-like fungus. Though objectional for appearance in some grades of lumber, it has little or no effect on strength.
8. Decay — disintegration of wood fibers due to fungi. Early stages of decay may be difficult to recognize. In advanced stages wood is soft, spongy, and crumbles easily.
9. Holes — caused by handling equipment or boring insects and worms. These will lower the lumber grade.
10. Warp — any variation from true or plane surface. Warp may include any one or combina-

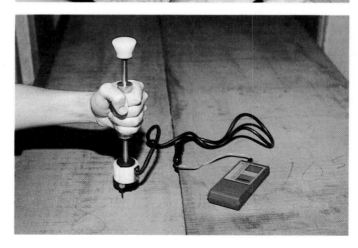

Fig. 1-14. Top. Modern moisture meter. LEDs indicate correct moisture percentage. Middle. Probes inserted into end cut. Bottom. Hammer probe provides readings up to 1 in. below surface. (Forestry Suppliers Inc.)

tions of the following: cup, bow, crook, and twist (or wind). See Fig. 1-17.

SOFTWOOD GRADES

Basic principles of grading lumber, set down by the American Lumber Standards Committee, are

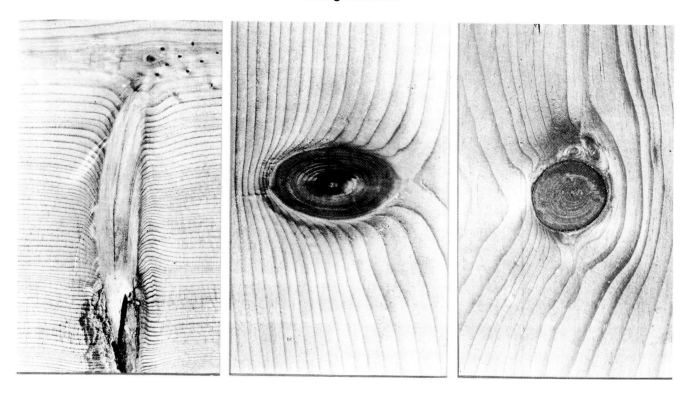

Fig. 1-15. Common kinds of knots. Left. Spike. Center. Intergrown. Right. Encased. Spike and intergrown knots will remain tight but an encased knot will usually loosen and fall out. (Forest Products Lab.)

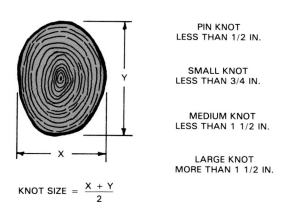

PIN KNOT
LESS THAN 1/2 IN.

SMALL KNOT
LESS THAN 3/4 IN.

MEDIUM KNOT
LESS THAN 1 1/2 IN.

LARGE KNOT
MORE THAN 1 1/2 IN.

$$\text{KNOT SIZE} = \frac{X + Y}{2}$$

Fig. 1-16. Knot size is determined by adding the width and length and then dividing by two.

published by the U.S. Department of Commerce. Detailed rules are developed and applied by the various associations of lumber producers—Western Wood Products Association, Southern Forest Products Association, California Redwood Association, and similar groups. These agencies publish grading rules for the species of lumber produced in their regions. They also have qualified personnel who supervise grading standards at sawmills.

Basic classifications of softwood grading include boards, dimension, and timbers. The grades within these classifications are shown in Fig. 1-18, next page.

Another classification called FACTORY and SHOP LUMBER is graded primarily for remanufacturing purposes. It is used by millwork plants in the fabrication of windows, doors, moldings, and other trim items.

The carpenter must understand that quality construction does not require that all lumber be of the best grade. Today, lumber is graded for specific

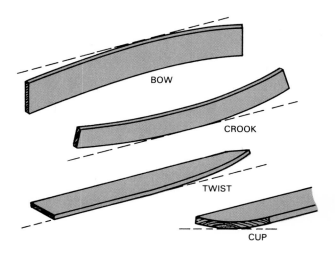

Fig. 1-17. Warp may occur in lumber across any dimension.

boards

APPEARANCE GRADES	**SELECTS**	B & BETTER (IWP—SUPREME) C SELECT (IWP—CHOICE) D SELECT (IWP—QUALITY)			**SPECIFICATION CHECK LIST** ☐ Grades listed in order of quality. ☐ Include all species suited to project. ☐ For economy, specify lowest grade that will satisfy job requirement. ☐ Specify surface texture desired. ☐ Specify moisture content suited to project. ☐ Specify Ⓦ grade stamp. For finish and exposed pieces, specify stamp on back or ends.
	FINISH	SUPERIOR PRIME E			
	PANELING	CLEAR (ANY SELECT OR FINISH GRADE) NO. 2 COMMON SELECTED FOR KNOTTY PANELING NO. 3 COMMON SELECTED FOR KNOTTY PANELING			
	SIDING (BEVEL, BUNGALOW)	SUPERIOR PRIME			

				WESTERN RED CEDAR	
			ALTERNATE BOARD GRADES	FINISH PANELING AND CEILING	**CLEAR HEART** A B
BOARDS SHEATHING	NO. 1 COMMON (IWP—COLONIAL) NO. 2 COMMON (IWP—STERLING) NO. 3 COMMON (IWP—STANDARD) NO. 4 COMMON (IWP—UTILITY)	SELECT MERCHANTABLE CONSTRUCTION STANDARD UTILITY		BEVEL SIDING	CLEAR — V.G. HEART A — BEVEL SIDING B — BEVEL SIDING C — BEVEL SIDING

dimension

LIGHT FRAMING 2" to 4" Thick 2" to 4" Wide	CONSTRUCTION STANDARD UTILITY ECONOMY	This category for use where high strength values are **NOT** required; such as studs, plates, sills, cripples, blocking, etc.
	STUD ECONOMY STUD	An optional all-purpose grade limited to 10 feet and shorter. Characteristics affecting strength and stiffness values are limited so that the "Stud" grade is suitable for all stud uses, including load bearing walls.
STRUCTURAL LIGHT FRAMING 2" to 4" Thick 2" to 4" Wide	SELECT STRUCTURAL NO. 1 NO. 2 NO. 3 ECONOMY	These grades are designed to fit those engineering applications where higher bending strength ratios are needed in light framing sizes. Typical uses would be for trusses, concrete pier wall forms, etc.
STRUCTURAL JOISTS & PLANKS 2" to 4" Thick 6" and Wider	SELECT STRUCTURAL NO. 1 NO. 2 NO. 3 ECONOMY	These grades are designed especially to fit in engineering applications for lumber six inches and wider, such as joists, rafters and general framing uses.

timbers

BEAMS & STRINGERS	SELECT STRUCTURAL NO. 1 NO. 2 (NO. 1 MINING) NO. 3 (NO. 2 MINING)	**POSTS & TIMBERS**	SELECT STRUCTURAL NO. 1 NO. 2 (NO. 1 MINING NO. 3 (NO. 2 MINING)

Fig. 1-18. Softwood lumber classifications and grades. Names of grades and their specifications will vary among lumber manufacturers' associations and among regions producing lumber.

uses. In a given structure, several grades may be appropriate. The key to good economical construction is the proper use of the lowest grade suitable for the purpose.

HARDWOOD GRADES

Grades for hardwood lumber are established by the National Hardwood Lumber Association. FAS (firsts and seconds) is the best grade. It specifies that pieces be no less than 6 in. wide by 8 ft. long

and yield at least 83 1/3 percent clear cuttings. The next lower grade is SELECTS and permits pieces 4 in. wide by 6 ft. long. A still lower grade is NO. 1 COMMON. Lumber in this group is expected to yield 66 2/3 percent clear cuttings.

LUMBER STRESS VALUES

In softwood lumber, all dimension and timber grades except Economy and Mining are assigned stress values. Slope of grain, knot sizes, and knot

product classification

BOARD MEASURE

The term "board measure" indicates that a board foot is the unit for measuring lumber. A board foot is one inch thick and 12 inches square.

The number of board feet in a piece is obtained by multiplying the nominal thickness in inches by the nominal width in inches by the length in feet and dividing by 12: $\frac{(T \times W \times L)}{12}$.

Lumber less than one inch in thickness is figured as one-inch.

	thickness in.	width in.		thickness in.	width in.
board lumber	1"	2" or more	beams & stringers	5" and thicker	more than 2" greater than thickness
light framing	2" to 4"	2" to 4"	posts & timbers	5" x 5" and larger	not more than 2" greater than thickness
studs	2" to 4"	2" to 4" 10' and shorter	decking	2" to 4"	4" to 12" wide
structural light framing	2" to 4"	2" to 4"	siding	thickness expressed by dimension of butt edge	
joists & planks	2" to 4"	6" and wider	mouldings	size at thickest and widest points	
Lengths of lumber generally are 6 feet and longer in multiples of 2'					

Standard Lumber Sizes / Nominal, Dressed, Based on WWPA Rules

Product	Description	Nominal Size Thickness In.	Nominal Size Width In.	Dressed Dimensions Surfaced Dry	Dressed Dimensions Surfaced Unseasoned	Lengths Ft.
FRAMING	S4S	2 3 4	2 3 4 6 8 10 12 Over 12	1-1/2 2-1/2 3-1/2 5-1/2 7-1/4 9-1/4 11-1/4 Off 3/4	1-9/16 2-9/16 3-9/16 5-5/8 7-1/2 9-1/2 11-1/2 Off 1/2	6 ft. and longer in multiples of 1'

Product	Description	Nominal Size	Dressed Dimensions	Lengths Ft.
		5 and Larger	Thickness In. Width In. 1/2 Off Nominal	
TIMBERS	Rough or S4S	5 and Larger	1/2 Off Nominal	Same

Product	Description	Nominal Size Thickness In.	Nominal Size Width In.	Dressed Dimensions Surfaced Dry Thickness In.	Dressed Dimensions Surfaced Dry Width In.	Lengths Ft.
DECKING Decking is usually surfaced to single T&G in 2" thickness and double T&G in 3" and 4" thicknesses	2" Single T&G	2	6 8 10 12	1 1/2	5 6 3/4 8 3/4 10 3/4	6 ft. and longer in multiples of 1'
	3" and 4" Double T&G	3 4	6	2 1/2 3 1/2	5 1/4	
FLOORING	(D & M), (S2S & CM)	3/8 1/2 5/8 1 1 1/4 1 1/2	2 3 4 5 6	5/16 7/16 9/16 3/4 1 1 1/4	1 1/8 2 1/8 3 1/8 4 1/8 5 1/8	4 ft. and longer in multiples of 1'
CEILING AND PARTITION	(S2S & CM)	3/8 1/2 5/8 3/4	3 4 5 6	5/16 7/16 9/16 11/16	2 1/8 3 1/8 4 1/8 5 1/8	4 ft. and longer in multiples of 1'
FACTORY AND SHOP LUMBER	S2S	1 (4/4) 1 1/4 (5/4) 1 1/2 (6/4) 1 3/4 (7/4) 2 (8/4) 2 1/2 (10/4) 3 (12/4) 4 (16/4)	5 and wider (4" and wider in 4/4 No. 1 Shop and 4/4 No. 2 Shop)	25/32 (4/4) 1 5/32 (5/4) 1 13/32 (6/4) 1 19/32 (7/4) 1 13/16 (8/4) 2 3/8 (10/4) 2 3/4 (12/4) 3 3/4 (16/4)	Usually sold random width	4 ft. and longer in multiples of 1'

ABBREVIATIONS

Abbreviated descriptions appearing in the size table are explained below.

S1S — Surfaced one side.
S2S — Surfaced two sides.

S4S — Surfaced four sides.
S1S1E — Surfaced one side, one edge.
S1S2E — Surfaced one side, two edges
CM — Center matched.

D & M — Dressed and matched.
T & G — Tongue and grooved.
EV1S — Edge vee on one side.
S1E — Surfaced one edge.

Fig. 1-19. Standard lumber sizes are set by government agencies and lumber associations. (Western Wood Products Assoc.)

locations are critical considerations.

There are two methods of assigning stress values:
1. Visual.
2. Machine rated.

In the machine rated method, lumber is fed into a special machine and subjected to bending forces. The stiffness of each piece (modulus of elasticity, E) is measured and marked on each piece. Machine stress-rated lumber (MSR) must also meet certain visual requirements.

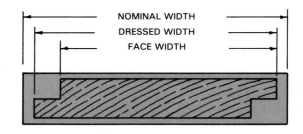

Fig. 1-20. Nominal size is greater than dressed size.

LUMBER SIZES

When listing and calculating the size and amount of lumber, the nominal dimension is always used. Fig. 1-19 illustrates the nominal and dressed sizes for various classifications of lumber used by the carpenter. Note that nominal sizes are sometimes listed in quarters. For example: 1 1/4 in. material is given as 5/4. This nominal dimension is its rough unfinished measurement, Fig. 1-20. The dressed size is less than the nominal size as a result of seasoning and surfacing. Dressed sizes of lumber, established by the American Lumber Standards, are applied consistently throughout the industry.

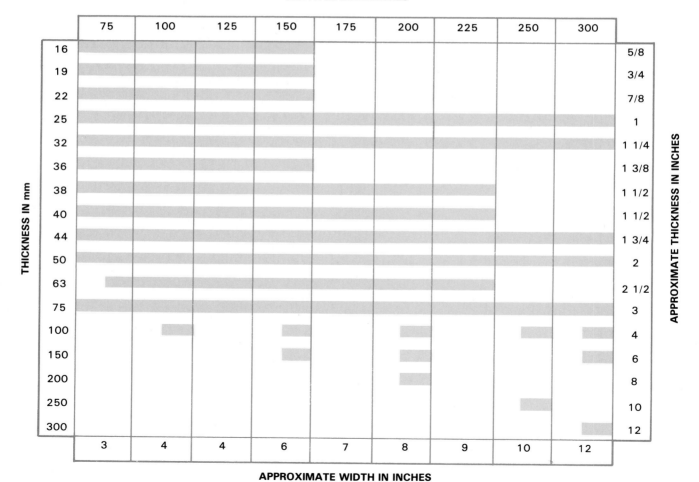

Fig. 1-21. Dimensions of metric lumber are given in millimetres. Lengths are always in metres and range from 1.8 m to 6.3 m in 0.3 m (about 1 ft.) increments.

FIGURING BOARD FOOTAGE

The unit of measure for lumber is the board foot. This is a piece 1 in. thick and 12 in. square or its equivalent (144 cu. in.).

Standard size pieces can be quickly calculated by visualizing the board feet included. For example: a board 1 x 12 and 10 ft. long will contain 10 bd. ft. If it were only 6 in. wide, it would be 5 bd. ft. If the original board had been 2 in. thick, it would have contained 20 bd. ft.

The following formula can be applied to any size piece where the total length is given in feet:

$$\text{Bd. ft.} = \frac{\text{No. pcs.} \times T \times W \times L}{12}$$

An example of the application of the formula follows: find the number of board feet in six pieces of lumber that measure $1'' \times 8'' \times 14'$:

$$\text{Bd. ft.} = \frac{6 \times 1 \times \overset{4}{\cancel{8}} \times 14}{\underset{2}{\underset{1}{\cancel{12}}}} = \frac{56}{}$$

$$= 56 \text{ bd. ft.}$$

Stock less than 1 in. thick is figured as though it were 1 in. When the stock is thicker than 1 in., the nominal size is used. When this size contains a fraction such as 1 1/4, change it to an improper fraction (5/4) and place the numerator above the formula line and the denominator below. For example: find the board footage in two pieces of lumber that measure $1 \ 1/4'' \times 10'' \times 8'$.

$$\text{Bd. ft.} = \frac{\overset{1}{\cancel{2}} \times \overset{5}{\cancel{5}} \times 10 \times 8}{\underset{2}{\underset{1}{\cancel{4}}} \ \underset{3}{\cancel{12}}} = \frac{50}{3}$$

$$= 16 \ 2/3 \text{ bd. ft.}$$

Use the nominal size of the material when figuring the footage. Items such as moldings, furring strips, and rounds are priced and sold by the lineal foot. Thickness and width are disregarded.

METRIC LUMBER MEASURE

Metric sized lumber gives thickness and width in millimetres (mm) and length in metres (m). There is little difference between metric and conventional dimensions for common sizes of lumber. For example, the common $1'' \times 4''$ board is 25 mm × 100 mm. Visually, they would appear to be about the same size. Metric lumber lengths start at 1.8 m (about 6 ft.) and increase in steps of 300 mm (about a foot) to 6.3 m. This is a little more than 20 ft. See Fig. 1-21 for a chart of standard sizes. Metric lumber is sold by the cubic metre (m³). See Reference Section.

PANEL MATERIALS

Wood panels for construction are manufactured in several different ways:
1. As plywood where thin sheets are laminated to various thicknesses.
2. As composite plywood where veneer faces are bonded to different kinds of wood cores.
3. As nonveneered panels including waferboard, particle board, and oriented strand board.

PLYWOOD

Plywood is constructed by gluing together a number of layers (plies) of wood with the grain direction turned at right angles in each successive layer. An odd number (3, 5, 7) of plies are used so they will be balanced on either side of a center core and so the grain of the outside layers will run in the same direction. The outer plies are called FACES or face and back. The next layers under these are called CROSS-BANDS and the other inside layer or layers are called the CORE. See Fig. 1-22. A thin plywood panel made of three layers would consist of two faces and a core.

There are two basic types of plywood:
1. EXTERIOR plywood which is bonded with waterproof glues. It can be used for siding, concrete forms, and other constructions where it will be exposed to the weather or excessive moisture.
2. INTERIOR plywood which is bonded with glues that are not waterproof. It is used for cabinets and other inside construction where the moisture content of the panels will not exceed 20 percent.

Plywood is made in thicknesses of 1/8 in. to more than 1 in. with the common sizes being 1/4, 3/8, 1/2, 5/8, and 3/4 in. A standard panel size is 4 ft. wide by 8 ft. long. Smaller size panels are available in the hardwoods.

Metric plywood panels are slightly smaller than the standard 4 by 8 ft. panel. Fig. 1-23 compares the two.

SOFTWOOD PLYWOOD GRADES

Softwood plywood for general construction is manufactured in accordance with U.S. Product Standard PS 1-74/ANSI A199.1. This standard provides a system for designating the species, strength, type of glue, and appearance.

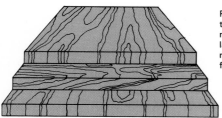

PLYWOOD: This staple of the construction industry is made by peeling logs and laying up the veneers at right angles to each other for rigidity and strength.

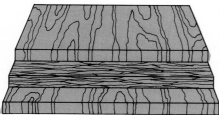

COMPOSITE PLYWOOD: This structural panel closely resembles plywood, except the middle ''ply'' is a core made of oriented wood fibers.

Fig. 1-22. Panel construction. Top. Standard plywood is made by peeling logs and laying up veneers at right angles for rigidity and strength. Bottom. composite plywood looks like regular plywood. However, the middle ''ply'' is a core made of oriented wood fibers glued together. (Georgia-Pacific)

Many species of softwood are used in making plywood. There are five separate plywood groups based on stiffness and strength. Group 1 includes the stiffest and strongest, Fig. 1-24.

GRADE-TRADEMARK STAMP

Construction and industrial panels are marked in two different ways as to quality:
1. A grade lettering system may be used to indicate the quality of the veneer used on the face and back of the panel. The letters and their meanings are given in Fig. 1-25.
2. A name indicating the panel's intended use or ''performance rating.''

The APA (American Plywood Association) has a rigid testing program based upon PS 1/74. Mills which are members of the association may use the official grade-trademark. It is stamped on each piece of plywood.

A typical stamp for an engineered grade of plywood is shown in Fig. 1-26. The span rating shows a pair of numbers separated by a slash mark (/). The number on the left indicates the maximum recommended span in inches when the plywood is used as roof decking (sheathing). The right-hand number applies to span when the plywood is used as subflooring. The rating applies only when the sheet is placed the long dimension across three or more supports. Generally, the larger the span the greater the panel's stiffness.

Fig. 1-27 lists some typical engineered grades of plywood. Included are descriptions and most common uses.

EXPOSURE RATINGS

The grade-trademark stamp gives an ''exposure durability'' classification to plywood. There are two basic types:
1. Exterior type which has 100 percent waterproof glueline.
2. Interior type with highly moisture resistant glueline.

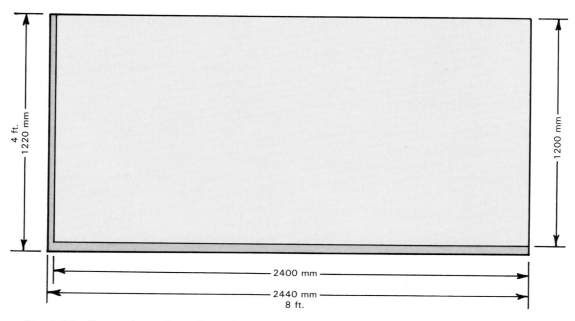

Fig. 1-23. Comparison of metric and conventional size plywood panels. Metric size is slightly smaller each way.

GROUP 1	GROUP 2	GROUP 3	GROUP 4	GROUP 5
Apitong Beech, American Birch Sweet Yellow Douglas Fir 1[a] Kapur Keruing Larch, Western Maple, Sugar Pine Caribbean Ocote Pine, South. Loblolly Longleaf Shortleaf Slash Tanoak	Cedar, Port Orford Cypress Douglas Fir 2 Fir California Red Grand Noble Pacific Silver White Hemlock, Western Lauan Almon Bagtikan Mayapis Red Lauan Tangile White Lauan Maple, Black Mengkulang Meranti, Red Mersawa Pine Pond Red Virginia Western White Spruce Red Sitka Sweetgum Tamarack Yellow- poplar	Alder, Red Birch, Paper Cedar, Alaska Fir, Subalpine Hemlock, Eastern Maple, Bigleaf Pine Jack Lodgepole Ponderosa Spruce Redwood Spruce Black Engelmann White	Aspen Bigtooth Quaking Cativo Cedar Incense Western Red Cottonwood Eastern Black (Western Poplar) Pine Eastern White Sugar	Basswood Fir, Balsam Poplar, Balsam

Fig. 1-24. Classification of softwood plywoods rates species for strength and stiffness. Group 1 represents strongest woods. (American Plywood Assoc.)

VENEER GRADES

N	Smooth surface "natural finish" veneer. Select, all heartwood or all sapwood. Free of open defects. Allows not more than 6 repairs, wood only, per 4 x 8 panel, made parallel to grain and well matched for grain and color.
A	Smooth, paintable. Not more than 18 neatly made repairs, boat, sled, or router type, and parallel to grain, permitted. May be used for natural finish in less demanding applications.
B	Solid surface. Shims, circular repair plugs, and tight knots to 1 inch across grain permitted. Some minor splits permitted.
C Plugged	Improved C veneer with splits limited to 1/8 inch width and knotholes and borer holes limited to 1/4 x 1/2 inch. Admits some broken grain. Synthetic repairs permitted.
C	Tight knots to 1 1/2 inch. Knotholes to 1 inch across grain and some to 1 1/2 inch if total width of knots and knotholes is within specified limits. Synthetic or wood repairs. Discoloration and sanding defects that do not impair strength permitted. Limited splits allowed. Stitching permitted.
D	Knots and knotholes to 2 1/2 inch width across grain and 1/2 inch larger within specified limits. Limited splits allowed. Stitching permitted. Limited to Interior, Exposure 1 and Exposure 2 panels.

Fig. 1-25. Description of softwood plywood veneer grades. (American Plywood Assoc.)

However, panels can be manufactured in three exposure durability classifications:
1. Exterior.
2. Exposure 1.
3. Exposure 2.

Panels marked "Exterior" can be used outdoors and may be continually exposed to weather and moisture. Panels marked "Exposure 1" can withstand moisture during extended periods but should only be used indoors. Panels designated as "Exposure 2" can be used in protected locations. They may be subjected to some water leakage or high humidity but, generally, should be protected from weather.

Most plywood is manufactured with waterproof exterior glue. However, interior panels may be manufactured with intermediate or interior glue.

HARDWOOD PLYWOOD GRADES

The Hardwood Plywood Institute uses a number system for grading the faces and backs of a panel. A grading specification of 1-2 would indicate a good face with grain carefully matched and a good back but without careful grain matching. A No. 3 back would permit noticeable defects and patching but would be generally sound. A special or PREMIUM grade of hardwood is known as "architectural" or "sequence-matched." This usually requires an order to a plywood mill for a series of matched plywood panels.

For either softwood or hardwood plywood, it is common practice to designate in a general way the grade by a symbol. G2S means good two sides. G1S means good one side.

In addition to the various kinds, types, and grades, hardwood plywood is made with different core constructions. The two most common are the

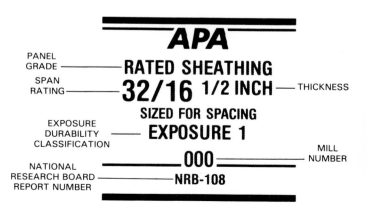

Fig. 1-26. Typical grade-trademark which is stamped on all of plywood manufactured in compliance with panels national plywood standard, PS 1/74. (American Plywood Assoc.)

	Grade Designation	Description & Common Uses	Typical Trademarks
PROTECTED OR INTERIOR USE	**APA RATED SHEATHING EXP 1 or 2**	Specially designed for subflooring and wall and roof sheathing, but can also be used for a broad range of other construction and industrial applications. Can be manufactured as conventional veneered plywood, as a composite, or as a nonveneered panel. For special engineered applications, including high load requirements and certain industrial uses, veneered panels conforming to PS 1 may be required. Specify Exposure 1 when long construction delays are anticipated. Common thicknesses: 5/16, 3/8, 7/16, 1/2, 5/8, 3/4.	**APA** RATED SHEATHING 32/16 1/2 INCH SIZED FOR SPACING EXPOSURE 1 000 NRB-108
	APA STRUCTURAL I & II RATED SHEATHING EXP 1	Unsanded all-veneer PS 1 plywood grades for use where strength properties are of maximum importance: structural diaphragms, box beams, gusset plates, stressed-skin panels, containers, pallet bins. Made only with exterior glue (Exposure 1). STRUCTURAL I more commonly available. Common thicknesses: 5/16, 3/8, 1/2, 5/8, 3/4. (3)	**APA** RATED SHEATHING STRUCTURAL I 24/0 3/8 INCH SIZED FOR SPACING EXPOSURE 1 000 PS 1-74 C-D INT/EXT GLUE NRB-108
	APA RATED STURD-I-FLOOR EXP 1 or 2	For combination subfloor-underlayment. Provides smooth surface for application of resilient floor covering and possesses high concentrated and impact load resistance. Can be manufactured as conventional veneered plywood, as a composite, or as a nonveneered panel. Available square edge or tongue-and-groove. Specify Exposure 1 when long construction delays are anticipated. Common thicknesses: 5/8 (19/32), 3/4 (23/32).	**APA** RATED STURD-I-FLOOR 24 oc 23/32 INCH SIZED FOR SPACING T&G NET WIDTH 47-1/2 EXPOSURE 1 000 NRB-108
	APA RATED STURD-I-FLOOR 48 oc (2-4-1) EXP 1	For combination subfloor-underlayment on 32- and 48-inch spans and for heavy timber roof construction. Provides smooth surface for application of resilient floor coverings and possesses high concentrated and impact load resistance. Manufactured only as conventional veneered plywood and only with exterior glue (Exposure 1). Available square edge or tongue-and-groove. Thickness: 1-1/8.	**APA** RATED STURD-I-FLOOR 48 oc 1-1/8 INCH (2-4-1) SIZED FOR SPACING EXPOSURE 1 T&G 000 INT/EXT GLUE NRB-108 FHA-UM-66
EXTERIOR USE	**APA RATED SHEATHING EXT**	Exterior sheathing panel for subflooring and wall and roof sheathing, siding on service and farm buildings, crating, pallets, pallet bins, cable reels, etc. Manufactured as conventional veneered plywood. Common thicknesses: 5/16, 3/8, 1/2, 5/8, 3/4.	**APA** RATED SHEATHING 48/24 3/4 INCH SIZED FOR SPACING EXTERIOR 000 NRB-108
	APA STRUCTURAL I & II RATED SHEATHING EXT	For engineered applications in construction and industry where resistance to permanent exposure to weather or moisture is required. Manufactured only as conventional veneered PS 1 plywood. Unsanded. STRUCTURAL I more commonly available. Common thicknesses: 5/16, 3/8, 1/2, 5/8, 3/4. (3)	**APA** RATED SHEATHING STRUCTURAL I 42/20 5/8 INCH SIZED FOR SPACING EXTERIOR 000 PS 1-74 C-C NRB-108
	APA RATED STURD-I-FLOOR EXT	For combination subfloor-underlayment under resilient floor coverings where severe moisture conditions may be present, as in balcony decks. Possesses high concentrated and impact load resistance. Manufactured only as conventional veneered plywood. Available square edge or tongue-and-groove. Common thicknesses: 5/8 (19/32), 3/4 (23/32).	**APA** RATED STURD-I-FLOOR 20 oc 19/32 INCH SIZED FOR SPACING EXTERIOR 000 NRB-108

(1) Specific grades, thicknesses, constructions and exposure durability classifications may be in limited supply in some areas. Check with your supplier before specifying.

(2) Specify Performance-Rated Panels by thickness and Span Rating.

(3) All plies in STRUCTURAL I panels are special improved grades and limited to Group 1 species. All plies in STRUCTURAL II panels are special improved grades and limited to Group 1, 2, or 3 species.

Fig. 1-27. Selected list of engineered grades of softwood plywood.
(American Plywood Assoc.)

veneer core and the lumber core. See Fig. 1-28. VENEER cores are the least expensive. They are fairly stable and warp resistant. LUMBER cores are easier to cut, the edges are better for shaping and finishing, and they hold nails or screws better. Plywood is also manufactured with a particle board core. It is made by gluing veneers directly to a particle board surface.

Fig. 1-29. Hardboard is much used in construction. Left. Samples of hardboard. Right. Particle board. (Weyerhaeuser Co.)

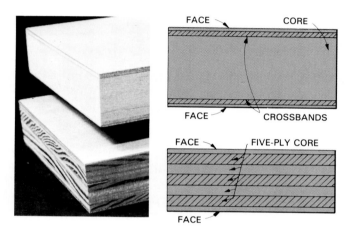

Fig. 1-28. Hardwood plywood is of two types: Top. Lumber core. Bottom. Veneer core.

COMPOSITE BOARD

Panels made up of a core of reconstituted wood with a thin veneer on either side are called composite board or composite panels. These materials are widely used in modern construction. They are good as sheathing, subflooring, siding, and interior wall surfaces.

In cabinetwork, hardboard and particle board, Fig. 1-29, serve as appropriate materials for drawer bottoms and concealed panels in cases, cabinets, and chests. They are manufactured by many different companies and sold under various trade names.

HARDBOARD

Hardboard is made of refined wood fibers, pressed together to form a hard, dense material (50 - 80 lb. per cu. ft.). There are two types: standard and tempered.

Tempered hardboard is impregnated (filled) with oils and resins. These materials make it harder, slightly heavier, more water resistant, and darker in appearance. Hardboard is manufactured with one side smooth (S1S) or both sides smooth (S2S). It is available in thicknesses from 1/12 in.

to 5/16 in. The most common thicknesses are 1/8, 3/16, and 1/4 in. Panels are 4 ft. wide and come in standard lengths of 8, 10, 12, and 16 ft.

PARTICLE BOARD

Particle board is made of wood flake, chips, and shavings bonded together with resins or adhesives. It is not as heavy as hardboard (about 40 lb. per cu. ft.) and is available in thicker panels. Particle board may be constructed of layers made of different size wood particles. Large ones in the center provide strength. Fine ones at the surface provide smoothness.

Wide use is made of particle board as a base for veneers and laminates. It is an important material in the construction of counter tops, cabinets, drawers and shelving, many types of folding and sliding doors, room dividers, and a variety of built-ins.

It is popular because of its smooth, grain-free surface and its stability. Its surface qualities make it a popular choice as a base for laminates. Doors made of it do not warp and require little adjustment following installation.

Particle board is available in thicknesses ranging from 1/4 in. to 1 7/16 in. The most common panel size is 4 x 8 ft.

WAFERBOARD

Waferboard, also called waferwood, is produced from high quality flakes of wood that are about 1 1/2 in. square. These flakes are bonded together under heat and pressure with phenolic resin, a waterproof adhesive. Both sides of waferboard have the same textured surface, Fig. 1-30. This surface has a natural slickness which can be minimized by special treatments. The density of waferboard is about 40 lb. per cu. ft. Standard panel size is 4 x 8 ft. in a thickness range from 1/4 in. to 3/4 in.

ORIENTED STRAND BOARD

Somewhat like waferboard in appearance, oriented strand board is also made up of wood fibers adhered to each other with suitable resins and glues. The fibers are put down in successive layers arranged at right angles to one another.

Fig. 1-30. Top. Oriented strand board. Bottom. Waferboard. These panels can be used as sheathing, subflooring, or as interior wall finish. (Georgia-Pacific)

WOOD TREATMENTS

Wood and wood products should be protected from attack by fungi, insects, and borers. Application of special chemicals or wood preservatives will accomplish this.

The degree of protection depends on the effectiveness of the chemical and how thoroughly it penetrates the material. Millwork plants employ extensive treatment processes in the manufacture of such items as door frames and window units.

There are two general classes of wood preservatives:

1. Oils, such as creosote, and petroleum solutions of pentachlorophenol.

2. Certain salts that can be dissolved in water.

When selecting a preservative you should consider its effectiveness in protecting the wood as well as any side effects that may result. Some products produce discoloration of painted surfaces or objectionable odors.

A number of commercial preservatives are available for on-the-job application. Study the manufacturer's directions and recommendations. Use special precautions in handling solutions. Some contain toxic and/or flammable chemicals.

HANDLING AND STORING

Building materials are expensive and every precaution should be taken to maintain them in good condition. After they are delivered to the construction site, this becomes the responsibility of the carpenter.

Piles of framing lumber and sheathing should be laid on level skids raised at least 6 in. above the ground. Be sure all pieces are well supported and are lying straight.

Cover the material with canvas or waterproof paper. Polyethylene film provides a watertight covering, Fig. 1-31.

If moisture absorption is likely, cut steel banding on panel materials to prevent edge damage when fibers expand. Keep coverings open and away from the sides and bottom of lumber stacks to promote good ventilation. Too tight coverings encourage mold growth.

Exterior finish materials, door frames, and window units should not be delivered until the structure is partially enclosed and the roof surfaced. In cold weather, the entire structure should be enclosed and heated before interior finish and cabinetwork are delivered and stored.

Fig. 1-31. If lumber or plywood is to be stored outdoors, cover it with a tarp or polyethylene film.

When finish lumber is received at a higher or lower moisture content than it will attain in the structure, it should be open-stacked with wood strips so air can circulate freely around each piece.

Plywood, especially the fine hardwoods, must be handled with care. Sanded faces become soiled and scarred if not protected. In storing, the panels should be laid flat as illustrated in Fig. 1-32.

Fig. 1-32. Plywood, especially fine hardwood veneers, should be stored indoors and stacked for good air circulation. (E.L. Bruce Co.)

NONWOOD MATERIALS

The carpenter works with a number of materials other than lumber and wood-base products. Some of the more common items include:
1. Gypsum lath.

Fig. 1-33. Asphalt and fiberglass shingles are important materials for modern residential construction. (Certainteed)

2. Wallboard and sheathing.
3. Insulation boards and blankets.
4. Shingles of asphalt, metal, and fiber glass, Fig. 1-33.
5. Metal flashing materials.
6. Caulking materials, Fig. 1-34.
7. Resilient flooring materials and carpeting.

METAL STUDS are another example of a non-wood material. Originally designed for commercial and institutional construction, their use is now being extended to residential structures.

A typical stud consists of a metal channel with openings through which electrical and plumbing lines can be installed. See Fig. 1-35. These are attached to base and ceiling channels with metal screws or clips. Wall surface material is attached to the stud using self-drilling drywall screws. Some web-type studs have a special metal edge into which nails can be driven. Metal stud systems are usually designed for nonload bearing walls and partitions.

Fig. 1-34. Caulking and sealing materials are designed to resist moisture and prevent air infiltration through cracks and joints. Be sure to follow the manufacturer's recommendations for a specific application.
(Beecham Home Improvement Products, Inc.)

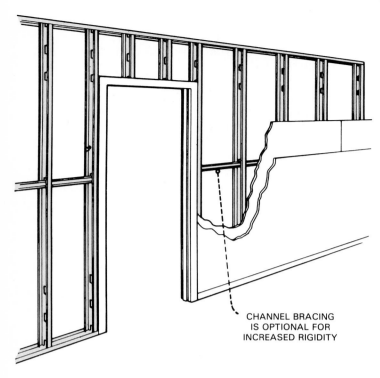

Fig. 1-35. Metal stud partition systems are now being used in residential construction. (National Gypsum Co.)

METAL FASTENERS

Nails, the metal fasteners commonly used by carpenters, are available in a wide range of types and sizes. Basic kinds are illustrated in Fig. 1-36.

The common nail has a heavy cross section and is designed for rough framing. The thinner box nail is used for toe nailing in frame construction and light work. The casing nail is the same weight as the box nail, but has a small conical head. It is used in finish carpentry work to attach door and window casings and other wood trim. Finishing nails and brads are quite similar and have the thinnest cross section and the smallest head.

The nail size unit is called a "penny" and is abbreviated with the lower case letter d. It indicates the length of the nail. A 2d (2 penny) nail is 1 in. long. A 6d (6 penny) nail is 2 in. long. See Fig. 1-37 and Fig. 1-38. This measurement applies to common, box, casing, and finish nails. Brads and small box nails are specified by their actual length and gauge number.

Fig. 1-39 shows a few of the many specialized nails. Each is designed for a special purpose. Annular or spiral threads greatly increase holding power. Some nails have special coatings of zinc, cement, or resin. Coating or threading increases its holding power. Nails are made from such materials as iron, steel, copper, bronze, aluminum, and stainless steel. Nails for power nailing are shown in Fig. 1-40.

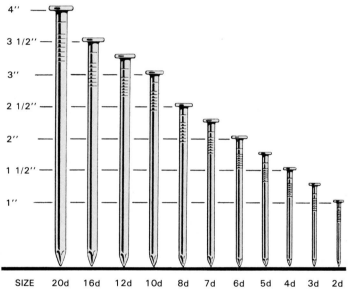

Fig. 1-37. Nail sizes are given in a unit called the "penny." It is written as "d." (United States Steel)

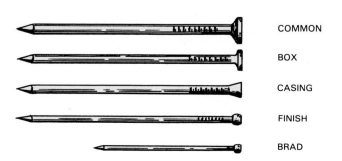

Fig. 1-36. There are five basic types of nails.

Size	Length''	COMMON		BOX	
		Diam.''	No./Lb.	Diam.''	No./Lb.
4d	1 1/2	.102	316	.083	473
5d	1 3/4	.102	271	.083	406
6d	2	.115	181	.102	236
7d	2 1/4	.115	161	.102	210
8d	2 1/2	.131	106	.115	145
10d	3	.148	69	.127	94
12d	3 1/4	.148	63	.127	88
16d	3 1/2	.165	49	.134	71
20d	4	.203	31	.148	52
30d	4 1/2	.220	24	.148	46
40d	5	.238	18	.165	35

Fig. 1-38. Nail sizes and approximate number in a pound. (Georgia-Pacific)

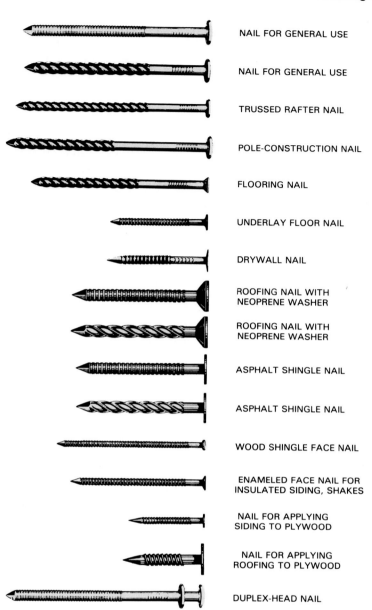

NAIL FOR GENERAL USE

NAIL FOR GENERAL USE

TRUSSED RAFTER NAIL

POLE-CONSTRUCTION NAIL

FLOORING NAIL

UNDERLAY FLOOR NAIL

DRYWALL NAIL

ROOFING NAIL WITH
NEOPRENE WASHER

ROOFING NAIL WITH
NEOPRENE WASHER

ASPHALT SHINGLE NAIL

ASPHALT SHINGLE NAIL

WOOD SHINGLE FACE NAIL

ENAMELED FACE NAIL FOR
INSULATED SIDING, SHAKES

NAIL FOR APPLYING
SIDING TO PLYWOOD

NAIL FOR APPLYING
ROOFING TO PLYWOOD

DUPLEX-HEAD NAIL

Fig. 1-39. Annular and spiral threaded nails are designed for special purposes.   (Independent Nail and Packing Co.)

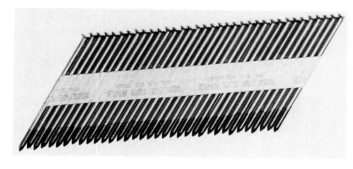

Fig. 1-40. Clip of nails for power nailer. Ends are coated for easier driving and greater holding power.
(Paslode Co., Div. of Signode)

the wood screw changes by 13 thousandths (.013) of an inch.

Most wood screws used today are made of mild steel with a zinc chromate finish. They are labeled as F. H., which stands for flat head. Nickel and chromium plated screws, also screws made of brass, are available for special work. Wood screws are usually priced and sold by the box. Each box contains 100 screws.

Additional fasteners the carpenter will find useful include lag screws, hanger bolts, carriage bolts (especially designed for woodwork), corrugated fasteners, and metal splines. Specialized metal fasteners are described in other sections of this book where their application is covered.

ADHESIVES

The adhesives the carpenter uses may be classified as glue and mastics. Research and development have created many new products in this area. Some are highly specialized, being designed for a specific material and/or application. Brief

Wood screws have greater holding power than nails and are often used for interior construction and cabinetwork. Their size is determined by the length and diameter (gauge number). Screws are classified according to the shape of head, surface finish, and the material from which they are made. See Fig. 1-41.

Wood screws are available in lengths from 1/4 to 6 in., and in gauge numbers from 0 to 24. The gauge number can vary for a given length of screw. For example, a 3/4 in. screw is available in gauge numbers of 4 through 12. The No. 4 is a thin screw, while the No. 12 has a large diameter. From one gauge number to the next, the size of

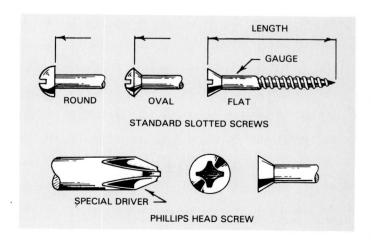

Fig. 1-41. Kinds of heads found on common wood screws.

descriptions of several of the commonly used glues follows:

POLYVINYL RESIN EMULSION GLUE (generally called polyvinyl or white glue) is excellent for interior construction. It comes ready to use in plastic squeeze bottles, Fig. 1-42, and is easily applied. This glue sets up rapidly, does not stain the wood or dull tools, and holds wood parts securely.

Polyvinyl glue hardens when its moisture content is removed through absorption into the wood or through evaporation. It is not waterproof and therefore is not suitable for assemblies that will be subjected to high humidities or moisture. The vinyl-acetate materials used in the glue are thermoplastic. Under heat they will soften. They should not be used in constructions where the temperature may rise above 165 °F.

UREA-FORMALDEHYDE RESIN GLUE (usually called urea resin) is available in a dry powder form which contains the hardening agent or catalyst. It is mixed with water to a creamy consistency before use.

Urea resin is moisture resistant, dries to a light brown color, and holds wood securely. It hardens through chemical action when water is added and sets at room temperatures in 4 to 8 hours.

CONTACT CEMENT is applied to each surface and allowed to dry until a piece of paper will not stick to the film. The surfaces are then pressed firmly together and bonding takes place immediately. The pieces must be carefully aligned for the initial contact because they cannot be moved after they touch. The bonding time is not critical and can usually be performed any time within one hour.

Contact cement is made with a neoprene rubber base and is an excellent adhesive for applying plastic laminates or joining parts that cannot be clamped together easily. It works well for applying thin veneer strips to plywood edges and can also be used to join combinations of wood, cloth, leather, rubber, and plastics.

Contact cement usually contains volatile, flammable solvents. The work area where it is applied must be well ventilated.

CASEIN GLUE is made from milk curd, hydrated lime, and sodium hydroxide. It is supplied in powder form and is mixed with cold water for use. After mixing, it should set for about 15 minutes before it is applied. It is classified as a water resistant glue.

Casein glue is used for structural laminating and works well where the moisture content of the wood is high. It has good joint filling qualities and is, therefore, often used on materials that have not been carefully surfaced. Casein is used for gluing oily woods such as teak, padouk, and lemon wood. Its main disadvantages are that it stains the wood, especially such species as oak, maple, and redwood, and has an abrasive effect on tool edges.

MASTICS

Mastics are a heavy, pasty type of adhesive that have revolutionized the methods used in the application of wallboards, wood paneling, and some types of floors. They vary in their characteristics and application methods, and are usually designed for a specific type of material. Some are waterproof. Others must be used where there is no excessive moisture.

One application method consists of placing several gobs on the surface of the material and

Fig. 1-42. A polyvinyl resin emulsion glue is being used to assemble stairs.

Fig. 1-43. Mounting gypsum wallboard on concrete block wall. Beads of adhesive or mastic were applied with a caulking gun. (National Gypsum Co.)

then pressing the unit firmly in place. This causes the mastic to spread over a wider area. Some mastics are spread over the surface with a notched trowel. Still others are designed for caulking gun application, Fig. 1-43.

Mastics are usually packaged in metal containers or gun cartridges ready for application. Always follow the directions of the manufacturer.

TEST YOUR KNOWLEDGE — UNIT 1

1. The natural cement that holds wood cells together is called _____.
2. New wood cells are formed in the _____ layer.
3. Which of the following kinds of wood are classified as a hardwood? Hemlock, redwood, willow, spruce.
4. When a softwood log is cut so the annular rings form an angle greater than 45 deg. with the surface of the boards, the lumber is called _____.
5. What is the moisture content of a board if a test sample that originally weighed 11.5 oz. was found to weigh 10 oz. after oven drying?
6. The fiber saturation point is about _____ M.C. for nearly all kinds of wood.
7. The letters E.M.C. are an abbreviation for the term _____ moisture content.
8. A large knot is defined as one that is over _____ in. in size.
9. Where should a plywood panel marked "Exposure 1" be used?
10. The best grade of "selects" softwood lumber is _____.
11. The best available grade of hardwood lumber is _____.
12. How many board feet of lumber are contained in a pile of 24 pieces of 2" x 4" x 8'?

OUTSIDE ASSIGNMENTS

1. Prepare a visual aid that shows various metal fasteners used in carpentry. Include nails, brads, screws, carriage bolts, staples (such as are used in power staplers), and other items. Try to include a good representation of spiral, ring groove, and coated nails. Label each item or group of items, giving the correct name, size, and other information.
2. Secure a group of softwood samples that will be representative of the species of lumber used in your locality. Instead of writing the proper name on each piece, use a number that corresponds with your master list. This will permit you to give a wood identification quiz to members of your class. As the samples are passed around, the students can record the numbers and their answers on a sheet of paper.
3. Visit a local building supply center and secure information and literature concerning the various grades and species of lumber normally carried in stock. Prepare written descriptions of the defects permitted in several of the grades commonly selected by builders in the area. Secure list prices of these grades to gain some understanding of the savings that can be gained by using a lower classification. Make a summary report to your class on your findings and conclusions.
4. After a study of reference materials, prepare a paper on stress-rated lumber. Include information on grade stamps and their interpretation. Also define the f (fiber stress in bending) and the corresponding E (stiffness) rating. Try to include a description of modern equipment and/or new techniques used in the grading process.

WOOD IDENTIFICATION

A key element in woodworking and in carpentry is the proper identification of the wood.

These samples placed alphabetically are intended as a guide and an aid to the student in learning to identify various woods. They show typical color and grain characteristics for 35 different species.

ASH, WHITE. Strong, stiff, and fairly heavy (42 lbs. per cu. ft.). Works fairly well with hand tools but splits easily. Heartwood is a pale tan with a texture similar to Oak. Used for millwork, cabinets, furniture, upholstered frames, boxes, and crates. Used extensively for baseball bats, tennis rackets, and other sporting equipment.

ASPEN. Soft, light, close-grained, and easy to work. White sapwood with light tan and brown streaked heartwood. Sources: Europe, Western Asia, and Middle Atlantic States. Used for furniture and interior paneling.

BASSWOOD. The softest and lightest (26 lbs. per cu. ft.) hardwood in commercial use. Fine, even texture with straight grain. Especially easy to work with hand tools and highly resistant to warpage. Heartwood is a light yellowish-brown. Used for drawing boards, food containers, moldings, woodenware, and core stock for plywood.

BEECH. Heavy, hard, and strong (44 lbs. per cu. ft.). A good substitute for Sugar Maple but somewhat darker in color, with a slightly coarser texture. Used for flooring, furniture, brush handles, food containers, and boxes and crates.

BIRCH. A hard, strong, wood (47 lbs. per cu. ft.). Works well with machines and has excellent finishing characteristics. Heartwood, reddish-brown with white sapwood. Fine grain and texture. Used extensively for quality furniture, cabinetwork, doors, interior trim, and plywood. Also used for dowels, spools, toothpicks, and clothespins.

BIRCH, WHITE. Selected sapwood rotary-cuttings of regular birch veneer. Same characteristics as birch except nearly white in color. Used to make plywood for installations requiring a very light, fine textured material.

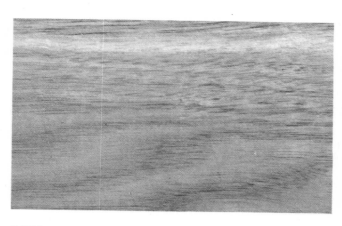

BUTTERNUT. Fairly soft, weak, light in weight (27 lbs. per cu. ft.) with a coarse texture. Grain patterns resemble walnut. Large open pores require a paste filler. Works easily with hand or machine tools. Sometimes used for interior trim, cabinetwork, and wall paneling.

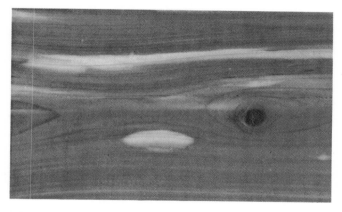

CEDAR, RED, EASTERN. Medium dense softwood, (34 lbs. per cu. ft.). Close-grained and durable. Heartwood is red; sapwood is white. It has an aroma that inhibits the growth of moths. Knotty wood is available only in narrow widths. Used mostly for chests and novelty items.

CEDAR, RED, WESTERN. A softwood, light in weight (23 lbs. per cu. ft.). Similar to redwood except for cedar-like odor. Pronounced transition from spring to summer growth (see edge-grain sample). Source: Western coast of North America, especially Washington. Used for shingles, siding, structural timbers, and utility poles.

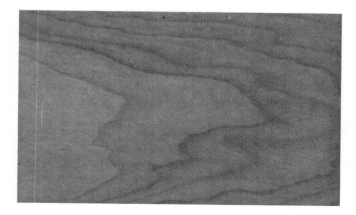

CHERRY, BLACK. Moderately hard, strong, and heavy (36 lbs. per cu. ft.). A fine, close-grained wood that machines easily and can be sanded to a very smooth finish. Heartwood is a reddish-brown with beautiful grain patterns. One of the fine furniture woods, however, there is a scarcity of good grades of lumber.

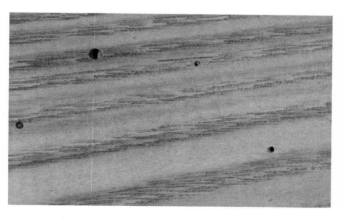

CHESTNUT. Coarse textured, open grain, and very durable. Reddish-brown heartwood. Easily worked with hand or machine tools. Source: Eastern United States. Available only in a wormy grade due to the "chestnut blight."

CYPRESS. Light in weight, soft, and easily worked. Fairly coarse texture with annual growth rings clearly defined (sample shows edge grain). Source: Southeastern Coast of the United States. Noted for its durability against decay. Used for exterior construction and interior wall paneling.

ELM, AMERICAN. Strong and tough for its weight (36 lbs. per cu. ft.). Fairly coarse texture with open pores. Annular ring growth is clearly defined. Bends without breaking and machines well. Used for barrel staves, bent handles, baskets, and some special types of furniture.

FIR, DOUGLAS. A strong, moderately heavy (34 lbs. per cu. ft.) softwood. Straight close grain with heavy contrast between spring and summer growth. Splinters easily. Used for wall and roof framing and other structural work. Vast amounts are used for plywood. Machines and sands poorly. Seldom used for finish.

GUM, SWEET. Also called Red Gum. Fairly hard and strong (36 lbs. per cu. ft.). A close-grained wood that machines well but has a tendency to warp. Heartwood is reddish-brown and may be highly figured. Used extensively in furniture and cabinetmaking. Stains well, often used in combination with more expensive woods.

HEMLOCK. A softwood. Light in weight and moderately hard. Light reddish-brown in color with a slight purplish cast. (Sample shows edge grain.) Source: Pacific Coast and Western States. Used for construction lumber and pulpwood; also for containers and plywood core stock.

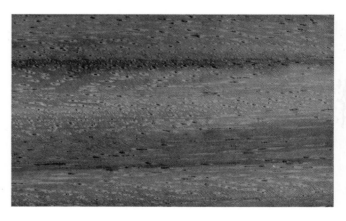

KELOBRA. Coarse texture with very large pores. Pronounced grain patterns are large and often have wavy lines. Source: Mexico, Guatemala, and British Honduras. Used for furniture and cabinetry mainly in the form of veneers.

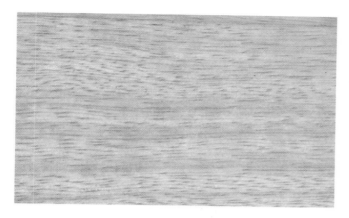

LIMBA. A light blond wood from the Congo, often sold under the trade name Korina. It has an open grain with about the same texture and hardness of Mahogany. Works easily with either hand or machine tools. Used for furniture and fixtures, especially where light tones are required.

MAHOGANY, AFRICAN. Characteristics are similar to American varieties. Slightly coarser texture and more pronounced grain patterns. Quartersawing or slicing produces a ribbon grain effect (as shown). Source: Ivory Coast, Ghana, and Nigeria. Used for fine furniture, interior finish, art objects, and boats.

MAHOGANY, PHILIPPINE. Medium density and hardness (37 lbs. per cu. ft.). Open grain and coarse texture. Works fairly well with hand or machine tools. Varies in color from dark red (Tanguile) to light tan (Lauan). Used for medium price furniture, fixtures, trim, wall paneling. Also, boat building and core stock in plywood.

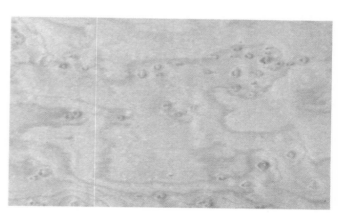

MAPLE, BIRDSEYE. Hard or Sugar Maple with tiny spots of curly grain that look like bird's eyes. The cause of this figure is not known. It may be distributed throughout the tree or located only in irregular stripes or patches. Used for highly decorative inlays and overlays.

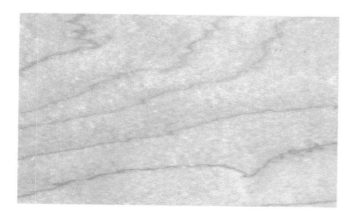

MAPLE, SUGAR. Also called Hard Maple. It is hard, strong, and heavy (44 lbs. per cu. ft.). Fine texture and grain pattern. Light tan color, with occasional dark streaks. Hard to work with hand tools but machines easily. Is an excellent turning wood. Used for floors, bowling alleys, woodenware, handles, and quality furniture.

OAK, QUARTERED. Sawing or slicing oak in a radical direction results in a striking pattern as shown. The "flakes" are formed by large wood rays that reflect light. Used where dramatic wood grain effects are desired.

Wood Identification

OAK, RED. Heavy (45 lbs. per cu. ft.) and hard with the same general characteristics as White Oak. Heartwood is reddish-brown in color. No tyloses in wood pores. Used for flooring, millwork, and inside trim. Difficult to work with hand tools.

OAK, WHITE. Heavy (47 lbs. per cu. ft.), very hard, durable, and strong. Works best with power tools. Heartwood is greyish-brown with open pores that are distinct and plugged with a hairlike growth called tyloses. Used for high quality millwork, interior finish, furniture, carvings, boat structures, barrels, and kegs.

PALDAO. A fairly hard wood with large pores that are partially plugged. Grain patterns are striking and beautiful and provide an excellent example of an ''exotic'' wood. Source: Philippine Islands. Sometimes selected by architects for special fixtures or built-ins for public or institutional buildings.

PINE, PONDEROSA. Lightweight (28 lbs. per cu. ft.) and soft. Straight grained and uniform texture. Not a strong wood but works easily and has little tendency to warp. Heartwood is a light reddish-brown. Change from springwood to summerwood is abrupt. Used for window and door frames, moldings, and other millwork; toys, models.

PINE, SUGAR. Lightweight (26 lbs. per cu. ft.) soft, and uniform texture. Heartwood, light brown with many tiny resin canals that appear as brown flecks. Straight grained and warp resistant. Cuts and works very easily with hand tools. Used for foundry patterns, sash and door construction, and quality millwork.

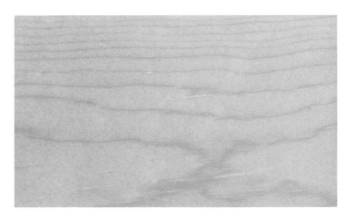

PINE, WHITE. Soft, light (28 lbs. per cu. ft.), and even texture. Cream colored with some resin canals but not as prevalent as in Sugar Pine. Used for interior and exterior trim and millwork items. Knotty grades often used for wall paneling. Works easily with hand or machine tools.

Wood Identification

POPLAR, YELLOW. Moderately soft, light in weight (34 lbs. per cu. ft.), and even textured. Heartwood is a pale olive-brown and sapwood is greyish-white. Works well with hand or machine tools and resists warping. Used in a wide variety of products including inexpensive furniture, trunks, toys, and core stock for plywood.

REDWOOD. Soft and light in weight (28 lbs. per cu. ft.). Texture varies but is usually fine and even grained. Easy to work and durable. Heartwood is reddish-brown. Used for structures, outside finish, and sometimes for interior paneling. Its durability makes it especially valuable for products exposed to water and moisture.

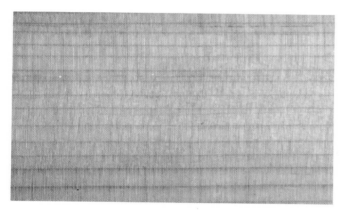

SPRUCE. A softwood, light in weight (24 lbs. per cu. ft.). Transition from spring to summer growth is gradual (see edge grain sample). There are several species; Sitka, Englemen, and a general classification called Eastern. Source: various parts of the United States and Canada. Used for pulpwood, light construction, and carpentry.

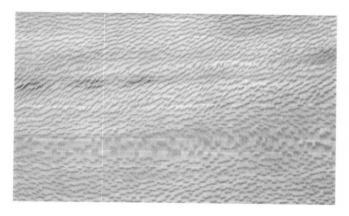

SYCAMORE. Medium density, hardness, and strength. A close-grained wood with a rather coarse texture. Easily identified by the flaky pattern of wood rays observed best in quartered stock. Source: Eastern half of United States. Used for drawer sides and lower priced furniture. Veneers are used for berry and fruit boxes.

WALNUT, BLACK. Fairly dense and hard. Very strong in comparison to its weight (38 lbs. per cu. ft.). Excellent machining and finishing properties. A fine textured open grain wood with beautiful grain patterns. Heartwood is a chocolate brown with sapwood near white. Used on quality furniture, gun stocks, fine cabinetwork, etc.

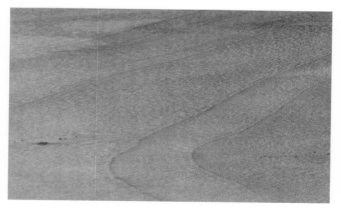

WILLOW, BLACK. Very soft and light in weight (27 lbs. per cu. ft.). Resembles basswood in workability, although there is some tendency for the machined surface to be fuzzy. Heartwood varies from light gray to dark brown. Used for some inexpensive furniture, core stock for plywood, wall paneling, toys, and novelty products.

35

Hands tools are essential for every step of construction. Above. Curved claw nail hammer is the most used tool on the construction job. (American Plywood Assoc.) Bottom. A framing square is another indispensible tool. Its most important uses include roof and stair framing. (Orem Research)

Unit 2

HAND TOOLS

have brought numerous improvements. Special tools have been developed to do specific jobs.

Experienced carpenters appreciate the importance of having good tools and select those that are accurately made from quality materials. They know that such tools last longer and will enable them to do better work.

A detailed study of the selection, care, and use of all the hand tools available is not possible. Only a general description of tools most commonly used by the carpenter will be included.

MEASURING AND LAYOUT TOOLS

Measuring tools must be handled with considerable care and kept clean. Only then can a high level of accuracy be assured.

The folding wood rule is indispensable. Most carpenters carry one at all times. A standard size rule is 6 ft. long. Some are equipped with a metal slide that is helpful in making inside measurements. Tape rules which range in size from 6 to 12 ft. are also handy, Fig. 2-1. Long tapes are available in lengths of 50 ft. and longer.

Fig. 2-1. Steel tape rules are available in lengths from 6 to 12 ft. They are useful for taking inside measurements. (Stanley Tools)

SQUARES

The framing square, also called a rafter square, is especially designed for the carpenter. Its uses are many and varied. A description of the square and how it is used in framing is included in Unit 9. It is available in steel, aluminum, or steel with a copper-clad or blued finish.

Try squares are available with blades 6 to 12 in. long. Handles are made of wood or metal. These are used to check the squareness of surfaces and edges and to lay out lines perpendicular to an edge, Fig. 2-2.

The combination square serves a similar purpose and is also used to lay out miter joints. The

Hand tools are essential to every aspect of carpentry work. A great variety of tools is required because residential construction covers a broad range of activities.

The carpenter, a skilled worker, carefully selects the kind, type, and size tools that best suit personal requirements. Tools become an important part of the carpenter's life, helping him or her to perform with speed and accuracy the various tasks of the trade.

Although the basic design of common woodworking tools has changed little in many years, modern technology and industrial "know-how"

Fig. 2-2. Try square is handy for laying out lines perpendicular to an edge.

adjustable sliding blade allows it to be conveniently used as a gauging tool.

The T-bevel has an adjustable blade making it possible to transfer an angle from one place to another. It is useful in laying out cuts for hip and valley rafters. Fig. 2-3 illustrates how the framing square is used to set the T-bevel at a specified angle of 45 degrees.

Wing dividers with locking legs are available in several sizes and serve a number of purposes. Dimensions can be stepped off along a layout or transferred from one position to another. Circles and arcs can also be scribed on surfaces.

MARKING GAUGE

A marking gauge is used to lay out parallel lines along the edges of material. It may also be used to:
1. Transfer a dimension from one place to another.
2. Check sizes of material.

The butt gauge can be used for about the same purposes. However, it is used mostly to lay out the gain (recess) for hinges.

Fig. 2-4. A chalkline is a fast, simple way to lay down straight lines over long distances on a building. (American Plywood Assoc.)

In layout work a scratch awl is used to scribe lines on the surfaces of materials. It is also used to mark points and to form starter holes for small screws or nails.

Using a chalk line is an easy way to mark long, straight lines. The line (covered with chalk) is held tight and close to the surface. Then it is snapped, Fig. 2-4. This action drives the chalk onto the surface forming a distinct mark. A special reel rechalks the line each time it is wound back into the case.

The level and plumb bob are important devices for laying out vertical and horizontal lines. A standard level is 24 in. long. The body of the level may be made of wood, aluminum, or special lightweight alloys. It is often used in connection with a straightedge when the span or height of the work is greater than the length of the level. (A straightedge is a straight strip of wood usually laminated for greater stability.) Some carpenters prefer a 4

to 6 ft. long level. It eliminates the need for a straightedge on such work as installing door and window frames. The plumb bob establishes a vertical line when attached to a line and suspended. Its weight pulls the line in a true vertical position for lay out and checking. The point of the plumb bob is always directly below the point from which it hangs. Fig. 2-5 shows the tools used for measuring and layout.

SAWS

The principal types of saws used by the carpenter are illustrated in Fig. 2-6 and Fig. 2-7. These are available in several different lengths as well as

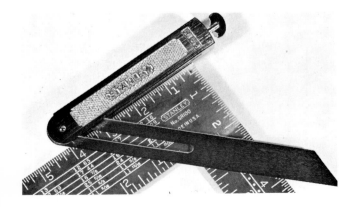

Fig. 2-3. Framing square can be used to set a T-bevel at a 45 deg. angle.

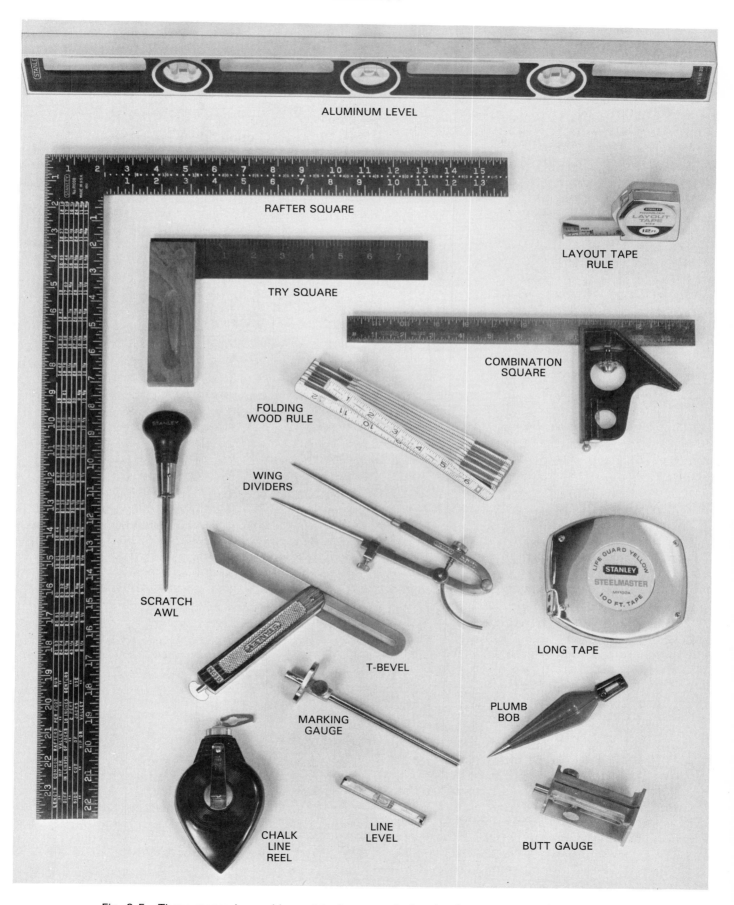

ALUMINUM LEVEL

RAFTER SQUARE

LAYOUT TAPE RULE

TRY SQUARE

COMBINATION SQUARE

FOLDING WOOD RULE

SCRATCH AWL

WING DIVIDERS

T-BEVEL

LONG TAPE

MARKING GAUGE

PLUMB BOB

CHALK LINE REEL

LINE LEVEL

BUTT GAUGE

Fig. 2-5. These measuring and lay out tools are used often by the carpenter. (Stanley Tools)

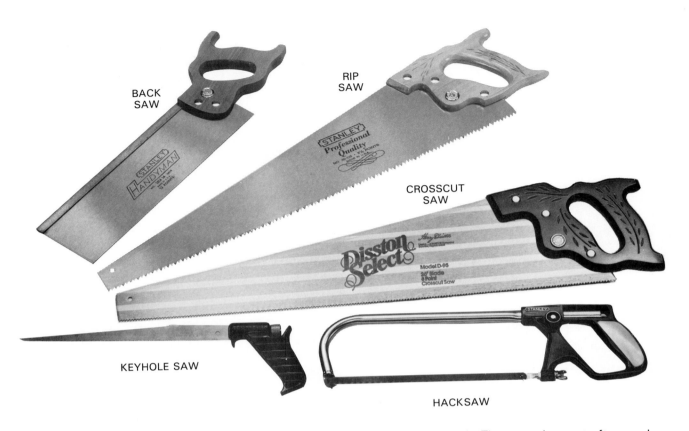

BACK
SAW

RIP
SAW

CROSSCUT
SAW

KEYHOLE SAW

HACKSAW

Fig. 2-6. Carpenters use many different kinds of saws in the course of their work. These are the most often used.

various tooth sizes (given as points per inch). Another saw, occasionally used for cutting curves, is the coping saw, Fig. 2-7.

Crosscut saws, as the name implies, are designed to cut across the wood grain. Their teeth are pointed. See view A, Fig. 2-8.

Ripsaws have chisel shaped teeth which cut best along the grain, view B, Fig. 2-8. Note that the teeth are set (bent) alternately from side to side so the kerf (opening cut in the wood) will be large enough for the blade to run freely. Some saws have a taper, ground from the toothed edge to the back edge. It eliminates the need for a large amount of set on the teeth.

Every carpenter needs a good crosscut saw with a tooth size ranging from 8 to 11 points. A crosscut saw for general use has 8 teeth per inch. A finishing saw used for fine cutting has 10 or 11 teeth per inch.

A ripsaw is no longer considered a "must." Most ripping operations are performed with power saws.

The backsaw has a thin blade reinforced with a steel strip along the back edge. The teeth are small (14-16 points). Thus, the cuts produced are fine. It is used mostly for interior finish work. A similar

type of saw is used in the miter box, Fig. 2-9. Slots or holders accurately guide the blade when forming miters and other types of joints.

Many compass and keyhole saws have a quick-change feature that permits the use of various

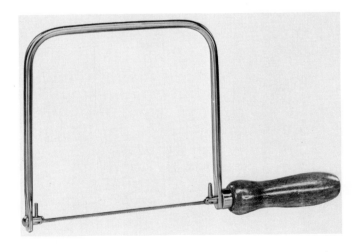

Fig. 2-7. Coping saw is used for cutting curves. It has about 15 teeth per inch.

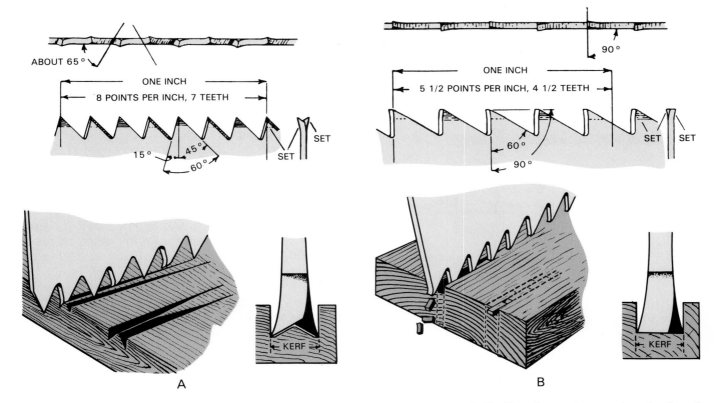

Fig. 2-8. Differences in teeth of crosscut and ripsaw. A—Crosscut teeth cut like a knife. Top shows shape and angle of teeth. Bottom shows how teeth cut. B—Ripsaw teeth cut like a chisel. top. Angle of 90 deg. is often increased to give negative rake. Bottom. How rip teeth cut.

sizes and kinds of blades. Although originally designed to cut keyholes, they now serve as general purpose saws for irregular cuts or work where working space is limited.

The sheetrock-drywall saw has large, specially designed teeth for cutting through paper facings, backings, and the gypsum core. Gullets are rounded to prevent their clogging from the gypsum material. See Fig. 2-10.

Because of the wide range of work that the carpenter must be prepared to handle, a hacksaw should be included in the hand tool assortment. This will be needed to cut nails, bolts, other metal fasteners, and various metal trim used on both exterior and interior work. Most hacksaws have an adjustable frame, permitting the use of several sizes of blades.

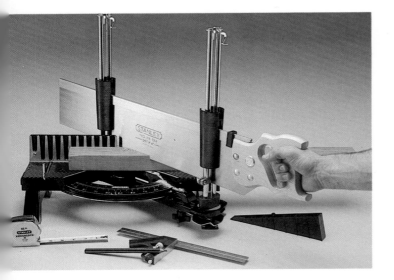

Fig. 2-9. Miter box and backsaw are useful for making fine, accurate cuts. (Stanley Tools)

Fig. 2-10. This saw is designed to cut drywall. (Stanley Tools)

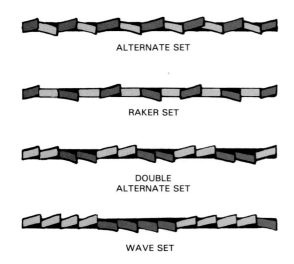

ALTERNATE SET

RAKER SET

DOUBLE
ALTERNATE SET

WAVE SET

Fig. 2-11. Enlarged view of hacksaw blades showing set of teeth.

HACKSAW BLADES

Hacksaw blades are made of high speed steel, tungsten alloy steel, molybdenum steel, and other special alloys. ALL HARD blades are heat treated all over. This makes them very brittle and easily broken if misused. FLEXIBLE BACK blades are hardened only around the teeth. They are usually preferred for all-around use on metals.

The blade's cutting edge may have anywhere from 14 to 32 teeth per inch. As a rule, you should use a blade with 14 teeth per inch for brass, aluminum, cast iron, and soft iron. For drill rod, mild steel, tool steel, and general work, 18 teeth per inch is recommended. For tubing and pipe, 24 teeth per inch is best. Generally, the thinner the metal being cut the finer the blade should be. At least two or three teeth should rest on the metal; otherwise the sawing motion may shear off teeth.

Like wood-cutting saws, hacksaw blades have a set. This provides clearance for the blade to slide through the cut. Edge views showing types of set are illustrated in Fig. 2-11.

Blades should be installed with teeth pointing forward, away from the handle. Use both hands to operate the saw. Apply enough pressure on forward strokes to allow each tooth to remove a small amount of metal. Remove pressure on the return stroke to reduce wear on the blade. Saw with long steady strokes paced at about 40 to 50 strokes per minute.

PLANING, SMOOTHING, AND SHAPING

The most important tools in this group are the planes and chisels. Several other tools that depend on a cutting edge to perform the work are included in the illustration, Fig. 2-12.

Standard surfacing planes include the smooth plane (8-9 in. long), jack plane (14 in. long), and the fore and jointer plane (18 to 24 in. long). Fig. 2-13 illustrates the parts and how they fit together. The jack plane is commonly selected for general purpose work.

Another surfacing plane, the block plane, is especially useful for the carpenter. It is small (6-7 in. long) and can be used with one hand. The blade is mounted at a low angle and the bevel of the cutter is turned up. This plane produces a fine, smooth cut making it suitable for fitting and trimming work.

Router and rabbet planes are designed to form dados, grooves, and rabbets. The need for these specialized hand planes has diminished in recent years due to the availability of power driven carpentry equipment.

Wood chisels are used to trim and cut away wood or composition materials to form joints or recesses, Fig. 2-14. They are also helpful in paring and smoothing small, interior surfaces that are inaccessible for other edge tools.

Width sizes range from 1/8 to 2 in. For general work, the carpenter usually selects 3/8, 1/2, 3/4, and 1 1/4 in. sizes. A soft-face hammer or mallet should always be used to drive the chisel when making deep cuts.

The scraper has an edge that is hooked. It produces thin shaving-like cuttings. The edge is usually filed and then a tool called a burnisher is used to form the hook.

There are a number of different types of scrapers. The carpenter generally prefers one with a blade that can be quickly replaced.

Two tools used for smoothing and shaping are the tungsten carbide coated file, and the multi-blade forming tool (Surform). Each cuts rapidly and will do a considerable amount of work before it becomes dull. The surface produced is rough and requires sanding to remove the tool marks, Fig. 2-15. The cornering tool, shown with the planes and chisels, is used to remove sharp corners from exposed wooden parts.

DRILLING AND BORING

Holes larger than 1/4 in. are made with auger bits or adjustable ones called EXPANSIVE bits. The operation is called boring.

Small holes are formed with hand or push drills. Tools in this group are illustrated in Fig. 2-16.

Auger bits vary in the shape and design of the twist. The size (diameter) of standard bits ranges from 3/16 in. to as large as 2 in. The more common range is from 1/4 to 1 in. The size is stamped on the tang or shank. Often it is a whole number which indicates the number in sixteenths.

"SURFORM" PLANE

ROUTER PLANE

UTILITY KNIFE

SCRAPER

CORNERING TOOL

PUTTY KNIFE

JACK PLANE

RABBET PLANE

FLOORING CHISEL

BLOCK PLANE

WOOD CHISELS

CABINET MAKERS PLANE

Fig. 2-12. Edge tools are used in carpentry for planing, smoothing, and shaping operations. (Stanley Tools)

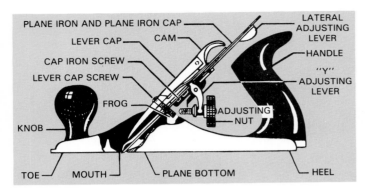

Fig. 2-13. Parts of a standard plane.

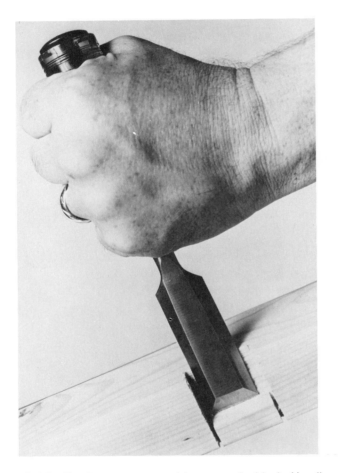

Fig. 2-14. Cutting a recess with a wood chisel. Handles should be able to withstand light hammer blows. (Stanley Tools)

Boring bits are mounted in a brace that holds and turns them into the wood. Fig. 2-17 shows the proper method for boring with a brace and bit. Other tools designed for the brace include:
1. The countersink which forms the recess for screw heads.
2. The screwdriver bit which is used to set regular wood screws.

An expansive bit is adjustable and can be used to bore large holes. When boring all the way through the stock, it is best to back-up the work with a scrap piece as shown in Fig. 2-18.

The size of a hand drill is determined by the capacity of its chuck. Usually the chuck capacity is 1/4 or 3/8 in. Regular twist drills are used in the hand drill. A practical set for the carpenter should range from 1/16 to 1/4 in. with increments (jumps) of 1/32 in.

Push drills are designed to form small holes quickly. When the handle is pushed down, the drill chuck revolves. A spring inside the handle forces it back out to its original position when pressure is released. Push drills use special fluted bits with sizes from 1/16 to about 3/16 in. Carpenters frequently use this type of drill to make holes for nails and screws. It can be operated with one hand. See Fig. 2-19 for two types of hand drills.

Fig. 2-15. A surform tool is made up of many tiny planing surfaces that cut quickly. Sanding operations are needed to smooth the surface afterwards.

FASTENING PARTS TOGETHER

In carpentry work, much of the work consists of fastening parts together. Nails, screws, bolts, and other types of connectors are used. Fig. 2-20 shows a group of hand tools commonly used to perform operations in this area.

Of the tools shown, the claw hammer is used most often. See Fig. 2-21. Carpenters usually carry a hammer in a holster attached to their belts. Two shapes of hammer heads are in common use:
1. The curved claw.
2. The ripping (straight) claw.

The curved claw is the most common and is best suited for pulling nails. The ripping claw which may be driven between fastened pieces is used somewhat like a chisel to pry them apart.

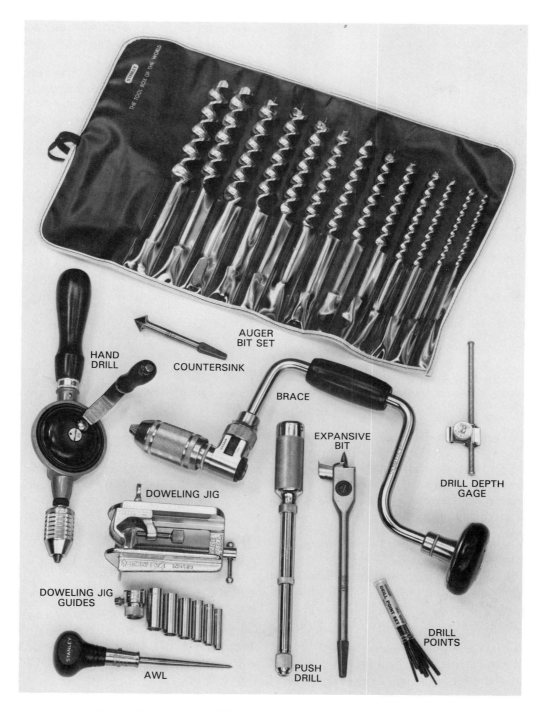

HAND DRILL

COUNTERSINK

AUGER BIT SET

BRACE

EXPANSIVE BIT

DRILL DEPTH GAGE

DOWELING JIG

DOWELING JIG GUIDES

DRILL POINTS

AWL

PUSH DRILL

Fig. 2-16. Tools for drilling and boring holes. (Stanley Tools)

Parts of a standard hammer are shown in Fig. 2-22. The face can be either flat or have a slightly rounded convex surface (bell face). The bell face is most often used. It will drive nails flush with the surface without leaving hammer marks on the wood. The hammer head is forged of high quality steel and is heat-treated to give the poll and face extra hardness.

The size of a claw hammer is determined by the weight of its head. It is available in a range from 7 to 20 oz. The 13 oz. size is popular for general purpose work. Carpenters generally use either the 16 or 20 oz. size for rough framing.

A hammer should be given good care. It is especially important to keep the handle tight and the face clean. If a wooden handle becomes loose, it can be tightened by driving the wedges deeper or by installing new ones. Fig. 2-23 illustrates the procedure for pulling nails to relieve strain on the handle and protect the work.

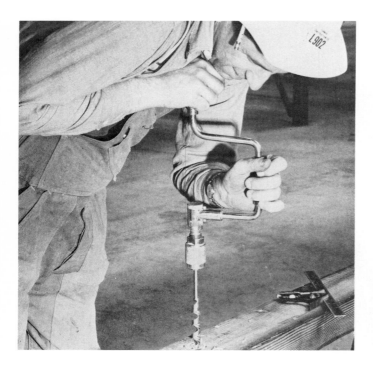

Fig. 2-17. Proper method for using the brace and bit. Use body to apply pressure. (Stanley Tools)

Hatchets are used for rough work on such jobs as making stakes and building concrete forms. Some are designed for wood shingle work, especially wood shakes. See Unit 10.

Ripping bars, also called wrecking bars, vary in length from 12 to 36 in. They are used to strip concrete forms, disassemble scaffolding, and other rough work involving prying, scraping, and nail pulling.

The nail set is designed to drive the heads of casing and finishing nails below the surface of the wood. Tips range in diameter from 1/32 to 5/32 in. by increments (steps) of 1/32 in. Overall length is usually about 4 inches.

Fig. 2-18. An expansive bit can be used to bore through wood.

A number of screwdriver sizes and styles are available. Sizes are specified by giving the length of the blade, measuring from the ferrule to the tip. The most common sizes for the carpenter are 3, 4, 6, and 8 in. These sizes are shown in the group illustration, Fig. 2-20.

The size of a Phillips screwdriver is given as a point number ranging from a No. 0, the smallest, to a No. 4. Size numbers 1, 2, and 3 will fit most of the screws of the type used in carpentry work.

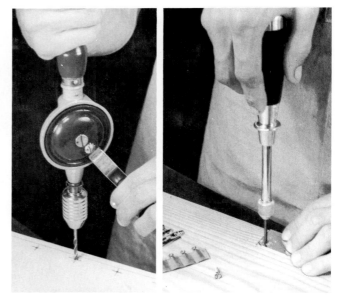

Fig. 2-19. Left. Drilling holes with a hand drill. Right. Using a push drill.

Tips of screwdrivers must be carefully picked for the job. For a slotted screw they must be square, of correct width and fit snugly into the screw. The width of the tip should be equal to the length of the bottom of the screw slot. The sides of the screwdriver tip should be carefully ground to an included angle of not more than 8 deg. and to a thickness that will fit the screw slot. Fig. 2-24 shows properly shaped tips for various types of screwdrivers.

A handy tool for the carpenter is the spiral ratchet screwdriver, Fig. 2-25. It operates somewhat like the push drill except the ratchet keeps the chuck turning in the same direction. Various bits to match screw sizes and types can be mounted in the chuck.

A kit of tools for the carpenter would not be complete without a pair or two of pliers and several wrenches. Sizes will vary depending on the construction needs.

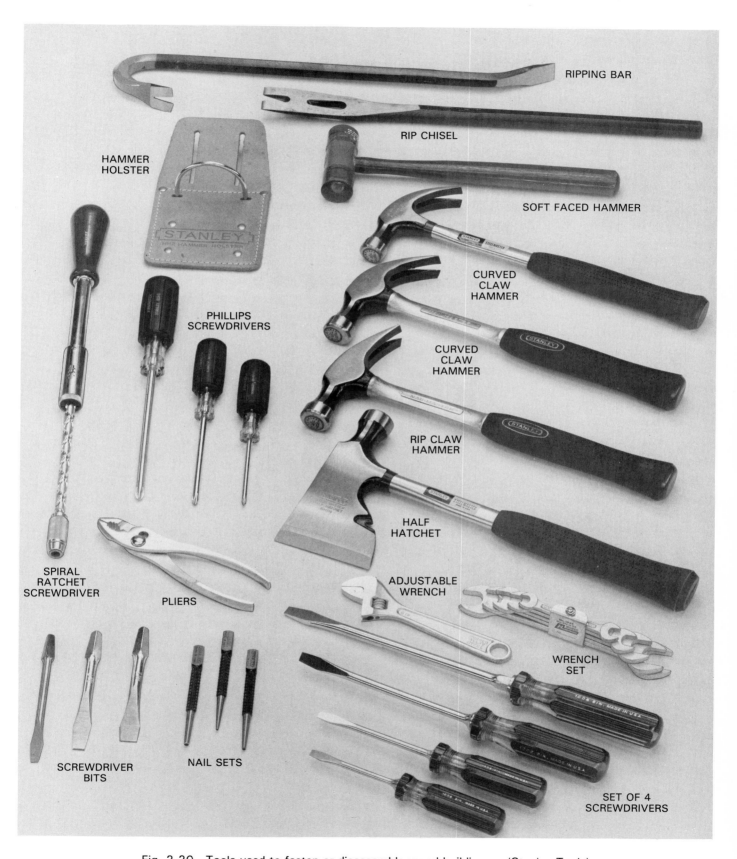

Fig. 2-20. Tools used to fasten or disassemble wood buildings. (Stanley Tools)

Fig. 2-21. Carpentry apprentice using 16 oz. hammer, nails temporary bracing to a wall frame. Hammer is most used of all fastening tools. (Waterloo Iowa Daily Courier)

Today, mechanical tackers, staplers, and nailers perform a variety of operations formerly done by hand nailing. See Fig. 2-26. They provide an efficient method of attaching insulation, roofing material, underlayment, ceiling tile, and many

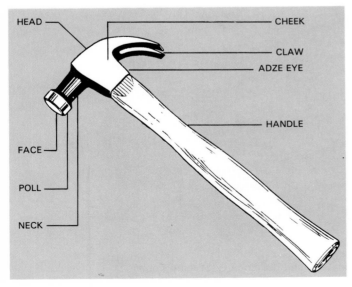

Fig. 2-22. Parts of a standard claw hammer.

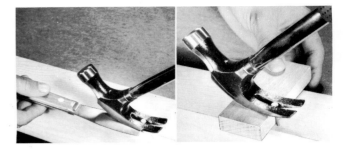

Fig. 2-23. Proper way to pull a nail. Left. Place a putty knife under the hammer to protect the surface. Right. A block of wood will increase leverage and protect the surface.

Using a spiral ratchet screwdriver to install hinges. (Stanley Tools)

Fig. 2-24. Screwdriver tips. Left. Phillips. Center. Standard. Right. Cabinet.

Fig. 2-25. Spiral ratchet screwdriver and bits. (Stanley Tools)

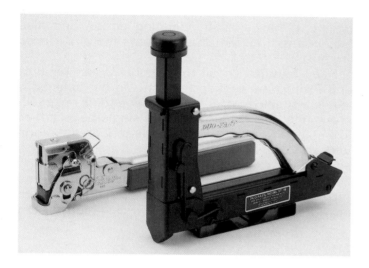

Fig. 2-26. Tackers are useful for many tacking and nailing jobs. Left. Hammer tacker. Right. Mallet operated nailer-tacker.

Fig. 2-27. This nailer-tacker is operated with a mallet. (Duo-Fast Corp.)

Fig. 2-28. C-clamps can be used to place temporary pressure on glued up parts. This type is also called a carriage clamp.

other products. An advantage of the regular stapler is that it leaves one hand free to hold the material. Fig. 2-27 illustrates a heavy duty stapler-nailer that is operated by striking it with a mallet.

CLAMPING TOOLS

There are several clamping tools that may prove helpful, especially in finish work. C-clamps, Fig. 2-28, are adaptable to a wide range of assemblies where parts need to be held together:
1. While metal fasteners are being applied.
2. Until an adhesive sets.

Sometimes they are used to clamp jigs or fixtures to machines for some special setup. C-clamps are available in sizes from 1 to 12 in. Sizes represent the tool's largest opening.

Hand screws, shown in Fig. 2-29, are ideal for woodworking because the jaws are broad and distribute the pressure over a wide area. Sizes (length of the jaw) range from 4 to 24 in.

TOOL STORAGE

Some type of chest or cabinet is needed to store and transport tools. In addition to being portable, the chest, or cabinet should protect the tools from weather damage, loss, and theft.

Folding tool panels provide a practical solution to transportation and storage. Fig. 2-30 shows a design that can be carried like a large suitcase. When opened it is a free-standing tool panel. Tool-holders, especially designed to hold a given item are preferred over nails, screws, or hooks. Attach the holder to a small subpanel and then mount the unit on a main panel as shown in Fig. 2-31. Special

Fig. 2-29. Handscrews are ideal for clamping wood because the pressure is distributed over a broad area.

locking devices may be required to hold the tools while the case is being transported.

One of the important advantages of an organized tool panel is the ease with which tools can

be checked. At the end of the day, you can determine if any tools are missing by checking to see if there are empty holders.

CARE AND MAINTENANCE

Experienced carpenters take pride in their tools and keep them in good working condition. They know that even high quality tools will not perform satisfactorily if they are dull or out of adjustment.

Tools should be wiped clean after being used. Occasionally saturate the wiping cloth lightly with oil. Some carpenters use a lemon oil furniture polish. It has a slight cleaning action and also leaves a thin oil film that protects metal surfaces from rust.

Keep handles on all tools tight. When handles and fittings are broken they should be replaced.

It is a simple matter to hone edge tools on an oilstone, Fig. 2-32. For tools with single-bevel edges like planes and chisels, place the tool on the stone with the bevel flat on the surface. Raise the back edge of the tool a few degrees so only the cutting edge is in contact. Then move the tool back and forth until a fine wire edge can be detected by pulling the finger over the edge. Now place the back of

Fig. 2-30. Folding tool panel may be closed and carried like a huge suitcase. (American Plywood Assoc.)

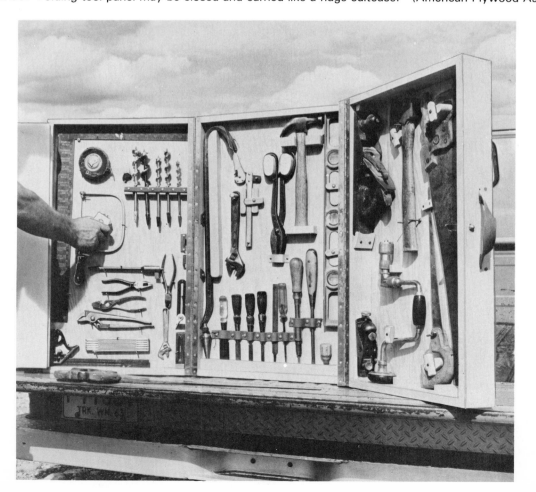

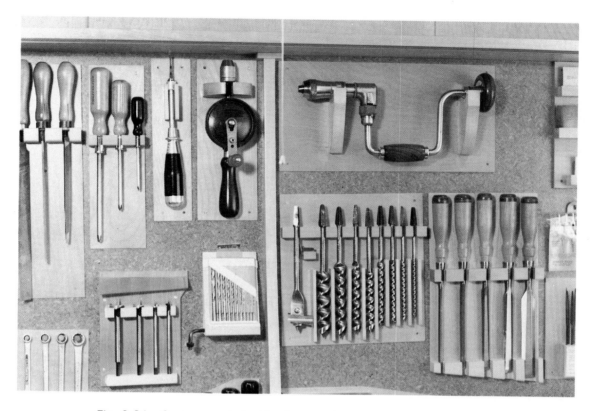

Fig. 2-31. Custom-made tool holders can be mounted on vertical panel.

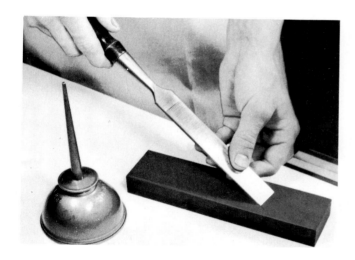

Fig. 2-32. Honing a chisel on an oilstone. Place a few drops of oil on the stone first.

The grinding angle for tools will vary somewhat depending on the work they must perform. Fig. 2-34 shows grinding and honing angles recommended for a plane iron.

Some tools may be sharpened with a file. For

the tool flat on the oilstone and stroke lightly several times. Turn the tool over and again stroke the beveled side lightly. Repeat this total operation several times until the wire edge has disappeared from the cutting edge.

If not damaged, a cutting edge can be honed a number of times before grinding is required. When the bevel becomes blunt, reshape it with a grinding operation, Fig. 2-33.

Fig. 2-33. Grinding a plane iron. Note the attachment that makes it possible to form an accurate bevel.

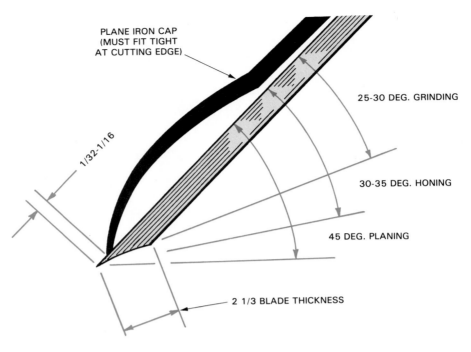

PLANE IRON CAP
(MUST FIT TIGHT
AT CUTTING EDGE)

25-30 DEG. GRINDING

30-35 DEG. HONING

45 DEG. PLANING

1/32-1/16

2 1/3 BLADE THICKNESS

Fig. 2-34. Grinding and honing angles for a plane iron.

Fig. 2-35. Sharpening an auger bit. Left. Filing the cutting lip.
Right. File the spur on the inside surface only.

Fig. 2-37. Setting the teeth. In the position shown, the saw
set will bend the tooth away from the operator.

Fig. 2-36. Filing a rip saw. The file should contact the back
of a tooth set away from the operator.

auger bits, it is best to use a special auger bit file. See Fig. 2-35. Sharpen the lips or cutters by stroking up through the throat. Do not file the underside. File the inside of the spurs as shown, being sure to keep them the same length.

Saws require filing as well as setting. Before filing they often require jointing. In this operation the height of the teeth are struck off evenly. Filing a saw is a tedious operation. Most carpenters prefer to send their saws to a shop where they can be machine sharpened by an expert.

Fig. 2-38. Correctly filed saw teeth. Above. Rip teeth. Bottom. Crosscut teeth.

When a saw is only slightly dull, the teeth can often be sharpened with a few file strokes, as illustrated in Fig. 2-36. Use a triangular saw file. Be sure to match the original angle of the teeth. File the back of one tooth and the front of an adjacent tooth in a single stroke. Fig. 2-37 shows a saw set in operation. Fig. 2-38 shows how correctly filed teeth should appear.

TEST YOUR KNOWLEDGE — UNIT 2

1. A standard folding wood rule is _____ ft. long.
2. The blade of a framing square is 24 in. long. How long is the tongue (other leg)?
   a. 10 in.
   b. 12 in.
   c. 16 in.
   d. 20 in.
3. When checking structural members to see if they are horizontal or vertical, what tool should be used?
4. A 10 point saw will have 12 teeth per inch. True or False?
5. A backsaw is used for fine work. Its teeth are usually spaced:
   a. 8 per inch.
   b. 8-10 per inch.
   c. 12-14 per inch.
   d. 14-16 per inch.
6. The bevel of a block plane blade is turned _____ (down, up).
7. An auger bit with the number 10 stamped on the tang or shank would bore a hole _____ in. in diameter.
8. Never strike a wood chisel with a mallet. True or False?
9. The size of a claw hammer is determined by the:
   a. Length of the handle.
   b. Length of the head.
   c. Weight of entire hammer.
   d. Weight of the head.
10. Large screws can be driven with a screwdriver bit mounted in a _____.
11. Spurs of an auger bit are filed on the _____ (inside, outside).
12. The operation of filing off the points of saw teeth until they are all level is called _____.

OUTSIDE ASSIGNMENTS

1. After a study of reference books and manufacturers' catalogs, prepare a list of tools you believe the carpenter will need for rough framing and exterior finish of a typical residential structure.
2. Prepare a report on the historical development of woodworking tools used by the carpenter. Reference books on woodworking and encyclopedias will contain helpful material. If you make a report to your class, use overhead projector or other visual aids to show students pictures and drawings of early tools. If possible, secure actual specimens from a collector.

Skilled carpenter laying out cut on finish stair tread. Note appropriate clothing worn which includes overalls, heavy work shirt with snug fitting cuffs, work shoes, hard hat, and safety glasses. The Occupational Safety and Health Administration (OSHA) specifies the wearing of approved hard hats on all construction sites. In addition to personal protective equipment, the law covers a wide range of standards to insure a safe and healthful environment in all branches of business and industry. (Stanley Tools)

GENERAL SAFETY RULES

Good carpenters recognize that safety is an important part of the job. They know that accidents are prevalent in building construction and that they often result in partial or total disability. Even minor cuts and bruises can be painful.

Safety is based on knowledge, skill, and an attitude of care and concern. Carpenters should know correct and proper procedures for performing the work. They should also be familiar with the potential hazards—how they can be minimized or eliminated.

Good attitudes toward safety are important. This includes belief in the importance of safety and willingness to give time and effort to a continuous study of the safest ways to perform work. It means working carefully and following the rules.

CLOTHING

Wear clothing appropriate for the work and weather conditions. Trousers or overalls should fit properly and have legs without cuffs. Keep shirts and jackets buttoned. Sleeves should also be buttoned or rolled up. Never wear loose or ragged clothing, especially around moving machinery.

Shoes should be sturdy with thick soles that will protect feet from protruding nails. Tennis or lightweight canvas shoes are not satisfactory. Never wear shoes with leather soles. They will not provide satisfactory traction on smooth wood surfaces, roofs in particular.

Construction work will require a hard hat (hat that will help prevent head injury from falling objects). Headgear should provide the necessary protection, be comfortable, permit good visibility, and shade your eyes. All clothing should be maintained in a satisfactory state of repair and not be permitted to become badly soiled.

PERSONAL PROTECTIVE EQUIPMENT

Safety glasses should be worn whenever the work involves even the slightest hazard to your eyes. Standard specifications state that a safety lens must withstand the blow of a 1/8 in. steel ball dropped from a height of 50 in.

Safety boots and shoes are required on heavy construction jobs. They consist of special reinforced toes that will withstand a load of 2500 lbs.

Hard hats should be worn whenever you are exposed to any possibility of falling objects. Standard specifications require that such hats with-

stand a certain degree of denting. They must be able to resist breaking when struck with an 80 lb. ball dropped from a 5 ft. height.

Wear gloves of an appropriate type when handling rough materials. Use a respirator when working in dusty areas, while installing insulation, or where finishing materials are being sprayed.

HAND TOOLS

Always select the correct type and size of tool for your work. Be sure it is sharp and properly adjusted. Guard against using any tool if the handle is loose or in poor condition. Dull tools are hazardous to use because force must be applied to make them cut. Oil or dirt on a tool may cause it to slip and cause an injury.

When using tools, hold them correctly. Most edge tools should be held in both hands with the cutting action away from yourself. Be careful when using your hand or fingers as a guide to start a cut.

Handle and carry tools with care. Keep edged and pointed tools turned downward. Carry only a few tools at one time unless they are mounted in a special holder. Do not carry sharp tools in pockets of your clothing. When not in use, tools should be kept in special boxes, chests, or cabinets.

POWER TOOLS

Before operating any power tool or machine you must be thoroughly familiar with the way it works and the correct procedures to follow. In general, when you learn to use equipment the correct way, you also learn to use it the safe way.

There are a number of general safety rules that apply to power equipment. In addition, special safety rules must be observed in the operation of each individual tool or machine. Those that apply directly to the power tools commonly used for modern carpentry are listed in Unit 3. Study and follow them carefully.

GOOD HOUSEKEEPING

This refers to the neatness and good order of the construction site. Maintaining a clean site contributes to the efficiency of the worker and is an important factor in the prevention of accidents.

Place building materials and supplies in neat piles. Locate them to allow adequate aisles and walkways. Rubbish and scrap should be placed in containers until disposal can be made. Do not permit blocks of wood, nails, bolts, empty cans, or pieces of wire to accumulate. They interfere with your work and constitute a tripping hazard.

Keep tools and equipment not being used in panels or chests. This will provide protection for the tools as well as the workers. In addition to improving efficiency and safety, good housekeeping helps maintain a better appearance at the construction project. This, in turn, will contribute to the morale of all workers.

DECKS AND FLOORS

To perform an operation safely, either with hand or power tools, the carpenter should stand on a firm, solid base. The surface should be smooth but not slippery. Do not attempt to work over rough piles of earth or on stacks of material that are unstable. Stay well away from floor openings, floor edges, and excavations as much as possible. Where this cannot be done, install adequate guardrails or barricades. In cold weather remove ice or cover it with sand or calcium chloride (salt).

EXCAVATIONS

Shoring and adequate bracing must be placed across the face of any excavation where the ground is cracked or caving is likely to occur. Inspect the excavation and shoring daily and especially after rains. Follow state and local regulations. Never climb into an open trench until proper reinforcement against cave-in has been installed or until the sides have been sloped to the "angle of repose" of the material being excavated.

Before beginning excavations determine whether there are underground utilities in the area. If so, locate and arrange protection for them during excavation operations.

Excavated soil and rock must be stored at least 2 ft. away from the edge of an excavation. Use ladders or steps to enter trenches which are more than 4 ft. deep.

SCAFFOLDS AND LADDERS

Scaffolds should have a minimum safety factor of four to one. This means that the scaffold will carry a load four times greater than the load it will probably be required to support. All scaffolding should be constructed under the direction of an experienced carpenter. Inspections should be made daily before use. Ladders should be checked at frequent and regular intervals. Their use should be limited to climbing from one level to another. Working while being supported on a ladder is

hazardous and should be kept to a minimum. There are many safety rules that must be observed in the use of scaffolds and ladders. These are covered in Unit 24.

FALLING OBJECTS

When working on upper levels of a structure, you should be especially cautious in handling tools and materials so there is no chance of them falling on workers below. Do not place tools on the edge of scaffolds, stepladders, window sills, or on any other surface where they might be knocked off.

If long pieces of lumber must be placed temporarily on end and leaned against the side of the structure, be sure they will not fall sideways. When moving through a building under construction, be aware of overhead work, and, wherever possible, avoid passing directly underneath. Stay clear of materials being hoisted. Wear an approved hard hat whenever there is a possibility of falling objects.

LIFTING AND CARRYING

Injuries may be caused by improper lifting or carrying heavy objects. When lifting, stand close to the load, bend your knees, and grasp the object firmly. Then lift by straightening your legs and keeping your body as nearly vertical as possible. To lower the object, reverse the procedure.

When carrying a heavy load, do not turn or twist your body but make adjustments in position by shifting your feet. If the load is heavy or bulky, secure help from others. Never underestimate the weight to be moved or overestimate your own ability. Always secure assistance when carrying long pieces of lumber.

FIRE PROTECTION

Carpenters should have a good understanding of fire hazards. Know their causes and methods of control.

Class A fires result from burning wood and debris. Class B fires involve highly volatile materials such as gasoline, oil, paints, and oil soaked rags. Class C fires are caused by electrical wiring and equipment. Any of these fires can occur on a typical construction site.

Approved fire prevention practices should be followed throughout the construction project. Good housekeeping is an important aspect. Special precautions should be taken during the final stages of construction when heating and wiring are being installed and when highly flammable surface finishes are being applied. Always keep containers of volatile materials closed when not in use and dispose of oily rags and combustible materials promptly.

Fire extinguishers should be available on the construction site. Be sure to use the proper kind for each type of fire. Study and follow local regulations.

FIRST AID

A knowledge of first aid is important. You should understand approved procedures and be able to exercise good judgment in applying them. Remember that an accident victim may receive additional injury from unskilled treatment by an unqualified person. Information of this nature can be secured from your local Red Cross.

As a preventative measure against infection, keep an approved first aid kit on the job site. Because of the nature of the material being handled and the dirty conditions of the work area, even superficial wounds should be treated promptly. Clean, sterilize, and bandage all cuts and nicks.

Always wear protective headgear in a hard-hat area of a construction site.   (Orem Research)

Safety tips for construction work. Left. When working with glass fibers, wear long sleeves, gloves, glasses, and a respirator. (Manville Building Materials Corp.) Center. Use extra care when handling large panels on windy days. Right. Ladders should extend above the roof edge at least 3 ft. See page 548. (Georgia-Pacific)

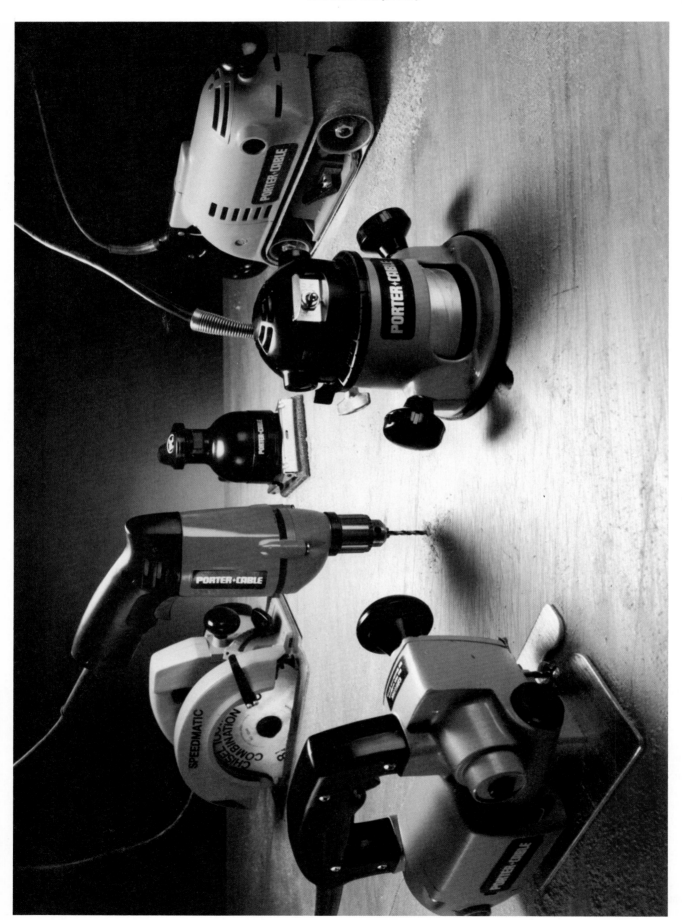

Portable power tools save time and energy in carpentry. They are used in nearly every stage of construction from rough framing to interior trim and cabinetwork. (Porter Cable Corp.)

Unit 3

POWER TOOLS

Modern power tools greatly reduce the time required to perform many of the operations in carpentry work. Heavy sawing, planing, and boring can be accomplished with far less human energy. Moreover, when proper tools are used in the correct way, high levels of accuracy can more easily be maintained.

There are two general types of power tools:
1. Portable.
2. Stationary.

Portable tools are light in weight, easily carried, and are held in the hands during operation. The tool is moved to the work. Portable tools are extensively used in nearly every stage of carpentry, especially rough framing.

Stationary tools (also called machines) are mounted on benches or stands which rest firmly on the floor. Work is brought to the machine.

Space permits only a brief description of the kinds of power tools most commonly associated with carpentry work. This should be supplemented with woodworking textbooks and reference books devoted to power tool operation. Manufacturer's bulletins and operator's manuals are also a good source of information.

POWER TOOL SAFETY

Safety must be practiced continually. Before operating any power tool, you must become thoroughly familiar with:
1. The way it works.
2. The correct way to use it.

You must be wide awake and alert. Never operate a power tool when tired or ill. Think through the operation before performing it. Know what you are going to do and what the tool will do. Make all adjustments before turning on the power. Be sure blades and cutters are sharp and are of the correct type for the work.

While operating a power tool, do not allow yourself to be distracted. See Fig. 3-1. Do not distract the attention of others while they are operating power tools. Keep all safety guards in position and wear safety glasses.

Feed the work carefully and only as fast as the tool will cut it easily. Overloading is hazardous to the operator and will likely damage the tool or work. When the operation is complete, turn off the power and wait until the moving parts have stopped before leaving the machine.

ELECTRICAL SAFETY

Always make sure that the source of electric power is the correct voltage and that the tool switch is in the "off" position before it is plugged into an electrical outlet.

Stationary power tools are factory equipped with magnetic starters. These are safety devices which will automatically turn the switch to the "off" position in case of power failure. This feature is important since personal injury or damage to the equipment could result if power is reestablished with the switch "on." Make sure these switches are in good working order.

The electrical cord and plug must be in good condition and must provide a ground for the tool. This means that extension cords should be the three-wire type. Make sure that the conducting wire is large enough to prevent excessive voltage drop.

Be careful in stringing electrical extension cords around the work site. Place them where they will not be damaged or interfere with other workers.

SHOCK PROTECTION

Electrical shock is one of the potential hazards of working with power tools. Always be sure that proper grounding is provided. Receptacles should be of the concealed contact type with a grounding terminal for continuous ground. Plugs and cords should be an approved type.

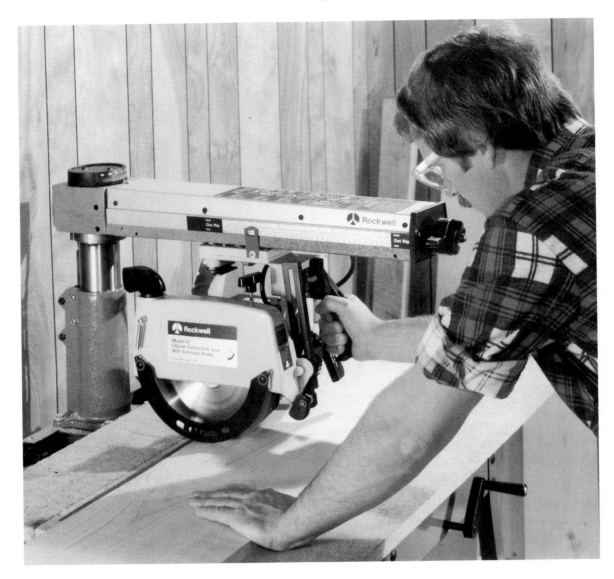

Fig. 3-1. When operating a power tool, give full attention to the work.

Portable power tools should be double insulated or otherwise grounded to protect the worker from dangerous electrical shock. Even though the circuit may be grounded, an operator of a portable power tool could be electrocuted should a bare conductor ground on a metal tool case. A ground fault circuit interrupter (GFCI) should be used on all construction job sites. These units can be installed in a circuit or can be plugged into an outlet which is grounded. These units ''sense'' when a short has occurred and will turn off power to the tool. See Fig. 3-2.

PORTABLE CIRCULAR SAWS

This power tool is also called an electric hand saw or builders' saw. Its size is determined by the

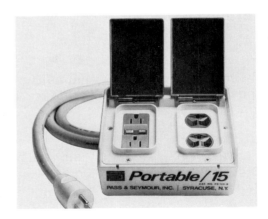

Fig. 3-2. Portable ground fault interrupter is used where permanent ground fault protection is not available. Power tools are simply plugged into the outlets. It will trip at around 5 milliamperes turning off power to the grounded tool. (Pass & Seymour, Inc.)

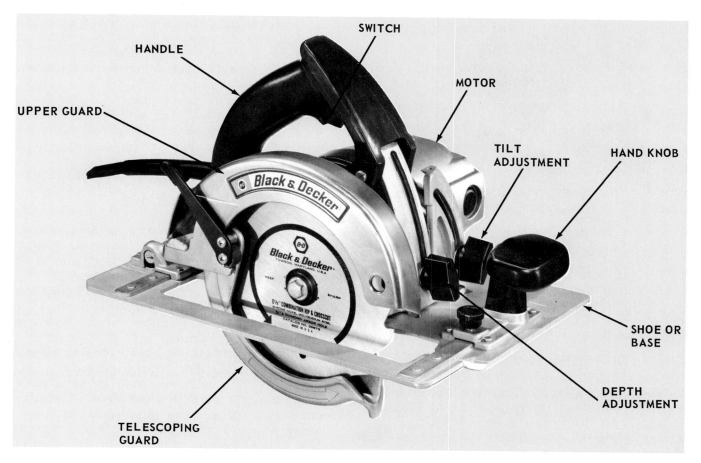

Fig. 3-3. Parts of a portable circular saw. During operation, telescoping guard is pushed back by the stock. Spring returns guard when the cut is completed.

diameter of the largest blade it will take. Most carpenters prefer a 7 or 8 in. saw. The depth of cut is adjusted by raising or lowering the position of the base or shoe. See Fig. 3-3. On many saws it is possible to make bevel cuts by tilting the shoe.

Portable saws are often guided along the layout line "free-hand." Therefore extra clearance in the saw kerf is required. To provide this clearance, teeth usually have a wide set. Fig. 3-4 shows standard types of blades. The rough-cut combination blade is popular because it is suitable for both ripping and crosscutting. Some carpenters prefer carbide tipped teeth, Fig. 3-5, because they usually stay sharp longer than teeth of a standard blade.

To use a portable saw, grasp the handle firmly in one hand with the forefinger ready to operate the trigger switch. The other hand should be placed on the stock, well away from the cutting line. Some saws require both hands on the machine.

Rest the base on the work and align the guide mark with the layout line. Turn on the switch, allow the motor to reach full speed, and then feed it smoothly into the stock as shown in Fig. 3-6. Release the switch as soon as the cut is finished. Hold the saw until the blade stops.

The portable saw may be used to make cuts in assembled work. For example, flooring and roofing boards are often nailed into place before ends are trimmed.

ROUGH CUT COMBINATION

RIP

CROSSCUT

STANDARD COMBINATION OR MITER

Fig. 3-4. Standard saw blades come in various types.

Fig. 3-5. Carbide tipped blade stays sharp longer. Handle blades carefully to prevent damage to points.

SAFETY FOR PORTABLE CIRCULAR SAWS

1. Stock must be well supported in such a way that the kerf will not close and bind the blade during the cut or at the end of the cut.
2. Thin materials should be supported near the cut. Small pieces should be clamped to a bench top or sawhorse.
3. Be careful not to cut into the sawhorse or other supporting device.
4. Adjust the depth of cut to the thickness of the stock, plus about 1/8 in.
5. Check the base and angle adjustment to be sure they are tight. Plug the cord into a grounded outlet and be sure it will not become tangled in the work.

6. Always place the saw base on the stock with the blade clear before turning on the switch.
7. During the cut, stand to one side of the cutting line. Never reach under the material being cut.
8. Some portable saws have two handles. In using such saws, be sure to keep both hands on the handles during the cutting operation.
9. Always unplug the machine to change blades or make adjustments.
10. Always use a sharp blade that is properly set.

SABER SAWS

The saber saw is also called a portable jig saw. It is useful for a wide range of light work. Carpenters, cabinetmakers, electricians, and home craftspeople use it.

A standard model and its basic parts are shown in Fig. 3-7. The stroke of the blade is about 1/2 in. The saw operates at a speed of approximately 2500 strokes per minute.

Blades for wood cutting have from 6 to 12 teeth per inch, Fig. 3-8. For general purpose work, a blade with 10 teeth per inch is satisfactory. Always select a blade that will have at least two teeth in contact with the edge being cut.

Saws will vary in the way the blade is mounted in the chuck. Follow directions in the manufacturer's manual. Also follow the lubrication schedule specified in this manual.

The saber saw can be used to make straight or bevel cuts as shown in Fig. 3-9. Curves are usually cut by guiding the saw along a layout line. How-

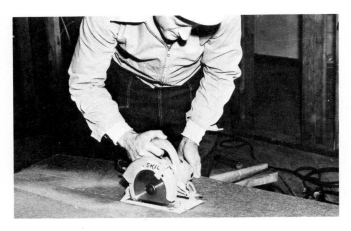

Fig. 3-6. Using a portable circular saw to cut a siding panel. Make sure work is well supported before cutting begins. (American Plywood Assoc.)

Fig. 3-7. Saber saws cut with an action much like a hand saw. It can make 1000 to 3000 no-load strokes per minute. (Bosch Power Tool Corp.)

Fig. 3-8. Saber saw blades are designed for cutting various kinds of material.

3. Disconnect the saw to change blades or make adjustments.
4. Place the base of the saw firmly on the stock before starting the cut.
5. Turn on the motor before the blade contacts the work.
6. Do not attempt to cut curves so sharp that the blade will be twisted.
7. Make certain the work is well supported. Check clearance so that you do not cut into sawhorses or other supports.

When cutting internal openings, a starting hole can be drilled in the waste stock, or the saw can be held on end so the blade will cut its own opening. See Fig. 3-10. This is called plunge cutting and must be undertaken with considerable care. Rest the toe of the base firmly on the work and turn on the motor. Then slowly lower the blade into the stock.

ever, circular cuts may be made more accurately with a special guide or attachment.

Since the blade cuts on the upstroke, splintering will take place on the top side of the work. This must be considered when making finished cuts, especially in fine hardwood plywood. Always hold the base of the saw firmly against the surface of the material being cut.

SAFETY RULES FOR SABER SAWS

1. Make certain the saw is properly grounded through the electrical cord. The switch must be in the ''off'' position before connecting to power source.
2. Select the correct blade for your work and be sure it is properly mounted.

Fig. 3-10. On internal cuts a saber saw will make its own opening. Be sure to rest the base on the workpiece.

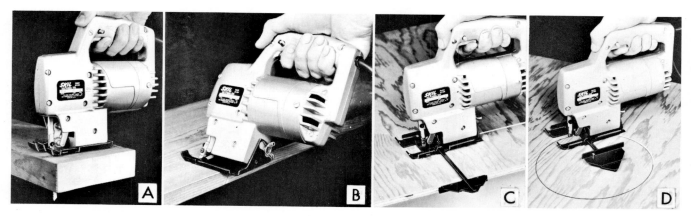

Fig. 3-9. How to use a saber saw. A—Straight cutting. B—Angle cutting. C—Using a fence. D—Using a circle cutting guide. (Skil Corp.)

Another portable power tool is the reciprocating saw shown in Fig. 3-11. Operation is similar to the saber saw.

PORTABLE ELECTRIC DRILLS

Portable electric drills come in a wide range of types and sizes. The size is determined by the chuck capacity; 1/4 and 3/8 in. generally being selected by the carpenter. Fig. 3-12 illustrates the basic parts and reduction gears of a typical model. Speeds of approximately 1000 rpm are best for woodworking. Bits designed for use in a portable electric drill are shown in Fig. 3-13.

A variable speed drill is shown in Fig. 3-14. The trigger switch has an adjusting knob for presetting desired speed.

SAFETY RULES FOR PORTABLE DRILL

1. Select the correct drill or bit for your work and mount it securely in the chuck.
2. Stock must be held so it will not move during the operation.
3. Connect the drill to a properly grounded outlet with switch in the "off" position.
4. Turn on the switch for a moment to see if the bit is properly centered and running true.
5. With the switch off, place the point of the bit in the punched layout hole.
6. Hold the drill firmly in one or both hands and at the correct drilling angle.
7. Turn on the switch and feed the drill into the work. The pressure required will vary with the size of the drill and the kind of wood being drilled. See Fig. 3-15.
8. During operation, keep the drill aligned with the direction of the hole.
9. When drilling deep holes, especially with a twist drill, withdraw the drill several times to clear the cuttings.

Fig. 3-11. Heavy-duty reciprocating saw operation is much like the saber saw. (Black & Decker Mfg. Co.)

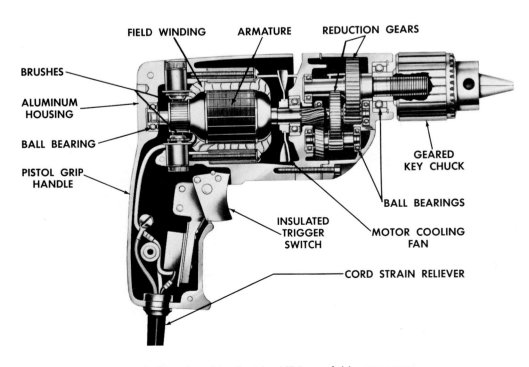

Fig. 3-12. Portable electric drill is useful in carpentry.

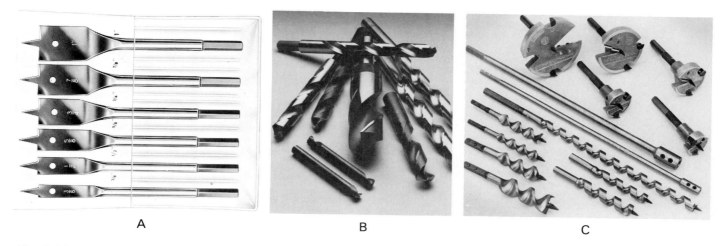

Fig. 3-13. Bits for portable electric drills come in several types. A—Spade bits for light duty. B—High speed twist drills. (Triumph Twist Drill Co.) C—From top to bottom: self feeding, large-hole boring bits; 18 in. extenders; double twist bits (left); ship augers (right). (Black & Decker Ind./Const. Div.)

10. Always remove the bit from the drill as soon as you have completed your work.

Cordless portable drills, Fig. 3-16, are handy for many jobs. Power is supplied by a small nickel-cadmium battery that can be recharged. Such drills are used for general maintenance work and on production jobs where there are no power lines.

POWER PLANES

The power plane produces finished wood surfaces with speed and accuracy, Fig. 3-17. The motor, which operates at a speed of about 20,000 rpm, drives a spiral cutter. The depth of

Fig. 3-15. Pressure required to operate a drill will vary with drill size, type of wood, and diameter of the drill bit. This drill is operating a bit cutting a large diameter hole. (Black & Decker Mfg. Co.)

Fig. 3-14. Variable speed electric drill. Direction of rotation can also be reversed. No-load speed can be varied from 0 to 1500 rpm. (Bosch Power Tool Corp.)

cut is adjusted by raising or lowering the front shoe. The rear shoe (main bed) must be kept level with the cutting edge of the cutterhead.

The power plane is equipped with a fence that is adjustable for planing bevels and chamfers. For surfacing operations, it is removed.

Hold and operate the power plane in about the same manner as a hand plane. The work should be rigidly supported in a position that will permit the

Fig. 3-16. Cordless drills are useful where power is not easily available. This one is drilling a hole in masonry. (Porter Cable)

Fig. 3-18. Power block plane. Its small size makes it useful for jobs such as cutting a chamfer.

operation to be easily performed. Start the cut with the front shoe resting firmly on the work and the cutterhead slightly behind the surface. Refer once more to Fig. 3-17. Be sure the electric cord is kept clear. Start the motor and move the plane forward with smooth, even pressure on the work. When finishing the cut, apply extra pressure on the rear shoe.

SAFETY RULES FOR POWER PLANES

1. Study the manufacturer's instructions for adjustment and operation.
2. Be sure the machine is properly grounded.
3. Hold the standard power plane in both hands before you pull the trigger switch. Continue to hold it in both hands until the motor stops after releasing the switch.
4. Always clamp the work securely in the best position to perform the operation.
5. Do not attempt to operate a power plane with one hand that was designed for two hands.

6. Disconnect the electric cord before making adjustments or changing cutters.

The power block plane, Fig. 3-18, can be used on small surfaces. It has about the same features and adjustments as the regular power plane. Being small, it is designed to be operated with one hand. When using this tool, the work should be securely held or clamped in place.

In planing small stock, kickbacks may occur. Be sure the hand not holding the plane is kept well out of the way.

PORTABLE ROUTERS

Routers are used to cut irregular shapes and to form various contours on edges, Fig. 3-19. When equipped with special guides, they can be used to

Fig. 3-19. Portable router with motor mounted in an adjustable base. Motor revolves in a clockwise direction when viewed from above. (Bosch Power Tool Corp.)

Fig. 3-17. Portable power plane is being used to plane a chamfer. Note arrow. (Bosch Power Tool Corp.)

Some heavy duty radial arm saws are equipped with a power feed. Control unit includes buttons for start and stop and control of feed drive. Knob (arrow) adjusts rate of feed.   (Black & Decker Mfg. Co.)

cut dados, grooves, mortises, and dovetail joints. Important uses in carpentry include the cutting of gains for hinges when hanging passage doors and routing housed stringers for stair construction. See Units 16 and 17.

When mounting bits in the router, the base is usually removed as illustrated in Fig. 3-20.

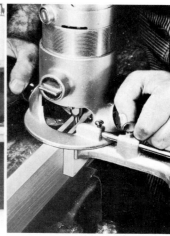

Fig. 3-21. Routers perform many shaping operations. Left. Making a dado with a series of cuts. Right. Cutting a groove in the edge of a board.

Straight bits are used when cutting dados and grooves. See Fig. 3-21.

Some bits for shaping and forming edges have a pilot tip that guides the router. The router motor revolves in a clockwise direction (when viewed from above) and should be fed from left to right when making a cut along an edge, as illustrated in Fig. 3-22. When cutting around the outside of oblong or circular pieces, always move in a

Fig. 3-20. Straight router bit is being mounted in a collet type chuck.

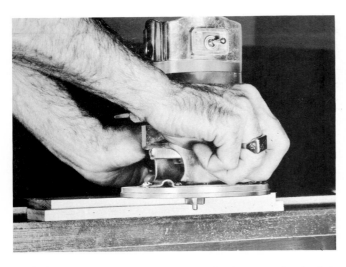

Fig. 3-22. Pilot tip on cutter controls the cut. Apply light pressure between tip and material. Too much pressure will cause burn marks.

counterclockwise direction.

Fixtures and templates are available that will guide the router through various decorative or blanking cuts. Fig. 3-23 shows a decorative cut being made on a cabinet door.

SAFETY RULES FOR PORTABLE ROUTERS

1. The bit must be securely mounted in the chuck and the base must be tight.
2. Be sure the motor is properly grounded.
3. Wear eye protection.
4. Be certain the work is securely clamped so it will remain stationary during the routing operation.
5. Place the router base on the work, template, or guide, with the bit clear of the wood, before turning on the power. Hold it firmly when turn-

ing on the motor. Starting torque could wrench the tool from your grasp.
6. Hold the router with both hands and feed it smoothly through the cut in the correct direction.
7. When the cut is complete, turn off the motor. Do not lift the machine from the work until the motor has stopped.
8. Always unplug the motor when mounting bits or making adjustments.

PORTABLE SANDERS

Portable sanders include three basic types:
1. Belt.
2. Disc.
3. Finish.

They vary widely in size and design. Manufacturer's instructions should be followed carefully in the mounting of abrasive belts, discs, and sheets. Also, follow the manufacturer's lubrication schedule.

The belt sander's size is determined by the width of the belt. Using the sander takes some skill. Support stock firmly. Switch must always be on "off" before plugging in the electric cord. Like all portable power tools, the sander should be properly grounded. Check the belt and make sure it is tracking properly.

Hold the sander over the work. Start the motor. Then, lower the sander carefully and evenly onto the surface. When using belt and finish sanders make sure to travel with the grain. Move it forward and backward over the surface in even strokes. At the end of each stroke, shift it sideways about one-half the width of the belt.

Continue over the entire surface, holding the sander level and sanding each area the same amount. Do not press down on the sander. Its weight is sufficient to provide the proper pressure for the cutting action. When work is complete, raise the machine from the surface and allow the motor to stop.

Finishing sanders, Fig. 3-24, are used for final sanding where only a small amount of material needs to be removed. They are also used for cutting down and rubbing finishing coats. There are two general types:
1. Orbital.
2. Oscillating.

STAPLERS AND NAILERS

A wide variety of power staplers and nailers is available. Most of them are air (pneumatic) powered. Those that are electrically operated should be properly grounded. Fig. 3-25 shows

Fig. 3-23. Router can be mounted in a variable arc attachment to produce a decorative design.

Stationary planer is small enough to take onto a construction job. This unit will plane stock up to 13 in. wide and nearly 6 in. thick. (Delta International Machinery Corp.)

Fig. 3-24. This orbital finishing sander is equipped with a dust bag. It operates at a speed of 10,000 orbits per minute. (Bosch Power Tool Corp.)

pneumatic and electrically powered tools that drive nails, staples, and screws.

SAFETY RULES FOR POWER STAPLERS AND NAILERS

1. Study the manufacturer's operating directions and follow them carefully.
2. Use the correct type and size of fastener recommended by the manufacturer.
3. For air-powered nailers, always use the correct pressure (seldom over 90 lb.). Be sure the compressed air is free of dust and excessive moisture.
4. Always keep the nose of the stapler or nailer pointed toward the work. Never aim it toward yourself or other workers.
5. Check all safety features and be sure they are working. Make a test by driving the staples or nails into a block of wood.
6. During use on the job, hold the nose firmly against the surface being stapled or nailed.
7. Always disconnect the power tool from the air or electrical power supply when it is not being used.

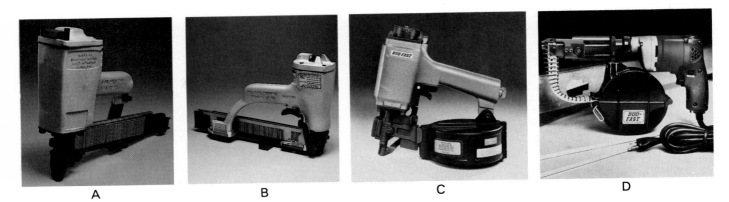

A   B   C   D

Fig. 3-25. Pneumatic and electrically powered fastening tools. A—Pneumatic stapler drives staples from 5/8 to 2 in. long. B—Roofing stapler installs staples up to 1 1/4 in. long. (Paslode Co.) C—Coil nailer can drive 25 different nails from 1 to 2 in. long. D—Automatic screw fastener is designed for driving screws into drywall to wood or metal studs. (Duo-Fast Corp.)

RADIAL ARM SAWS

Motor and blade of the radial arm saw are carried by an overhead arm. The stock is supported on a stationary table. The arm is attached to a vertical column at the back of the table. The depth of cut is controlled by raising or lowering the overhead arm. Fig. 3-26 shows the parts of a typical radial arm saw.

The motor is mounted in a yoke and may be

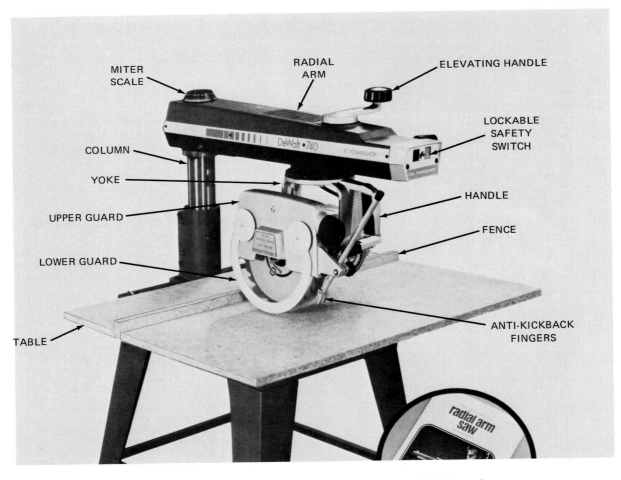

Fig. 3-26. Radial arm saw makes most cuts with material held stationary. (Black & Decker Mfg. Co.)

tilted for angle cuts. The yoke is suspended from the arm on a pivot which permits the motor to be rotated in a horizontal plane. Adjustments make it possible to perform many sawing operations.

When crosscutting, mitering, beveling, and dadoing, the work is held firmly on the table and the saw is pulled through the cut, Fig. 3-27. For ripping and grooving, the blade is turned parallel to the table and locked into position. Stock is then fed into the blade in somewhat the same manner as a table saw, Fig. 3-28.

Fig. 3-28. Ripping operation. Fence, not visible, is set in the table to guide the work.

Fig. 3-27. Crosscutting on a radial arm saw. Lower guard is in place.

For regular crosscuts and miters, first be sure the saw is against the column. Then place your work on the table and align the cut. Hold the stock firmly against the table fence with your hand at least 6 in. away from the path of the saw blade. Turn on the motor. Grasp the saw handle pulling the saw firmly and slowly through the cut. See Fig. 3-29. The saw may tend to "feed itself." You must control the rate of feed. When the cut is

completed, return the saw to the rear of the table and shut off the motor.

The radial arm saw is especially useful in cutting compound miters. It is also a good tool for cutting

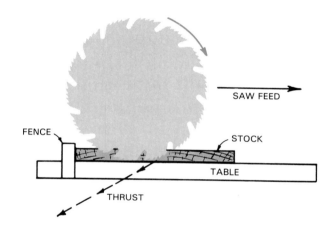

Fig. 3-29. Blade rotation is in direction of the saw feed.

Fig. 3-30. Using a radial arm saw to cut dimension lumber. Material of this size would be difficult to slide through a table saw. (Delta International Machinery Corp.)

larger dimension lumber which is difficult to slide across a saw table. See Fig. 3-30. The proper saw setup for cutting a large sheet of plywood is shown in Fig. 3-31.

SAFETY RULES FOR RADIAL ARM SAWS

1. Stock must be held firmly on the table and against the fence for all crosscutting operations. The ends of long boards must be supported level with the table.
2. Before turning on the motor, be sure clamps and locking devices are tight. Check depth of cut and table slope. It must be slightly lower at back than front to prevent blade from ''running'' forward.
3. Keep the guard and anti-kickback device in position.
4. Always return the saw to the rear of the table after completing a crosscut or miter cut. Never remove stock from the table until the

Fig. 3-31. Setup for cutting large sheet material on radial saw. Blade is turned at right angle to arm and material is pushed through.

saw has been returned.
5. Maintain a 6 in. margin of safety. Keep your hands this distance away from the path of the saw blade.

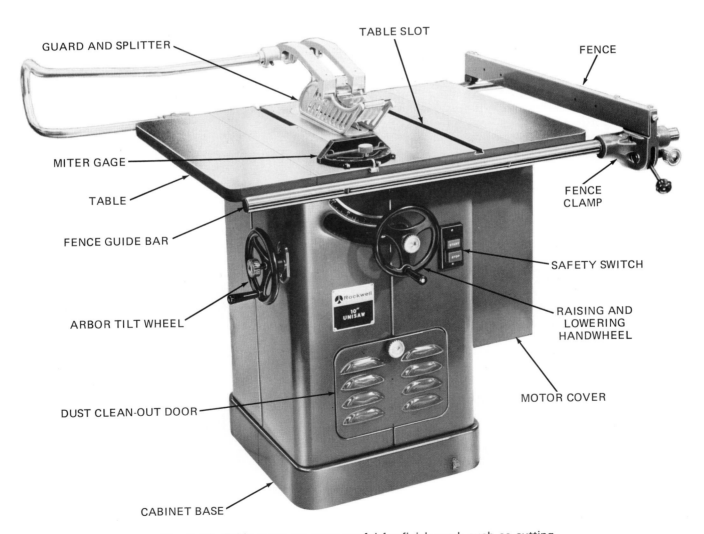

Fig. 3-32. Table saws are most useful for finish work such as cutting moldings and components for built-ins.

6. Shut off the motor and wait for the blade to stop before making any adjustments.
7. Do not leave the machine before the blade has stopped.
8. Keep the table clean and free of scrap pieces and excessive amounts of sawdust. Do not attempt to clean off the table while the saw is running.
9. In crosscutting, always pull blade toward you.
10. Stock to be ripped must be flat and have one straight edge to guide it along the fence.
11. When ripping, always feed stock into the blade so that the bottom teeth are turning toward you. This will be the side opposite the anti-kickback fingers.

TABLE SAWS AND JOINTERS

These power tools are basic machines used in cabinetmaking. They are frequently used by carpenters on projects which include on-the-job built cabinets and built-ins. Carpenters use them to some extent for cutting and fitting moldings and other inside trim work.

When used for carpentry, the smaller sizes (4 to 6 in. jointers and 8 to 10 in. table saws) are usually selected because they can be easily moved from one job to another.

Space is this book does not permit more than a brief introduction to this equipment. Woodworking textbooks provide a complete description of the wide variety of work they will do along with instruction on how to use them.

The table saw, also called a circular saw, is used for ripping stock to width and cutting it to length. It also will cut bevels, chamfers, and tapers. Properly set up, the table saw can be used to produce grooves, dados, rabbets, and other forms basic to a wide variety of joints. The size of the saw is determined by the largest blade it will take. Fig. 3-32 shows a typical model with the parts identified.

Fig. 3-34. Crosscutting. Above. Squaring stock to a marked line. Below. Guard has been removed to show operation.

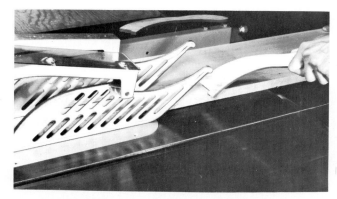

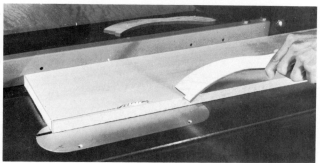

Fig. 3-33. Ripping stock with table saw. Guard should always be in place for safe operation. Top. Guard in position. Bottom. Guard removed to show operation.

Stock to be ripped must have at least one flat face to rest on the table and one straight edge to run along the fence. Fig. 3-33 shows correct procedure for making a ripping cut. Be sure to follow the safety rules.

Fig. 3-34 shows a standard crosscutting operation. A line is squared across the stock to show where the cut is to be located. For accurate work make a check mark on the side of the line where

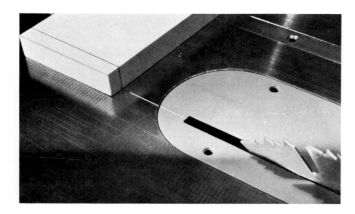

Fig. 3-35. Line scribed on saw table will help you align stock for more accurate cutting to a line.

the saw kerf will be located. The guard tends to hide the blade; it is helpful, when aligning the cut, to use a line scribed in the table surface. Since most of the work will be located to the left, it should extend back from the left side of the blade as shown in Fig. 3-35.

Fig. 3-36 shows a compound angle being sawed on a table saw. To make these cuts both the miter gage and the saw blade must be set at an angle. Tables of compound angles are available to determine the proper angles.

SAFETY RULES FOR TABLE SAWS

1. Be certain the blade is sharp and right for the job at hand.
2. Make sure the saw is equipped with a guard and use it.
3. Set the blade so it extends about 1/4 in. above the stock to be cut.
4. Stand to one side of the operating blade and do not reach across it.
5. Maintain a 4 in. margin of safety. (Do not let your hands come closer than 4 in. to the operating blade even though the guard is in position.)
6. Stock should be surfaced and at least one edge jointed before being cut on the saw.
7. Use the fence or miter gauge to control the stock. Do not cut stock free hand.
8. Always use push sticks when ripping short, narrow pieces.
9. Stop the saw before making adjustments.
10. Do not let small scrap cuttings accumulate around the saw blade. Use a push stick to move them away.
11. Resawing setups and other special setups must be carefully made and checked before the power is turned on.

12. Remove the dado head or any special blades after use.
13. Other workers, helping to "tail-off" the saw, should not push or pull on the stock but only support it. The operator must control the feed and direction of the cut.
14. As work is completed, turn off the machine and remain until the blade has stopped. Clear the saw table and place waste in a scrap box.

Principal parts of a jointer are shown in Fig. 3-37. The cutter head holds three knives and revolves at a speed of about 4500 rpm. The size of the jointer is determined by the length of these knives.

The three main adjustable parts are:
1. The infeed table.
2. The outfeed table.
3. The fence.

The outfeed table must be the same height as the knife edges at their highest point of rotation. This is a critical adjustment. See Fig. 3-38. If the table is too high the stock will be gradually raised out of the cut and a slight taper will be formed. If it is too low, the tail end of the stock will drop as it leaves the infeed table and cause a "bite" in the surface or edge.

The fence guides the stock over the table and knives. When jointing an edge square with a face, it should be perpendicular to the table surface, Fig. 3-39. The fence is tilted when cutting chamfers or bevels.

SAFETY RULES FOR JOINTERS

1. Before turning on the machine, make adjustments for depth of cut and position of fence.

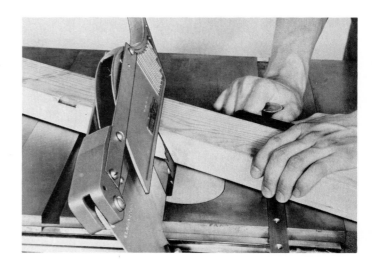

Fig. 3-36. Cutting a compound angle on 2 x 4 stock. Both blade and miter gage are set at an angle. Rockwell International, Power Tool Div.)

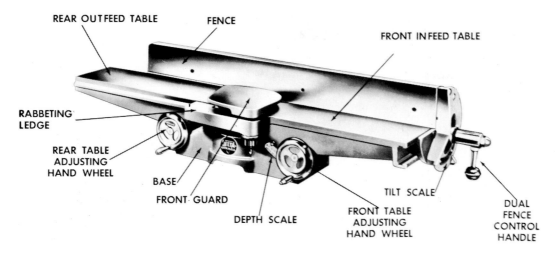

Fig. 3-37. Small 4 to 6 in. jointers can be moved easily from one job site to another.

Be sure the guard is in place and is operating properly.

2. The maximum cut for jointing on a small jointer is 1/8 in. for an edge and 1/16 in. for a flat surface.
3. Stock must be at least 12 in. long. Stock to be surfaced must be at least 3/8 in. thick unless a special feather board is used.
4. Feed the work so the knives will cut "with the grain." Use stock that is free from knots, splits, and checks.
5. Keep your hands away from the cutter head even though the guard is in position. Maintain at least a 4 in. margin of safety.
6. Use a push block when planing a flat surface. Do not apply pressure directly over the knives with your hand.
7. Do not plane end grain unless the board is at least 12 in. wide.
8. The jointer knives must be sharp. Dull knives will vibrate the stock and may cause a kickback.
9. When work is complete, turn off the machine. Stand by until the cutter head has stopped.

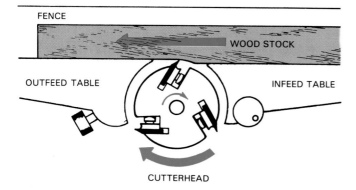

Fig. 3-38. How a jointer works. Note direction of the wood grain. Avoid working against the grain.

SPECIAL SAWS

Special saws have been developed for certain uses in working with wood construction. If light weight and compact size are important factors, carpenter may use a power miter box or a frame and trim saw for accurate crosscuts and mitering. The motor and blade of the power miter saw are supported on a pivot. To operate it, the carpenter sets the angle from a scale marked off in degrees. The cut is made by pulling downward on the handle. A trigger control in the handle turns the motor on and off. See Fig. 3-40.

Fig. 3-39. Jointing an edge. "Step" the hands along the stock so they will not bear down on the stock while it passes over the cutter head.

The frame and trim saw is supported on a pair of overhead shafts or guides. The support rotates left and right a little more than 45 deg. to make crosscuts and miter cuts. It is capable of all sawing operations except ripping. Its capacity is 16 in. width on crosscuts and 12 in. on miter cuts of 45 deg. An extension table allows one worker to cut long stock alone. See Fig. 3-41.

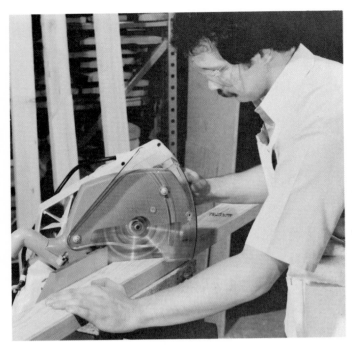

Fig. 3-40. Using a power (motorized) miter box to make an angle cut on a 2 x 4. Cuts can be made either to the right or left as with a regular miter box.

SAW SAFETY

1. Keep guards in place while operating.
2. Wear safety glasses or a face shield to protect eyes from sawdust and other debris.
3. Lock the saw securely at the angle of the cut.
4. Hold stock firmly against the fence.
5. Keep free hand clear of the cutting area.
6. Work only with a sharp saw blade.

Fig. 3-41. Sawbuck frame and trim saw being used to make a compound miter cut. It crosscuts, miters, bevels, and makes compound cuts on any stock up to 2 x 12. (Delta International Machinery Corp.)

Fig. 3-42. Skilled worker sharpens carbide tipped circular saw blade. (Foley Mfg. Co.)

POWER TOOL CARE AND MAINTENANCE

Care of power tools is especially important if they are to function properly while giving long service. Sharp blades and cutters ensure accurate work and make the tool much safer to operate. The good carpenters take pride in their tools' condition and appearance.

Most power tools are equipped with sealed bearings that seldom need attention. Follow the manufacturer's recommendations for lubrication schedules. Gear mechanisms for portable power tools usually require a special lubricant. All equipment will require a few drops of oil on controls and adjustment of bearings from time to time.

Clean and polish bare metal surfaces with 600 wet-or-dry abrasive paper when required. These surfaces can be kept smooth and clean by wiping them occasionally with light oil or furniture polish. Some carpenters apply a coat of paste wax to protect the surface and reduce friction.

Some power tools, especially those with a number of accessories, can be purchased with a case. While making transport easier, such cases keep the accessories organized and protected.

Cutters and blades require periodic sharpening. Most carpenters are too busy and usually do not have the equipment to accurately grind cutters or completely fit saw blades. They usually send these items to a saw shop where an expert job can be performed. Carpenters may, however, lightly hone cutters before grinding operations are required. Also they may prefer to file saw blades several times before sending them in for a complete fitting. A fitting includes jointing, gumming, setting, and filing.

Carbide tipped tools will stay sharp at least 10 times longer than those with regular steel edges. A special diamond grinding wheel is used to sharpen carbide edges, Fig. 3-42.

TEST YOUR KNOWLEDGE — UNIT 3

1. What is a ground fault interrupter and why should it be used in carpentry?
2. The size of a portable circular saw is determined by the _____.
3. For general purpose work, a saber saw blade should have about _____ teeth per in.
4. When the base of the saber saw rests on a horizontal surface, the blade cuts on the _____ (up, down) stroke.
5. When drilling deep holes do not withdraw a twist drill until the hole is completed. True or False?
6. The depth of cut of a power plane is adjusted by raising or lowering the _____.
7. A standard router bit is held in a _____ type chuck.
8. The size of a belt sander is determined by the _____.
9. To adjust the depth of cut of a radial arm saw, the _____ is raised or lowered.
10. When crosscutting with the radial arm saw, the blade is _____ (pushed away, pulled toward) the operator.
11. For regular work, the _____ of the jointer should be perfectly aligned with the knife edges at their highest point.
12. What cuts can be performed with a frame and trim saw?

OUTSIDE ASSIGNMENTS

1. Visit a builder's supply center and study the various portable circular saws on display. Also, secure descriptive literature concerning the various sizes. After careful consideration, select a brand, model, and size that you believe would be best for rough framing and sheathing work on residential structures. Give your reasons and report to your class. Include specifications and prices of your selection.
2. Visit with a carpenter in your locality and learn what procedures are followed in maintaining and sharpening tools. If he or she uses standard saw blades, learn how they are kept sharp. Secure a reaction to the use of hardened tooth and carbide tipped blades. If he or she has some tools sharpened at a saw shop, find approximate prices. If a tool maintenance center is located in your area, find out what services are available and what they cost. Prepare your notes carefully, then make a report to your class.

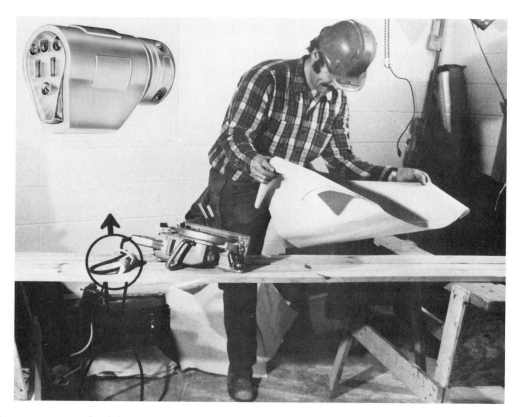

For electrical safety on construction job sites the ground continuity monitor is a constant check that an extension cord is properly wired and grounded. The monitor (inset) is wired into the extension cord. A light on the monitor will glow when the cord is grounded and wired properly. (Daniel Woodhead Co.)

Leveling instruments are handy tools on a building site. They are useful in finding building lines and in assuring that footings, foundations, and walls are square and plumb. The instrument shown here is called an automatic level. It is accurate to 1/16 in. at 150 ft.

Worker checks level of grade stake using an electronic device that generates an infrared laser beam. The beam, automatically maintained in an absolutely level position, can rotate over the entire building site at 420 rpm. A special detector is mounted on the rod to locate the exact level of the beam. Any number of rodmen can lay out or check construction levels using the same beam generator. The same equipment can be used to establish vertical lines and planes.
(David White Instruments, Div. of Realist, Inc.)

Unit 4

LEVELING INSTRUMENTS

Before construction of a foundation or a slab for a building can begin, the carpenter must know where the structure will be located on the property. Most communities have strict requirements. Buildings must be set back certain distances from the street and must maintain minimum clearances from other property. The local code must be checked carefully for these requirements before layout begins.

PLOT PLAN

Many communities require the builder or owner to furnish a plot plan before a building permit is issued, Fig. 4-1. This plan, as well as a survey of the lot (property), may already have been provided by a surveyor.

If there is a plot plan, it will indicate the location of the structure and indicate distances to property lines. There will be stakes or markers on the property to indicate property lines.

Surveyors should locate the property lines, in any event. They should also draw up the plot plan if one is required. The help of an engineer or surveyor will protect the owner and builder against costly errors in measurement.

ESTABLISHING BUILDING LINES

Building lines are the lines marking where the walls of the structure will be. They are the lines that must conform to the code requirements on distance of the structure from boundary lines.

Once the property lines are known and marked by the surveyor, the building lines can be found by measuring off the distance with tapes. Be sure to observe proper setbacks and clearances. See Fig. 4-2. Check the local code carefully for compliance.

MEASURING TAPES

For measurements and layouts involving long distances, steel tapes, called measuring tapes,

may be used. They are housed in winding reels like the one shown in Fig. 4-3. Tapes are available in lengths of 50 to 300 ft. There are various types and the graduations are different from one to the other.

The carpenter will usually select one that is marked off in feet, inches, and eighths like two that are shown. Surveying, on the other hand, requires a tape graduated in feet and decimal parts of a foot.

USING TAPES TO LOCATE LINES

To use measuring tapes for finding building lines first find the boundary lines. Then measure off the required setback and clearances. Make sure the measurement is taken with the tape perpendicular to the boundary line. There are two ways of checking this:
1. Swinging an arc from two dimensions.
2. Using the "6—8—10" method.

In the arc method, the tape is extended from one property line to the dimension being measured. Then make a mark on the ground while swinging the tape back and forth. The marking instrument will make a curved line (arc) on the ground. Measure off the other dimension from an intersecting property line and again make an arc at the correct length. The point where the arcs intersect (cross) marks the spot where the intersecting lines are perpendicular.

In the "6—8—10" method, measure off 6 ft. along one line and 8 ft. on the intersecting line. Then measure diagonally from one mark to the other. When the corner formed by the intersecting lines is 90 degrees (perpendicular) the diagonal line will measure 10 ft. Fig. 4-4 shows both methods.

LAYING OUT WITH LEVELING INSTRUMENTS

In residential construction, especially in built-up areas of a community, it is important that building

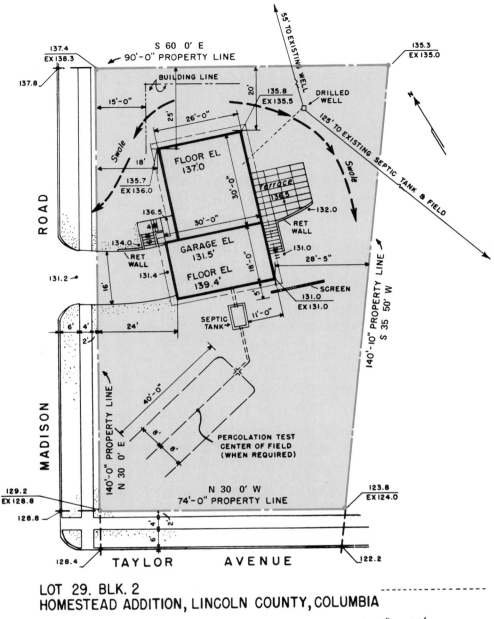

LOT 29. BLK. 2
HOMESTEAD ADDITION, LINCOLN COUNTY, COLUMBIA

Scale 1" = 20'

Fig. 4-1. Plot plan. Many communities require one of the
builder or owner before a building permit will be issued.

lines be accurately established in relation to lot
lines. It is also important that footings and foundation walls be:
1. Level.
2. Square.
3. The correct size.
If the building is small, the carpenters' level,
framing square, and a rule are accurate enough for
laying out and checking the building lines. But, as
size increases, special leveling instruments are
needed for greater accuracy and efficiency. They

should be used for single family dwellings as well
as for commercial structures and institutional
building.

LEVELING INSTRUMENTS

The level and level-transit are commonly used in
laying out and checking construction work. These
instruments can also be used for surveying and
other space and land-layout jobs. When the job is
too large for the chalk line, straightedge, level, and

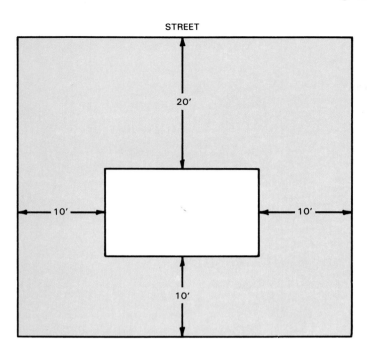

Fig. 4-2. Simple rectangular structure laid out by taking measurements using lot lines as reference points.

square, leveling instruments should be used.

The instruments include an optical device which operates on the basic principle that a LINE OF SIGHT is a straight line that neither dips, sags, nor curves. Any point along a level line of sight will be the same height as any other point. Through the use of these instruments, the line of sight replaces the chalk line and straightedge.

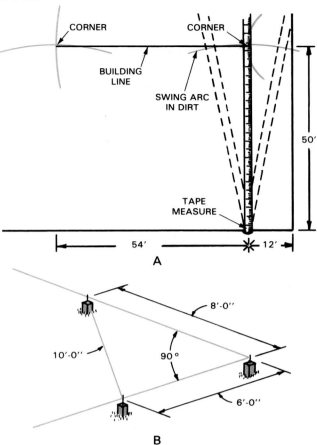

Fig. 4-4. Methods for checking that intersecting lines are perpendicular (create an angle of 90 degrees when they meet). A—Swinging an arc. B—Marking off 6 ft. on one line and 8 ft. on the intersecting line will result in a measurement of 10 ft. across the marks if the lines are perpendicular to each other.

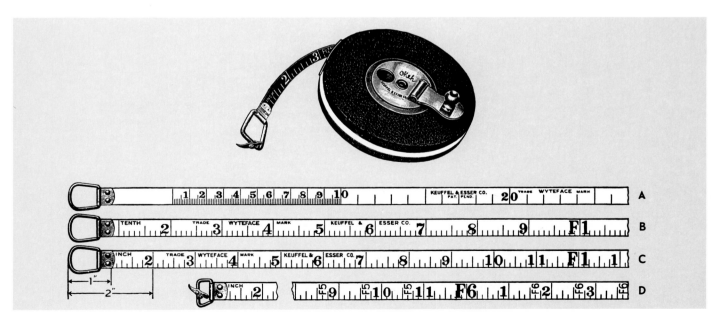

Fig. 4-3. Measuring tape. Top. Graduations should be in feet, inches, and eighths. (Keuffel and Esser Co.) Bottom. Measuring tapes are made with different systems and graduations. A—Metric. B—Feet and decimal graduations. C—Feet, inch, and eighth inch graduations. D—Feet and inches with feet repeated at each inch mark.

Fig. 4-5. Builders' level is used to sight level lines and lay out or measure horizontal lines. A—Telescope lens. B—Focusing knob. C—Instrument level vial. D—Eyepiece. E—Index vernier. F—Horizontal tangent screw. G—Horizontal clamp screw. H—Tripod mounting stud. I—Four leveling screws. J—Horizontal graduated circle.

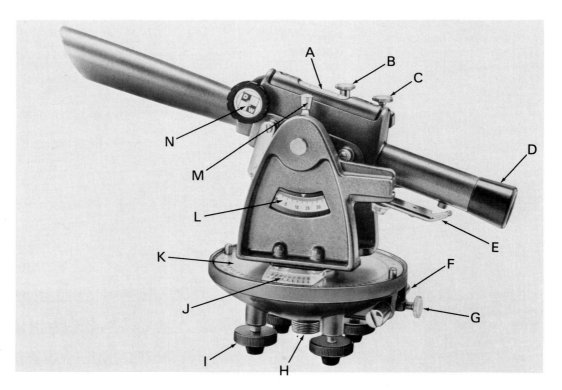

Fig. 4-6. Level-transit can be used to lay out or check level and plumb lines. It can also be used to measure angles in either horizontal or vertical planes. A—Instrument level vial. B—Vertical clamp screw. C—Vertical tangent screw. D—Eyepiece. E—Telescope lock lever. F—Horizontal tangent screw. G—Horizontal clamp screw. H—Tripod mounting stud. I—Four leveling screws. J—Index vernier. K—Horizontal graduated circle. L—Vertical arc. M—Vertical arc pointer. N—Focusing knob.
(David White Instruments, Div. of Realist, Inc.)

The builders' level, also called a dumpy level or an optical level, is shown in Fig. 4-5. It consists of an accurate spirit level and a telescope assembly. These are attached to a circular base. Leveling screws are used to adjust the base after the instrument has been mounted on the tripod. The telescope rotates on the base so that any angle in a horizontal plane can be laid out or measured.

The level-transit, Fig. 4-6, works like the builders' level. An additional feature permits the telescope to be pivoted up and down in a vertical plane. Using this instrument, it is possible to accurately measure vertical angles or determine if a wall is perfectly plumb (vertical). Its vertical movement also simplifies the operation of aligning a row of stakes, especially when they vary in height.

In use, both the builders' level and level-transit are mounted on tripods, Fig. 4-7. Some models, like the one shown, have legs whose length is adjustable making the tripod easier to use on sloping

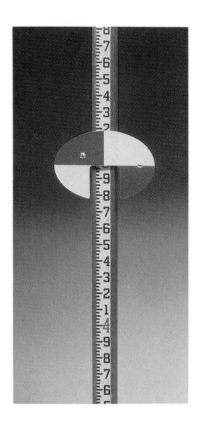

Fig. 4-8. Leveling rod with target. The target can be moved up or down to match the line of sight of the level-transit. (David White Instruments, Div. of Realist Inc.)

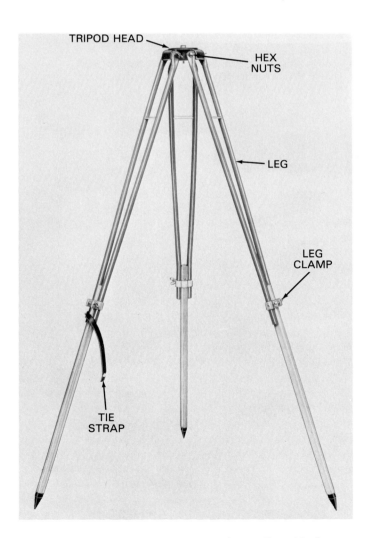

Fig. 4-7. Tripod legs hinge at top and are adjustable for use on uneven terrain.
(David White Instruments, Div. of Realist, Inc.)

ground. This feature also permits the legs to be shortened for handling and storing.

When it is necessary to sight over long distances, a leveling rod is used. See Fig. 4-8. It is designed so that differences in the elevation between the position of the level and various positions where the rod is held can be easily read. The rod is especially useful for surveying. Readings can be made by the person operating the level or the target can be adjusted up and down to the line of sight and then the rodworker (one holding the rod) can make the reading.

The rod shown has graduations in feet and decimal parts of a foot. This is the type used for regular surveying work. Rods are also available with graduations in feet and inches.

When sighting short distances (100 ft. or less) a regular folding rule can be held against a wood strip and read through the instrument. This procedure will be satisfactory for jobs such as setting grade stakes for a footing. Always be sure to hold the strip and rule in a vertical position.

CARE OF LEVELING INSTRUMENTS

Leveling instruments are more delicate than most other tools and equipment. Special precautions must be followed in their use so they will continue to provide accurate readings over a long

A level-transit can be used to check the plumb of walls in new structures. The horizontal clamp screw is locked and the telescope can swing in a vertical arc.   (David White Instruments, Div. of Realist, Inc.)

Instrument worker on a building site is shooting a plumb line with the level-transit. Note he is giving a hand signal to a rodworker.   (David White Instruments, Div. of Realist, Inc.)

Leveling instruments and equipment will vary somewhat depending on the manufacturer. Always read and study carefully the instructions for a given brand.

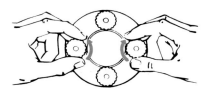

Fig. 4-9. Adjust leveling screws to center bubble in level vial.

period of time. Some suggestions follow:

1. Keep the instrument clean and dry. Store it in its carrying case when it is not in use.
2. When the instrument is set up, have a plastic bag or cover handy to use in case of rain. Should it become wet, dry it before storing.
3. It is best to grip the instrument by its base when moving it from the case to the tripod.
4. Never leave the instrument unattended when it is set up near moving equipment.
5. When moving a tripod-mounted instrument, handle it with care. Hold it upright, never carry it in a horizontal position.
6. Never over-tighten leveling screws or any of the other adjusting screws or clamps.
7. Always set the tripod on firm ground with the legs spread well apart. When set up on floors or pavement, take extra precautions to insure that the legs will not slip.
8. For precision work, permit the instrument to reach air temperature before making readings.
9. When the lenses collect dust and dirt, clean them with a camel's hair brush or special lens paper.
10. Never use force on any of the adjustments. They should turn easily by hand.
11. Have the instrument cleaned, oiled, and checked yearly by a qualified repair station or by the manufacturer.

SETTING UP THE INSTRUMENT

Set up the tripod so it will be a firm and stable base for the instrument. Make sure the points are well into the ground. On hard surfaces, be sure the points will not slip.

Check the wing nuts on the adjustable legs. They should be tight. Tighten the hex nuts holding the legs to the head to the tension desired. Use a 1/2 in. open end wrench.

The legs should have a spread of about 3 1/2 ft. Adjust them so the head appears to be level.

Lift the instrument carefully from its case by the base plate. Before mounting the instrument, loosen the clamp screws. On some instruments the leveling screws must be turned up so the tripod cup assembly can be hand tightened to the instrument mounting stud. The telescope lock lever of the level-transit should be in the closed position.

Now, attach the instrument to the tripod. If it is to be located over an exact point, such as a bench mark, attach the plumbing bob and move the instrument over the spot. Do this before the final leveling.

LEVELING THE INSTRUMENT

Leveling the instrument is a very important operation in preparing it for use. None of the readings taken or levels sighted will be accurate unless the instrument is level throughout the work. First, release the horizontal clamp screw and line up the telescope so it is directly over a pair of the leveling screws. Grasp the two screws between the thumb and forefinger as shown in Fig. 4-9. Turn both screws uniformly with your thumbs moving toward each other or away from each other. Keep turning until the bubble of the level vial is centered between the graduations. You will find that on most instruments the BUBBLE WILL TRAVEL IN THE DIRECTION THAT YOUR LEFT THUMB MOVES. See Fig. 4-10. Leveling screws should bear firmly on the base plate. Never tighten the screws so much that they bind.

When the bubble is centered, turn the telescope 90 deg. over the other set of leveling screws and repeat the leveling operation. Now check and recheck the instrument over each pair of screws. When the instrument is level, the telescope can be

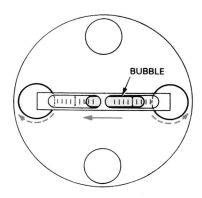

Fig. 4-10. The bubble of the level vial will generally move in the same direction as the left thumb.

turned in a complete circle without any change in the bubble.

SIGHTING

The telescope will magnify or enlarge the image (object being sighted). Most builders' levels have a telescope with a power of about 20X. This means that the object will appear to be only one-twentieth of its actual distance.

First, line up the telescope by sighting along the barrel and then look into the eyepiece, Fig. 4-11. Adjust the focusing knob until the image is clear and sharp. When the cross hairs are in approximate position on the object, Fig. 4-12, tighten the horizontal-motion clamp. Make the final alignment by turning the tangent screw.

USING THE INSTRUMENTS

Leveling instruments can be used by the carpenter to prepare the building site for excavation and/or leveling. Jobs that can be done with them include:
1. Locating the building lines and laying out horizontal angles (square corners).
2. Finding grade levels and elevations.

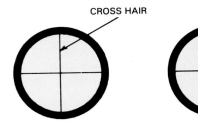

Fig. 4-12. View through the telescope. Left. Cross hairs split the image area in half vertically and horizontally. Right. Object in view should be centered on cross hairs.

3. Determine plumb (vertical) lines.

For layout, the builders' level or level-transit must start from a reference point. This can be a stone marker in the ground or a point on a manhole cover or a mark on a permanent structure nearby. The point where the instrument is located is called the STATION MARK. It may be the bench mark or the corner of the property or a previously marked point which is to be a corner of the building.

THE HORIZONTAL GRADUATED CIRCLE

Laying out corners with the transit requires an understanding of how the horizontal graduated

Fig. 4-11. Sighting a level line with a builders' level. Note that both eyes are kept open during sighting. This reduces eyestrain and provides the best view. (David White Instruments, Div. of Realist, Inc.)

circle is marked. It is divided into spaces of 1 degree of a circle, Fig. 4-13. When you swing the telescope of the builders' level or level-transit, the graduated circle remains stationary but another scale called the vernier scale moves. It is marked off in 15 minute intervals. When laying out or measuring angles where there are fractions of degrees involved you will use this vernier scale.

The upper half of Fig. 4-14 shows a section of the graduated circle and the scale. It reads 75 degrees. Notice that the zero mark on the vernier lines up exactly with the 75 degree mark. Now look at the lower half of Fig. 4-14. The zero mark has moved past the mark for 75 degrees but is not on 76 degrees. You need to read along the vernier scale until you find a mark that is closest to being directly over a circle mark. That number is 45. The reading is 75 degrees plus the number on the vernier, 45 minutes.

Vernier scales will not be the same on all instruments You should study the operator's manual for instructions about the particular model you are using.

LAYING OUT AND STAKING A HOUSE

Staking out usually proceeds after one building line has been established using a property line as reference and measuring in the proper distance. Referring to Fig. 4-15, this would be line AB. The

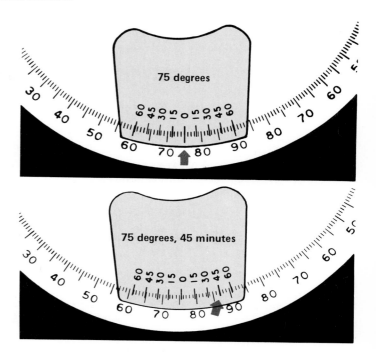

Fig. 4-14. Reading the horizontal circle of a transit. Top. When zero mark of vernier is right on a degree mark, the reading is an even degree. Bottom. When zero mark is between degrees, read across vernier to find minute mark that aligns with a degree mark.
(David White Instruments, Div. of Realist, Inc.)

next step is to attach a plumb bob to the center screw or hook on the underside of the instrument. Shift the tripod about until the point of the plumb bob is directly over the point marking the corner. This is usually marked on a stake about 2 in. square. It may have a nail or tack marking the exact spot.

Level the instrument before proceeding further. Check the plumb bob and then turn the telescope so the vertical cross hair is directly in line with the edge of the rod held at station B, Fig. 4-15, Left. Lay out the distance to the second corner with the measuring tape.

Set the horizontal circle on the instrument at zero to align with the vernier zero and swing the instrument 90 degrees (or any required angle). Position the rod along line AC so it aligns with the cross hairs. The distance to station C is also laid out with the measuring tape.

Move the instrument to station C, sight back to station A and then turn 90 degrees to locate the line of sight to station D. Measure the required distance to D. If the shape is a perfect rectangle or square, you will have completed the layout. However, you may want to move the instrument to station D to check your work.

In practice, you will find that it is difficult to locate a stake in a single operation. This is especially true when using a builders' level where the

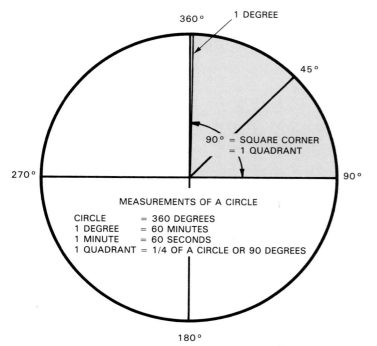

Fig. 4-13. Graduated circle of a transit corresponds to the 360 degrees of a full circle. Ninety degrees represents a quadrant which would give you a square corner for a building.

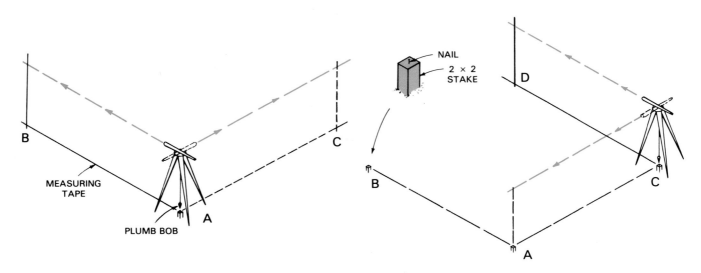

Fig. 4-15. Steps for laying out building lines. Left. Locate instrument over stake marking corner and line up "0" on the instrument circle with building line AB. Swing the instrument 90 degrees to establish line AC. Right. Move instrument to point C to establish point D. Rod must be kept vertical with plumb line or carpenters' level.

line of sight must be "dropped" with a plumb line to ground level. Usually it is best to set a temporary stake as in Fig. 4-16. Mark it with a line sighted from the instrument. Then with the measuring tape pulled taut and aligned with the mark, drive the permanent stake and locate the exact point as shown.

All major rectangles and squares of a building line can be laid out using leveling instruments in the manner just described. After batter boards are set and lines attached, the carpenters' level and square can be used to locate stakes for small projections and irregular shapes.

FINDING GRADE LEVEL

Finding the difference in the grade level between several points or transferring the same level from

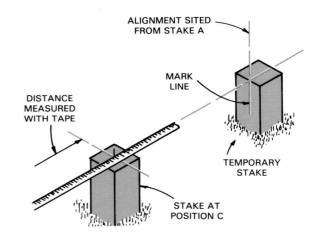

Fig. 4-16. Temporary stake may be used to establish an exact point. First set the temporary stake and mark the line of sight on it. Drive second stake and transfer mark from stake A.

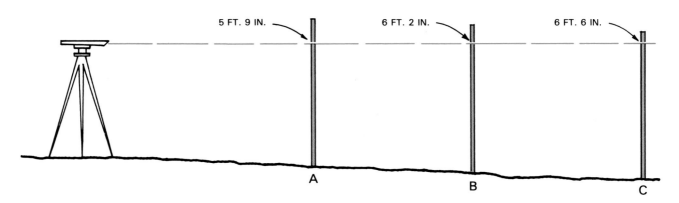

Fig. 4-17. Establishing a grade level. Point A is 9 in. higher than point C.

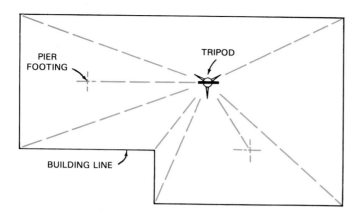

Fig. 4-18. A central location for the instrument will speed up finding and setting grade stakes.

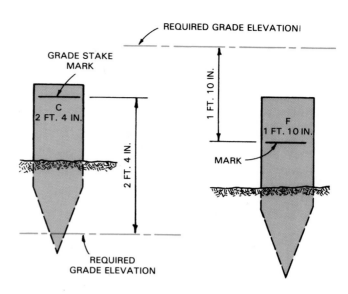

Fig. 4-19. Cut and fill stakes are used to tell excavator how much fill to remove or add to reach grade level. Letter "C" means cut (remove). Letter "F" means fill (add).

one point to another is called grade leveling. With the instrument level, the line of sight will also be level and the readings, shown in Fig. 4-17, can be used to calculate the difference in elevation. When there is a great amount of slope on the building site, the instrument can be set up between the points. The reading is taken with the rod in one position and then the instrument is carefully rotated 180 degrees to get the reading at a second position.

When setting grade stakes for a footing or erecting batter boards, the instrument should be set in a central location as shown in Fig. 4-18. The distances will be about equal and it will reduce the need for changing focus on each corner. An elevation established at one corner can be quickly transferred to other corners or points in between.

SETTING FOOTING STAKES

Grade stakes for footings are usually set to the approximate level by "eye" judgment. They are then carefully checked with the rod and level as they are driven deeper. The top of each stake should be at the required elevation.

Sometimes reference lines are drawn on construction members or stakes near the work. Then the carpenter transfers them to the forms with a carpenters' level and rule.

There may be situations where the existing grade will not permit the setting of a stake or reference mark at the actual level of the grade. In such cases a mark is made on the stake with the information on how much fill to add or remove. The letters "C" and "F" standing for CUT and FILL, are generally used. See Fig. 4-19 for an example of how stakes are marked.

When laying out sloping building plots or "carrying" a benchmark (an offically established eleva-

tion) to the building site, it will likely be necessary to set up the instrument in several locations. Fig. 4-20 shows how reading from two positions is used to calculate or establish grade levels.

CONTOUR LINES

Contour lines are lines that run through points of equal level. Such lines can be laid out on an area of ground by setting up the instrument at an appropriate point and directing the rodworker to hold the rod at the beginning point. Sight the rod and set the target on the horizontal cross hair. The rodworker then moves to the next required location and moves the rod up and down the slope until the target again aligns with the scope. Set a stake and repeat the procedure as many times as required.

RUNNING STRAIGHT LINES WITH TRANSIT

Although the builders' level can be used to line up stakes, fence posts, poles, and roadways, more accuracy is gained with the level-transit, especially when different levels are involved.

Set the instrument directly over the reference point. Level the instrument and then release the lock that holds the telescope in the level position. Swing the instrument to the required direction or until a stake is aligned with the vertical cross hair. Tighten the horizontal circle clamp so the telescope can move only in a vertical plane. Now by pointing the telescope up or down, any number of points can be located in a perfectly straight line. See Fig. 4-21.

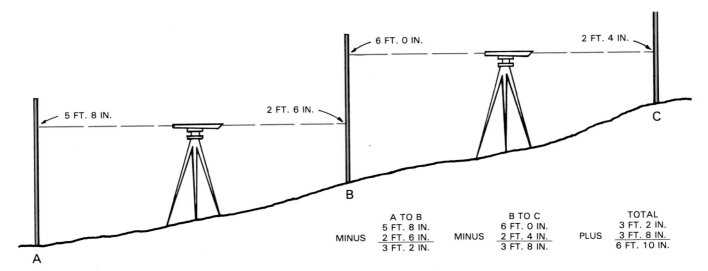

Fig. 4-20. When there is a great deal of slope on the property or when long distances are involved, the transit or level-transit will need to be set up in two or more locations.

	A TO B		B TO C		TOTAL
	5 FT. 8 IN.		6 FT. 0 IN.		3 FT. 2 IN.
MINUS	2 FT. 6 IN.	MINUS	2 FT. 4 IN.	PLUS	3 FT. 8 IN.
	3 FT. 2 IN.		3 FT. 8 IN.		6 FT. 10 IN.

VERTICAL PLANES AND LINES

The level-transit can be used to:
1. Measure vertical angles.
2. Lay out and check building walls, flagpoles, or TV antenna masts.

To establish or check vertical lines and planes, first level the instrument, then release the lever that holds the telescope in a horizontal position. Swing the instrument vertically and horizontally until a reference point in the required plane or line is sighted and then lock the horizontal clamp screw. As you tilt the telescope up and down, all of the points sighted will be located in the same vertical plane, Fig. 4-22.

Plumb lines can be checked or established by first operating the instrument as shown. Then move it to a second position, usually 90 degrees either to the right or left and repeat the procedure.

To measure or lay out angles in a vertical plane, follow the same general procedure that was used for horizontal angles.

A plumb bob and line may often be the most practical way to check vertical planes and lines. For layouts inside a structure where a regular builders' level or level-transit is impractical, use a plumb line as shown in Fig. 4-23.

TEST YOUR KNOWLEDGE — UNIT 4

1. What are building lines?
2. Explain how to check perpendicularity of intersecting lines.
3. In the use of leveling instruments, the_____ replaces the chalk line and straightedge.
4. The builders' level consists of a telescope assembly that is mounted on a_____ base.
5. For surveying work, a measuring tape with graduations reading in feet and _____ is usually selected.
6. The most important operation in setting up a builders' level or level-transit is the _____.

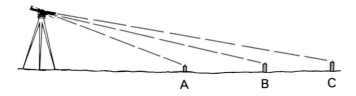

Fig. 4-21. How to use the level-transit to align a row of stakes.

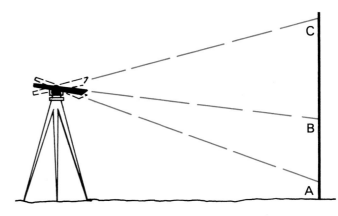

Fig. 4-22. A level-transit can be used to lay out or check points in a vertical plane.

Fig. 4-23. Carpenter uses plumb line (arrow) to make a layout on the floor exactly below a specified point on an open stair. (Stanley Tools)

7. When sighting through the telescope, you should adjust the _____ until the image is sharp and clear.
8. When setting grade stakes for a building footing, the instrument should be set up in a _____ location.
9. To position a leveling instrument directly over a given point, a _____ _____ is used.
10. A circle is divided into degrees, _____, and _____.

OUTSIDE ASSIGNMENTS

1. Study the catalog of a supplier or manufacturer and develop a set of specifications for a builders' level. Be sure it includes a good carrying case. Also select a suitable tripod and measuring tape. Secure list prices for all of the items.
2. Through drawings and a written description, tell how you would proceed to lay out a baseball diamond using a level-transit.
3. Make a study of the procedures you would follow and calculations you would make to determine the height of a flagpole, tall building, or mountain, using the level-transit and trigonometric functions. Prepare a report for your class.

Fig. 5-1. Beautiful renderings in full color help buyers visualize how a new dwelling will look. Plans for this house are shown throughout this unit. (L.F. Garlinghouse Co.)

Unit 5

PLANS, SPECIFICATIONS, CODES

In building construction, a good plan and a well defined contract are important. "Early understandings make long friendships."

Every carpenter must know how to read and understand architectural drawings (plans) and correctly interpret the information found in written specifications. Simply put, the plans tell you how to build and the specifications tell you what materials must be used.

Copies made of original drawings of the architect are usually called blueprints. At one time all copies were made on chemically treated paper so that the lines were white on a blue background. The term blueprint is still used even though prints may have dark lines on a white background.

SET OF PLANS

Vast amounts of information are needed to build a modern house. Since a single drawing could not possibly hold all that information, many sheets of drawings are required. When bound together, these sheets are known as a set of plans.

The carpenter and building contractor will not be the only persons needing a set of plans. The owner receives a set. Tradespeople such as the electrician, plumber, and heating contractor need sets so they can install their systems. Lumber dealers and other suppliers will use them to determine what and how much materials are needed. The carpenter uses only parts of the plan.

WHAT SET OF PLANS INCLUDES

A set of house plans usually includes:
1. A plot plan.
2. A foundation or basement plan.
3. Floor plans.
4. Elevation drawings. These show the front, rear, and sides of the building.
5. Drawings of the electrical, plumbing, heating, and air conditioning layouts.

STOCK PLANS

When a set of plans is mass produced to be sold to many clients it is called a stock plan. Today, a wide range of such ready-drawn plans are available for people who want to build. This unit shows a number of drawings taken from a set of stock plans.* They will be used for reference as we discuss the house plan in more detail. See Fig. 5-1.

SCALE

Drawings must be reduced so they will fit on the drawing sheet. This must always be done in such a way that the reduced drawing is an exact proportion of the actual size. Such drawings are said to be drawn to scale. Residential plan views are generally drawn to 1/4 in. scale (1/4'' = 1' - 0''). This means that for each 1/4 in. on the plan, the building dimension will be 1 ft.

When certain parts of the structure need to be shown in greater detail, they are drawn to a larger scale such as 1'' = 1' - 0''. Others, such as framing plans, are often drawn to a smaller scale (1/8'' = 1' - 0''). Fig. 5-2 is a partial list of conventional inch-foot scales.

While carpenters are not responsible for making drawings, many find it helpful if they are able to make a simple sketch for a building or remodeling job. The sketch is often good enough to be used as a guide during the building. A pictorial sketch is a drawing showing three dimensions much like a photograph. It is often needed so the customer can visualize what the completed job will look like. See Fig. 5-3.

FLOOR AND FOUNDATION PLANS

Floor plans show the size and outline of the building and its rooms. They also give much additional information which will be useful to the

*Plan No. 10372 L.F. Garlinghouse, Topeka, KS

SCALE	WHERE USED	RATIO	RELATIONSHIP TO ACTUAL SIZE
1/32''	SITE PLANS	1:384	1/32'' = 1'0''
1/16''	PLOT PLAN	1:192	1/16'' = 1'0''
1/8''	PLOT PLAN	1:96	1/8'' = 1'0''
1/4''	ELEVATION PLANS	1:48	1/4'' = 1'0''
3/8''	CONSTRUCTION DETAILS	1:32	3/8'' = 1'0''
1/2''	CONSTRUCTION DETAILS	1:24	1/2'' = 1'0''
3/4''	CONSTRUCTION DETAILS	1:16	3/4'' = 1'0''
1''	CONSTRUCTION DETAILS	1:12	1'' = 1'0''

Fig. 5-2. Common inch-foot scales used in residential drawings. Construction details may also be shown in 1 1/4'' scale.

carpenter and other workers in the construction trades.

Floor plans will have many dimension lines to show the location and size of inside partitions, doors, windows, and stairs. Dimension lines will be explained later.

Plumbing fixtures, as well as appliance and utility installations, can also be shown in this view. Fig. 5-4 shows the floor plan of the residence pictured in Fig. 5-1.

Foundation plans are similar to floor plans and are often combined with basement plans. When shown, the footings are represented as a dotted line. It is assumed that the basement floor is in place and that the grade (ground) covers the footings on the outside. See Fig. 5-5.

A complete set of architectural drawings usually includes a plot plan. This is a drawing of the location of the structure on the building site (lot or acreage). It includes the lot lines and the outside lines of the building. See Fig. 5-6.

ELEVATIONS

Elevations are drawings showing the outside walls of the structure. These drawings are scaled so that all elements will appear in their true relationship. Generally, the various elevations are related to the site by listing them according to the direction they face. However, when plans are not designed for a certain location, the names "front," "rear," "left side," and "right side" are used. Figs. 5-7 and 5-8 are typical elevation drawings.

By studying the elevation views, the carpenter can determine:
1. Floor levels.
2. Grade lines.
3. Window and door heights.
4. Roof slopes.
5. Kinds of materials used on wall and roof surfaces.

FRAMING PLANS

Sometimes a house plan will also have drawings showing the size, number, and location of the structural members of a building's frame. These are known as framing plans. See Figs. 5-9 through 5-12. Separate plans may be drawn for the floors, ceilings, walls, and roof. These plans will specify the sizes and spacing of the framing members. The members are each drawn in as they will be positioned in the building. Openings needed for

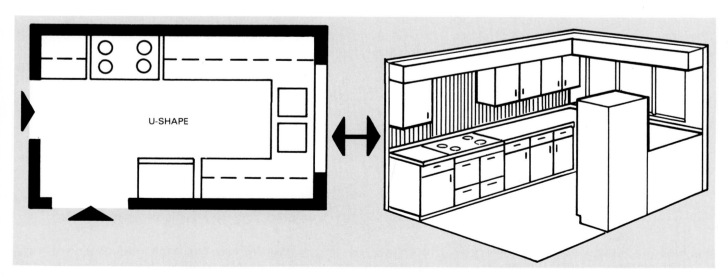

U-SHAPE

Fig. 5-3. It is easier for the buyer to visualize how the kitchen will look if shown in a picture-like sketch as shown at right. This can be done easily using isometric sketch paper.

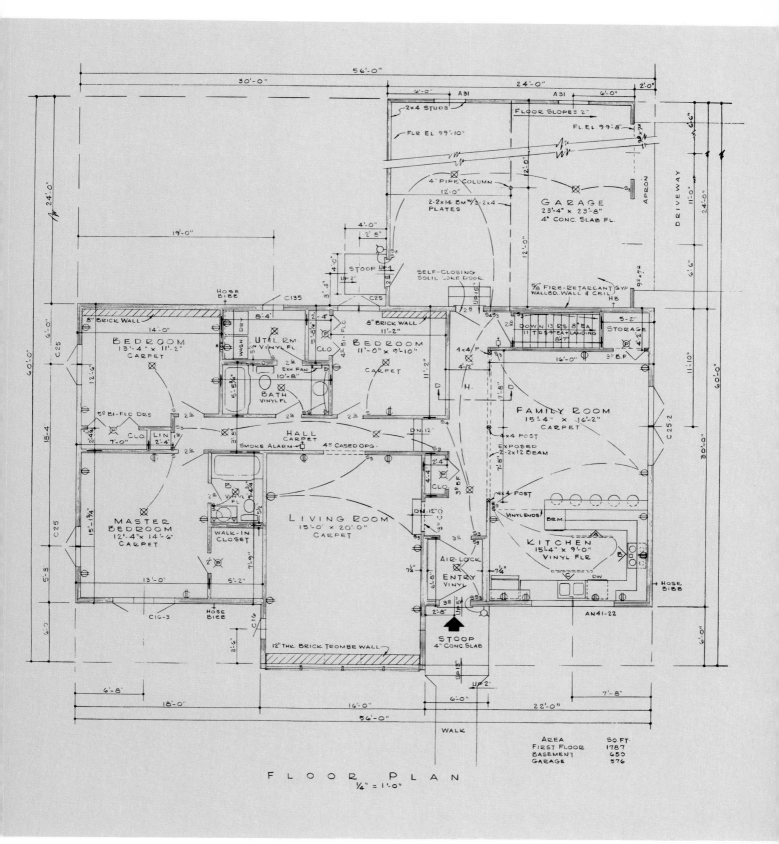

Fig. 5-4. Floor plan for residence shown in Fig. 5-1. Such plans always include shape and size of rooms, and location of plumbing, wiring, and cabinets. Original was drawn to scale of 1/4'' = 1'-0''. This reproduction is made much smaller.

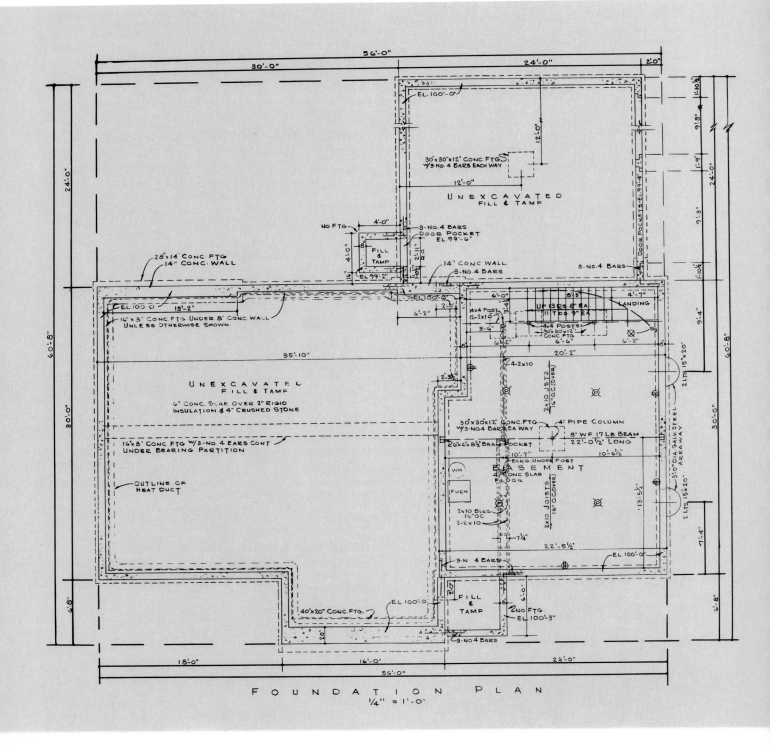

Fig. 5-5. Foundation plan shows basement and footings.

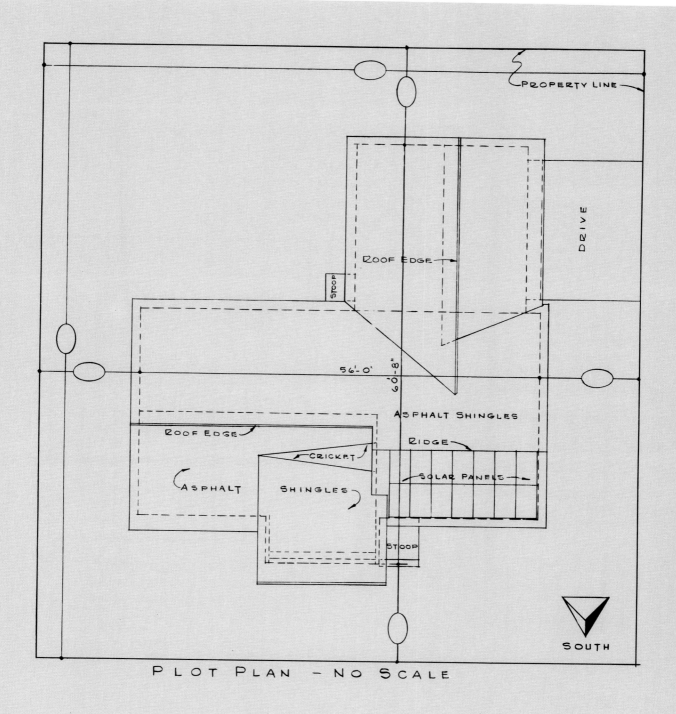

Fig. 5-6. Plot plan. Stock plans leave spaces for filling in lot dimensions.

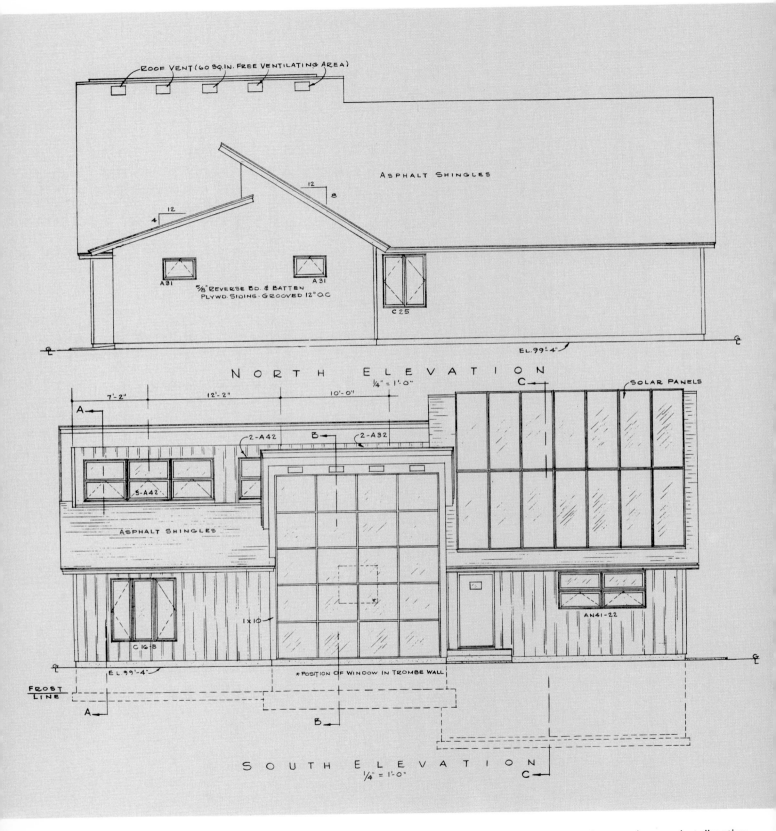

Fig. 5-7. In stock plans, elevations are usually marked ''front'' and ''back'' because the architect does not know what direction the house will be facing. In this solar house, however, the front of the house must face south to catch the sun.

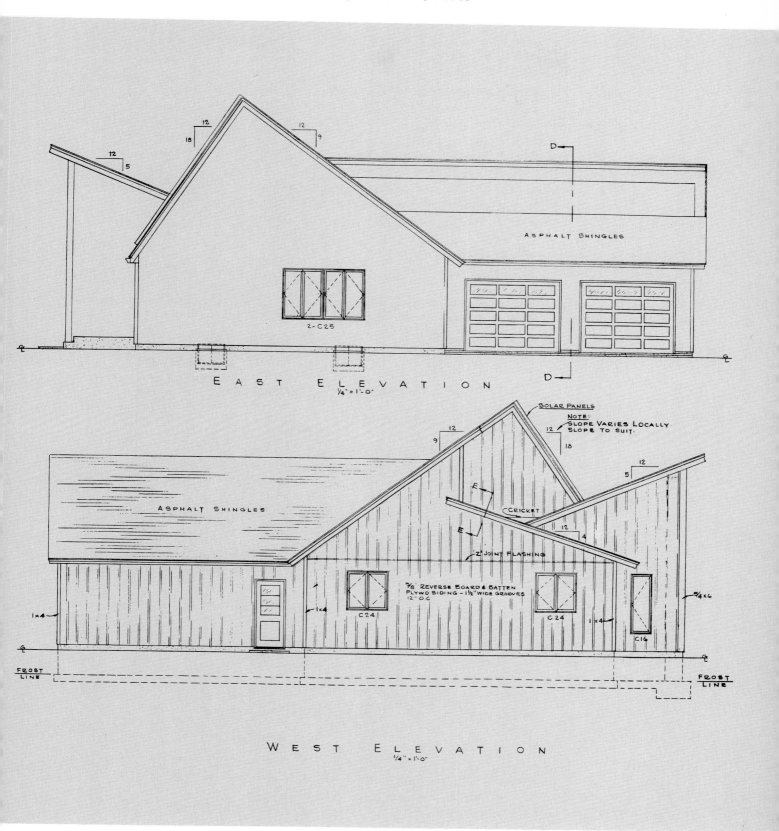

Fig. 5-8. Right and left elevation drawings. All parts appear in their true relationship to one another.

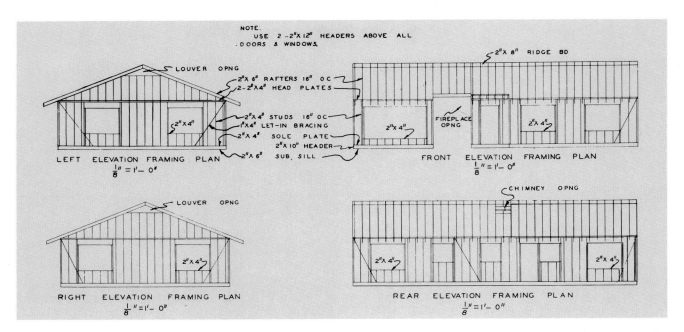

Fig. 5-9. Sometimes plans will include drawings for framing. Here elevation drawings help the carpenter with respect to dimension lumber needed for studs, rafters, headers, plates, and sills. Note: this drawing is not for the house in Fig. 5-1.

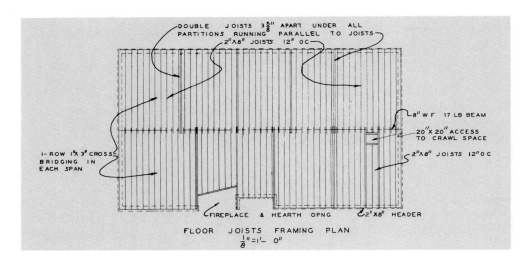

Fig. 5-10. Framing plan for floor in a simple ranch home.

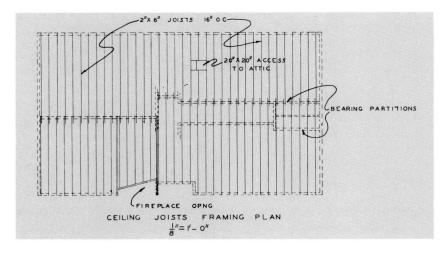

Fig. 5-11. Framing plan for ceiling joists in ranch home.

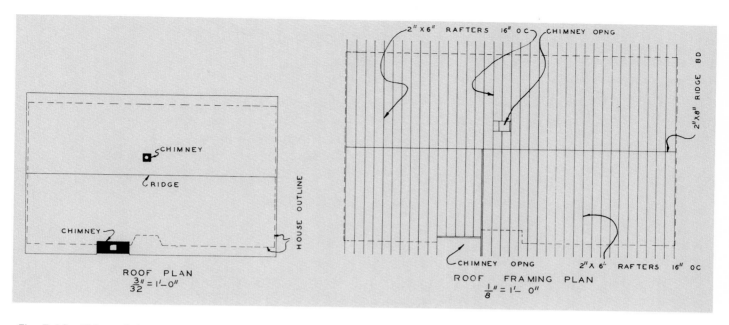

Fig. 5-12. This roof plan and framing plan belongs to the ranch home of Figs. 5-9, 5-10, and 5-11. It helps the carpenter visualize how the rafters must be laid out and where openings will need special framing.

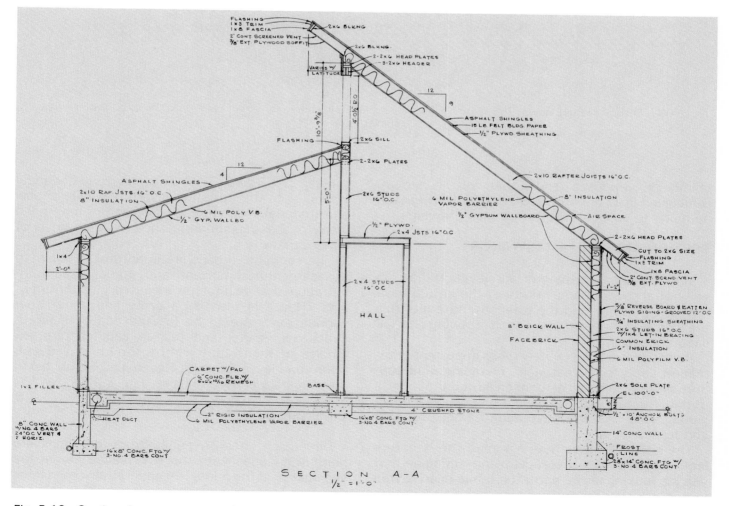

Fig. 5-13. Section views are a way to show details of some part of a structure. It will often be labeled with a double letter which refers to a cutting line in a larger drawing. This one refers to the South elevation drawing in Fig. 5-7.

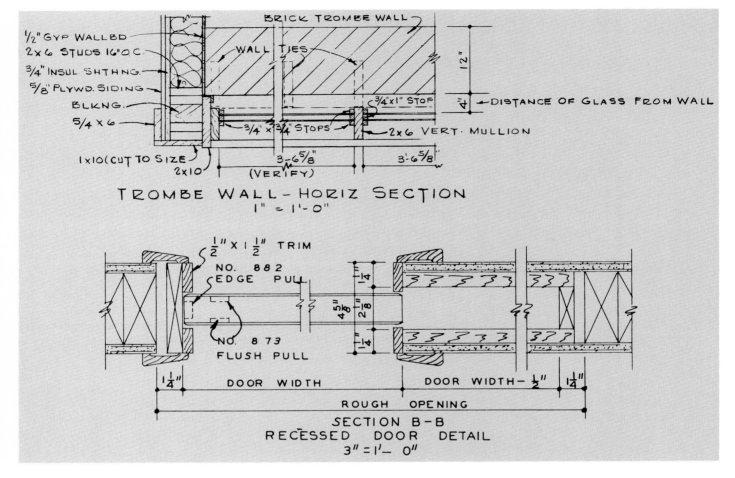

Fig. 5-14. Several section views. Top. Trombe wall cross section is shown for structural details. Bottom. Section view of pocket door.

chimneys, windows, and doors will be shown. Dimension lines are not used but the drawings will be made to scale as with other plan drawings.

SECTION AND DETAIL DRAWINGS

Looking at a floor plan or an elevation drawing will not show you small parts of the structure or how the parts fit into the total structure. For this the carpenter needs to consult drawings called sections and details.

A section drawing or view gives important information about size, materials, fastening, and support systems as well as concealed features. Parts of the structure likely to have a section drawing include walls, window and door frames, footing, and foundations.

The section shows how a part of the structure looks when cut by a vertical plane. (Imagine that you are looking at the part after it had been sawed in two and you are facing the cut edge.) Because of the need to show many details, section drawings are made to a large scale.

Figs. 5-13 and 5-14 show typical sectional views. Note the attention given to details of size and material. A complicated structure may need many of these sections to show all the details of construction.

Like sections, details are important in showing the carpenter important and complicated construction that cannot be included in plan drawings. They are also large scale and show how various parts are to be located and connected. Fireplaces, stairs, and built-in cabinets are examples of items that will be shown in a detail. Refer to Fig. 5-15 and Fig. 5-16. Some detail drawings are shown full size, Fig. 5-17.

DIMENSIONS

Dimension lines on architectural drawings are continuous lines with the size placed above the line near or at the middle. In general, all dimensions over one foot are expressed in feet and inches. For example: standard ceiling height is given as 8' - 0'' rather than 96''.

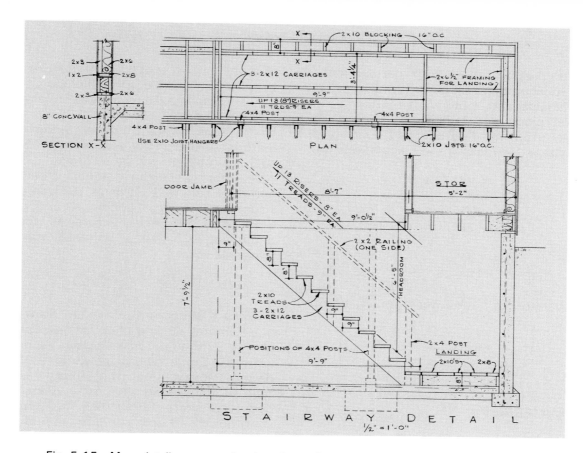

Fig. 5-15. More details may need to be given of complicated parts of the construction.

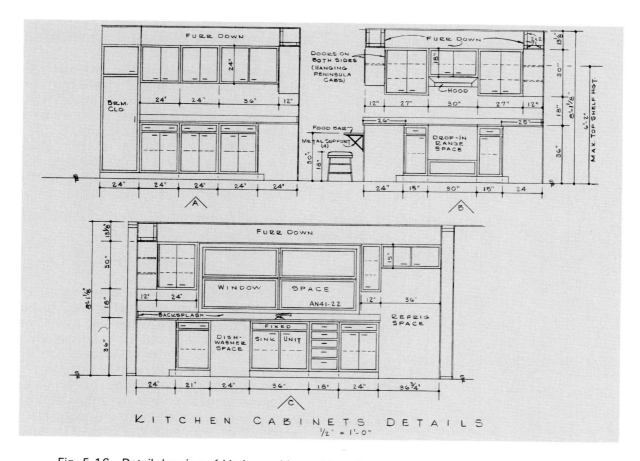

Fig. 5-16. Detail drawing of kitchen cabinets. Note the amount of detail given on dimensions.

Carpenters prefer to work with feet and inches since the measurement in this form is usually easier to visualize and apply. When working with architectural plans and laying out various distances, they often need to add and subtract dimensions. Steps for making the calculations are:

Addition

$$6' - 8''$$
$$+4' - 6''$$
$$+2' - 4''$$
$$+1' - 2''$$
$$=13' - 20'' \text{ or } 14' - 8''$$

Subtraction

$$8' - 4''$$
$$-6' - 10''$$

(since 10 cannot be subtracted from 4, borrow 12'' from 8')

Thus:

$$7' - 16''$$
$$-6' - 10''$$
$$=1' - 6''$$

LISTS OF MATERIALS

Sets of plans will also include a materials list, Fig. 5-18. It is known by other names as well: "bill of materials," "lumber list," or "mill list." Whatever the list is called, it will include all of the materials and assemblies needed to build the structure.

A materials list usually will include the number of the item, its name, description, size, and the material of which the item is made. Built-in items such as cabinets will be included.

Another part of the materials list is the window and door schedule. It gives the quantity needed, sizes of the rough openings, and descriptions. Sometimes the manufacturer is specified. Fig. 5-19 shows a door schedule taken from a materials list.

SYMBOLS

Since architectural plans are drawn to a small scale, materials and construction can seldom be shown as they actually appear. Also it would require too much time to produce drawings of this nature. The architect, therefore, uses symbols to represent materials and other items and certain approved short-cuts (called conventional representations). These simplify the illustration of assemblies and other elements of the structure. Generally accepted symbols are illustrated in Figs. 5-20 through 5-23.

Abbreviations are commonly used on plans to save space. Refer to the appendix for a listing.

HOW TO SCALE A DRAWING

Architectural plans include dimension lines that show many distances and sizes, but carpenters may require a dimension that is not shown. To get this dimension, the carpenter will need to scale the drawing. An architect's scale, Fig. 5-24, may be used for this purpose. Each division of an architect's scale represents 1 ft. The foot division is divided into 12 major parts. Each is equal to 1 in.

Another way to scale a plan is to use a regular folding rule. Calculate the distance as shown in Fig. 5-25.

CHANGING PLANS

Minor changes in plans, desired by the owner as the job progresses—such as changing the size or location of a window or making a revision in the design of a built-in cabinet—can usually be handled by the carpenter. Sketches or notations should be recorded on each set of plans so there will be no misunderstandings.

Major changes such as the relocation of a load-bearing wall, or stairs, may generate a "chain-reaction" of problems and should be undertaken only after the necessary plan changes have been made by an architect and approved by the owner.

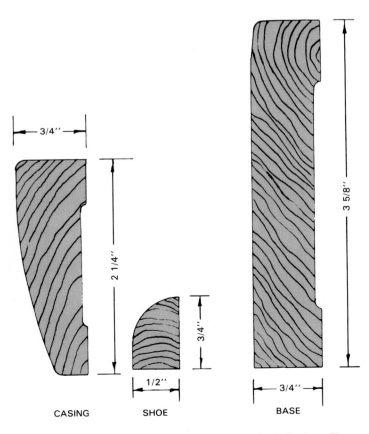

Fig. 5-17. Some detail drawings are made full size. The above are cross sections of special millwork needed for finishing a house.

```
                    MATERIALS LIST
                    PLAN NO. 10372


 4900   Face bricks for interior walls
 2950   Common bricks for interior walls (**bedrooms**) & **center** of trombe wall
 7 1/4  Cu. yds. concrete for basement floor slab
   20   Cu. yds. concrete for 1st. floor slab (6" thick)
 7 1/4  Cu. yds. concrete for garage floor slab and apron
    1   Cu. yd. concrete for entrance platforms and steps
   14   Cu. yds. concrete for footings
45 3/4  Cu. yds. concrete for foundation walls
  224   Lin. ft. 4" plastic drainage tubing
   45   Tons 3/4" to 1" crushed stone -- under slabs & around drainage tubing

                   STRUCTURAL STEEL

    1   8WF17 steel beam 22'-01/2" long
    1   4" steel pipe columns 7-4 long with plates
    1   4" steelpipe columns 8'-3" long with plates (verify)
    8   3/8" x 11" x 24" steel plate lintels over trombe wall openings
    1   Complete gas vent
 3150   Lin. ft. No. 4 reinforcing bars
 2500   Sq. ft. 6" x 6" x 10/10 gauge reinforcing mesh
    2   Galvanized steel areaways 36" diameter 24" high
   64   1/2" anchor bolts 10" long
    8   16 x 8 screened foundation vents w/dampers

                  CARPENTER'S LUMBER
```

**Note - Recommended framing lumber shall have a mimimum allowable extreme fiber bending stress of 1450 PSI and a mimimum modulus of elasticity of 1,400,000 PSI, except rafter joists over living room have F_b of 1500 and a E of 1,400,000.

```
                              Floors -- 40 PSF
Design Live Loads -           Ceilings -- 20 PSF
                              Roof -- 40 PSF

       First Floor Joists & Headers
  5    2 x 10 x 8
 19    2 x 10 x 12
 21    2 x 10 x 14
  2    2 x 10 x 16

       Sub Sills
  1    2 x 6 x 10
  3    2 x 6 x 12
  1    2 x 6 x 14
  2    2 x 6 x 16
  2    2 x 6 x 20
  1    2 x 3 x 16
```

Fig. 5-18. A materials list includes all of the materials and assemblies, such as doors, windows, and cabinets, that will be used in the building. Only part of the list is shown.

Doors:	Frames	Openings	Jambs
1	Outside entrance	3-0 x 6-8 x 1 3/4	
1	Outside solid core service	2-8 x 6-8 x 1 3/4	
1	Outside service	2-8 x 6-8 x 1 3/4	
1	Inside	2-8 x 6-8 x 1 3/8	
5	Inside	2-6 x 6-8 x 1 3/8	
2	Inside	2-0 x 6-8 x 1 3/8	
1	Inside	1-6 x 6-8 x 1 3/8	
2	Garage	9-0 x 7-0 x 1 3/8	
1	Combination	2-8 x 6-8 x 1 1/8	
1	Combination	3-0 x 6-8 x 1 1/8	
1	Bi-fold door	5-0 x 6-8 x 1 3/8	
1	Bi-fold door	4-0 x 6-8 x 1 3/8	
2	Bi-fold doors	3-0 x 6-8 x 1 3/8	
2	Sides of door trim	5-0 x 6-8, 1/2 x 2 1/4	
4	Sides of door trim	4-0 x 6-8, 1/2 x 2 1/4	
7	Sides of door trim	3-0 x 6-8, 1/2 x 2 1/4	
5	Sides of door trim	2-8 x 6-8, 1/2 x 2 1/4	
10	Sides of door trim	2-6 x 6-8, 1/2 x 2 1/4	
4	Sides of door trim	2-0 x 6-8, 1/2 x 2 1/4	
2	Sides of door trim	1-6 x 6-8, 1/2 x 2 1/4	

Windows: All *Andersen Perma-Shield casement and awning windows are to be complete with frames, sash, interior trim, exterior trim, screens, storm sash and hardware.

Quan.	No.	
1	C16-3	Perma-shield casement window
5	C25	Perma-shield casement window
1	C16	Perma-shield casement window
1	C135	Perma-shield casement window
1	AN41-22	Perma-shield awning window
5	A42	Perma-shield awning window
2	A32	Perma-shield awning window
2	A31	Perma-shield awning window

*Andersen Corporation
Bayport, MN 55003

2 Steel basement units 2 lites 15 x 20 with screens

Custom Window Material - Trombe Wall

50	Lin. ft. 2 x 10 Surround material
15	Lin. ft. 2 x 8 Surround material
15	Lin. ft. 2 x 4 Blocking
165	Lin. ft. 2 x 6 Mullion Material
588	Lin. ft. 3/4" x 3/4" Stop material
294	Lin. ft. 3/4" x 1" Stop material
36	16 Ga 1 1/2" x 8" Galvanized wall ties
500	Sq. ft. 1/4" Plate glass

Fig. 5-19. Part of the door and window schedule from a materials list. Sometimes a manufacturer will be specified.

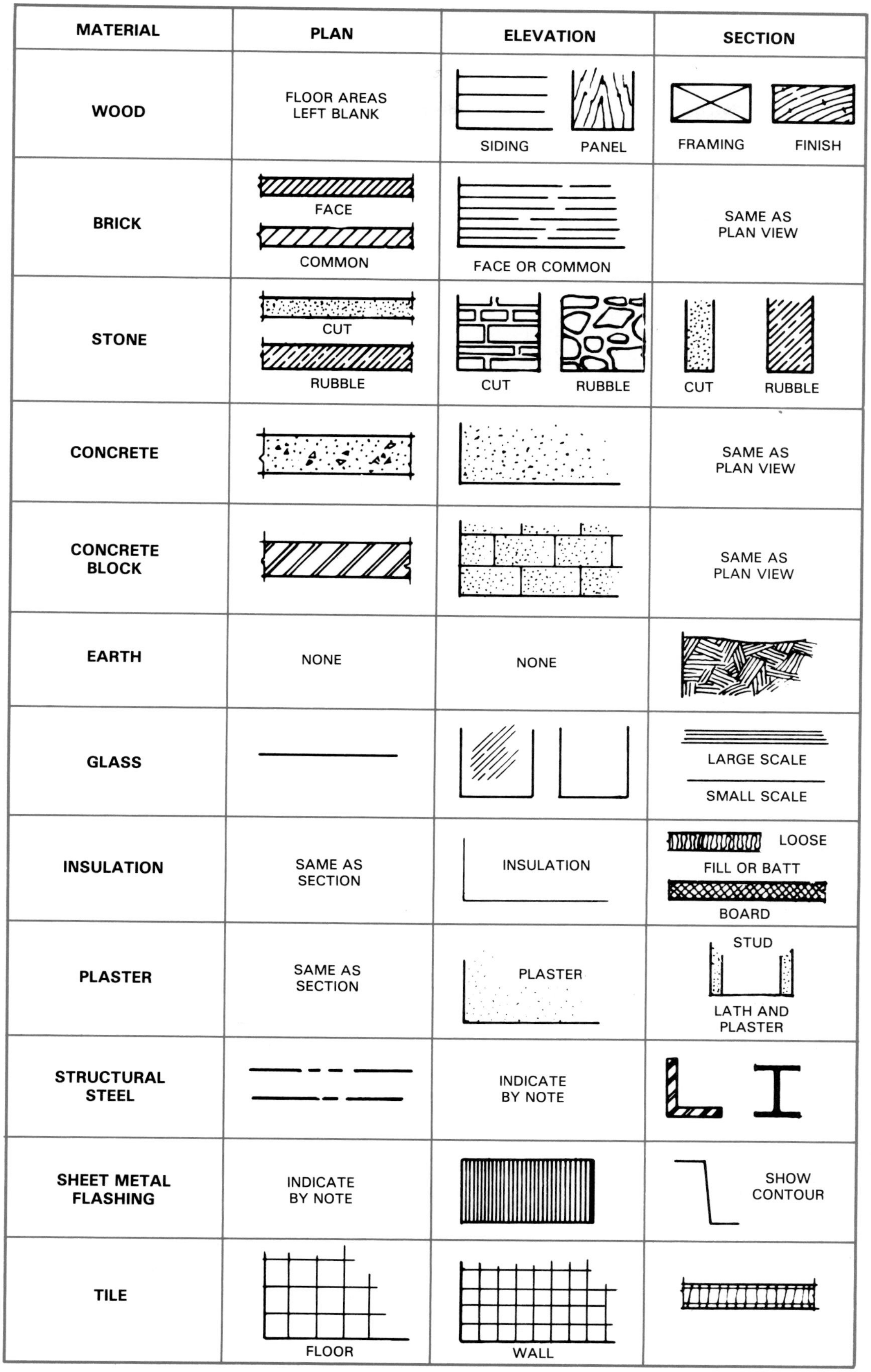

MATERIAL	PLAN	ELEVATION	SECTION
WOOD	FLOOR AREAS LEFT BLANK	SIDING · PANEL	FRAMING · FINISH
BRICK	FACE / COMMON	FACE OR COMMON	SAME AS PLAN VIEW
STONE	CUT / RUBBLE	CUT · RUBBLE	CUT · RUBBLE
CONCRETE			SAME AS PLAN VIEW
CONCRETE BLOCK			SAME AS PLAN VIEW
EARTH	NONE	NONE	
GLASS			LARGE SCALE / SMALL SCALE
INSULATION	SAME AS SECTION	INSULATION	LOOSE FILL OR BATT / BOARD
PLASTER	SAME AS SECTION	PLASTER	STUD / LATH AND PLASTER
STRUCTURAL STEEL		INDICATE BY NOTE	
SHEET METAL FLASHING	INDICATE BY NOTE		SHOW CONTOUR
TILE	FLOOR	WALL	

Fig. 5-20. Symbols are used to represent things that are impractical to draw.

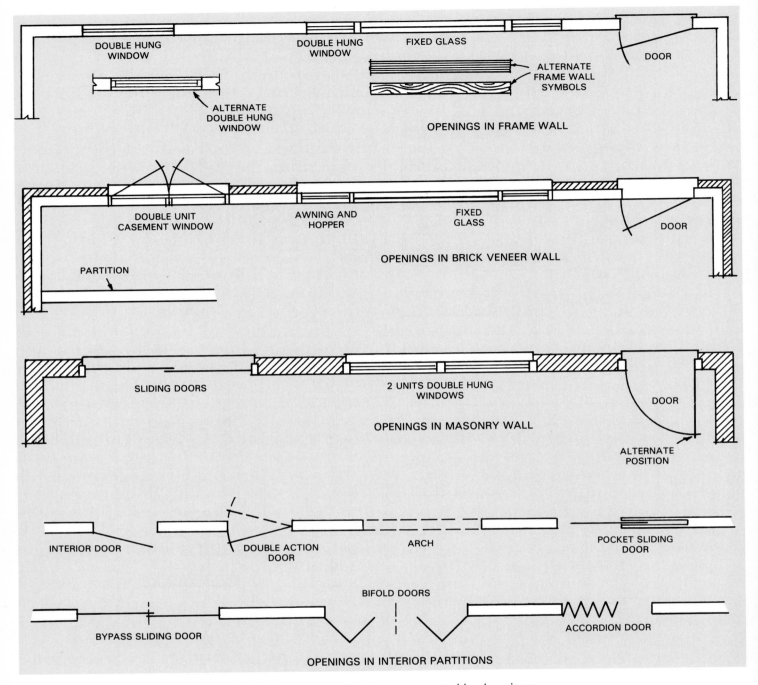

Fig. 5-21. How openings are represented in plan views.

SPECIFICATIONS

Although the working drawings show many of the requirements for a structure, certain supplementary information is best presented in written form or specifications (commonly called Specs). See Fig. 5-26. The carpenter should check and carefully follow these specs.

Headings generally included in specifications for a residential structure include:

1. Basic information for General Requirements, Conditions, and Information.
2. Excavating and Grading.
3. Masonry and Concrete Work.
4. Sheet Metal Work.
5. Rough Carpentry and Roofing.
6. Finish Carpentry and Millwork.
7. Insulation, Caulking, and Glazing.
8. Lath and Plaster or Drywall.
9. Schedule for Room Finishes.
10. Painting and Finishing.
11. Tile Work.

12. Electrical Work.
13. Plumbing.
14. Heating and Air Conditioning.
15. Landscaping.

Under each heading the content is usually divided into: scope of work, specifications of materials to be used, application methods and procedures, and guarantee of quality and performance.

Carefully prepared specifications are valuable to the contractor, estimator, tradespeople, and the building supply dealer. They protect the owner and help to insure quality work. In addition to the items previously described, the specifications may include information and requirements regarding building permits, contract payment provisions, insurance and bonding, and provisions for making changes in the original plans.

MODULAR CONSTRUCTION

The modular coordination (construction) concept is based on the use of a standard grid divided into 4 in. squares. See Fig. 5-27. Actually each individual square (module) should be considered to be the base of a cube so it can be applied to elevations as well as horizontal planes.

All dimensions are based on multiples of 4 in. including 16 in., 24 in., and 48 in. The last two dimensions are sometimes called the minor and major module. Many building materials and fabricated units are manufactured to coordinate with modular dimensions. This helps to eliminate costly cutting and fitting during construction. A good example of this system is illustrated in standard concrete blocks. They are manufactured in nominal sizes of 8 x 8 x 16. The actual size is 3/8 in. less in each dimension to allow for bonding (mortar joints). See Unit 6.

Modular dimension standards for manufactured components have been developed by the National Lumber Manufacturers Association. The system is called Unicom which stands for ''uniform manufacture of components.'' The Unicom system helps to make it possible to apply modern mass

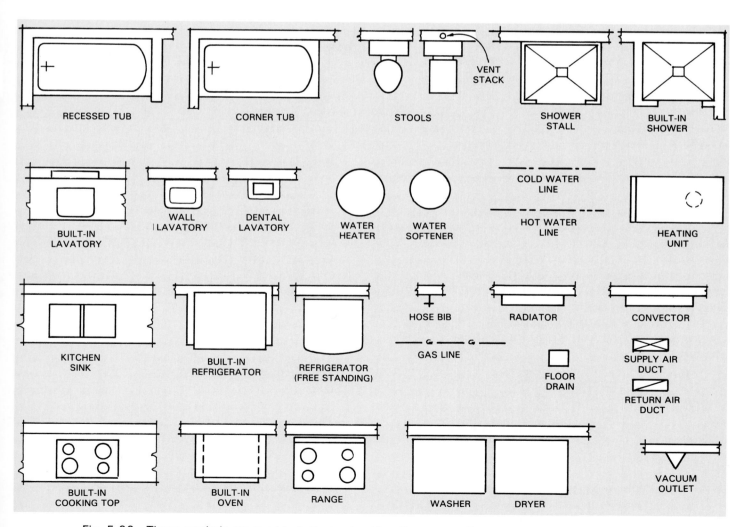

Fig. 5-22. These symbols are used to indicate plumbing fixtures, appliances, and mechanical equipment.

ELECTRICAL SYMBOLS

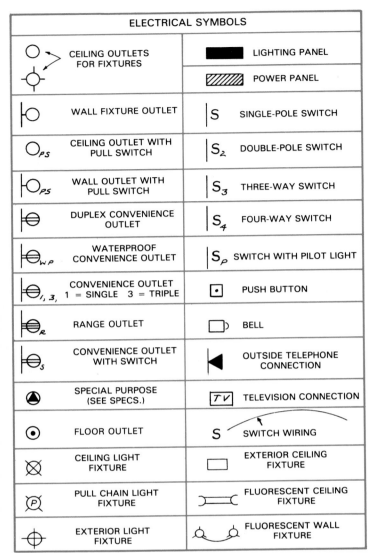

Symbol	Description	Symbol	Description
◯	CEILING OUTLETS FOR FIXTURES	■	LIGHTING PANEL
		▨	POWER PANEL
◖O	WALL FIXTURE OUTLET	S	SINGLE-POLE SWITCH
◯PS	CEILING OUTLET WITH PULL SWITCH	S₂	DOUBLE-POLE SWITCH
◖O PS	WALL OUTLET WITH PULL SWITCH	S₃	THREE-WAY SWITCH
⊖	DUPLEX CONVENIENCE OUTLET	S₄	FOUR-WAY SWITCH
⊖WP	WATERPROOF CONVENIENCE OUTLET	Sₚ	SWITCH WITH PILOT LIGHT
⊖₁,₃	CONVENIENCE OUTLET 1 = SINGLE 3 = TRIPLE	⊡	PUSH BUTTON
⊖R	RANGE OUTLET	▭	BELL
⊖S	CONVENIENCE OUTLET WITH SWITCH	◀	OUTSIDE TELEPHONE CONNECTION
⬤	SPECIAL PURPOSE (SEE SPECS.)	TV	TELEVISION CONNECTION
⊙	FLOOR OUTLET	S	SWITCH WIRING
⊗	CEILING LIGHT FIXTURE	▭	EXTERIOR CEILING FIXTURE
Ⓟ	PULL CHAIN LIGHT FIXTURE		FLUORESCENT CEILING FIXTURE
⊕	EXTERIOR LIGHT FIXTURE		FLUORESCENT WALL FIXTURE

Fig. 5-23. Electrical symbols. Although carpenters will not be responsible for the wiring, they should be able to recognize the symbols.

production methods to building construction. For example a sheet of plywood fits the module exactly at 48 by 96 in.

A modular system also exists in the SI metric system. It is based on a grid made up of 100 mm squares.

The 100 mm module is about, but not exactly, 4 in. The ISO standard also recommends that the submultiples of 25, 50, and 75 mm be used as well as the multiples of 300, 400, 600, 800, and 1200. The 600 and 1200 multiples become the minor and major modules. See the reference to metrics in the Technical Information section.

BUILDING CODES

A building code is a collection of laws listed in booklet form that apply to a given community. The

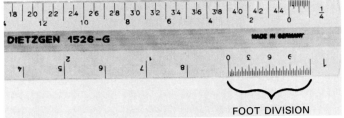

Fig. 5-24. Architects scale may be used to check the dimensions on a scaled drawing. Each division represents 1 ft. Fractions of a foot are checked against the foot division.

code covers all important aspects of the erection of a new building and also the alteration, repair, and demolition of existing buildings. The basic purpose is to provide for the health, safety, and general welfare of the occupants of the home being built and other people in the community.

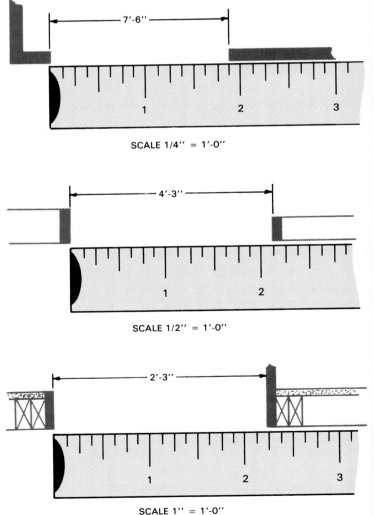

Fig. 5-25. How to use a folding rule to scale a plan.

PLAN NO. 12275

DIVISION 5

CARPENTRY, MILLWORK, & HARDWARE

General Conditions:

This contractor shall read the General Conditions and Supplementary General Conditions which are a part of these Specifications.

Scope of Work:

Furnish and install all rough lumber, millwork, rough and finished hardware, including all grounds, furring, and blocking, frames and doors, etc., as shown on the drawings or hereinafter specified.

Rough Lumber:

All lumber used for grounds, furring, blocking, etc., shall be #1 Common Douglas Fir or Pine.

Grounds and Blocking:

Furnish and install suitable grounds wherever required around all openings for nailing metal or wood trim such as casings, wainscoting, cap, shelving, etc.; also building into masonry all blocking as required. All wood grounds must be perfectly true and level and securely fastened in place.

Doors:

Interior doors to be plain sliced red oak for stain finish. Solid core doors to be laminated 1 3/4" thick, or as called for in the door schedule. See plan for Label doors.

Temporary Doors and Enclosures:

Provide temporary board enclosures and batten doors for all entrance openings. Provide suitable hardware and locks to prevent access to work by unauthorized persons.

At any openings used for ingress and egress of material, provide and maintain protection at jambs and sills as long as openings are so used.

Rough Hardware:

hardware such as nails, spikes, anchors, bolts, rods, etc., in connection with rough car- by this contractor. Use aluminum non-staining nails for all exterior wood.
All finishing
around openings shall be
edges of trim shall be lightly sandpape.
smoothed and machine sanded at the factory, and
carpenters at the building in a first class manner before being set in p
applied.

Finishing Hardware:

An allowance of $1,000.00 shall be included for finishing hardware, which will be selected by the architect and installed by this contractor. Locksets shall match existing design and finish. Adjustment of cost above or below this allowance shall be made according to authentic invoice for hardware for this building from hardware supplier. Hardware supplier shall furnish schedule and templates to frame manufacturer.

Caulking:

Caulk all exterior windows, doors, etc. and openings throughout building with polysulfide based sealant, see specifications under cut stone.

Wood Paneling:

All wood paneling shall be Weyerhauser Forestglo, Orleans Oak 1/4" prefinished. Apply over 1/2" drywall on steel studs and furring.

Fig. 5-26. One sheet of the specification for the house in Fig. 5-1. Other topics included in specs are plumbing, heating, and wiring.

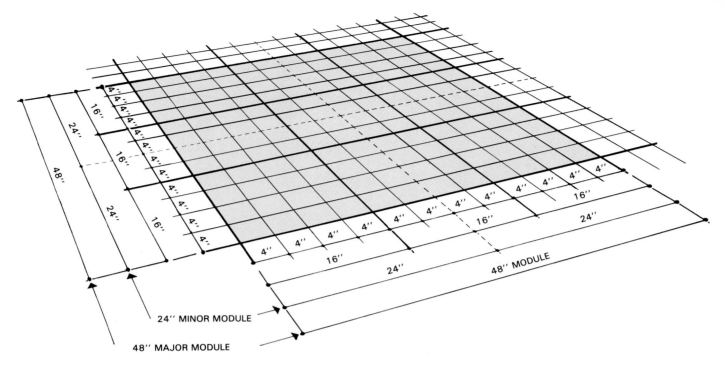

Fig. 5-27. Plans for modular structures, to be built with components or by conventional framing methods, are designed to exact grid sizes. (National Forest Products Assoc.)

Every carpenter should become thoroughly familiar with local building codes. Work which does not conform must be done over and can add considerably to the expense of construction. See Fig. 5-28.

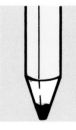

In doing any kind of carpentry work, the importance of closely following all building codes applicable to the job cannot be overemphasized.

A code sets the minimum standards that are acceptable in a community for design, quality of materials, and quality of construction. It also sets requirements concerning such design factors as:
1. Sizes, heights, and bulk of buildings.
2. Room sizes.
3. Ceiling heights.
4. Lighting and ventilation.

Many local codes contain detailed directions regarding installation of a building's systems. These instructions govern the methods and materials used in installing plumbing, wiring, and heating systems.

A carpenter should be aware that building codes may vary from community to community. What is common practice in one area may not be allowed in another. For example, some communities' codes will require one or more sumps in basement floors and perimeter drainage around the footings.

Another community may not mention them. These differences should be carefully noted.

Some items in a code necessarily must be adjusted to local conditions. In northern climates, footings need to be deeper than in southern states; structures in "hurricane belts" require extra bracing.

MODEL CODES

Modern research and development have resulted in so many improvements in building construction that it now becomes a tremendous task to prepare and continually up-date building codes. Because of this, many communities have adopted model codes. Today, four major organizations provide a service of this nature.

The *Uniform Building Code,* published by the International Conference of Building Officials, is widely accepted. This organization provides annual revisions and the entire code is republished every three years. A short form is available which covers buildings not over two stories in height and containing less than 6000 sq. ft. of ground floor area.

Another organization, the Building Officials and Code Administrators International, Inc. has developed the BOCA - *Basic Building Code.* An abridged form, designed for residential construction, includes plumbing and wiring standards. It, too, is revised every three years.

One of the first model codes was introduced by The American Insurance Association (successor to the National Board of Fire Underwriters). This publication is now known as the *National Building Code.* An abbreviated edition is also available.

A model building code called the *Standard Building Code,* used in Southern States, covers problems in this region. It is prepared under the direction of the Southern Building Code Congress International, Inc.

In addition to building codes adopted by the local community (cities, towns, counties), the carpenter must be informed of certain laws at the state level that govern buildings. Several states have developed building codes for adoption by their local communities. However, for the most part, state codes deal mainly with fire protection and special needs for public buildings.

A carefully prepared and up-to-date code is not sufficient in itself to insure safe and adequate buildings. All codes must be properly administered by officials that are experts in the field. Under these conditions, the owner can be assured of a well constructed building and the carpenter will be protected against the unfair competition of those who are willing to sacrifice quality for an excessive margin of profit.

Communities have inspectors who enforce the building code. They will make periodic inspections during construction or remodeling. The inspectors are persons who have worked in the construction trades or who are otherwise knowledgeable about construction.

13.1.2   Required Room Sizes

No dwelling unit shall be erected or constructed which does not comply with the following minimum room sizes:

(a)	Living Room:	250 sq. feet
(b)	Dining Room:	100 sq. feet
(c)	Kitchen:	90 sq. feet
(d)	Bath Room (Three Fixtures)	40 sq. feet
(e)	Powder Room (Two Fixtures)	24 sq. feet
(f)	First Bedroom:	150 sq. feet
(g)	Each Additional Bedroom:	110 sq. feet
(h)	Den or Library, etc.	100 sq. feet

(Permitted only when dwelling unit includes two (2) or more bedrooms of the required sizes.)

(i)	Garage:	253 sq. feet

(Minimum dimensions shall be 11 feet x 23 feet.)

13.1.3   Required Rooms

No dwelling unit shall be erected or constructed which does not contain one each of the following rooms: Living Room, Dining Room, Kitchen, Bath Room, Bedroom and Garage. Each room is to be separated from each other room by full height partitioning and doors, except that Living Room, Dining Room and Kitchen need not be fully separated from each other, and the Garage may be a detached building. Required rooms and all additional bedrooms, den and library shall not be located in a basement.

Fig. 5-28. Sample of a local building code. Codes will vary from community to community. (Village of Flossmoor, Illinois)

VILLAGE OF FLOSSMOOR, ILLINOIS
BUILDING PERMIT

Bldg. Permit No._____   Elec. Permit No. _____

Plbg. Permit No._____   Date Issued_____

Street Address_____ Lot_____ Block_____

General Contractor's Name_____ Phone_____

Address _____

	REJECTED	ACCEPTED
BUILDING		
1st Inspection; Foundations		
2nd Inspection; Drywall		
3rd Inspection; For Occupancy Permit		
For Building Inspections Call ___-____		
PLUMBING		
Plumbing Contractor:		
1st Inspection; Exterior Rough-In		
2nd Inspection; Drywall		
3rd Inspection; Final		
For Plumbing Inspections Call ___-____		
ELECTRICAL		
Electrical Contractor		
1st Inspection; Drywall		
2nd Inspection for Occupancy Permit		
For Electrical Inspections Call ___-____		
PUBLIC WORKS INSPS. CALL ___-____		
SEWER TAP—SEWER LINES & DRAIN LINES INSPS. CALL ___-____		
WATER TAP—CALL ___-____		
HEATING—For Inspections Call ___-____		
REFRIGERATION (Permit Required) **For Inspections Call** ___-____		

NOTICE!

(1) The approval of the drawings will not sanction nor permit any violation of village zoning or building code.
(2) A complete set of approved drawings along with permit must be kept on the premises during construction.
(3) The permit will become null and void in the event of any deviation from the accepted drawings.
(4) No foundation, structural, electrical, nor plumbing work shall be concealed without approval.
(5) THE BUILDING MAY NOT BE OCCUPIED OR USED FOR STORAGE WITHOUT FIRST OBTAINING AN OCCUPANCY PERMIT.

No work shall be done on any part of the building beyond the point indicated in each successive inspection without acceptance. No structural framework of any part of any building or any underground work shall be covered or concealed without acceptance.

THIS PERMIT MUST BE PROMINENTLY DISPLAYED ON BUILDING AT ALL TIMES

636

Fig. 5-29. Construction cannot begin until a building permit is obtained from building officials of the community. Permit and inspection card must be displayed at the building site. This permit is combined with the card.

STANDARDS

Building codes are based on standards developed by manufacturers, trade associations, government agencies, professionals, and tradespeople, all of whom are seeking a desirable level of quality through efficient means. A particular material, method, or procedure is technically described through specifications. Specifications become standards when their use is formally adopted by broad groups of manufacturers and builders and/or recognized agencies and associations.

Organizations devoted to the establishment of standards, many of which are directly related to the field of construction, include:
1. The American Society for Testing and Materials (ASTM).
2. American National Standards Institute (ANSI).
3. Underwriters' Laboratories, Inc. (UL).

Commercial standards are developed by the Commodity Standards Division of the U.S. Department of Commerce. The chief purpose of the agency is to establish quality requirements and approved methods of testing, rating, and labeling. These standards are designated by the initials CS, followed by a code number and the year of the latest revision.

BUILDING PERMITS AND INSPECTIONS

Steps for securing building permits, Fig. 5-29, will vary from community to community. Usually the contractor or building owner will file a formal application with the village or city clerk. The application, Fig. 5-30, with one or two sets of plans. is given to the clerk. Usually the drawings submitted must include:
1. Floor plans.

VILLAGE OF FLOSSMOOR

APPLICATION FOR BUILDING PERMIT

Type of Work:
Erection _____ Remodel _____ Addition _____ Repair _____ Demolish _____
Construction: Brick _____ Frame _____ Other _____
Color: Roof _____ Brick _____ Trim _____ Siding _____
Proposed use _____
Livable floor area _____

Required Service Entrance Conductors — Full Rated
100 Amp. Service and 100 Amp. Service Switch— .... Wire Size ....
150 Amp. Service and 150 Amp. Service Switch— .... Wire Size ....
200 Amp. Service and 200 Amp. Service Switch— .... Wire Size ....
400 Amp. or Larger Service and 400 Amp. Service Switch— ....
Wire Size ....
Underground Service ☐

MAJOR APPLIANCE CIRCUITS REQUIRED BY CODE

☐ Range................ Amps.   ☐ 2 Laundry Circuits—20..... Amps.
☐ Built-in Oven .... Amps.   ☐ Heating Plant .... Amps.
☐ Water Heater .... Amps.   ☐ Lighting Circuits......... Amps.
☐ Dish Washer .... Amps.   ELECTRICAL HEATING CIRCUITS
☐ Garbage Disposal .... Amps.   ☐ Cable .... Amps.
☐ Sump Pump .... Amps.   ☐ Baseboard Units .... Amps.
☐ Clothes Dryer........... Amps.   ☐ Electrical Furnace .... Amps.
☐ Bathroom Heater .... Amps.   COMMERCIAL WIRING
☐ Fixed Air Conditioner ..... Amps.   ☐ Lighting .... Amps.
☐ Window Air Conditioner ... Amps.   ☐ Motors .... Amps.
☐ Electric Door Opener ..... Amps.   ☐ Appliances .... Amps.
☐ 2 Kitchen Circuits—20 ..... Amps.   ☐ Other .... Amps.

*NOTE — No Lighting or Other Current Consuming Device Shall Be Connected to the 2 Kitchen Circuits, the 2 Laundry Circuits, Heating Plant, Sump Pump or Air Conditioner Circuits.

Street Address: _____
Date _____ Township _____
Real Estate Index No. _____ — _____ — _____
Lot: _____ Block _____
Subdivision: _____

Property Owner _____
 Present Address _____
 Telephone _____
Architect _____
 Address _____
 Telephone _____
General Contractor: _____
 Address _____
 Telephone _____ Bond Expir. Date _____
Air Conditioning & Heating Contractor: _____
 Address _____
 Telephone _____
Electrical Contractor: _____
 Address _____
 Telephone _____ Licensed _____
Plumbing Contractor: _____
 Address _____
 Telephone _____ Bond Expir. Date _____
 State License No. _____

As owner of the property, for which this permit is issued and/or as the applicant for this permit, I expressly agree to conform to all applicable ordinances, rules and regulations of the Village of Flossmoor.

Contractor or Owners Signature: _____

Estimated cost of building complete, including all materials and labor $ _____

Building Permit No. _____
Electrical Permit No. _____
Plumbing Permit No. _____

Fig. 5-30. An application for a building permit must be accompanied by information about the structure that is to be built. In addition to the information given on the form, plans must be submitted. (Village of Flossmoor, Illinois)

2. Specifications.
3. Site plan.
4. Elevation drawings.

Sometimes a filing fee and plan review fee are required. These are in addition to the fee required for the building permit itself.

The plans are examined by building officials to determine if they meet the requirements of the local code. Some communities have an architectural committee which will determine if the plans are satisfactory.

Sometimes it will be necessary for the builder or owner to submit supporting data to show how the building design will meet the code. When the plan meets all of the requirements of the building code, a building permit will be issued. Permit fees are based on cost of construction. Usually the range is from $30 to $255 or higher for structures costing more than $50,000.

When construction is started, the building permit and an inspection card are posted on the building site. Sometimes the two are combined as in Fig. 5-29.

As work progresses the building inspector will make inspections and fill out the inspection card for approval of work completed. It is important that the permit and card always be attached to the building or somewhere on the construction site.

Work on the structure should not proceed beyond the point indicated in each successive inspection. Carpenters on the job must pay close attention to this record. Mechanical work (heating, plumbing, and electrical wiring) may never be enclosed before the installations have been approved by the building inspector. In some communities, a final inspection must be made and an occupancy permit issued. Until then, the building may not be occupied.

IMPORTANT TERMS

Architectural drawings, blueprints, building code, building permit, detail drawing, dimension lines, elevation drawings, framing plan, modular construction, plot plan, scale, section drawing, specification, stock plan, Unicom system.

TEST YOUR KNOWLEDGE — UNIT 5

1. A set of house plans usually includes what drawings?
2. Residential plan views are usually drawn to a scale of _____ in. = _____ ft. _____ in.
3. Floor plans show the _____ and outline of the building and its rooms.
4. The plot plan shows the _____ _____.

5. Elevation drawings show the _____ walls of the structure.
6. A section view shows how a _____ of a structure looks when _____ _____ _____ _____.
7. Dimension lines are _____ lines with the size being placed _____ the line near the _____.
8. Draw symbols which represent these materials and items:
   a. Concrete.
   b. Double hung window.
   c. Interior door.
   d. Refrigerator.
   e. Wall lavatory.
   f. Three-way switch.
   g. Range outlet.
   h. Wall fixture outlet.
9. To obtain a plan dimension not shown, an _____ scale may be used.
10. Working drawings (plans) provide much information required by the builder. Supplementary information is supplied by written _____, commonly called _____.
11. The modular construction concept is based on the use of a standard grid divided into _____ in. squares.
12. A building code covers all important aspects of the erection of a building. True or False?

OUTSIDE ASSIGNMENTS

1. Secure a complete set of plans for an average size residence and make a careful study of the views shown. Try to borrow a set from a local builder. Make a list of symbols, notes, and abbreviations that you do not understand. Then go to reference books and architectural standards books to secure the information. Ask your instructor for help if you have difficulty with some of the views.
2. Study the building code in your community. Become familiar with the various sections that are covered and especially note requirements that apply to residential work. Submit a general outline of the material you feel is most important.
3. Make a trip to your city offices and visit with the commissioner or director of building. Be sure to call for an appointment in advance. During your visit secure information concerning building permits and inspection procedures. Learn the cost and what plans and specifications need to be submitted. Also get information about zoning restrictions and other public ordinances that apply to residential construction. Prepare carefully organized notes and make an oral report to your class.

Architectural drafter prepares a set of plans for a complete structure. Every part of the building from foundation to roof is drawn carefully to scale. This must be completed before beginning of foundation work which is taken up in the next unit. (Stenson-Warm-Grimes-Port Architects, Inc.)

On some construction jobs, carpenters will build the forms for footings and foundation walls and may help pour concrete. In any event, they must know standards and practices in concrete work. Where all-weather foundation systems are used, carpenters will build the foundation entirely of wood. (Portland Cement Assoc. and Osmose)

Unit 6

FOOTINGS AND FOUNDATIONS

In the construction of single family dwellings and other smaller structures, carpenters must work with other tradespeople. They also work closely with the architect and owner in carrying out the total building plan.

On some jobs the carpenters may be required to lay out the building lines and supervise the excavation. They build the forms for footings and foundation walls. Anyone working in the carpentry trade needs a working knowledge of standards and practices in concrete work. In this unit, some of the material presented deals with masonry. It is included because of its relationship to carpentry.

CLEARING THE SITE

Preparation of the building site may require grading and/or removal of trees. Grading may be needed before the building lines are laid out. This

Fig. 6-1. Using a transit to establish lot lines. It is a good idea to have an engineer or surveyor do this for you.

may require the placement of grade level stakes. The proper establishing of grade is explained in Unit 4. It will usually require the use of a transit.

If the property is wooded, great care ought to be used in deciding what trees should be removed. Much depends on where the trees are located and their type. In general, evergreens should be used as protection against the cold winter winds. Deciduous (leaf dropping) trees are best used as shade against hot summer sun. Try to place the house to take advantage of the protection offered by trees already on the property.

Mark trees that are to be taken down so the person responsible for their removal takes the right ones. Avoid digging trenches through the root system of trees being retained. It could cause them to die. Likewise, trees usually will not tolerate more than a foot of additional fill around their root systems.

LAYING OUT BUILDING LINES

After the site is cleared, someone must locate and check lot lines. To protect the owner and builder, this should be done by or with the help of a registered engineer or licensed surveyor, Fig. 6-1. Their assistance may include the establishment of building lines and grade levels. The carpenter should be familiar with local building codes.

It is best to locate building lines with leveling instruments. Follow the procedures described in Unit 4. Lines can, however, be transferred from lot markers. In such cases, it is important that distances be laid out perpendicular to existing lines. Building lines, likewise, must be square. To establish a right angle, the 6-8-10 method can be used. Refer to Unit 4, Fig. 4-4.

Following procedures described in Unit 4, locate corners formed by the intersection of the outsides of foundation walls. Mark the positions by driving stakes. Set tacks in the stake tops at the exact spot.

After locating all building lines, check them carefully. Measure the length and, even though they were laid out with a transit, measure the diagonals of squares and rectangles, Fig. 6-2. An out-of-square foundation will cause problems throughout construction.

BATTER BOARDS

Batter boards, Fig. 6-3, are set up around the building layout stakes. Use 2 x 4 pieces for the stakes and 1 x 6, or wider, pieces for the ledgers.

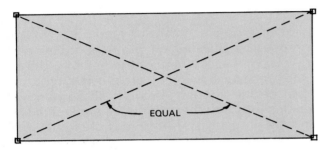

Fig. 6-2. Diagonals of a square or rectangle will always be equal. Always measure to diagonal corners to see if building lines are square.

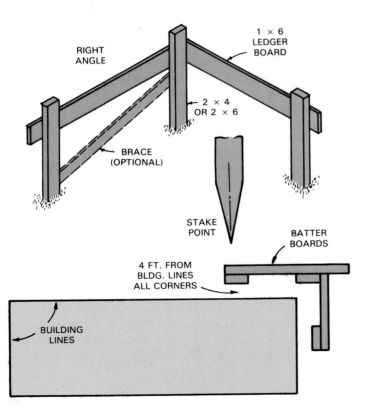

Fig. 6-3. Set up batter boards about 4 ft. from building lines on all four corners. In loose soil or when boards are more than 3 ft. off the ground, use braces.

Locate the batter boards 4 ft. or more away from the corners created by the building lines.

Nail the ledger boards to the stakes. They should be level and at a convenient working height, preferably slightly above the top of the foundation. The batter boards should be roughly level with each other. Also, be sure that the ledger boards are long enough to extend well past each corner.

Using lines and a plumb bob, pull the lines so they pass directly over the layout stakes. Mark the top of the ledger boards where the lines cross. Make a shallow saw kerf or drive a nail as shown in Fig. 6-4. Pull the lines tight and fasten them.

If you are using saw kerfs, drive nails into the backs of the ledger boards. You may prefer to wrap them around the ledger and run them through the saw kerf several times.

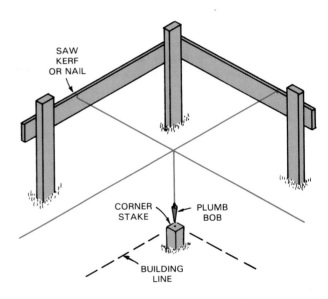

Fig. 6-4. Using batter boards and plumb line to establish building lines. Lines must intersect over the tack in the corner stake.

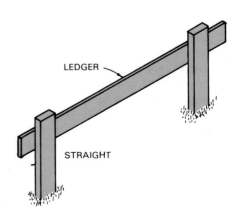

Fig. 6-5. Straight batter board is used in some situations as shown in lower half of Fig. 6-6.

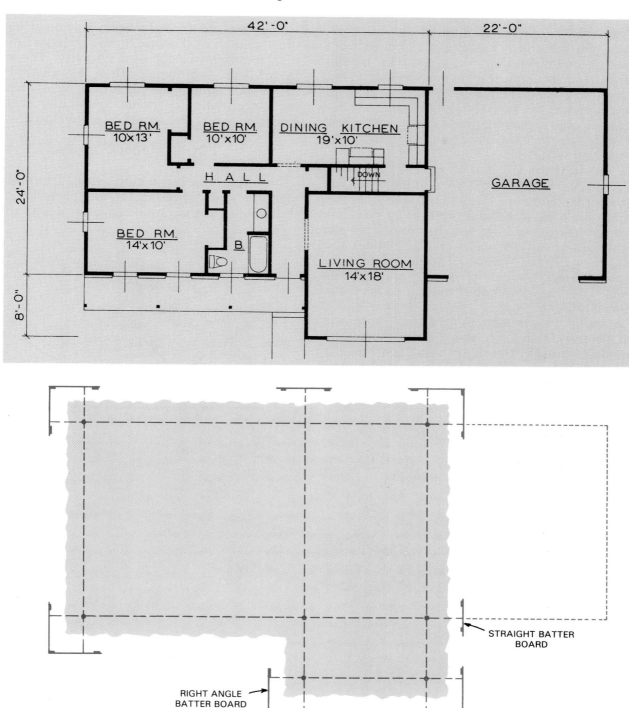

Fig. 6-6. Top. Plan showing overall dimensions. Bottom. Sketch of building lines set up around area to be excavated. Actual excavation (shown in gray) should be at least 2 ft. outside of building lines.

Sometimes a straight batter board, Fig. 6-5, is used when it will not be located at a corner.

EXCAVATION

Building sites on steep slopes or rugged terrain should be rough graded before the building is laid out. Top soil should be removed and piled where it will not interfere with construction. This can be used for the finished grade after the building is complete.

Where no grading is needed, the site can be laid out and batter boards erected. See the lower drawing, Fig. 6-6.

Stakes marking the outer edge of the rough excavation are set and the lines are removed from the batter boards during the work. For regular basement foundations, the excavation should extend beyond the building lines by at least 2 ft. to allow clearance for form work. Foundations for structures with a slab floor or crawl space will need little excavating beyond the trench for footings and walls.

The depth of the excavation can be calculated from a study of the vertical section views of the architectural plans.

In cold climates, it is important that foundations be located below the frost line. If footings are set too shallow, the moisture in the soil under the footing may freeze. This could force the foundation wall upward causing cracks and serious damage. Local building codes usually cover these requirements.

It is common practice to establish both the depth of the excavation and the height of the foundation by using the highest elevation on the perimeter of the excavation. This is then known as the control point, Fig. 6-7.

Foundations should extend about 8 in. above the finished grade. Then the wood finish and framing members will be adequately protected from moisture. The finished grade should be sloped away from all sides of the structure so surface water will run away from the foundation.

The depth of the excavation may be affected by the elevation of the site. It may be higher or lower than the street or adjacent property. The level of sewer lines also has an effect. Normally, solving these problems is the responsibility of the architect. Information on grade, foundation, and floor levels is usually included in the working drawings.

FOUNDATION SYSTEMS

All structures settle somewhat. A properly designed and constructed foundation will distribute the weight to the ground in such a way that the

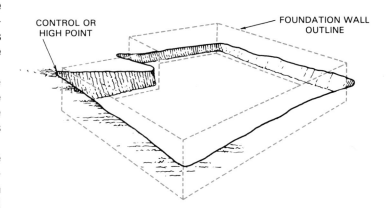

Fig. 6-7. Highest elevation outside the excavation is used to establish depth of excavation.

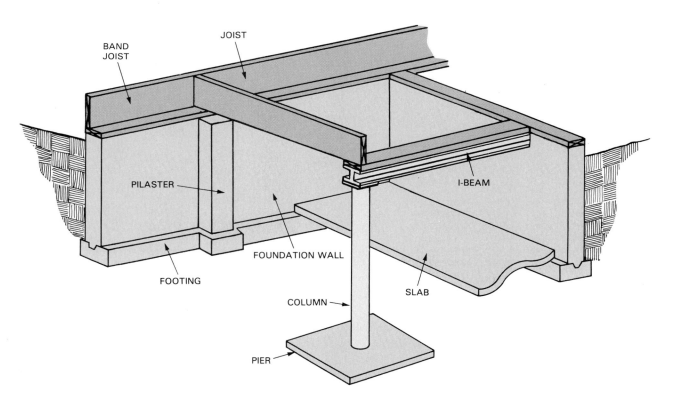

Fig. 6-8. Elements of a typical foundation. Footings and piers spread the load over a wider area. This type foundation is often used when owner wants a basement.

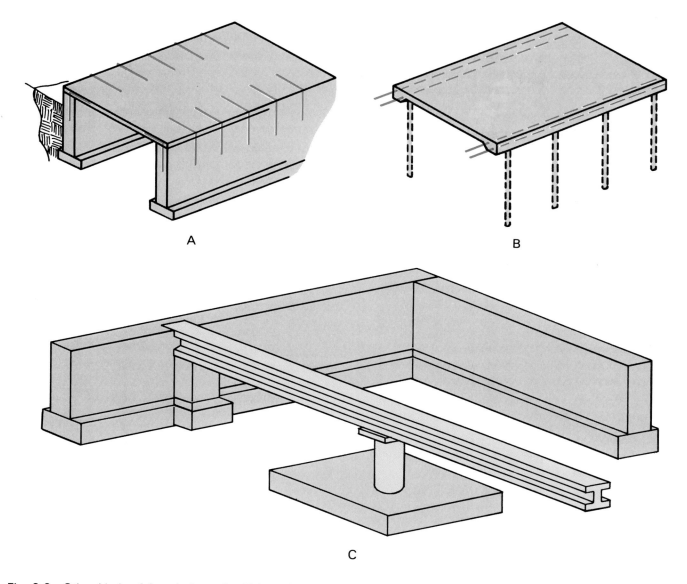

Fig. 6-9. Other kinds of foundations. A—Slab on foundation is used in cold climates. B—Slab-on-ground is popular in warm climates. Piers (dotted lines) can be used as support in unstable soils. C—Crawl space foundation is similar to basement foundations.

settling will be negligible or at least uniform. Fig. 6-8 shows a typical foundation. Fig. 6-9 presents several foundation types in simple form.

For light construction, such as residential, the spread foundation, illustrated in Fig. 6-8, is most common. It transmits the load through the walls, pilasters, columns, or piers. These elements rest on a footing which is nothing more than an enlarged base.

FOOTINGS

PLAIN footings carry light loads and usually do not need reinforcing. REINFORCED footings have steel rebar in them for added strength against cracking. They are used when the load must be

spread over a larger area or when the load must be bridged over weak spots such as excavations or sewer lines.

A STEPPED footing is one that changes grade levels at intervals to accommodate a sloping lot. See Fig. 6-10. Vertical sections should be at least 6 in. thick. Horizontal distance between steps should be at least 2 ft. If masonry units are to be used over the footing, distances should fit standard brick or block modules.

SLABS

Slab foundations take several forms. The slab can be used with other elements such as walls, piers, and footings, as shown in View B of Fig.

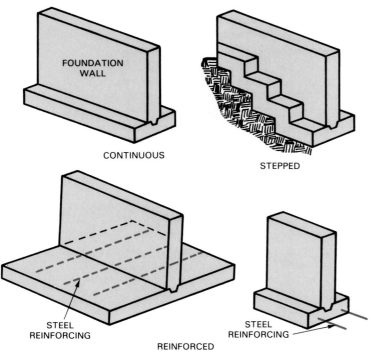

Fig. 6-10. Different types of footings are needed for different slope and soil conditions. Vertical runs of stepped footing should not exceed 3/4 of horizontal run between steps.

6-9. This is called s STRUCTURALLY SUPPORTED slab. A second type is laid directly on top of the ground like those shown in Fig. 6-11. These are referred to as GROUND SUPPORTED.

Some slabs, particularly in warmer climates, are constructed in one continuous pour. There are no joints or separately poured sections. This is called MONOLITHIC concrete or a monolithic pour. This type of construction is appropriate over soils with low bearing capacity, Fig. 6-12.

FOOTING DESIGN

Footings must be wide enough to spread the load over sufficient area. Load-bearing capacities of soils vary considerably. See Fig. 6-13.

In residential and smaller building construction, the usual practice is to make the footing twice as wide as the foundation wall. See Fig. 6-14. The average thickness of a footing is about 8 in.

Footings under columns and posts carry heavy, concentrated loads and are usually from 2 to 3 ft. square. The thickness should be about 1 1/2 times the distance from the face of the column to the edge of the footing.

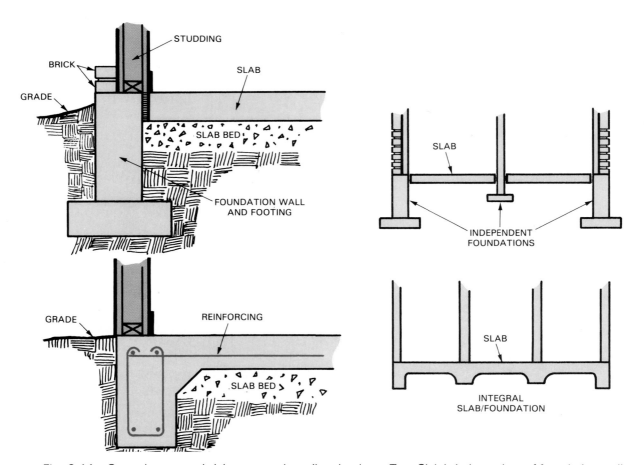

Fig. 6-11. Ground supported slabs rest on the soil under them. Top. Slab is independent of foundation walls. Bottom. Ground supported slab is monolithic structure incorporating a grade beam.

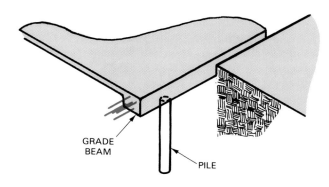

Fig. 6-12. Grade beam is the thickened section of slab. It usually is supported by piles. Note reinforcement in beam.

Type of soil	Capacity, tons per sq. ft.
Soft clay	1
Wet sand or firm clay	2
Fine, dry sand	3
Hard, dry clay or coarse sand	4
Gravel	6

Fig. 6-13. Load carrying capacity of different soil types.

W = WALL THICKNESS

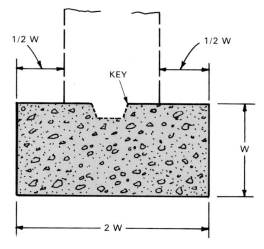

Fig. 6-14: Standard footing design for residential construction. It should be twice the width of the wall.

Reinforced footings are used:
1. In regions subject to earthquakes.
2. Where the footings must extend over soils containing poor load-bearing material.

Some structural designs may also require the use of reinforcing. The common practice is to use two No. 5 (5/8 in.) bars for 12 x 24 in. footings. At least 3 in. of concrete should cover the reinforcing at all points.

In a single story dwelling where they are independent of other footings, the chimney footings should have a minimum projection of 4 in. on each side. For a two-story house, chimney footings should have a minimum thickness of 12 in. and a minimum projection of 6 in. on each side. Exact dimensions will vary according to the weight of the chimney and the nature of the soil. Where chimneys are a part of outside walls or inside bearing walls, chimney footings should be constructed as part of the wall footing. Concrete for both chimney and wall footings should be placed at the same time.

Footings that must support cast-in-place concrete walls may be formed with a recess forming a keyed joint as illustrated in Fig. 6-15. A number of typical footing designs are shown in Fig. 6-16. These are not working drawings and will need to be adapted to local conditions and existing code requirements.

FORMS FOR FOOTINGS

After the excavation is completed, the footings are laid out and forms are constructed. Lines are replaced on the batter boards and corner points are dropped with a plumb bob to the bottom of the excavation.

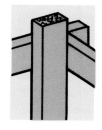

After the excavation is complete, check the batter boards carefully. They may have been disturbed by the excavating equipment. Make necessary adjustments before proceeding with the footing layout.

Fig. 6-15. When a concrete foundation is to be poured, the footing should be keyed. Key is formed by placing a wooden strip in the form during or immediately after the pour. (Portland Cement Assoc.)

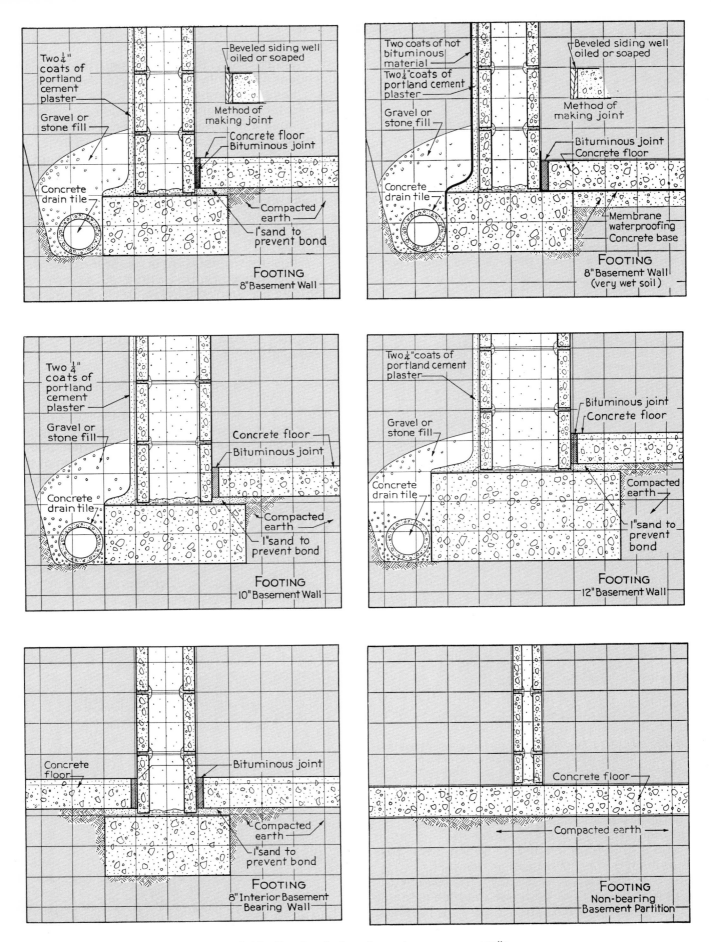

Fig. 6-16. Footing designs for various masonry walls.

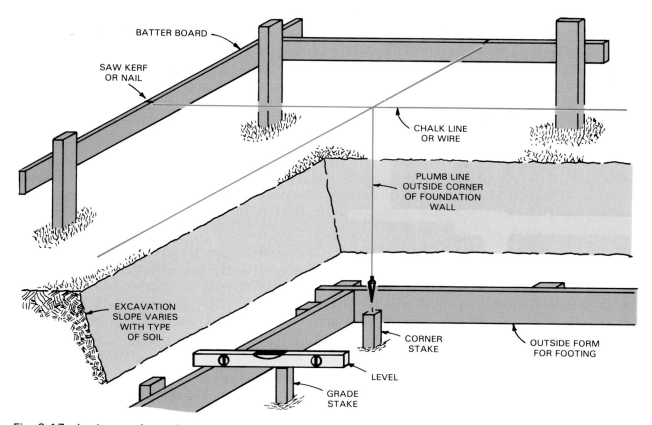

Fig. 6-17. Laying out forms for footings. The outside forms are built first. They must be level with grade stakes.

Drive stakes and establish points at the corners of the foundation walls. Set up a builders' level at a central point in the excavation and drive a number of grade stakes (level with the top of the footing) along the footing line and at approximate points where column footings are required. Corner stakes can also be driven to the exact height of the top of the footing. Connect the corner stakes with lines tied to nails in the top of the stakes.

Working from these building lines, construct the outside form for the footing. The form boards will be located outside the building lines by a distance equal to the footing extension (usually 4 in. for an 8 in. foundation wall). See Fig. 6-17. The top edge of the form boards must be level with the grade stakes.

Transfer the measurement from the grade stakes to the form. Use a carpenter's level. After the outside form boards are in place, it will be relatively easy to set the inside sections. See Fig. 6-18.

Forms constructed of 1 in. boards should be supported with stakes placed 2 to 3 ft. apart. Stakes may be placed farther apart when 2 in. material is used. Spacers or spreaders used to locate the inside form will save measuring time, Fig. 6-19.

Bracing of footing forms may sometimes be desirable. Usually it is not necessary unless 1 in. lumber is used for forming. Attach the brace to the top of a form stake and to the bottom of a brace stake located about a foot away. Refer to Fig. 6-19 again.

Stepped footings will require some additional formwork. Vertical blocking must be nailed to the form to contain the concrete until it sets. See Fig. 6-20.

Column footings are pads of concrete which will support stringers resting on columns. They carry weight transmitted to the column from the center beam of a building, Fig. 6-21.

Fig. 6-18. Measure distance and set the inside form board. (Portland Cement Assoc.)

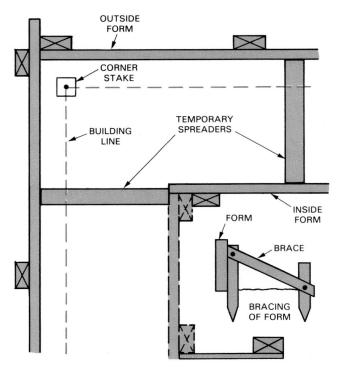

Fig. 6-19. Properly located footing form. Note that it extends beyond corner stake and building line. Precut spreaders were used to locate the inside form boards.

Forms for column footings are usually set after the wall footing forms are complete. These are located by direct measurements from the building lines. Forms are leveled to grade level stakes previously set.

Some hand digging and leveling of the excavation will probably be necessary as forms are set. *Loose dirt and debris must always be removed from the ground that will be located under a footing. This is necessary even though the resulting depth will be greater than required.*

The top of the footing must be level. The bottom may vary as long as the minimum thickness is maintained.

Form boards are temporarily nailed to stakes and to each other. Duplex (double-headed) nails may be used. If regular nails are used, they should be driven only partway into the wood.

Consider problems in form removal. Nail through stakes into the form boards. Do not nail from the inside.

When the forms are completed, check for sturdiness and accuracy. Remove the line, line stakes, and grade stakes. The concrete can now be placed.

CONCRETE

Concrete is made by mixing:
1. Cement.
2. Fine aggregate (sand).
3. Coarse aggregate (gravel or crushed stone).
4. Water.

To prepare concrete, the aggregate and cement are first mixed together. Then the water is added. The water causes a chemical action (called hydration) to take place and the mass hardens. The hardening process is not a result of drying. The concrete should be kept moist during the initial hydration process.

The compressive strength of concrete is high, but its tensile strength (stretching, bending, or twisting) is relatively low. For this reason, when concrete is used for beams, columns, and girders, it must be reinforced with steel. When it must resist compression forces only, reinforcement is usually not added.

CEMENT

Most cement used today is Portland cement. It is usually manufactured from limestone mixed with shale, clay, or marl. Each sack of Portland cement holds 94 lb. This is equal to one cubic foot in volume. Cement should be a free-flowing powder. If

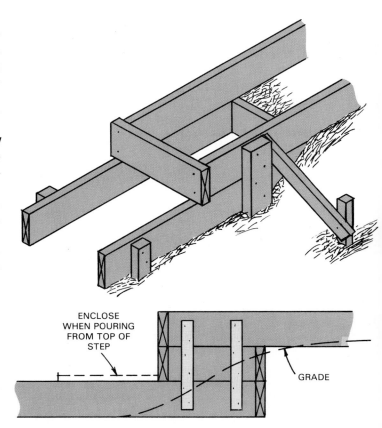

Fig. 6-20. Form constructed for a vertical section in a stepped footing. Lower level is poured first and will usually be allowed to set up slightly before step is poured. Top. Basic form. Bottom. Alternate design when more height is needed.

it contains lumps that cannot be pulverized easily between thumb and fingers, it should not be used.

AGGREGATES

Aggregate may consist of sand, crushed stone, gravel, or lightweight materials such as expanded slag, clay, or shale.

The large coarse aggregate (seldom over 1 1/2 in. in diameter) forms the basic structure of the concrete. The voids between these particles are filled with smaller particles; the voids between these smaller particles are filled with still smaller particles. Together they form a dense mass.

Today, practically all concrete is delivered to the building site in ready-mix trucks. Ready-mix concrete is purchased by the cubic yard (27 cu. ft.) and is available in a number of psi ratings. A minimum order is usually 1 cu. yd. Fractional parts (1/4, 1/2, 1/3) can be furnished beyond this amount.

ERECTING WALL FORMS

Many different types of wall forming systems are available. There are certain basic considera-

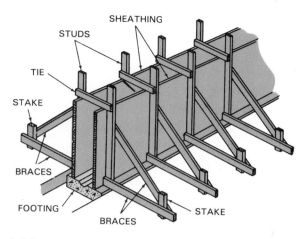

Fig. 6-22. Design for wall forms up to 3 ft. high. Use plywood sheathing and space studs 2 ft. apart.

tions that should be understood and applied to all systems.

For quality work the forms used must be tight, smooth, defect-free, and properly aligned. Joints between form boards or panels should be tight. This prevents the loss of the cement paste which tends to weaken the concrete and cause honeycombing.

Wall forms must be strong and well braced to resist the side pressure created by the plastic concrete. This pressure increases tremendously as the height of the wall is increased. Regular concrete weighs about 150 lb. per cu. ft. If it were immediately poured into a form 8 ft. high, it would create a pressure of about 1200 lb. per sq. ft. along the bottom side of the form.

In practice, this amount of pressure is reduced through compaction and hardening of the concrete. It tends to support itself. Thus, the lateral pressure will be related to:
1. The amount of concrete placed per hour.
2. The outside temperature.
3. The amount of mechanical vibration.

Low wall forms, up to about 3 ft. in height, can be assembled from 1 in. sheathing boards or 3/4 in. plywood, supported by two-by-four studs spaced 2 ft. apart. The height can be increased somewhat if the studs are closer together. Refer to Fig. 6-22.

For walls over 4 ft. in height, the studs should be backed with wales to provide greater strength, Fig. 6-23.

REESTABLISHING THE BUILDING LINE

Before setting up the outside foundation wall form you will need to mark the building line on top of the footing. Set up your lines on the batter boards once more. Then drop a plumb line from

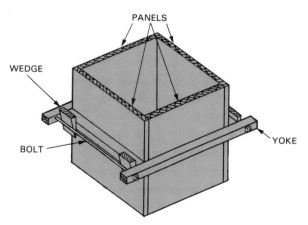

Fig. 6-21. Poured column footing. Top. Typical reusable form. Bottom. Actual footing.

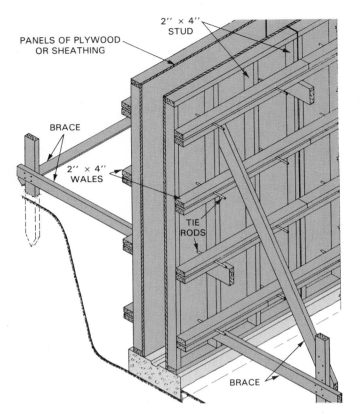

PANELS OF PLYWOOD OR SHEATHING

2″ × 4″ STUD

BRACE

2″ × 4″ WALES

TIE RODS

BRACE

Fig. 6-23. Prefabricated panels for wall form. For walls over 4 ft. high, wales, the horizontal doubled 2 × 4s, provide greater strength to the form. Note bracing to top and bottom of form.

the intersections (corners) of the building lines to the footing. Mark the corners on the footing. Snap a chalk line from corner to corner on the footing. This line will be the outside face of the foundation wall. As you set up the foundation forms, align the face of the outside form with the chalk line.

FORM HARDWARE

Wire ties and wooden spreaders have been largely replaced with various manufactured devices. Fig. 6-24 shows three types of ties. The rods go through small holes in the sheathing and studs. Holes through the wales can be larger for easy assembly or the wales can be doubled as shown.

The snap tie is designed so that a portion of it remains in the wall. Tapered ties are removed and should be coated with release. Bolts in the coil type ties should also be coated with release.

Corners of concrete forms can be secured by corner locks. Two types are shown in Fig. 6-25.

After the concrete has set, the clamps can be quickly removed and the forms stripped. A special wrench is used to break off the outer sections of the snap tie rod. The rod breaks at a small indentation located about 1 in. beneath the concrete surface. The hole in the concrete is patched with grout or mortar.

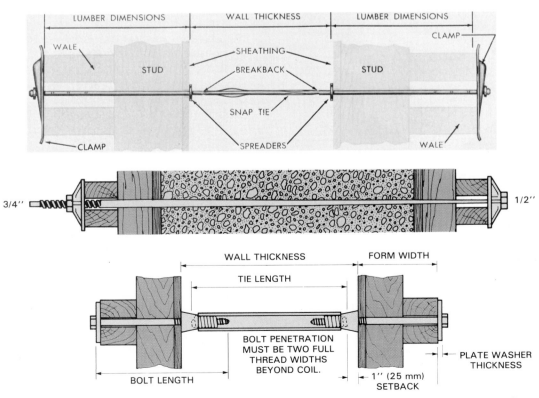

LUMBER DIMENSIONS — WALL THICKNESS — LUMBER DIMENSIONS

WALE
STUD
CLAMP
SHEATHING
BREAKBACK
STUD
SNAP TIE
SPREADERS
CLAMP
WALE

3/4″
1/2″

WALL THICKNESS
TIE LENGTH
FORM WIDTH
BOLT PENETRATION MUST BE TWO FULL THREAD WIDTHS BEYOND COIL.
BOLT LENGTH
1″ (25 mm) SETBACK
PLATE WASHER THICKNESS

Fig. 6-24. Form ties combine the functions of ties and spreaders. Top. Snap tie. (Universal Form Clamp Co.) Center. Taper tie. Bottom. Heavy duty coil tie. (The Burke Co.)

130

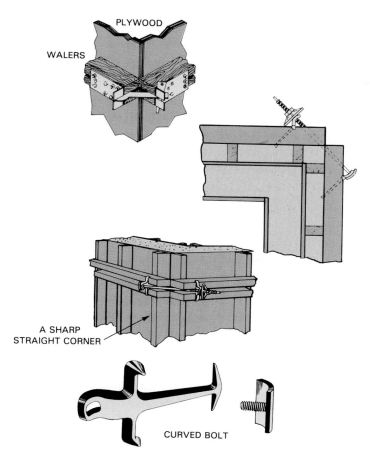

Fig. 6-26. Most wall forming is now done with prefabricated forms. Plywood panels are reinforced with angle iron. (Portland Cement Assoc.)

Fig. 6-25. Corners of wall forms must be carefully fastened so they will withstand pressure of poured concrete. This can be done by interlocking the wales or using patented locks. Top. A high speed corner lock. Clamps are attached with 6 to 8 penny duplex (doubleheaded) nails and lag screws. Bottom. A corner lock requiring no nails or screws. (The Burke Co.)

Other types of patented wall ties were shown in Fig. 6-24. The coil type spreader is assembled with a cone of wood, plastic, or metal and a lag screw. The cone provides smooth contact with the form and leaves a recess that is easy to fill.

Taper type ties are threaded at both ends for easy removal. The threaded plate is removed from the small end and the rod is pulled free from the other end.

PANEL FORMS

Today, prefabricated panels are used for most wall forming. The panels, made from a special grade of plywood, are attached to wood or metal frames. See Fig. 6-26.

Carpenters can build their own panels, using 3/4 in. plywood and 2 x 4 studs to form 2 ft. or 4 ft. by 8 ft. units. For standard columns and other units, prefabricated forms, as shown in Fig. 6-27, will save time.

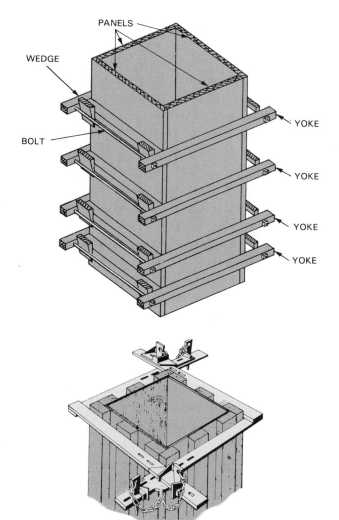

Fig. 6-27. Prefabricated column forms save construction time. Top. Yoke and wedge arrangement. Bottom. Scissor clamps. (The Burke Co.)

Tubular fiber forms may be purchased. These are usually found in heavy construction.

Carpenter-built prefabricated forms can be fastened as shown in Fig. 6-28. Be sure to add wales and bracing. Select straight lumber for wales.

Since panel forms are designed to be used many times, they should be treated to prevent the concrete from sticking to the surfaces. Use special form release coatings which are available.

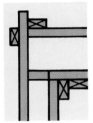

Check form work carefully before placing the concrete. A form that fails during the pouring will waste material and cause extra work.

Care should be exercised in removing form components. They should be thoroughly cleaned and then carefully sorted and stacked for movement to the next job or storage.

Manufacturers have developed many forming systems to replace or supplement panel forms built by the carpenter. For residential work they usually consist of steel frames and exterior grade plywood panels. Sometimes the plywood is coated with a special plastic material to create a smooth finish on the concrete and prevent it from sticking to the surface. The panel units are light for easy handling and transporting from one building site to another. Specially designed devices are used to assemble and space the components quickly and accurately. See Fig. 6-29.

WALL OPENINGS

Several procedures are followed in forming openings in foundation walls for doors, windows, and other holes. In poured walls, forms or stops are built into the regular forms. Nailing strips may be attached to the form and cast into the concrete. Frames are then secured to these strips after the forms are removed. Fig. 6-30 shows several methods of framing openings.

Special framing must also be attached inside the form for pipes or voids for carrying beams. As with windows and doors, form work for these structures must be attached to the outside wall form before the inside form is erected.

Tubes of fiber, plastic, or metal can be used for small openings. They are held in place by wood

Fig. 6-29. Manufactured forms are easier to erect and strip. Top. Form designed for residential foundation. Wedge-bolt connectors are tightened or loosened with a light blow of the hammer. (Symons Corp.) Bottom. Close-up of system having patented corners and form ties. (Universal Form Clamp Co.)

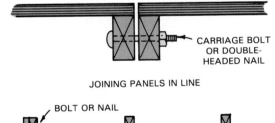

JOINING PANELS IN LINE

CARRIAGE BOLT
OR DOUBLE-
HEADED NAIL

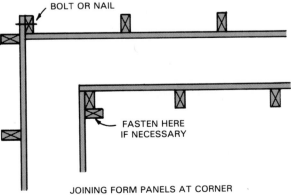

BOLT OR NAIL

FASTEN HERE
IF NECESSARY

JOINING FORM PANELS AT CORNER

Fig. 6-28. Methods of fastening form panels made by the carpenter.

blocks or plastic fasteners which are attached to the form. Larger forms made of wood can be attached with duplex nails driven through the form from the outside.

In concrete block construction, door and window frames are set in place. The masonry units are constructed around them. The outside surface of the frames have grooves into which the mortar flows, forming a key. Basement windows are usually located level with the top of the foundation wall. The sill carries the weight of the structure across the opening, Fig. 6-31.

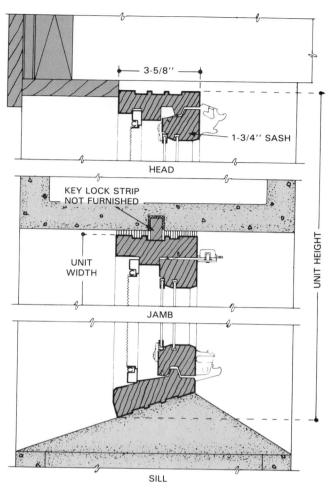

SCALE: 3" EQUALS 1'-0"

Detail above shows typical basement installation in concrete block wall with poured sill. Note key lock strip at side jamb to secure unit in opening. Strip is nailed to back of jamb in slot provided before installation.

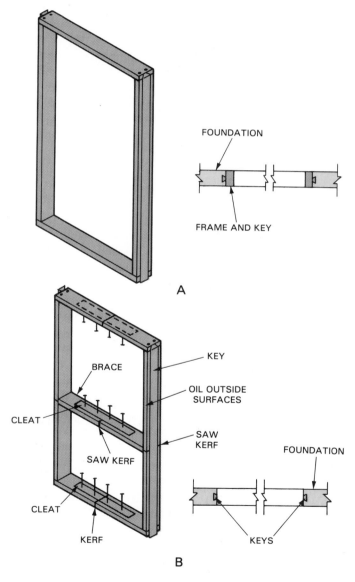

Fig. 6-30. Forms used to frame openings in foundation wall can be constructed by the carpenter. A—Permanent frame which will be left in the wall. B—Frame which will be removed is called a buck. Members are cut partway through for easy removal. Cleats and braces reinforce members at saw cuts. Bucks and frames are nailed to the form with duplex nails from the outside.

Fig. 6-31. Installing windows in basement walls. Top. Detail of basement window unit that can be placed in a concrete or masonry foundation wall. Bottom. Unit installed. (Andersen Corp.)

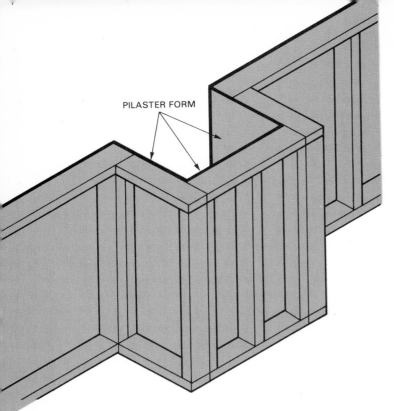

Fig. 6-32. Section of foundation form set up for a pilaster. Often the studs at the sides are omitted.

Fig. 6-33. Concrete being placed in a wall form from a concrete bucket. Avoid moving the concrete long distances in the form.

PILASTERS

Long walls may have pilasters. A pilaster is a thickened section of the wall which strengthens the wall or provides extra support for beams. Fig. 6-32 shows a form set up for pouring a pilaster.

PLACING CONCRETE

Usually most of the concrete can be poured directly from the ready-mix truck into the forms. To move the concrete to other areas not accessible to the truck, a wheelbarrow or bucket is generally used, Fig. 6-33.

Place concrete near to where it will rest. Never allow it to run or be worked over long distances. To do so could cause segregation. (This is a condition in which large aggregates get separated from cement paste and smaller aggregates.) Place concrete in forms promptly after mixing.

In general, concrete for walls should be placed in the forms in horizontal layers of uniform thickness not exceeding 6 to 12 in. As the concrete is placed, spade or vibrate it enough to compact it thoroughly. This produces a dense mass.

Working the concrete next to the form tends to produce a smooth surface. It prevents honeycombing along the form faces. A spade or thin board may be used for this purpose. Large aggregates are forced away from the forms and any air trapped along the form face is released. Mechani-

cal vibrators are effective in consolidating concrete.

However, vibrators create added pressure on the forms. This factor must be considered in the form design. The vibrator should not be held in one location long enough to draw a pool of cement paste from the surrounding concrete.

ANCHOR

Wood plates are fastened to the top of foundation walls with 1/2 in. anchor bolts or straps. They are spaced not more than 4 ft. apart. See Fig. 6-34. In concrete walls, they are set in place as soon as the pour is completed and leveled off.

Anchor bolts are set in the cores of a concrete block wall. They should be about 18 in. long. A piece of metal lath is placed in the second horizontal joint below the top of the wall to hold the grout or mortar. Bolts are installed after the wall is completed. Anchor clips are installed in the same way. See Fig. 6-35.

Wall forms help protect concrete from drying too fast. They should not be removed until the concrete is strong enough to carry the loads that will be placed on it. The material should be hard enough so the surface is not damaged by the stripping operation. Hardening of concrete will normally take a day or two.

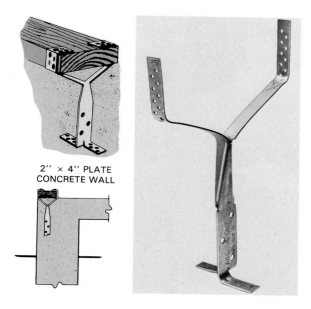

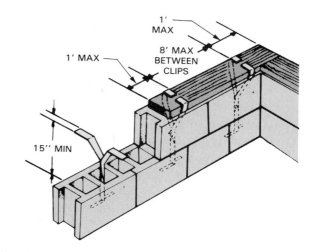

Fig. 6-34. Patented anchor clip is embedded into top of concrete wall. It secures wall plate to foundation. (The Panel Clip Co.)

CONCRETE BLOCK FOUNDATIONS

In some localities, concrete blocks are used for foundation walls and other masonry construction. The standard block is made from Portland cement and aggregates such as sand, fine gravel, or crushed stone. It weighs about 40 or 50 lb.

Lightweight units are made from Portland cement and natural or manufactured aggregates. Among these are volcanic cinders, pumice, and foundry slag. A lightweight unit weighs between 25 and 35 lb. It usually has a much lower U factor. (This is a measurement of the heat flow or heat transmission through materials.)

Blocks should comply with specifications provided by the American Society for Testing Materials. ASTM specifications for a Grade A load-bearing unit requires that the compression strength equal 1000 lb. per sq. inch. Thus, an 8 x 8 x 16 unit must withstand about 128,000 lb. or 64 tons.

SIZES AND SHAPES

Blocks are classified as solid or hollow. A solid unit is one in which the core (hollow) area is 25 percent or less of the total cross-sectional area. Blocks are usually available in 4, 6, 8, 10, and 12 in. widths and 4 and 8 in. heights. Fig. 6-36 shows some of the sizes and shapes with names that indicate use.

Sizes are actually 3/8 in. shorter than their nominal (name) dimensions to allow for the mortar

Fig. 6-35. Anchor bolts and straps are installed about the same way. Top. Anchor strap is embedded in the next-to-top block. Bottom. Embedding an anchor bolt in concrete or mortar.

joint. For example: the 8 x 8 x 16 block is actually 7 5/8 x 7 5/8 x 15 5/8. With a standard 3/8 in. mortar joint, the laid-in-the-wall height will be 8 in. and the width 16 in.

LINTELS

Masonry that is carried across the top of openings is supported by a structural unit called a lintel. This can be:
1. A precast concrete unit that includes metal reinforcing bars.
2. Steel angle irons, Fig. 6-37.

Another method is to lay a course of lintel blocks across the opening (supported by a frame). Then add reinforcing bars and fill the blocks with concrete, Fig. 6-38.

INSULATING FOUNDATION WALLS

In northern climates, when residential plans include a finished basement, insulating the foundation walls may be desirable. There are several ways to reduce the heat flow.

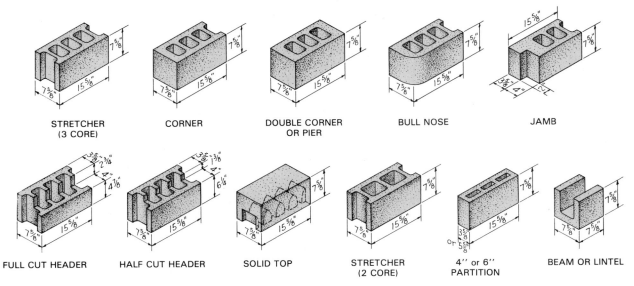

STRETCHER
(3 CORE)

CORNER

DOUBLE CORNER
OR PIER

BULL NOSE

JAMB

FULL CUT HEADER

HALF CUT HEADER

SOLID TOP

STRETCHER
(2 CORE)

4'' or 6''
PARTITION

BEAM OR LINTEL

Fig. 6-36. Concrete blocks are manufactured for many different purposes. These are typical.
(Portland Cement Assoc.)

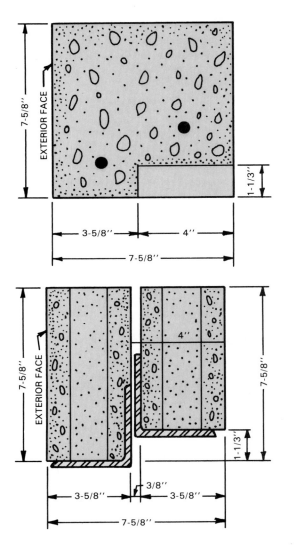

Fig. 6-37. Lintels for block walls. Top. Precast unit. Bottom.
Standard block units supported by steel angle irons. Notches
are for window or door frames.

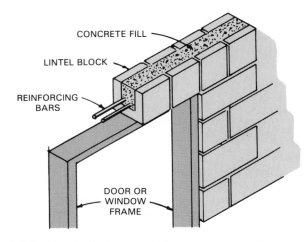

Fig. 6-38. Lintel blocks are laid over a supporting frame.
Then the cavity is filled with reinforced concrete.

Fig. 6-39. These blocks were insulated by filling the cores
with styrofoam insulation.

1. Lightweight masonry units that have a lower U factor may be used.
2. The cores of regular or lightweight blocks can be filled with an insulating material, Fig. 6-39.
3. Other methods include cavity wall construction or the use of various forms of insulation applied to the interior or exterior surface. Additional information about insulation is included in Unit 13.

WATERPROOFING

In most localities, the outside of poured concrete or concrete masonry basement walls should be waterproofed below the finished grade and drain tile installed. See Fig. 6-40.

Masonry (block) walls may be waterproofed by an application of cement plaster followed by several coats of an asphaltic material. The wall surface should be clean and dampened with a water spray just before the first coat of plaster is applied. The plaster can be made of cement and sand (1 to 2 1/2, mix by volume) or mortar may be used. When the first coat has partially hardened, it should be roughened with a scratcher to provide better bond for the second coat.

After the first coat has hardened at least 24 hours, the second coat is applied, Fig. 6-40. Again, dampen the surface just before applying the plaster. Both the plaster coats should extend from about 6 in. above the finished grade to the footing.

A cove of plaster should be formed between the footing and wall. This precaution prevents water from collecting and seeping through the joint. The second coat should be kept damp for at least 48 hours.

In poorly drained soils, or when it is important to secure added protection against moisture penetration, the plaster may require coating with asphalt waterproofing or hot bituminous material. Fig. 6-41 shows a method of waterproofing basements.

To waterproof poured walls, bituminous waterproofing is generally used without cement plaster. Polyethylene sheeting is often used as a waterproofing material. It should cover wall and footing in one piece.

Fig. 6-42, top, shows a drain tile being laid along the side of the footing. The drain should lead away from the foundation to an outlet that always remains open. In some localities, perimeter drains may be connected to a sump pump. Fig. 6-42, bottom, shows a tile placed through the footing for this purpose. The drain tile is usually placed at a slope of about 1 in. in 20 ft. and spaced 1/4 in. apart. Strips of tar paper cover the joints. Tiles should be covered with a 6 to 8 in. layer of coarse gravel or crushed stone.

As an energy conservation measure, plastic foam board may be added to the outside face of the foundation wall. In such cases, no back plastering or waterproofing should extend any higher than 2 in. below the final grade. More information on this type of insulation will be found in Unit 13.

Fig. 6-40. Applying plaster to a block foundation. First coat was roughened before completely dry.

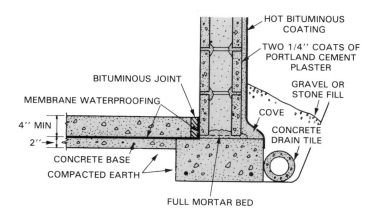

Fig. 6-41. Sectional view shows method of waterproofing basement. It is especially effective in very wet soil. Plaster is omitted on outside wall in poured foundations.

BACKFILLING

After the foundation has cured and waterproofing has been completed, the excavation outside the walls is filled with earth. Walls should be carefully braced before backfilling begins. This is especially important with block foundation. See Fig. 6-43 and Fig. 6-44.

If tile lines have been installed, care must be taken to protect the lines from movement. When

Fig. 6-42. Placing drain tile. Top. Laying drain tile along the footing. Perforated plastic pipe is also used for this purpose. Bottom. Bleeder tile will carry water from perimeter tile through the footing. A tile line will connect the bleeder tile with a sump pump in the basement.

Fig. 6-43. Walls should be adequately braced. This will help them withstand the pressure of backfilling and grade settling.

heavy power equipment is used to backfill, the operator must be careful not to damage the walls.

SLAB-ON-GROUND CONSTRUCTION

Today, many commercial and residential structures are built without basements. The main floor is formed by placing concrete directly on the ground. Footings and foundations are somewhat

Fig. 6-44. Backfilling operation is partly completed. Final grade level is marked by the waterproofing. Pressed steel areaways must be attached around windows before the backfilling is begun. (Portland Cement Assoc.)

similar to those for basements. However, they need to extend down only to solid soil and below the frost line. See Fig. 6-45.

In slab-on-ground construction, insulation and moisture control are essential. The earth under the floor is called the subgrade and must be firm and completely free of sod, roots, and debris.

A coarse fill, at least 4 in. thick, is placed over the finished subgrade. The fill should be brought to grade and thoroughly compacted.

This granular fill may be slag, gravel, or crushed stone, preferably ranging from 1/2 in. to 1 in. in diameter. The material should be uniform without fines to insure maximum air space in the fill. Air spaces will add to insulating qualities and reduce capillary attraction of subsoil moisture (action by which moisture passes through fill).

In areas where the subsoil is not well drained, a line of drain tile may be required around the outside edge of the exterior wall footings.

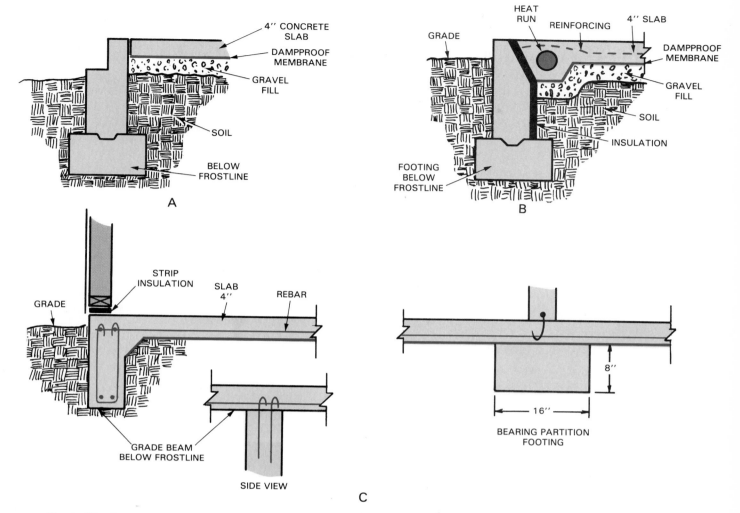

Fig. 6-45. Three types of slab-on-ground. A—Unreinforced slab with loads supported by footing and wall. Used where soil is coarse and well drained. B—Slab is reinforced with welded wire fabric. Inside foundation wall is insulated because of perimeter heat duct in slab. C—Monolithic slab. This type is used over problem soils. Loads are carried over a large area of the slab.

While preparing the subgrade and fill, various mechanical installations should be made. Underfloor ducts, where used, are usually embedded in the granular fill. Water service supply lines, if placed under the floor slab, should be installed in trenches deep enough to avoid freezing. Connections to utilities should be brought above the finished floor level before pouring the concrete.

After the fill has been compacted and brought to grade, a vapor barrier should be placed over the sub-base. Its purpose is to stop the movement of water into the slab.

Among materials widely used as vapor barriers are 55 lb. roll roofing, 4-mil polyethylene film, and asphaltic-impregnated kraft papers. Strips should be lapped 6 in. to form a complete seal. A vapor barrier is essential under every section of the floor. Instructions supplied by the manufacturer should be carefully followed.

Perimeter insulation is important. It reduces heat loss from the floor slab to the outside. The insulation material must be rigid and stable while in contact with wet concrete. Fig. 6-46 shows a foundation design with perimeter insulation typical of residential frame construction.

Thickness of the insulation varies from 1 to 2 in., depending on outside temperature and type of heating. The insulation can be placed either horizontally or vertically along the foundation wall. Refer to Unit 13 for additional information.

When the insulation, vapor barrier, and all mechanical aspects are complete, reinforcing mesh, if used, is laid. Check local building codes for requirements. Usually a 6 x 6 x 10 ga. mesh is sufficient for residential work. This should be located from 1 to 1 1/2 in. below the surface of the concrete. Fig. 6-47 shows the concrete being placed for a typical slab-on-ground residential floor.

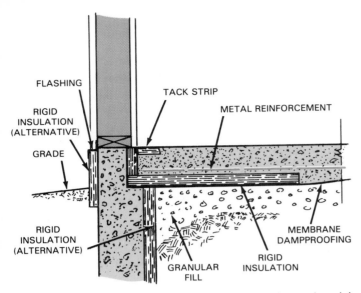

FLASHING
RIGID INSULATION (ALTERNATIVE)
GRADE
TACK STRIP
METAL REINFORCEMENT
RIGID INSULATION (ALTERNATIVE)
GRANULAR FILL
RIGID INSULATION
MEMBRANE DAMPPROOFING

Fig. 6-46. Perimeter insulation is important when using slab-on-ground construction in cold climates.

Terrazzo, ceramic tile, asphalt tile, wood flooring, linoleum, and wall-to-wall carpeting are coverings appropriate for use on concrete floors. When linoleum, asphalt tile, or similar resilient-type flooring materials are to be applied, the concrete surface is usually given a smooth steel-troweled finish. Information on finished wood floors is provided in Unit 15.

BASEMENT FLOORS

Many of the considerations previously listed for slab-on-ground floors also apply to basement floors. Basement floors are poured later in the building sequence; sometime after the framing and roof are complete and after the plumber has installed waste plumbing and water service lines, Fig. 6-48.

The concrete may be brought in through basement windows or stair openings. Sometimes the rough opening for a fireplace located on an outside wall will provide easy access. (Fig. 6-8 shows a cutaway of a poured concrete basement.)

Fig. 6-48. Basement floors are usually poured after building is erected. Workers are screeding off a section of the newly poured floor. (Portland Cement Assoc.)

ENTRANCE PLATFORMS AND STEPS

Entrance platform foundations should be constructed as a part of the main foundation or firmly attached to the main foundation. Reinforcing bars, placed in the wall when it is constructed, can provide a solid connection. Fig. 6-49 shows a method of forming special support brackets for entrance platforms and steps.

Steps may be included and poured as part of the platform, Fig. 6-50. When the steps are more than 3 ft. wide, 2 in. stock should be used to prevent the risers from bowing. Detailed information on stair construction is included in Unit 16. The 2 x 8 in. riser boards are set at an angle of about 15 deg. to provide a slight overhang (nosing). Also, the bottom edges of the boards are beveled to permit the mason to trowel the entire surface of the tread.

Some concrete steps must be poured against a wall or between two existing walls. A form can be constructed like the one shown in Fig. 6-51.

Fig. 6-47. Placing concrete for residential floor. Heating ducts, plumbing, water service lines, and vapor barrier are placed before the pour. (Portland Cement Assoc.)

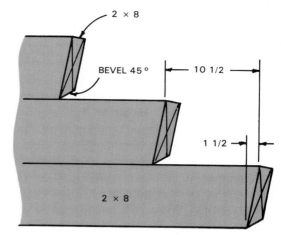

BEVELING FORM TO PROVIDE
TOE ROOM ON TREADS

Fig. 6-49. Entrance steps can be supported by special brackets cast in the wall when it is poured. Top. Brackets are formed by placing sloping 2 × 8s between side forms. Bottom. Resulting brackets support steps preventing settling. Loose backfill normally cannot support the weight of steps or platform.

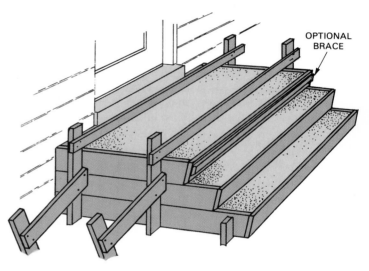

Fig. 6-50. Constructing forms for steps. Nosing is formed by tilting form boards inward at the bottom as shown at top.

SIDEWALKS AND DRIVES

Usually sidewalks and drives are laid after the finished grading is completed. If there is extensive fill, wait until the grade has settled.

Main walks leading to front entrances should be at least 4 ft. wide. Those to secondary entrances may be 3 ft. or slightly less.

In most areas, sidewalks are 4 in. thick and the formwork is constructed with 2 x 4 lumber. Walks and drives are usually laid directly on the soil. If there is a moisture problem and frost action, a coarse granular fill should be put down first.

When joining two levels of sidewalks with steps, it is usually best to pour the top level first. If retaining walls are used, these should be constructed next along with a segment of the lower sidewalk. Finally the steps are formed and poured as shown in Fig. 6-52.

When setting sidewalk forms, Fig. 6-53, provide a slope to one side of about 1/4 in. per foot. Increase thickness to 6 in. or add reinforcing where there will be heavy vehicular traffic. The concrete

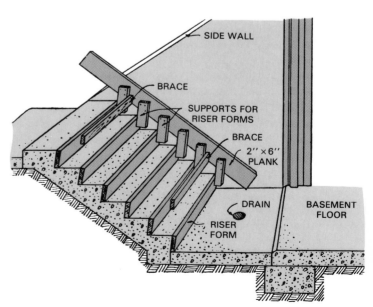

Fig. 6-51. Method of building forms when steps are located between two walls already in place.

Fig. 6-52. Form in place for steps between two levels of sidewalk. Sloping 2 × 4s support risers.

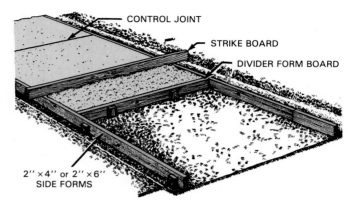

Fig. 6-53. Pouring a sidewalk. Wood strike board levels concrete between the forms. It is worked back and forth in a sawing motion as it is gradually moved ahead.

should not be permitted to bond against foundation walls or entrance platforms. It should be permitted to "float" on the ground. A 1/2 in. thick strip of asphalt impregnated composition board is commonly used to form this separation.

Driveways should be 5 or 6 in. thick with reinforcing mesh included. A single driveway should be at least 10 ft. wide and a double driveway a minimum of 16 ft. Minimum crossway slope should be 1/4 in. per foot.

Concrete should be placed between the forms so it will be close to its final position. Do not overwork the concrete while it is still plastic. This tends to bring excess water and fine material to the surface. It will cause scaling and dusting after the concrete has cured.

SCREEDING

After the concrete is roughly spread between the forms, screed it immediately. This is done by moving a straightedge back and forth in a saw-like motion across the top of the forms. A small amount of concrete should always be kept ahead of the straightedge.

When the screeding operation is complete, move a float made of wood, aluminum, or magnesium over the surface. When skillfully performed, this operation removes high spots, fills depressions, and smooths irregularities.

As the concrete stiffens and the water sheen disappears from the surface, finish edges and cut control joints. These joints should extend to a depth of at least one-fifth of the thickness of the concrete.

For sidewalks and driveways, the distance between the control joints is usually about equal to the width of the slab. Joints can be formed with a groover and straightedge or cut with a power saw. The saw must have a masonry blade. Sawing is done 18 to 24 hours after the concrete is poured.

EDGING

Edges of walks and driveways should be rounded by working an edger tool along the forms. The edging tool can also be used to finish the control joints.

Surface finishing operations should be performed after the concrete has hardened enough to become somewhat stiff. For a rough finish that will not become slick during rainy weather, the surface can be stroked with a stiff bristle broom. When a finer texture is desired, the surface should be steel troweled and then lightly brushed with a soft bristle broom. Several key steps in the pouring and finishing of a concrete drive are shown in Figs. 6-54 to 6-56.

For proper curing of concrete protect it against moisture loss during the early stages of hardening (hydration). Covers of waterproof paper or polyethylene film are commonly used. A convenient alternative is to spray the concrete with a plastic-based curing compound. The sprayed material forms a continuous membrane over the surface.

WOOD FOUNDATIONS

The All-Weather Wood Foundation system is so named because it can be installed in almost any weather. It provides comfortable living space in basement areas because the stud wall can be fully insulated. All wood parts are pressure treated with a solution of chemicals that make the fibers use-

Fig. 6-54. Place concrete near its final resting place. These forms are patented steel channels supported with steel pins driven into the ground.

Fig. 6-55. Polyethylene film underlayment prevents ground from soaking up water from the concrete. Hook (arrow) is being used to raise reinforcing into concrete.

less as a food for insects and the fungus growth that causes decay.

Foundation sections of 2 in. lumber and exterior plywood can be panelized in fabricating plants or constructed on the building site. Pressure treated wood foundations, Fig. 6-57, have been approved by major code groups and accepted by FHA, HUD, and FmHA (Farmers Home Adm.).

For a regular basement, the site is excavated to the required depth. Plumbing lines are installed and provisions made for foundation drainage according to local requirements.

Some soils will require a sump (pit for water collection) which is connected to a storm sewer, pump, or other positive drainage. The subgrade is then covered with a 4 to 6 in. layer of porous

gravel or crushed stone and carefully leveled, Fig. 6-58. Footing plates of 2 x 6 or 2 x 8 material are installed directly on this base and the wall sections erected.

Nails and other fasteners should be made of either silicon bronze, copper, or hot-dipped zinc coated steel. Special caulking compounds are used to seal all joints in the plywood sheathing.

Before pouring the basement floor, the porous

Fig. 6-56. Concrete surfacing tools. Left. Power screed strikes off and surfaces concrete. Engine vibrates the wood frame. Center. Lightweight aluminum float smooths screeded concrete. Right. Power troweler produces smooth, hard concrete surface.

Fig. 6-57. Wood foundations can be built during any kind of weather. All components are pressure treated against rot and insect damage. Outer walls below grade are protected by a waterproof membrane. (American Plywood Assoc.)

Fig. 6-58. Preparation of subgrade to receive wood foundation. Four to six inches of porous gravel are needed. (American Plywood Assoc.)

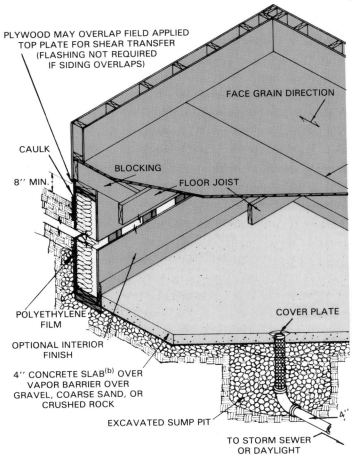

PLYWOOD MAY OVERLAP FIELD APPLIED TOP PLATE FOR SHEAR TRANSFER (FLASHING NOT REQUIRED IF SIDING OVERLAPS)

FACE GRAIN DIRECTION

CAULK

BLOCKING

8" MIN.

FLOOR JOIST

POLYETHYLENE FILM

COVER PLATE

OPTIONAL INTERIOR FINISH

4" CONCRETE SLAB(b) OVER VAPOR BARRIER OVER GRAVEL, COARSE SAND, OR CRUSHED ROCK

EXCAVATED SUMP PIT

TO STORM SEWER OR DAYLIGHT

4"

(b)A WOOD BASEMENT FLOOR SYSTEM IS UNDER DEVELOPMENT, WRITE APA FOR PRELIMINARY DETAILS.

Fig. 6-59. Typical wood foundation in cutaway. Note drainage sump which keeps the subsoil dry around the foundation. (American Plywood Assoc.)

gravel or crushed stone base is covered with a polyethylene film (6 mil. thick) and a screed board is attached to the foundation wall. See Fig. 6-59.

The first floor frame is installed on the double top plate of the foundation wall with special attention given to methods of attachment so that inward forces will be transferred to the floor structure. See Fig. 6-60. Where joists run parallel to the wall, blocking should be installed between the outside joist and the first interior joist.

Before backfilling, attach a 6-mil polyethylene moisture barrier to sections of the wall below grade. Bond the top edge of the barrier to the wall at grade level with a special adhesive. Install a treated wood strip over this and caulk it. (Later it will serve as a guide for backfilling.)

Lap vertical joints in the polyethylene film at least 6 in. Seal joints with the same adhesive.

Do not backfill until basement floor and first floor are installed. Let the basement floor to cure.

As with any foundation system, satisfactory performance demands full compliance with recommended standards covering design, fabrication, and installation. Standards for wood foundations are contained in a Manual published by the National Forest Products Association, 1619 Massachusetts Avenue, N.W., Washington, D.C. 20036.

Carpenters installing wood foundations should make certain that each piece of treated lumber and plywood carries the mark "AWPB-FDN." This assures them that the materials meet requirements of code organizations and federal regulatory agencies.

Fig. 6-60. Truss joists being installed over wood foundation. Use of 10d nail is recommended so inward pressure on wall is transferred to the floor system. (Osmose)

COLD WEATHER CONSTRUCTION

Cold weather may call for some changes in the way concrete and masonry materials are handled and placed. It may be necessary to:
1. Heat the materials.
2. Cover freshly placed concrete or masonry.
3. Erect an enclosure and keep the construction area heated.

When temperatures fall below 40 °F, concrete should have a temperature of 50 °F to 70 °F when it is placed. Since most concrete is ready-mix, only shelter will be needed on the building site. However, when blocks are being placed, the materials will need shelter and heat.

Shelter can be arranged with scaffolding sections, lumber, and tarpaulins. Then the materials can be stored and adequately heated. If shelters cannot be built, the bagged materials and masonry units should be wrapped with canvas or polyethylene tarpaulins when the temperature is below 40 °F. Be sure the material is stored so ground moisture cannot reach it. Fig. 6-61 lists recommendations for handling of masonry materials in cold weather.

PROTECTING CONCRETE

Freshly placed concrete should be protected with a covering. Hydration (chemical reaction of cement and water during hardening) creates heat. The covering will help hold the heat in until the concrete cures.

Avoid pouring concrete on frozen ground. When the ground thaws, uneven settling may crack the concrete. Before pouring, make sure that reinforcing, metals, embedded fixtures, and the insides of forms are free of ice.

MORTAR TEMPERATURES

Temperatures of materials used in mortar are important. Water is the easiest to heat. It can also store more heat and will help bring cement and aggregate up to temperature. Generally, it should not be hotter than 180 °F. There is a danger that much hotter water could cause the mortar to set instantly.

When the air temperature is lower than 32 °F, sand should be heated to thaw frozen lumps. If desired, the sand temperature can be raised as high as 150 °F. A 50 gal. drum, open on one end, or a metal pipe works well for containing fire. Heap the sand over and around the container.

ADMIXTURES

Admixtures are materials other than cement, water, and aggregate which are added to concrete or mortar to change its properties. Cold weather admixtures used for concrete or mortar include:
1. Antifreeze to lower the freezing point of the mixture.
2. Accelerators to speed up curing. They do not lower freezing point.

WHEN AIR TEMPERATURE REACHES	DO THIS TO MATERIALS	DO THIS TO PROTECT PLACED MASONRY
Below 40 °F	Heat mixing water. Keep mortar temperatures 40 °F to 120 °F	Cover walls and masonry materials to protect from moisture and freezing. Use canvas or plastic.
Below 32 °F	Do all of above but also: Heat sand to thaw frozen clumps. Heat wet masonry units to thaw ice.	Provide windbreak for workers when wind speed is above 15 mph. Cover walls and materials after workday to protect against wetness and freezing. Keep masonry temperature above 32 °F, using heaters or insulated blankets for 16 hours after placing of units.
Below 20 °F	Besides above: Heat dry masonry units to 20 °F.	Enclose structure and heat the enclosure to keep temperature above 32 °F for 24 hours after placing masonry units.

Fig. 6-61. Recommendations of Portland Cement Association should be followed for cold-weather construction.

3. Air-entraining agents that improve workability and freeze-thaw durability of mortar as it ages.
4. Corrosion inhibitors. When reinforcing is placed in winter construction these materials are thought to prevent rust formation.

According to the Portland Cement Association, accelerators and air-entraining agents have proved successful for winter use and are recommended. Other admixtures may help but are not recommended by the Association. Accelerators include: calcium chloride, soluble carbonates, silicates, and fluosilicates, calcium aluminate, and organic compounds such as triethanolamine.

For mixing of mortar, admixtures must be obtained by the carpenter or contractor. They are added on the job. Ready-mix companies will include them in the mix upon specifications provided by the carpenter, contractor, or architect.

ESTIMATING MATERIALS

Concrete is measured and sold by the cubic yard. A cubic yard is 3 ft. square and 3 ft. high. It contains 27 cu. ft. (3 x 3 x 3 = 27). To determine the number of cubic yards needed for any square or rectangular area, use the following formula. All dimensions should be converted to feet and fractions of feet:

$$\text{cubic yards} = \frac{\text{width} \times \text{length} \times \text{thickness}}{27}$$

For example, to find the concrete needed to pour a basement floor that is 30 ft. × 42 × 4 in.:

$$\text{cu. yd.} = \frac{30 \times 42 \times 1/3}{27}$$

$$= \frac{30 \times 42 \times 1}{27 \quad 3} = \frac{140}{9}$$

$$= 15.56 \text{ cu. yd.}$$

You should allow extra concrete for waste or slight variations in the cross sections of the form. An additional 5 to 10 percent is usually added.

The number of concrete masonry units needed can be estimated by:
1. Determining the number of units required in each course.
2. Multiplying by the number of courses between the footing and plate.

For example: find the number of 8 × 8 × 16 in. blocks required to construct a foundation wall with a total perimeter (distance around the outside) of 144 ft. and laid 11 courses high.

$$\text{Total number} = \frac{\text{perimeter}}{\text{unit length}} \times \text{number of courses}$$

$$= \frac{144}{1\ 1/3\ (16'')} \times 11$$

$$= \frac{144}{\frac{4}{3}} \times 11$$

$$= \frac{144}{4} \ 3 \times 11$$

$$= \frac{144}{4}\ 3 \times 11$$

$$= 108 \qquad \times 11$$

$$= 1188 \text{ blocks}$$

Another method is to figure the face area of the wall in square feet and divide by 100. This figure is then multiplied by 112.5 which is the number of 8 × 8 × 16 in. blocks required to construct 100 sq. ft. of wall. This figure can be multiplied by 2.6 to find the cu. ft. of mortar required. See the table in Fig. 6-62.

Wall thickness	For 100 sq. ft. of wall		For 100 concrete block
in.	Number of block*	Mortar** cu. ft.	Mortar** cu. ft.
8	112.5	2.6	2.3
12	112.5	2.6	2.3

*Based on block having an exposed face of 7 5/8 × 15 5/8 in. and laid up with 3/8-in. mortar joints.

**With face shell mortar bedding—10 percent wastage included.

Fig. 6-62. Quantities of concrete block and mortar can be calculated with this chart. (Portland Cement Assoc.)

TEST YOUR KNOWLEDGE — UNIT 6

1. Grading must (always, sometimes) be done before building lines are laid out.
2. Carpenters always locate lot lines. True or False?
3. Batter boards should be located _____ feet or more away from the building lines.
4. Building sites on steep slopes or rugged terrain should be rough graded before the building is laid out. True or False?
5. In cold climates, foundations should be

located below the _____ line.

6. In residential construction, a safe design is usually obtained by making the width of the footing _____ as wide as the foundation wall.

7. A _____ footing is one that changes grade levels at intervals to accommodate a sloping lot.

8. What is a grade beam?

9. Foundation forms constructed of 1 in. boards should be held in place with stakes placed _____ to _____ ft. apart.

10. Loose dirt and debris (should or should not) be removed from the ground under a footing.

11. Concrete is made by mixing _____, _____, _____, and water in proper proportions.

12. Concrete hardens by a chemical action called _____.

13. Each sack of Portland cement holds _____ lb.

14. Ready-mix concrete is purchased by the _____.

15. What is a pilaster?

16. When placing concrete in forms, working the concrete next to the forms tends to produce a _____ surface along the form faces.

17. Wood sill plates are fastened to the top of a foundation wall with _____, _____, or _____.

18. A concrete block specified as an 8 × 8 × 16 block is actually _____ × _____ × _____.

19. A concrete basement wall may be water-proofed by using an application of _____ _____, _____, or _____.

20. List three types of slab-on-ground.

21. In slab-on-ground construction, a _____ _____ should be laid over the sub-base to stop the movement of _____ _____ into the concrete slab.

22. In most areas, sidewalks are _____ in. thick.

23. A wood foundation is installed on a layer of _____ _____ that is _____ to _____ in. thick.

24. In cold weather it is accepted practice to pour concrete over frozen ground. True or False?

25. How many 8 × 8 × 16 concrete blocks are required to lay 100 square feet of wall surface?

OUTSIDE ASSIGNMENTS

1. Visit a ready-mix concrete plant and study the operations. Secure information about the following: source of aggregates; handling and storing cement and aggregates; equipment used to measure and proportion mixtures; size of truck-mounted mixers; distance trucks can travel without extra charge; cost of a cubic yard of concrete in various psi ratings and fractional parts of a cubic yard that can be ordered. Prepare a written report.

2. Find a set of house plans that includes a fireplace located on an outside wall. Prepare a scaled (1 1/2" = 1'-0") drawing of the formwork you would use for the footings under the fireplace wall. Show individual form boards and stakes. Include one or more section views to describe the shape of the footing.

3. Study reference materials and booklets prepared by such organizations as the Portland Cement Association or the Perlite Institute. Secure information about air-entrained concrete, slump tests, lightweight aggregates, ultra-lightweight concrete, thermal conductivity, compression tests, reinforcing, and prestressed concrete units. Prepare an outline and report to the class.

4. Study the building code in your area and learn about such requirements as: building setbacks from property lines, design of footings for residential structures, minimum depths for footings and foundations, and basic construction of concrete and masonry foundation walls. Summarize your findings in a written report for your class.

A

B

D

C

Both modern and traditional floor framing members are used in residential construction today. Solid lumber is most often used. However, special designs such as the truss joist, the web joist, the the steel joist are frequently installed by the carpenter. They have the advantage of allowing open sections for running ducts, pipes, and wiring. A—Conventional floor framed with 2 x 10 joists. (National Forest Products Assoc.) B—The floor truss or truss joist is made of 2 x 4s. Chords (top and bottom pieces) are strengthened by web (angled) members. (TrusWal Systems Corp.) C—Metal web truss system. (Gang-Nail Systems, Inc.) D—Steel joists system. (U.S. Gypsum Co.)

Unit 7

FLOOR FRAMING

When the foundation is completed and the concrete has properly set up, work can be started on the floor framing. Before starting with the framing, most carpenters like to have the area outside the walls backfilled. The ground can be brought to rough-grade level. This makes it easier to deliver and stack lumber on the building site. It also provides the carpenter with easier access to the building.

TYPES OF FRAMING

Framing methods used in a structure will be determined, to a large extent, by the basic design.

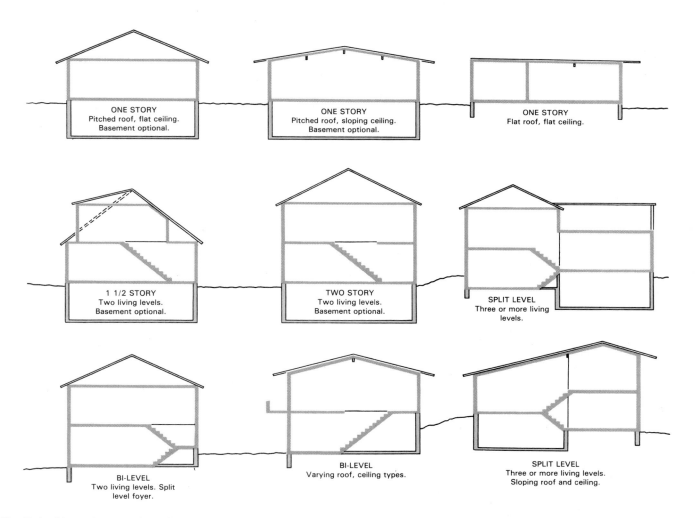

Fig. 7-1. How a house will be framed depends on the type, how spaces are tied together, and the kinds of materials used. These are basic types.

See Fig. 7-1. In addition to this, methods may vary because of conditions found in a certain locality, materials available, and the personal experience and preference of the builder.

In some parts of the country, buildings must be constructed with special resistance to wind and rain. In other parts, earthquakes may be the greatest hazard. In cold climates, heavy loads of damp snow may require special roof designs.

All structures should be built to reduce the effects of shrinkage and warping. They must also resist the hazard of fire.

Two basic types of framing are:
1. Platform framing, also called western framing.
2. Balloon framing. This type is seldom used any more.

Joists, studs, plates, and rafters are the common structural members in both types of framing. Material with a nominal 2 in. thickness is used. Plank and beam, also called post and beam, framing is different. Its heavy structural members are 4 in. or more thick. This kind of construction is covered in Unit 20.

PLATFORM FRAMING

Most modern residential construction has platform framing. The first floor is built on top of the foundation walls as though it were a "platform." The floor provides a work area upon which the carpenter can assemble and raise wall sections safely and accurately. The wall sections are one story high. They and partitions support a platform for the second floor. Each floor is framed separately. See Fig. 7-2.

Platform framing is satisfactory for both one story and multi-story structures. Settlement, due to shrinkage, occurs in an even and uniform manner throughout the structure.

Typical construction methods used at first and second floor levels are illustrated in Fig. 7-3. The only firestopping needed is built into the floor frame at the second floor level. It prevents the spread of fire in a horizontal direction. Here it also serves as solid bridging, holding the joists in a plumb (vertical) position. To save energy a strip of insulation is usually placed under the sill plate.

The type of framing is usually specified in the architectural plans. There will be sectional views of floors, walls, and ceilings. A typical detail drawing of first floor framing would include not only the type of construction but also the size and spacing of the various members. See Fig. 7-4.

BALLOON FRAMING

In balloon framing, now seldom used, the studs are continuous from the sill to the rafter plate. Ends of the second floor joists are supported on a ribbon. They are spiked to the stud as well. See Fig. 7-5. Firestopping must be added to the space between the studs. This space, which also occurs in load-bearing partitions, permits easy installation of service pipes and wiring.

In balloon framing, shrinkage is reduced because the amount of cross-sectional lumber is low. Wood shrinks across its width but practically no shrinkage occurs lengthwise. Thus the high vertical stability of the balloon frame makes it adaptable to two-story structures, especially where masonry veneer or stucco is used on the outside wall.

Fig. 7-2. Example of platform framing on single family dwelling. Platform (arrow) supports wall for next level. (American Plywood Assoc.)

GIRDERS AND BEAMS

Joists are the supports of the floor frame. They rest on top of the foundation walls. Usually the span (distance between the walls) is so great that additional support must be provided. Girders, also called beams, resting on the foundation walls and on posts or columns, provide the needed support. Girders may be solid timbers, built-up lumber, or steel beams. Sometimes a load-bearing partition replaces a girder or beam.

To determine the size of a girder:
1. Find the distance between girder supports.
2. Find the girder load width. A girder must carry the weight of the floors on each side to the mid-point of the joists which rest upon it.
3. Find the "total floor load" per sq. ft. carried by joists and bearing partitions to girder. This will

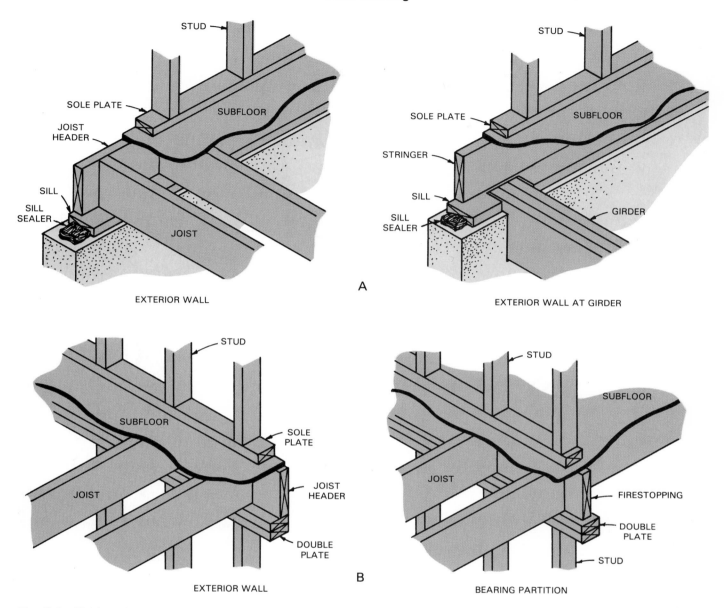

Fig. 7-3. Platform framing details. A—First floor. Joist header also acts as a firestop. B—Second floor framing is similar to first. Note solid bridging also acts as firestop.

be the sum of loads per sq. ft. listed in the diagram, Fig. 7-6. This does not include roof loads. These are carried on the outside walls unless braces or partitions are placed under the rafters. Then a portion of the roof load is carried to the girder by joists and partitions.

4. Find the total load on the girder. This is the product of girder span × girder load width × total floor load.
5. Select proper size of girder according to the code in your area. The table in Fig. 7-7 is typical. It indicates safe loads on standard size girders for spans from 6 ft. to 10 ft. Shortening the span is usually the most economical way to increase the load a girder will carry.

Built-up girders can be made of three or four pieces of 2 in. lumber nailed together with 20d nails. See Fig. 7-8. Joints should rest over columns or posts.

STEEL BEAMS

In many localities, steel beams are used instead of wood girders. Sizes depend on the load. It can be calculated in the same way as for wood girders.

Two types of steel beams are illustrated in Fig. 7-9. The W (wide-flange) is the type generally used in residential construction. Wood beams vary in depth, width, species, and grade. Steel beams vary in depth, width of flange, and weight.

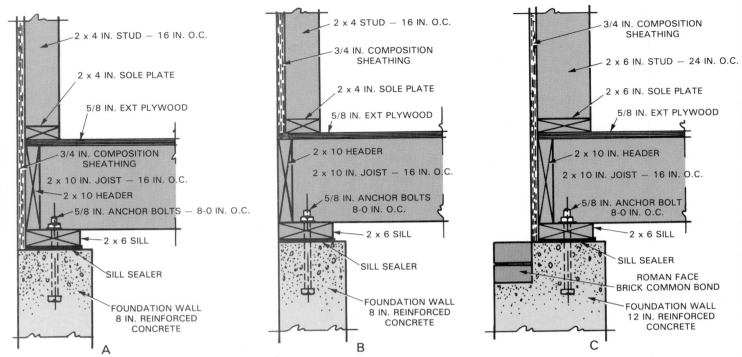

Fig. 7-4. Architectural detail drawings show methods of construction as well as materials to use. A—Sheathing brought to foundation. B—Sheathing brought to sole plate. C—Brick veneer construction.

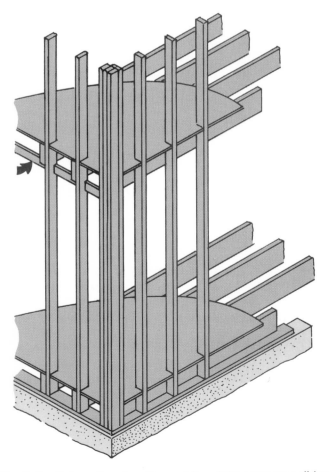

Fig. 7-5. Balloon framing. Second floor joists rest on a ribbon (arrow) set into the studs.

After the approximate load on a steel beam has been determined, the correct size can be selected from the table, Fig. 7-10. This table lists a selected group of steel beams commonly used in residential structures. For example, if the total load on the beam (evenly distributed) is 15,000 lb. and the span between supports is 16'-0'', then a W8-18 beam should be used. This specifies an 8 in. beam weighing 18 lb. per lineal foot. The width of the flange is 5 1/4 in.

POSTS AND COLUMNS

For ordinary wood posts (not longer than 9 ft. or smaller than 6 by 6 in.), it is safe to assume that a post whose greatest dimension is equal to the width of the girder it supports will carry the girder load. For example, a 6 by 6 in. post would be suitable for a girder 6 in. wide. For a girder 8 in. wide, a 6 by 8 in. or 8 by 8 in. post should be used.

Adequate footings must be provided for girder posts and columns. Wood posts should be supported on footings which extend above the floor level, as shown in Fig. 7-11. To make sure the posts will not slide off their footing, pieces of 1/2 in. diameter reinforcing rod or iron bolts of that size should be embedded in the footing before the concrete sets. They should project about 3 in. into holes bored in the bottoms of the posts.

A post anchor, Fig. 7-12, holds the wood post securely in place. It supports the bottom of the post above the floor, protecting the wood from

Floor Framing

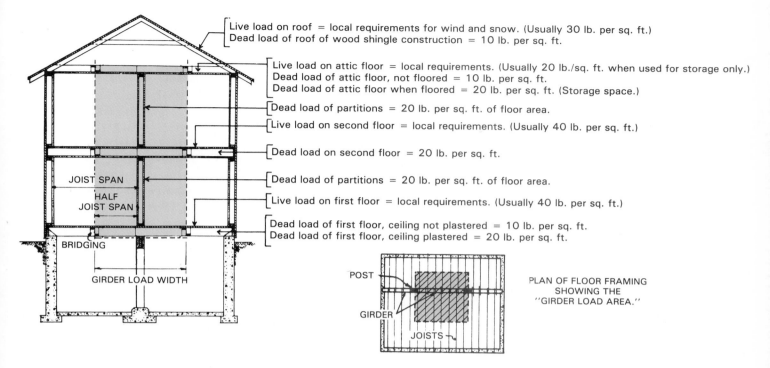

Fig. 7-6. Diagram shows method of figuring loads for frame of a two-story home.

GIRDERS	SAFE LOAD IN LB. FOR SPANS FROM 6 TO 10 FEET				
SIZE	6 FT.	7 FT.	8 FT.	9 FT.	10 FT.
6 x 8 SOLID	8,306	7,118	6,220	5,539	4,583
6 x 8 BUILT-UP	7,359	6,306	5,511	4,908	4,062
6 x 10 SOLID	11,357	10,804	9,980	8,887	7,997
6 x 10 BUILT-UP	10,068	9,576	8,844	7,878	7,086
8 x 8 SOLID	11,326	9,706	8,482	7,553	6,250
8 x 8 BUILT-UP	9,812	8,408	7,348	6,544	5,416
8 x 10 SOLID	15,487	14,782	13,608	12,116	10,902
8 x 10 BUILT-UP	13,424	12,768	11,792	10,504	9,448

Fig. 7-7. Table indicates typical safe loads for standard size wood girders.

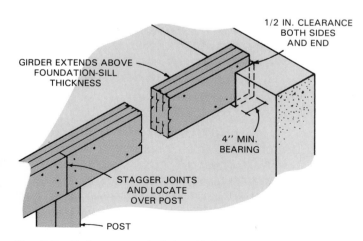

Fig. 7-8. Built-up wood girder. Nails should not be spaced farther apart than 32 inches along top and bottom edges. Metal bearing plate should be placed under girder at foundation wall.

dampness. The bracket can be adjusted for plumb if the anchor bolt was improperly placed.

When a wood column supports a steel girder, fitting the end of the column with a metal cap is desirable. If wood supports wood, a metal cap should be provided to give an even bearing surface. The metal will also prevent end grain of the post from crushing the horizontal grain of the wood girder.

Be sure the tops of posts and columns and also the seats in foundation walls are flat so the girder or beam is well supported with its sides plumb.

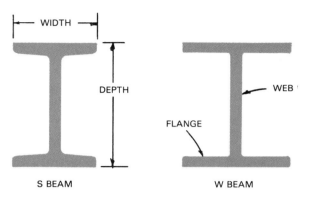

S BEAM   W BEAM

Fig. 7-9. Steel beams are commonly used in residential construction. "S" means standard; "W" means wide-flange.

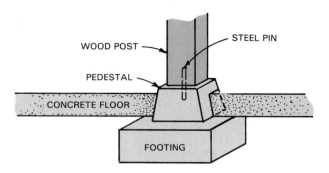

Fig. 7-11. Footings for columns must extend above the floor level for moisture protection.

A built-up wood post may be made by spiking together three 2 by 6s. The pieces should be free from defects and securely nailed together. Otherwise excessive loading may cause the members to buckle away from each other and fail.

Steel posts are most popular for girder and beam support. The post should be capped with a steel plate to provide a good bearing area. A steel post designed especially for this purpose is shown in Fig. 7-13. This has a threaded area inside the top end. A heavy stem threads into the top so the length becomes adjustable. As wooden structural members shrink, the post can be lengthened to provide needed adjustment.

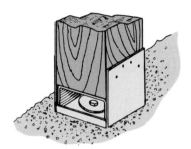

Fig. 7-12. Steel post anchor permits lateral adjustment. (Timber Engineering Co.)

FRAMING OVER GIRDERS AND BEAMS

A common method of framing joists over girders and beams is shown in Fig. 7-14. The steel beam

is placed level with the top of the foundation wall. The 2 in. wood pad then carries the joists level with the sill. When a wooden girder is used, it is usually set so the top is level with the sill.

If ceiling height under joists needs to be moved down, the joists can be notched and carried on a

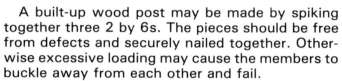

DESIGNATION WT./FT.	NOMINAL SIZE DP. x WD.	SPAN IN FEET									
		8'	10'	12'	14'	16'	18'	20'	22'	24'	26'
W8x10	8x4	15.6	12.5	10.4	8.9	7.8	6.9	—	—	—	—
W8x13	8x4	19.9	15.9	13.3	11.4	9.9	8.8	—	—	—	—
W8x15	8x4	23.6	18.9	15.8	13.5	11.8	10.5	—	—	—	—
W8x18	8x5 1/4	30.4	24.3	20.3	17.4	15.2	13.5	—	—	—	—
W8x21	8x5 1/4	36.4	29.1	24.3	20.8	18.2	16.2	—	—	—	—
W8x24	8x6 1/2	41.8	33.4	27.8	23.9	20.9	18.6	—	—	—	—
W8x28	8x6 1/2	48.6	38.9	32.4	27.8	24.3	21.6	—	—	—	—
W10x22	10x5 3/4	—	—	30.9	26.5	23.2	20.6	18.6	16.9	—	—
W10x26	10x5 3/4	—	—	37.2	31.9	27.9	24.8	22.3	20.3	—	—
W10x30	10x5 3/4	—	—	43.2	37.0	32.4	28.8	25.9	23.6	—	—
W12x26	12x6 1/2	—	—	—	—	33.4	29.7	26.7	24.3	22.3	20.5
W12x30	12x6 1/2	—	—	—	—	38.6	34.3	30.9	28.1	25.8	23.8
W12x35	12x6 1/2	—	—	—	—	45.6	40.6	36.5	33.2	30.4	28.1

Fig. 7-10. Allowable uniform loads for W steel beams. Loads are given in kips (1 kip = 1000 lb.). (Grosse Steel Co.)

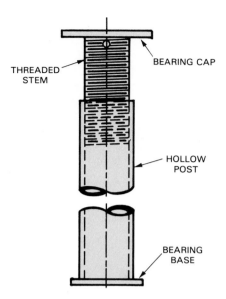

Fig. 7-13. Steel posts with threaded top section are easy to install and adjust.

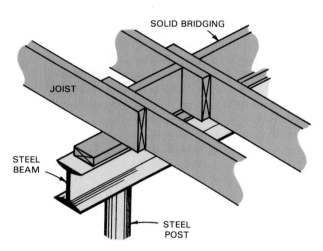

Fig. 7-14. Joists supported on top of a steel beam. Top of beam is set flush with top of foundation wall.

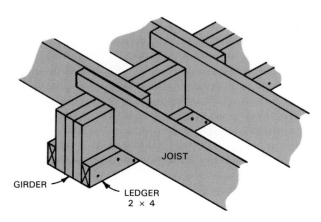

Fig. 7-15. This arrangement is used when girder is raised for extra headroom or when ceiling is lowered. Joists should rest on the ledger strip, not on top of the girder.

ledger, Fig. 7-15. When it is necessary for the underside of the girder to be flush with the joists to provide an unbroken ceiling surface, the joists should be supported with hangers or stirrups. See Figs. 7-16 and 7-17.

Framing joists to steel beams at various levels can be accomplished with special hangers in somewhat the same manner as suggested for wood girders. You must make allowance for the

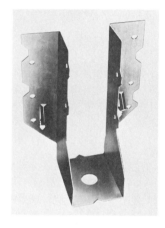

Fig. 7-16. Joist and beam hanger is used when bottom of girder must be flush with the bottoms of the joists. (The Panel Clip Co.)

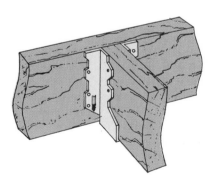

Fig. 7-17. Joist attached with a hanger.

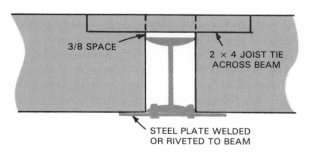

Fig. 7-18. Edge view of joists supported on an S-beam. Allow 3/8 in. space above beam for shrinkage.

fact that joists will likely shrink while the steel beam will remain the same size. For average work with a 2 x 10 in. joist, an allowance of 3/8 in. above the top flange of the steel girder or beam is usually sufficient.

A method of attaching joists is shown in Fig. 7-18. Notching the joists so they rest on the lower flange of an S-beam is not recommended because the flange surface does not provide sufficient bearing surface. Wide-flanged beams, however, do provide sufficient support surface for this method of construction. Fig. 7-19 shows butt methods of framing over girders.

Fig. 7-20. Sill sealers are sold in 50 ft. rolls. Nominally 1 in. thick, the seal compresses to as little as 1/32 in. (Owens-Corning Fiberglas Corp.)

SILL CONSTRUCTION

After girders and beams are set in place, the next step is to attach the sill to the foundation wall. This is the part of the side walls or floor frame that rests horizontally on the foundation. It is also called the sill plate or the mudsill. The latter term originates from the procedure of correcting irregularities in the masonry work by embedding the sill in a layer of fresh mortar or grout.

Sills usually consist of 2 × 6 in. lumber; however the width may vary depending on the type of construction. The sills are attached to the foundation wall with anchor bolts or straps. The size and spacing of anchors is specified in local building codes. Fig. 7-20 shows a resilient waterproof material called sill sealer which may be used under the sill before it is bolted in place. Its purpose is to seal the joint against drafts.

TERMITE SHIELDS

If termites are a problem in your locality, special shields should be provided. Termites live underground and come to the surface to feed on wood. They may enter through cracks in masonry or build earthen tubes on the sides of masonry walls to reach the wood.

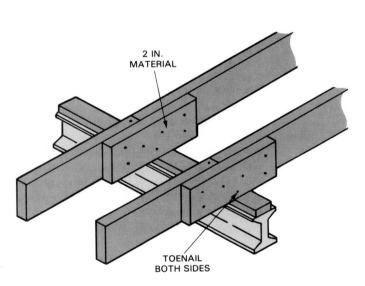

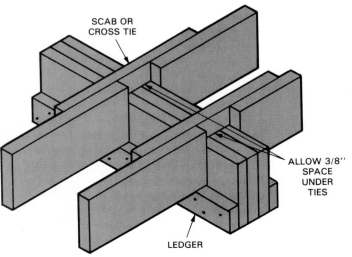

Fig. 7-19. Some carpenters like to butt joists over the girder as shown above.

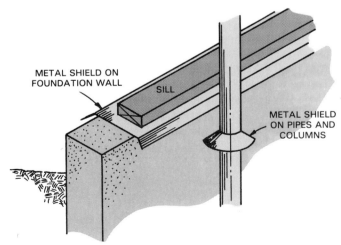

Fig. 7-21. Termite shields. Use galvanized sheet iron or other suitable metal.

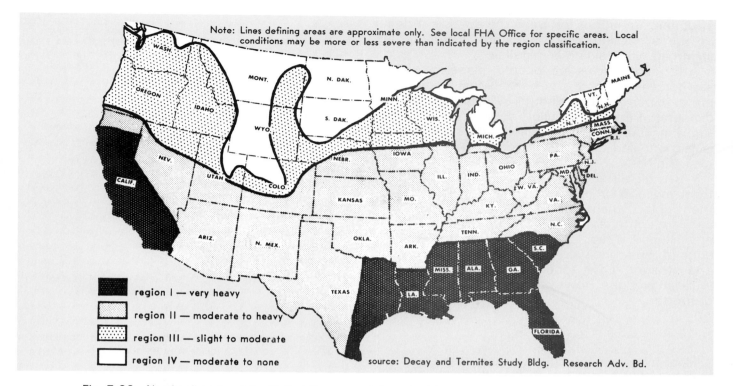

Fig. 7-22. Nearly all parts of the United States are infested with termites which can damage buildings. (Forest Products Lab.)

The wood sill should be at least 8 in. above the ground. A protective metal shield, not less than 26 gauge, should extend out over the foundation wall as shown in Fig. 7-21.

In areas where termite damage is great, additional measures should be taken. Sometimes it is necessary to use lumber for lower framing members that has been treated with chemicals, or to poison the soil around the foundation and under the structure.

Fig. 7-22 is a map of the United States locating various levels of termite infestation. Canada and Alaska are considered to be in region IV. Hawaii and Puerto Rico are in region I.

INSTALLING SILLS

Fig. 7-23 shows two types of sill anchor. With the strap type, position the sill and attach the straps with nails. Some types must be bent over the top of the sill; others are nailed on the sides.

When anchor bolts are used, remove the washers and nuts. Lay the sill along the foundation wall. Remember, the edge of the sill will be set back from the outside of the foundation a distance equal to the thickness of the sheathing.

Draw lines across the sill on each side of the bolts as shown. Measure the distance from the center of the bolt to the outside of the foundation and subtract the thickness of the sheathing. Use

this distance to locate the bolt holes. You will probably need to make separate measurements for each anchor bolt.

Since foundation walls are seldom perfectly straight, many carpenters prefer to snap a chalk line along the top where the outside edge of the sill should be located. This will insure an accurate floor frame which is basic to all additional con-

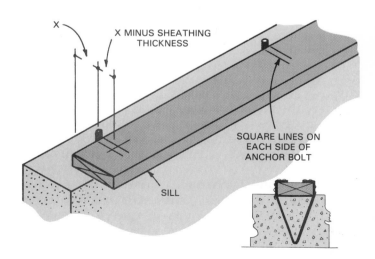

Fig. 7-23. Laying out anchor bolt holes. Anchor strap, right, needs no layout. Sill is positioned and strap is nailed to sill. (TECO)

struction. Variations between the outside surface of the sheathing and foundation wall can be shimmed when siding is installed.

After all the holes are located, place the sill on sawhorses and bore the holes. Most carpenters prefer to bore the hole about 1/4 in. larger than the diameter of the bolts to allow some adjustment for slight inaccuracies in the layout. As each section is laid out and holes bored, position the section over the bolts.

When all sill sections are fitted, remove them from the anchor bolts. Install the sill sealer and then replace them. Install washers and nuts. As nuts are tightened, see to it that the sills are properly aligned. Also, check the distance from the edge of the foundation wall. The sill must be level and straight. Low spots can be shimmed with wooden wedges. However, it is better to use grout or mortar.

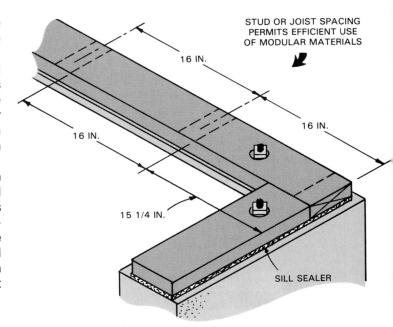

Fig. 7-24. After sill is attached to foundation, locations for studs or joists may be marked.

JOISTS

Floor joists are framing members that carry the weight of the floor between the sills and girders. In residential construction, they are generally nominal 2 in. lumber placed on edge. In heavier construction, steel bar joists and reinforced concrete joists are used.

The most common spacing of wooden joists is 16 in. O.C. (on center). However, 12, 20, and 24 in. centers are also used under certain conditions. The table in the Technical Information section lists safe spans for joists under average loads. For floors this is usually figured on a basis of 50 lb. per sq. ft. (10 lb. dead load and 40 lb. live load). Joists must not only be strong enough to carry the load that rests on them; they must also be stiff enough to prevent undue bending or vibration. Building codes usually specify that the deflection (bending downward at the center) must not exceed 1/360th of the span with a normal live load. This would equal 1/2 in. for a 15'-0" span.

LAYING OUT JOISTS

Study the plans carefully. Note the direction the joists are to run. Also, become familiar with the location of posts, columns, and supporting partitions. The plans may also show the center lines of girders.

The position of the floor joists can be laid out directly on the sill, Fig. 7-24. On platform construction, the joist spacing is usually laid out on the joist header rather than the sill. The position of an intersecting framing member may be laid out by marking a single line and then placing an X to indicate the position of the part, Fig. 7-25. Instead of measuring each individual space around the

perimeter of the building, it is more accurate and efficient to make a master layout on a strip of wood (called a rod). Use it to transfer the layout to headers or sill. The same rod is then used to make the joist layout on girders and the opposite wall. (When the joists are lapped at the girder, the "X" [location of the joists] is marked on the other side of the layout line for the opposite wall.) In this case the spacing between the stringer and first joist will be different than the regular spacing, Fig. 7-26.

Joists are doubled where extra loads must be supported. When a partition runs parallel to the joists, a double joist is placed underneath. Partitions which are to carry plumbing or heating pipes

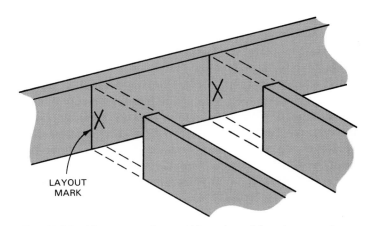

Fig. 7-25. How to mark actual location of framing members. Lines indicate location of an edge. The "X" indicates on which side of the line the member is positioned. Crowns are always turned upward.

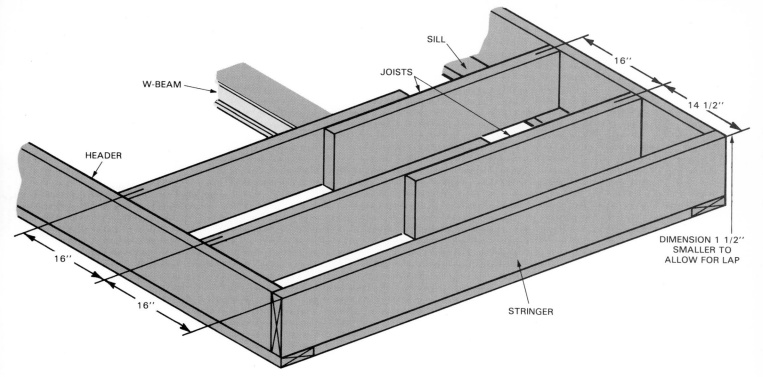

Fig. 7-26. When joists are lapped over the girder, spacing between first joist and the stringer is different.

are usually spaced far enough apart to permit easy access, Fig. 7-27.

Joists must also be doubled around openings in the floor frame for stairways, chimneys, and fireplaces. These joists are called trimmers. They support the headers which carry the tail (short) joists. See Fig. 7-28. The carpenter must become thoroughly familiar with the plans at each floor level so adequate support can be provided.

Select straight lumber for the header joist and lay out the standard spacing along its entire length, Fig. 7-29. Add the position for any doubled joists and trimmer joists that will be required along openings. Where regular joists will become tail joists, change the X mark to a T as shown.

INSTALLING JOISTS

After the header joists are laid out, toenail them to the sill. Position all full length joists with the crown (slight warpage called crook) turned up. Hold the end tightly against the header and along the layout line so the sides will be plumb. Attach the joists to the header using a nailing pattern con-

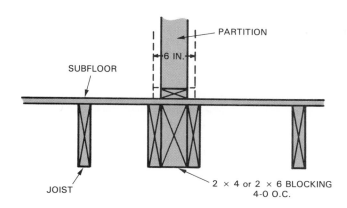

Fig. 7-27. Joists under partitions are doubled and spaced to allow access for heating or plumbing runs. If the wall must hold a plumbing stack (vent to roof), the wall will be framed with 2 x 6s.

sisting of three 16d nails, Fig. 7-30.

Now fasten the joists along the opposite wall. If the joists butt at the girder (join end to end without overlapping), they should be joined with a scarf or metal fastener. If they lap, they can be nailed together using 10d nails. Also use 10d nails to toenail the joists to the girder.

To increase the accuracy of the floor frame, some carpenters first nail the joists to the headers. The headers are then carefully aligned with the sill or a chalk line on the foundation. Then the assembly is toenailed to the sill, Fig. 7-31.

Nail doubled joists together using 12 or 16d nails spaced about 1 ft. along the top and bottom edge. First, drive several nails straight through to

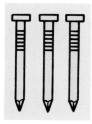

The Uniform Building Code specifies that, in standard framing, nails should not be spaced closer than one-half their length; nor closer to the edge of a framing member than one-fourth their length.

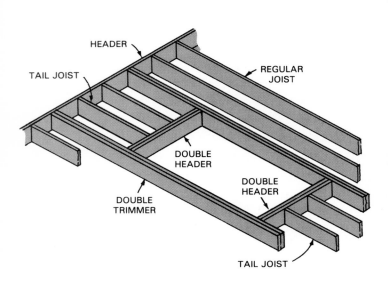

Fig. 7-28. Framing members are doubled around floor openings.

Fig. 7-30. Pneumatic nailer is used to attach header to joists. (Senco Products, Inc.)

pull the two surfaces tightly together and clinch the protruding ends. Finish the nailing pattern by driving the nails at a slight angle. This will prevent them from going all the way through while increasing their holding power.

FRAMING OPENINGS

Place boards or sheets of plywood across the joists to provide a temporary working deck to install header and tail joists. First set the trimmer joists in place. Sometimes a regular joist will be located where it can serve as the first trimmer. Fig. 7-32 is a plan view of how the finished assembly will appear.

The length of the headers can be figured from the layout on the main header joist. Cut headers and tail joists to length. Make the cuts square and true. Considerable strength will be lost in the finished assembly if the members do not fit tightly together. Lay out the position of the tail joists on the headers by transferring the marks made on the main header in the initial layout.

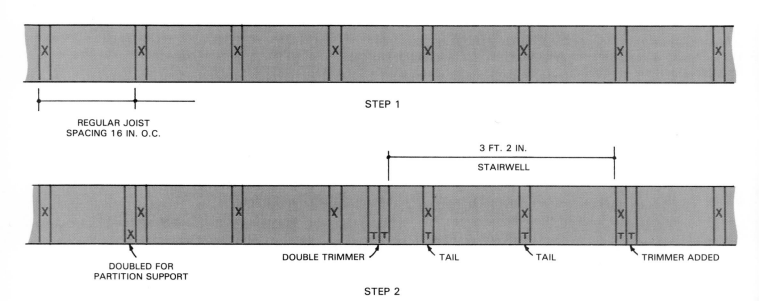

Fig. 7-29. Header laid out with joist positions. In step 1, rod was used to mark regular spacing. In step 2 double and trimmer joist positions have been added.

Fig. 7-31. Some carpenters attach joists to header then align the assembly on the sill and toenail it to the sill.

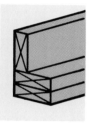

Be accurate in laying out and in cutting floor framing members. The strength of the assembly depends on all the parts fitting tightly together.

When the assembly of tail joists and first headers is small, they are sometimes nailed together and then set in place. Usually, however, the headers are installed and then the tail joists are attached. One of the tail joists can be temporarily nailed to each trimmer to accurately locate the header and hold it while it is being nailed.

Fig. 7-33 illustrates the procedure for fastening tail joists, headers, and trimmers. After the first header and tail joists are in position between the first trimmers, nail the second or double header in place. Be sure to nail through the first trimmer into the second header using three 16d nails at each end. Finally, the second trimmer is nailed to the first trimmer. A good nailing pattern for the entire assembly is shown in Fig. 7-34.

This nailing pattern will support a concentrated load of 300 lb. at any point on the floor. It will also hold a uniformly distributed load of 50 lb. per sq. ft. with any spacing and span of tail beams ordinarily used in residential construction, provided the long dimension of the floor opening is parallel to the joist. If the long way of the opening is at right angles to joists, excessive loading may be carried to the junction of headers with trimmers. Anticipated loads should be checked and more nails or additional supports should be provided at these junctions when needed.

Today, metal framing anchors are often used to assemble headers, trimmers, and tail joists, Fig. 7-35. They are manufactured from 18 gauge zinc coated sheet steel in a variety of sizes and shapes. Special nails for attaching the anchors are also available. The National Forest Products Association recommends the use of framing anchors or

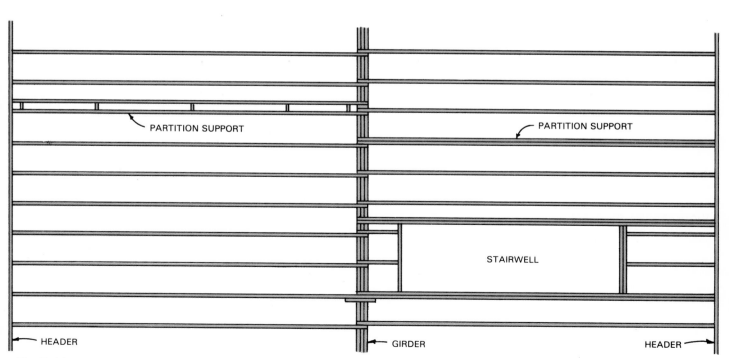

PARTITION SUPPORT

PARTITION SUPPORT

STAIRWELL

HEADER

GIRDER

HEADER

Fig. 7-32. Plan view of floor framed for stair opening and partition support. Single header is used next to girder since the distance between header and girder is so short.

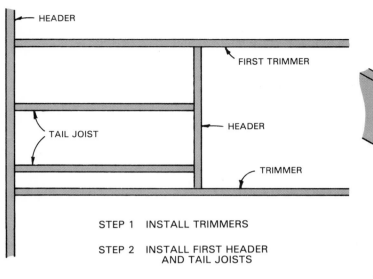

STEP 1   INSTALL TRIMMERS

STEP 2   INSTALL FIRST HEADER
AND TAIL JOISTS

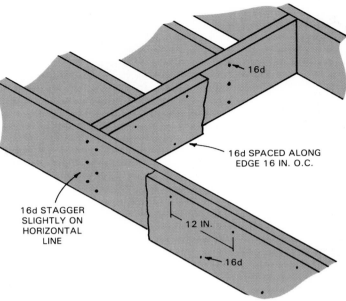

Fig. 7-34. Nailing pattern for attaching floor opening members.

ledger strips to support tail joists that are over 12 ft. long.

BRIDGING

Some recent studies have shown that bridging may be eliminated:
1. Where joists are properly secured at the ends.
2. Where subflooring is adequate and carefully nailed.
However, many local building codes list requirements in this area and general standards suggest

STEP 3   INSTALL SECOND HEADER

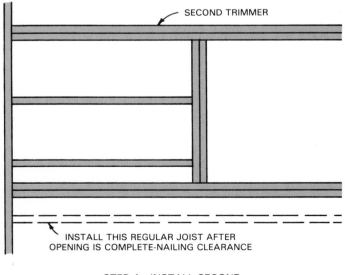

STEP 4   INSTALL SECOND
TRIMMER JOIST

Fig. 7-33. These steps can be followed in assembling frames for floor openings.

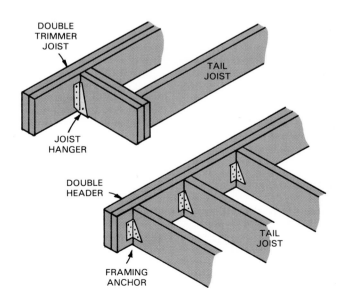

Fig. 7-35. Joist hangers and framing anchors are good for assembling floor framing members.

Floor Framing

that bridging be installed at intervals of no more than 8 ft.

Regular bridging, sometimes called herringbone or cross bridging, is composed of pieces of lumber set diagonally between the joists to form an X. Its purpose is to:
1. Hold the joists in a vertical position.
2. Transfer the load from one joist to the next.

Fig. 7-36 shows how the carpenters framing square can be used to lay out a pattern for bridging. Pieces can be cut rapidly on the radial arm saw or a jig can be set up to use the portable electric saw or a hand saw.

To install bridging, first determine the position of the run and then snap a chalk line across the tops of the joists as shown in Fig. 7-37. Start two 8d nails into the ends of all the bridging and then attach a piece to each side of every joist. Alternating the position of the two pieces, first on one side of the chalk line and then on the other side at the next joist, will make the installation easy and correct, Fig. 7-38. The lower ends of the bridging are not nailed until the subflooring is complete or until the under surfaces of the floor are to be enclosed.

Solid bridging, as the name implies, consists of solid blocks set between the joists. This is often used in odd-size spaces in a run of cross bridging. Solid bridging, also called blocking, improves an installation like the one shown in Fig. 7-39. Here the chief purpose is to keep the joist vertical. However, it also adds rigidity to the floor.

Several types of prefabricated steel bridging are available that can be installed quickly. The type shown in Fig. 7-40 is manufactured from sheet steel. A V-shaped cross section makes it rigid. No nails are required and it is driven into place with a

regular hammer. This design meets FHA Minimum Property Standards, and is approved by the Uniform Building Code.

After the bridging is installed, the floor frame should be checked carefully to see that nailing patterns have been completed in all members. After this is done, the frame is ready to receive the subflooring.

SPECIAL FRAMING PROBLEMS

In modern residential construction, the design may include a section of floor that overhangs (sticks out beyond) a lower floor or basement level. When the floor joists run perpendicular to the walls, the framing is comparatively easy. It is only necessary to use longer joists. If, however, the floor joists run parallel to the wall, the construction must be framed with cantilevered joists as illustrated in Fig. 7-41.

The exact spacing and length of the members will depend on the weight of the outside wall. Usually, cantilevered joists should extend inward at least twice as far as they stick out over the supporting wall. Note that since the load at the inside double header is upward, the ledger strip must be positioned at the top.

Entrance halls, bathrooms, and other areas are often finished with tile or stone that is installed on a concrete base. To provide room for this base, the floor frame must be lowered. When the area is not large, this can be done by doubling joists of a smaller dimension, Fig. 7-42. Additional support can be secured by reducing the spacing. When area is large, steel or wood girders and posts should be added.

Bathrooms must support unusually heavy loads, heavy fixtures, and often the additional weight of a tile floor. The fixed dead load imposed by a tile floor will average around 30 lb. per sq. ft. The load from bathroom fixtures adds from 10 to 20 lb. per

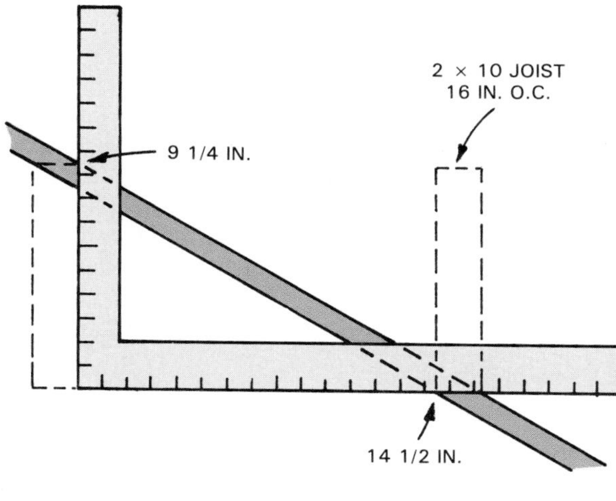

Fig. 7-36. Carpenter's square can be used to lay out bridging. Line for lower cut can be secured by shifting the tongue of the square to the 14 1/2 in. mark on the stock.

Fig. 7-37. Mark position of bridging by snapping a chalk line across the tops of the joists.

2 x 10 JOIST 16 IN. O.C.

9 1/4 IN.

14 1/2 IN.

163

Fig. 7-39. Solid bridging is usually installed over a girder.

Fig. 7-38. Installing wood bridging. First bridging is attached at the top. Pieces must not project beyond the top of the joists. Top. Carpenter works on top to attach bridging. Note use of solid bridging over girders and bearing walls. (Weyerhaeuser Co.) Bottom. Close-up view of solid and cross bridging. Lower end of cross bridge is nailed later.

In the usual rectangular joist, this point is assumed to be midway between the top and bottom. Variations in the quality of lumber and other conditions may shift the point slightly; still, this assumption is accurate enough.

If there is neither compression nor tension at the center, it is obvious that a hole—provided it is not

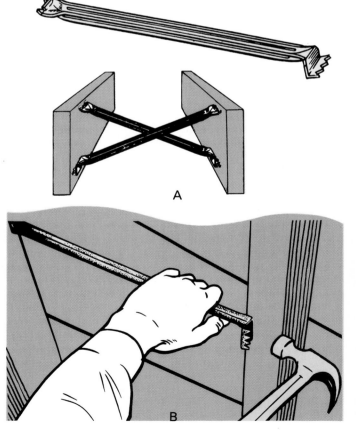

Fig. 7-40. Top. Steel bridging. Bottom. Installation method. (Timber Engineering Co.)

sq. ft., for a total of 40 to 50 lb. dead load. In addition, it is frequently necessary to cut joists to bring in water service and waste pipes. Special precautions must, therefore, be taken in framing bathroom floors to provide adequate support.

CUTTING FLOOR JOISTS

Before cutting joists to install plumbing it is useful to know how stress affects flooring joists. This knowledge will help you determine where to make holes and cut notches.

When the top of a joist is in compression and the bottom in tension, there is a point at which the stresses change from one to the other. At this point, there is neither tension nor compression.

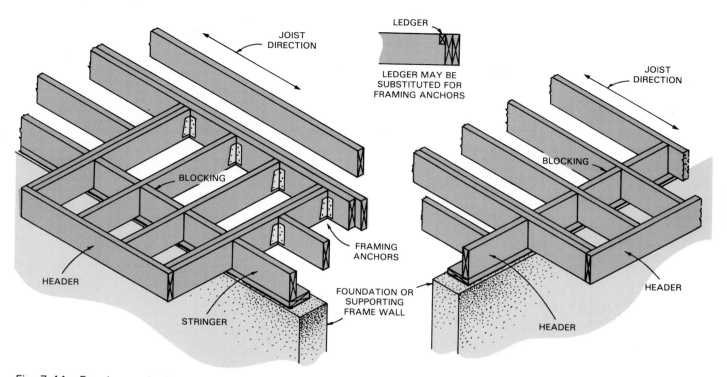

Fig. 7-41. Framing methods on overhangs depends on direction the joists run. Blocking holds the joists vertical, adds rigidity, and closes up the space.

larger than one-fourth the total depth of the joist— would have little effect on the strength.

Weight produces the greatest bend if it is at the center of the span. Therefore a weakness is more likely to reduce the strength of a joist or beam if it is near the center of the span. Considering this, follow these precautions in cutting joists:

1. When possible, holes should be cut at or close to the center. If limited to one-fourth of the total depth, no material reduction in strength will result.
2. Where it is necessary to cut joists, the cuts should be made from the top. For example: if a

2 x 8 joist is cut to a depth of 4 in., its strength will be reduced to that of a 2 x 4. If the cut from the top is 2 in., it will be equivalent to a 2 x 6. When a joist is cut, the loss in strength must be compensated for by providing headers and trimmers or by adding extra joists. Another way of solving the problem for large plumbing pipes is shown in Fig. 7-43.

3. If the cut is made elsewhere than at the center of span, the weakening effect will not be as

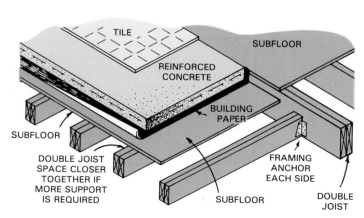

Fig. 7-42. Smaller joists are used when a concrete base is needed for tile or stone surfaces.

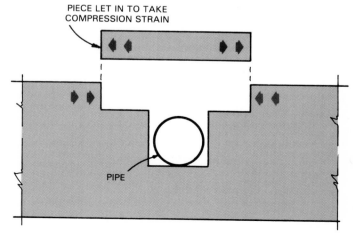

Fig. 7-43. Fit a block of wood into a deep notch to restore strength of joist.

great. Even so, it is advisable to provide fully as much compensating strength as is lost by the cut.

LOW-PROFILE FLOOR FRAMES

Many home buyers prefer a house with a low silhouette. Standard wood floor construction, whether over a basement or a crawl space, requires adequate distance between the framing members and the ground. This places the first floor level well above the finished grade.

Various framing systems have been devised to bring the floor level closer to the outside grade. Basically, these consist of designing the foundation in such a way that the floor frame is surrounded and protected by the wall, Fig. 7-44.

In construction of this type, special precautions must be taken to assure that the joists have adequate bearing surface. Allowance should also be made for shrinkage of the wood members.

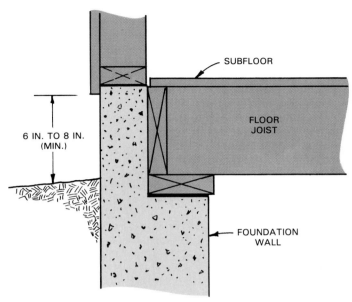

Fig. 7-44. Offset in foundation wall reduces distance between the floor level and the finished grade.

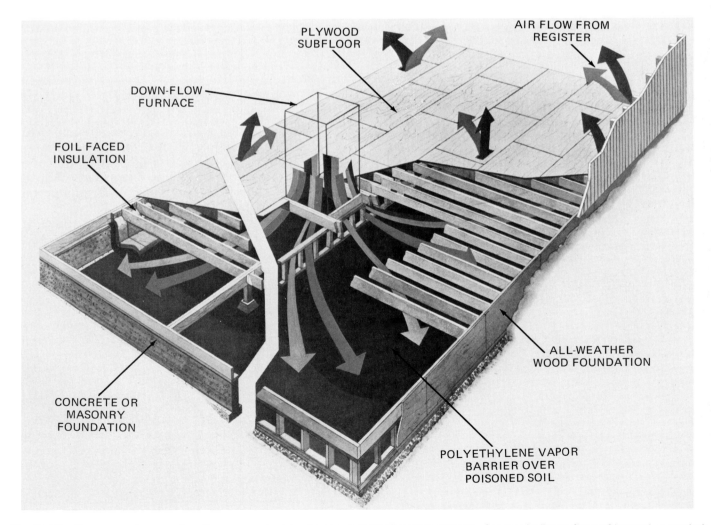

Fig. 7-45. Energy saving construction uses crawl space as a heating/cooling plenum. Arrows indicate flow of heated or cooled air. Either concrete or wood foundation can be used. Design data can be secured from American Plywood Association.

Fig. 7-45 shows a floor framing system over a crawl space. The underfloor space serves as a plenum (enclosed space for air under slightly greater pressure than that surrounding the enclosure) for heating and cooling. The system is easy to construct and permits considerable design flexibility. Money is saved by eliminating ductwork and by using smaller joists and beams. With proper insulation, the system can save a great deal of energy.

Foundations for low-profile floor framing may be constructed of poured concrete, masonry units such as concrete block, or wood. Whichever type of building material is used, proper barriers against moisture and heat or cold must be installed.

Foil-faced insulation should be laid around the entire perimeter with the foil facing upward or inward. It is suggested that the insulation have an insulating factor of about R 11. However, local codes and practice should be followed in all cases.

The plastic moisture barrier is put down across the entire floor area of the crawl space. It deflects moisture downward to a layer of gravel or crushed stone. Outside walls should be well insulated as shown in Fig. 7-46.

FLOOR TRUSSES

Floor trusses are widely used in modern construction. They provide long clear spans with a minimum of depth. The open webs result in a lightweight assembly that is easy to handle. At the same time, they reduce transmission of sound through floor/ceiling assemblies. Open webs make it easy to install plumbing, heating, and electrical systems. See Fig. 7-47.

Modern trusses and truss systems are designed with the aid of computers. These designs insure that loading requirements are met through the use of a minimum amount of material. Engineered jig hardware used in the assembly "builds-in" proper camber in each unit.

Some trusses are fabricated with lumber chords and patented galvanized steel webs. The webs have metal teeth which are pressed into the sides of the chords. They also have a reinforcing rib that withstands both tension and compression forces. See Fig. 7-48.

SUBFLOORS

The laying of the subfloor is the final step in completing the floor frame. Either plywood, shiplap, tongue-and-groove flooring, or common boards can be used. The subfloor serves three purposes:

1. It adds rigidity to the structure.
2. It provides a base for finish flooring material.
3. It provides a surface upon which the carpenter can lay out and construct additional framing.

Subflooring of the board or shiplap type is nailed at each joist with 8d nails. Use two nails in each board when the width is under 6 in. and three nails for widths of 6 in. and over.

If subfloor is tongued and grooved on ends and edges, end joints need not be made over joists. Subfloor is preferably laid without cracks between boards. If accumulation of water on the subfloor during construction is likely, it may be desirable to leave cracks to permit drainage.

PLYWOOD

In most modern construction, plywood is used for subflooring. It provides a smooth, even base and acts as a horizontal diaphragm that adds strength to the building. Plywood can be installed rapidly and usually insures a squeak-free floor.

Although 1/2 in. plywood over joists spaced 16 in. O.C. meets the minimum FHA requirements, many builders use 5/8 in. plywood. The long dimension of the sheet should run perpendicular to the joists. Joints should be broken in successive courses, Fig. 7-49. For 5/8 or 3/4 in. plywood, use 8d nails spaced 6 in. along edges and 10 in. along intermediate members. See Fig. 7-50.

Combined subfloor-underlayment systems utilize a special plywood panel with tongue and groove edges. This single layer provides adequate structural qualities and a satisfactory base for

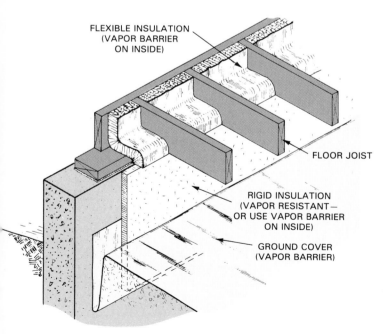

FLEXIBLE INSULATION (VAPOR BARRIER ON INSIDE)

FLOOR JOIST

RIGID INSULATION (VAPOR RESISTANT— OR USE VAPOR BARRIER ON INSIDE)

GROUND COVER (VAPOR BARRIER)

Fig. 7-46. Approved method of insulating heated crawl space. When weather permits, the installation should be made before subfloor is laid.

Fig. 7-47. Truss joists being installed on a residence. They are made of 2 x 4s fastened with special connector plates. Note notches on ends. They will receive a continuous band of 2 x 4s.   (TrusWal Systems Corp.)

direct application of carpet, tile, and other floor finishes.

Subfloor-underlayment panels are available for joists or beam spacing of 16, 20, 24, or 48 in. Maximum support spacing is stamped on each panel. A 3/4 in. thickness is used for 24 in. O.C. spacing. Be sure to follow instructions supplied by the manufacturer.

OTHER SHEET MATERIALS

Other sheet materials such as composite board, waferboard (also called waferwood), oriented strand board, and structural particleboard are also approved for use as subflooring. These products have been rated by the American Plywood Assoc. and meet all standards for subflooring. The speci-

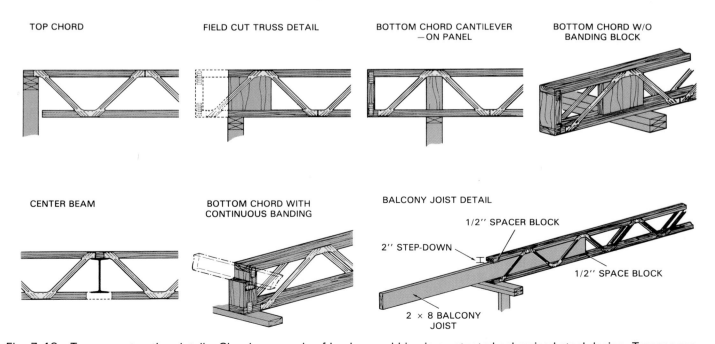

TOP CHORD FIELD CUT TRUSS DETAIL BOTTOM CHORD CANTILEVER —ON PANEL BOTTOM CHORD W/O BANDING BLOCK

CENTER BEAM BOTTOM CHORD WITH CONTINUOUS BANDING BALCONY JOIST DETAIL

1/2'' SPACER BLOCK
2'' STEP-DOWN
1/2'' SPACE BLOCK
2 x 8 BALCONY JOIST

Fig. 7-48. Truss construction details. Chords are made of lumber; webbing is a patented galvanized steel design. Trusses provide wide nailing surface because chord is laid flat.   (TrusWal Systems Corp.)

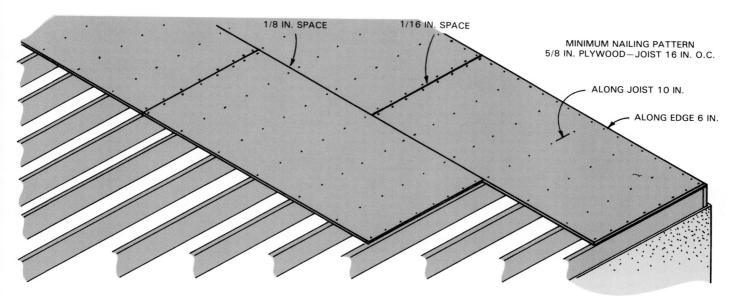

Fig. 7-49. Using plywood as a subfloor. Note nailing pattern and how joints are broken for added strength.

Fig. 7-50. Using a pneumatic-powered nailer to install 5/8 in. plywood subflooring. Floor joists are spaced 16 in. on center. (Paslode Corp.)

fications for application are the same as for plywood, Fig. 7-51. For additional information refer to pages 23 and 24.

GLUED FLOOR SYSTEM

In a glued floor system the subfloor panels are glued and nailed to the joists. Structural tests have shown that stiffness is increased about 25 percent with 2 x 8 joists and 5/8 in. plywood. In addition the system insures squeak-free construction, eliminates nail-popping, and reduces labor costs.

Before each panel is placed, a 1/4 in. bead of glue is applied to the joists as shown in Fig. 7-52. Spread only enough glue to lay one or two panels. Two beads of glue are applied on joists where panel ends butt together. All nailing must be completed before glue sets.

When laying tongue and groove panels, Fig. 7-53, apply glue along the groove, either continuous or spaced. Use a 1/8 in. bead so excessive squeeze-out will be eliminated. A 1/8 in. space must be provided at all end and edge joints. Material specifications and application procedures are provided in a booklet published by the American Plywood Association.

Fig. 7-51. Waferwood (or waferboard) panels are engineered and approved for use as subflooring. Same application methods are used as with plywood. For non-textile resilient floor applications a 1/4 in. underlayment is recommended. (Louisiana-Pacific Corp.)

169

Fig. 7-52. Applying adhesive for a glued floor system. A 1/4 in. width of glue to the top of the joist is sufficient. Two beads are applied where panels butt.   (American Plywood Assoc.)

STEEL JOISTS

Various kinds of cold-formed steel joists are available for floor frames. They are especially useful in buildings where long spans are required. Their loadbearing capacity is greater than wood joists of equal size.

Steel joists are attached to steel headers with special end clips and screws. They can also be used with regular wood headers and sills. In the latter construction, wood blocks are inserted into the joist ends which are then nailed to the header. Attachment can also be made with a special joist connector. Steel joists are sometimes set in pockets formed in poured concrete foundations. Openings in the floor frame are framed with a

variety of clips and brackets. Plywood subflooring is attached to the joist with a 3/8 in. bead of special adhesive and self-drilling screws or flooring nails. Joists are spaced 24 in. O.C. See Fig. 7-54.

ESTIMATING MATERIALS

If you are required to estimate the number and size of floor joists on a job, first scale the plan and determine the lengths that will be needed. Be sure to allow sufficient length for full bearing on girders and partitions. Average residential structures will require several different lengths.

Multiply the length of the wall that carries the joists by 3/4 for spacing 16'' O.C. (3/5 for 20, 1/2 for 24) and add one more. Also add extra pieces for doubled joists under partitions and trimmer joists and headers at openings.

No. of joists = length of wall × 3/4 + 1 + extras

Some carpenters figure one joist for every foot of wall upon which the joists rest (16'' O.C. spacing). The over-run allows for extra pieces needed for doubles, trimmers, and headers.

Header joists are usually figured separately and added to the above figures. Your listing should include the cross section size and the number of pieces of each length. For example,

To figure floor joists:

40 pcs. — 2 × 10 × 16'-0''
36 pcs. — 2 × 10 × 14'-0''

To figure header joists:

4 pcs. — 2 × 10 × 16'-0''
2 pcs. — 2 × 10 × 12'-0''

Fig. 7-53. Installing a tongue and groove panel in a glued floor system.   (American Plywood Assoc.)

Fig. 7-54. Carpenters are installing plywood subfloor to steel joists. At right, bead of adhesive is being laid down. Worker at left is using a grounded electric drill to drive self-drilling, self-tapping screws.   (U.S. Gypsum Co.)

Procedures for estimating the subflooring will vary depending on the type of material used. Usually the area is figured by multiplying the overall length and width, and then subtracting major areas that will not be covered. These include breaks in the wall line and openings for stairs, fireplaces, and other items.

This will give the net area and the basic amount of material needed. To this must be added waste and certain other extras. For example: when 8 in. shiplap is used, multiply the basic figure by 1.15 and then add another 15 percent for waste. If the shiplap is laid diagonally to the joist, another 5 percent should be added. Individual boards are not specified. The amount is simply listed in board feet (equal to the square footage needed) along with a description of the material.

When using sheet materials, there is practically no waste and the basic figure is divided by 32 (sq. ft. in a 4 x 8 sheet) and rounded out to the next whole number. This will be the number of pieces of plywood required. Be sure to specify the type of sheet material and its span rating.

A newer and more complete method of specifying plywood as recommended by the American Plywood Association is included in Unit 1.

TEST YOUR KNOWLEDGE — UNIT 7

1. The type of framing used in most one-story construction is _____.
2. When requirements call for the joists to be framed flush with the underside of a wood girder, it is best to use _____.
3. Standard construction usually requires that the sill be spaced back from the foundation wall a distance equal to the _____ _____.
4. The studs of a balloon type frame run continuously from the _____ to the rafter plate.
5. Name the two types of steel beams used in residential construction.
6. In residential construction, the deflection of first floor joists under normal live loads should not exceed _____ of the span.
7. A member of the floor frame that runs from the main header to a header for an opening is called a _____ joist.
8. When framing a floor opening, the double header should be nailed in place before the second _____ is installed.
9. Cantilevered joists should extend inward at least _____ times the distance that they overhang the supporting wall.
10. Large holes bored through joists for pipes or wiring should be made at the_____(top, bottom, center).
11. Shiplap of a nominal width of 8 in. should be applied with _____ (2, 3, 4) 8d nails at each joist.
12. When sheet material is used for the subflooring, the short dimension of the panel should run _____ (parallel, perpendicular) to the joist.

OUTSIDE ASSIGNMENTS

1. Secure a set of architectural plans for a house with a conventional basement. Study the methods of construction specified in the sections and detailed drawings. Then prepare a first-floor framing plan. Start by tracing the foundation walls and supports shown in the basement or foundation plans and then add all joists, headers, and other framing members. Your drawing should be similar to the one in Fig. 7-32.
2. Working from a set of architectural plans for a single-story house, develop a list of materials required to frame the floor. Select the type of subflooring, if not specified, and estimate the amount of material needed.
3. From the local building code in your area get the requirements for floor framing. Prepare a list of the requirements along with sketches that might clarify complicated written descriptions. Make an oral report to your class.

In modern residential construction, walls and ceiling joists form one structural system. The walls support the joists which form the ceiling or the next floor level. Ceiling joists or trusses are supported by the walls. The walls are stiffened and held plumb by the addition of the joists. Above. Carpenter places trusses along outside wall of a two-story house. Note metal bracing. (Truswall Systems Corp.) Left. Pneumatic nailer is used to attach plywood sheathing to outside wall.
(Plaslode Co.)

Unit 8

WALL AND CEILING FRAMING

Wall framing includes assembling of vertical and horizontal members that form outside and inside walls of a structure. This frame supports upper floors, ceilings, and roof. It also serves as a nailing base for inside and outside wallcovering materials. Inside walls are called PARTITIONS.

The term "system" commonly means methods and materials of construction. It is used in connection with floors, ceilings, and roofs as well as walls. Included are the design of the framework as well as the surface-covering materials and the methods for applying them. For example, a floor system includes:

1. The details of the sill construction.
2. Size and spacing of joists.
3. The kind of subflooring.
4. Application requirements.

PARTS OF THE WALL FRAME

The wall-framing members used in conventional construction include sole plates, top plates, studs, headers (also called lintels), and sheathing. Studs and plates are made from 2 x 4 lumber while headers usually require heavier material. Bracing made of 1 x 4 stock or steel strips must be built into the wall when the sheathing does not provide enough stiffness.

Fig. 8-1. In modern construction, studs are often placed on 24 in. centers. Plywood sheathing provides enough stiffening so that corner bracing is not needed. (Western Wood Products Assoc.)

In one-story structures, studs are sometimes placed 24 in. O.C. (on center), Fig. 8-1. However, 16 in. spacing is more common. Fig. 8-2 shows a typical wall frame with openings for a window and door. Note the extra studs used:

1. At the corner.
2. At the sides of the openings.
3. Where an interior partition joins the outside wall.

Conventional stud spacing of 16 in. or 24 in. has evolved from years of established practice. It is based more on accommodating the wall-covering materials than on the actual calculation of imposed loads.

Fig. 8-3 illustrates, in more detail, various parts and how they fit together. Full length studs become CRIPPLE studs when they end because of an opening. TRIMMER studs stiffen the sides of the opening and bear the direct weight of the header. The regular stud spacing is even and continuous along the wall regardless of openings. Modular sheet materials can, therefore, be applied with little cutting.

Wall-framing lumber must be strong and straight with good nail-holding power. Warped lumber will not do the job, especially if the interior finish is drywall. Stud and No. 3 grades are approved and used throughout the country. Species such as Douglas fir, larch, hemlock, yellow pine, and spruce are satisfactory. See Unit 1 for additional information.

CORNERS

Any of several methods can be used to form the outside corners of the wall frame. In platform con-

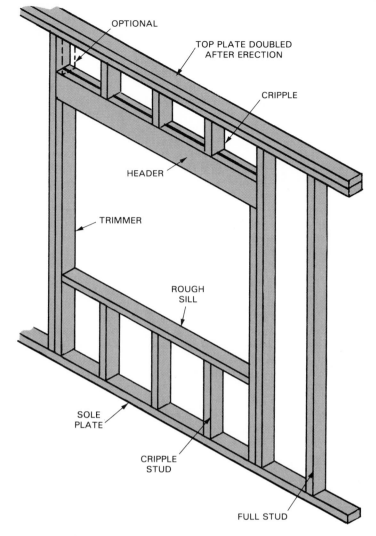

Fig. 8-3. Names of parts for a typical wall frame. In modern construction, headers usually go all the way to the top plate. Door frames may be framed in the same way except that lower cripple studs and sole plate are left out.

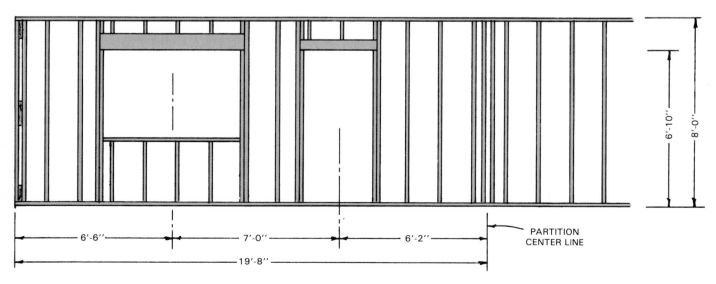

Fig. 8-2. View of a typical wall frame section. Eight foot height from sole plate to wall plate is extended by adding second top plate. It allows extra room for floor and ceiling coverings.

174

struction the wall frame is usually assembled in sections on the rough floor and then raised. Corners are formed when a side wall and end wall are joined.

Usually a second stud is included in the side wall frame. It should be spaced inward from the end stud with three or four blocks. When the end wall is raised into position, the complete corner is formed, Fig. 8-4. An alternate method is to turn the extra stud as shown.

Only straight studs should be selected for corners. Assemble with 10d nails spaced 12 in. apart. Stagger them from one edge to the other as shown. Include extra nails to attach the filler blocks.

Some carpenters prefer to build the corners for platform construction separately. They are set in place, carefully plumbed, and braced, before the wall sections are raised. This makes it easier to plumb and straighten the wall sections but does not permit the application of sheathing while the frame is still on the deck.

In balloon construction, the corners are made up and installed as a separate unit. See Fig. 8-5.

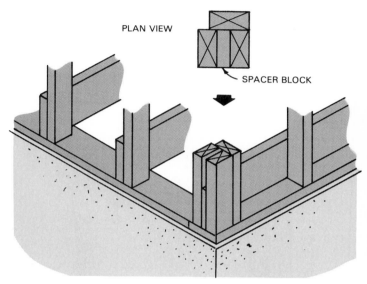

Fig. 8-5. Corner framing in balloon construction is similar to platform construction.

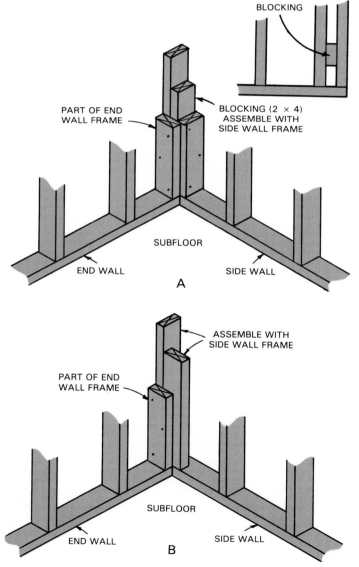

Fig. 8-4. Placement of studs to form corners in platform construction. A—Corner built from three full studs and blocking. B—Corner built with three full studs and no blocking.

PARTITION INTERSECTIONS

Where partitions meet outside walls, it is essential that they be solidly fastened. This requires extra framing.

The framing must also be arranged so inside corners provide a nailing surface for wall-covering materials. Several methods can be used to tie walls together and provide the nailing surfaces needed:

1. Install extra studs in the outside wall. Attach the partition to them.
2. Insert blocking and nailers between the regular studs.
3. Use blocking between the regular studs and attach patented back-up clips to support inside wall coverings at the inside corners. Fig. 8-6 shows all three methods.

ROUGH OPENINGS

Study the house plans to learn the size and location of the rough openings (R.O.). Plan views with dimension lines. Usually the measurement is taken from corners and/or intersecting partitions to the center lines of the openings.

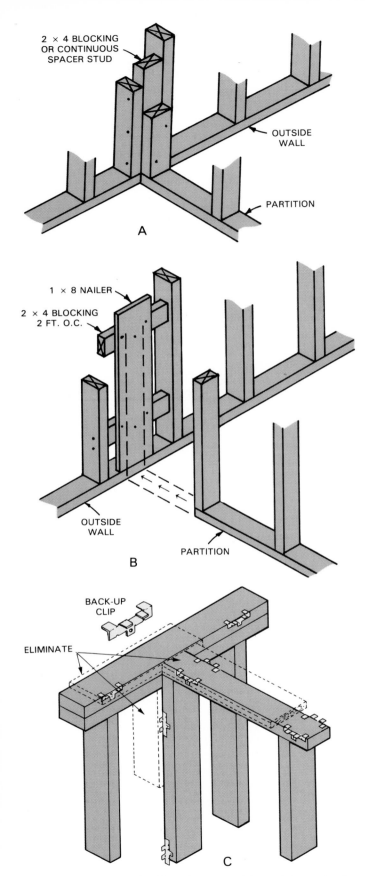

2 × 4 BLOCKING
OR CONTINUOUS
SPACER STUD

OUTSIDE
WALL

PARTITION

A

1 × 8 NAILER

2 × 4 BLOCKING
2 FT. O.C.

OUTSIDE
WALL

PARTITION

B

BACK-UP
CLIP

ELIMINATE

C

Fig. 8-6. Framing details where partitions intersect. A—Using extra studs. Nail studs to blocking with 16d nails in 12 in. O.C. Use 10d nails to attach partition stud. B—Blocking installed between studs. Use 16d nails. Backing board is attached with 8d nails. C—Back-up clips are sometimes used and take place of some framing studs. (TECO)

The height of rough openings can be secured from elevations and sectional views. Their size will be shown on the plan view or listed in a table called a door and window schedule. See Unit 5.

The size of the rough openings is always listed. Width is given first and the height second. Study Unit 11 for additional requirements for window and door openings.

Headers carry the weight of the ceiling and roof across door and window openings. They are formed by nailing two members together and placing them on edge. A 1/2 in. plywood spacer is inserted between the pieces to make the thickness equal to that of the wall, Fig. 8-7.

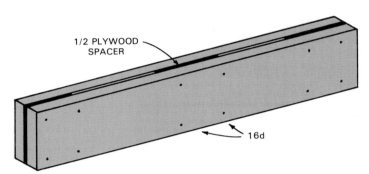

1/2 PLYWOOD
SPACER

16d

Fig. 8-7. Header construction. Plywood spacers are placed 16 to 24 in. apart on centers.

The length of the header will be equal to the rough opening plus two trimmers (3 in.). The size of the lumber used in the header depends on the width of the opening. Local building codes may include requirements for headers. The table in Fig. 8-8 gives the size of headers normally required for various R.O. widths under several load conditions.

Headers are also required across openings in load-bearing partitions. When the load is especially heavy or the span is unusually wide, a truss may be used, Fig. 8-9. The design of special trusses is usually included in the plans. They should be based on careful calculations by a competent architect or engineer.

In modern platform construction, extra studs are included around rough opening as shown in the standard assembly, Fig. 8-10. The studs and trimmers support the header and provide a nailing surface for window and door casing. Some carpenters also double the rough sill to add a nailing base for window stools and aprons.

ALTERNATE HEADER CONSTRUCTION

In large window openings the size of the header will reduce the length of the upper cripple studs to

MATERIAL ON EDGE	SUPPORTING ONE FLOOR, CEILING, ROOF (IN FT. & IN.)	SUPPORTING ONLY CEILING AND ROOF (IN FT. & IN.)
2 × 4	3-0	3-6
2 × 6	5-0	6-0
2 × 8	7-0	8-0
2 × 10	8-0	10-0
2 × 12	9-0	12-0

Fig. 8-8. Recommended header spans. Be sure to check local codes.

a point where they cannot be easily assembled. They should be replaced with flat blocking.

Another solution is to increase the header size to completely fill the space to the plate. Most builders follow this practice and extend it to include all openings, regardless of the span. They have found that the cost of labor required to cut and fit the cripple studs is usually greater than the cost of the larger headers.

A disadvantage of such construction is the extra shrinkage which, without special precaution in the application of interior wall finish, may cause cracks above doors and windows.

In balloon construction, which was once popular, studs extended from the sole plate to the roof plate. It was common practice to extend headers beyond the rough opening to the next regular stud, Fig. 8-11.

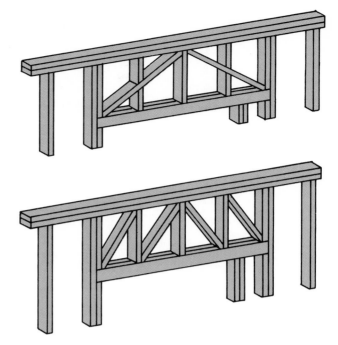

Fig. 8-9. Trussed header was used in older construction but required additional labor to build.

PLATE LAYOUT

Use only straight 2 x 4 stock for plates. Select two of equal length and lay them side by side along the location of the outside wall. Length should be

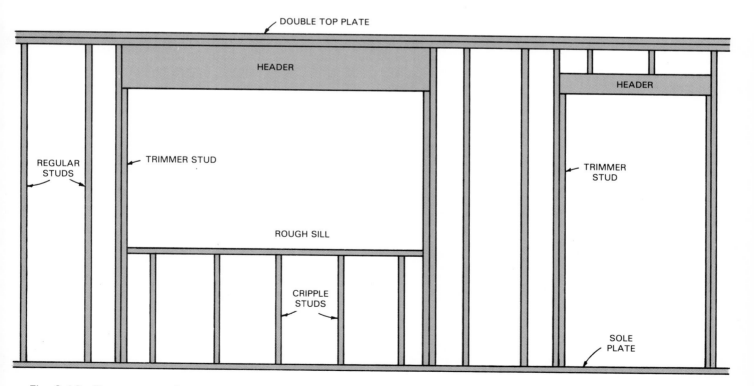

Fig. 8-10. Proper way to frame openings in walls. Trimmer studs carry the weight of the header. Header is wider than it needs to be. However, it is done this way to fill the space. It avoids labor of cutting and installing cripple studs and braces.

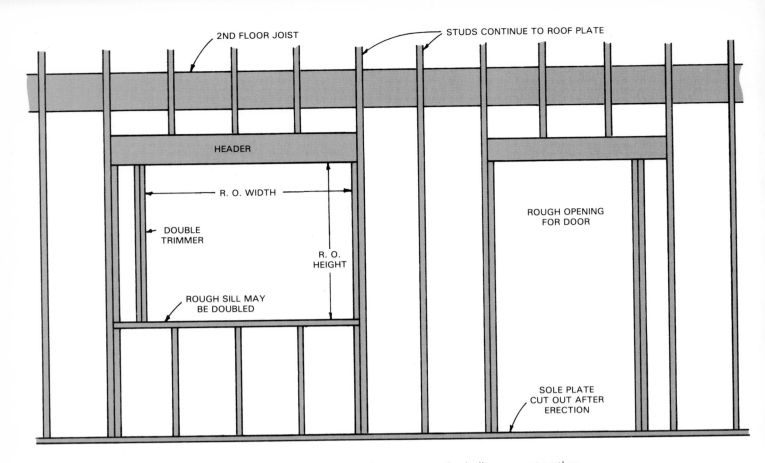

Fig. 8-11. Framing of openings for once popular balloon construction.

determined by what can easily be lifted off the floor and into a vertical position after it is assembled. Remember that the weight may include all the framing for rough openings, bracing, and sheathing. If wall jacks or a forklift are available, sections can be made larger. Where they must be lifted by hand sheathing may be attached after the wall is up.

Lay out the plates along the main side walls first. Align the ends with the floor frame and then mark the regular stud spacing all the way along both plates, Fig. 8-12. Some carpenters tack the pieces to the floor with several nails so they will not move while the layout is being made.

Study the architectural plans and lay out the centerline for each door and window opening.

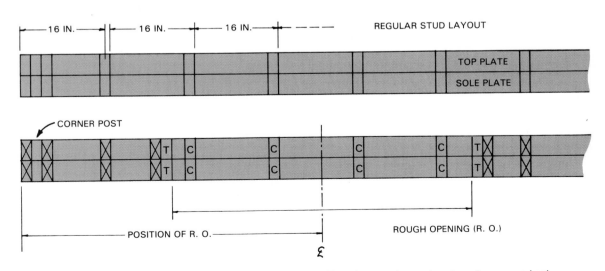

Fig. 8-12. Layout of sole and top plates. Top. Regular stud spacing has been marked. Bottom. Layout is converted for a window opening.

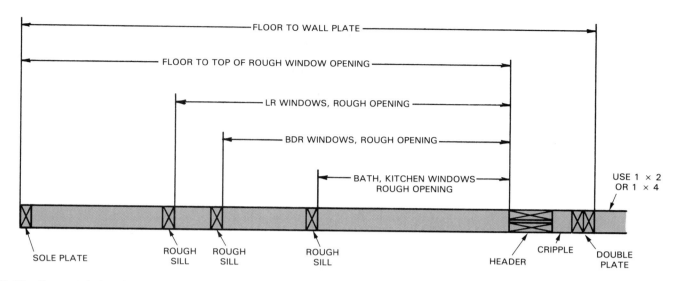

Fig. 8-13. Story pole is a handy guide which marks height of every horizontal member and length of every vertical member of the wall frame. It usually extends one story but may include more than one.

Fig. 8-14. A split level home with some short studding such as pictured will require the use of a story pole for checking heights.

Measure off one-half the width of the opening on each side of the centerline. Mark the plate for trimmer studs outside of these points. On each side of the trimmer stud include marks for a full length stud. Identify the positions with the letter "T" for trimmer studs and "X" for full length studs. Now mark all of the stud spaces located between the trimmers with the letter C. This designates them as cripple studs.

Lay out the centerlines where intersecting partitions will butt. Add full length studs if required by

the method of construction. When blocking between regular studs is used, the centerline will be needed as a guide for positioning the backing strip. Plan the layout of wall corners carefully so they will fit together correctly when the wall sections are erected.

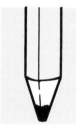

Check over your rough opening layouts carefully. Errors may be difficult to recognize at this point.

STORY POLE

A story pole is a long measuring stick made up by the carpenter on the job. It represents the actual wall frame with markings made at the proper height for every horizontal member of the wall frame. See Fig. 8-13.

While it is possible to go back to the framing plans every time and read the information needed, it is not very practical. Constant checking consumes a great deal of time and errors have more chance to creep in. It is much easier to read off the dimensions from the plan and mark them once and for all on the story pole. The carpenter is always careful to double check the accuracy of the measurements on the pole.

While very useful on any residential construction site the pole is a must for split-level homes, Fig. 8-14, where there are a number of stud lengths and many heights to remember.

179

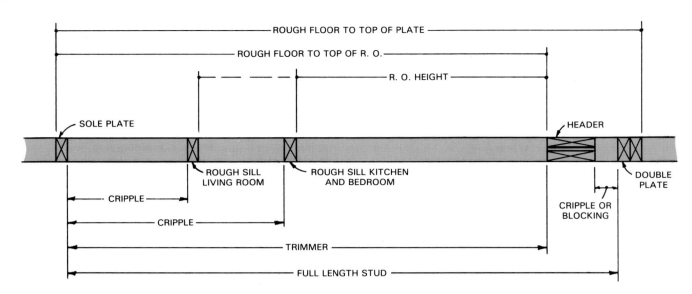

Fig. 8-15. Master stud pattern is like a story pole. Stud lengths can be secured from it.

MASTER STUD LAYOUT

A master stud pattern is like a story pole but does not include as much information. The layout can be made on either a straight 1 x 4 or 2 x 4. First lay out the distance from the rough floor to the ceiling. This dimension can be taken directly from the story pole or from the plans. Mark off the position of the sole plate and double top plate, Fig. 8-15. Now lay out the header. When several header sizes are used, they can be marked on top of each other.

Lay out the height of the rough openings, measuring down from the bottom side of the header. Then draw in the rough sill.

The length of the various studs (regular, trimmer, and cripple) can now be taken directly from this full-size layout.

When the header height of the doors is different from that of the windows, use the other side of the pattern to keep them separate. In multi-story or split-level structures, a master stud layout will probably be required for each level.

CONSTRUCTING WALL SECTIONS

Working from the master stud layout, cut the various stud lengths. In modern construction, it is seldom necessary to cut standard full length studs. These are usually precision end trimmed (P.E.T.) at the mill and delivered to the construction site ready to assemble.

Cut and assemble the headers. Their length, and also the length of the rough sill, can be taken directly from the plate layout.

The sequence to be followed in assembling wall sections, especially the rough opening, will vary among carpenters. Following is one of several pro-

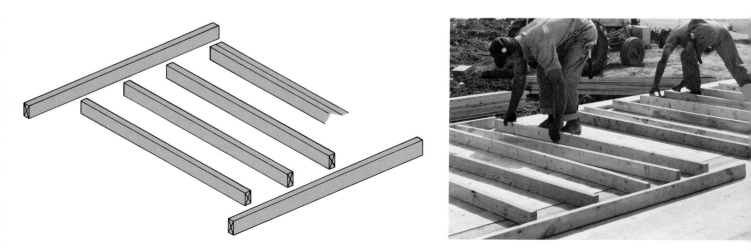

Fig. 8-16. Assembling the wall studs and plates. Left. Place a stud at every position marked on the plate. Right. Checking that studs are turned crown up. (National Forest Products Assoc.)

180

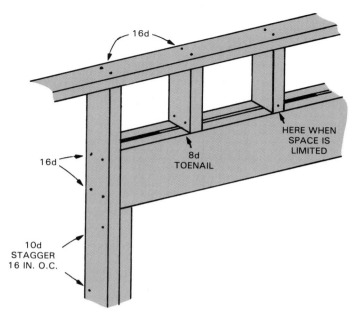

Fig. 8-17. Older method of installing headers with cripple studs filling space above. Note nailing pattern through full-length stud into header.

cedures that could be used:

Move the top plate away from the sole plate about a stud length. Turn both plates on edge with the layout marks inward. Place a full-length stud, crown up, at each position marked on the plates. See Fig. 8-16. Nail the top plate and sole plate to the studs using two 16d nails through each end of each stud.

Set the trimmer studs in place on the sole plate and nail them to the full-length studs. Now place the header so it is tight against the end of the trimmer and nail through the full-length stud into the header using 16d nails, Fig. 8-17. The upper cripples can be installed after the header is placed. It is now common for carpenters to omit these cripple studs and run the header all the way to the top plate, Fig. 8-18.

For window openings, transfer the position of the cripple studs from the sole plate to the rough sill and then make the assembly using 16d nails, Fig. 8-19. Some carpenters prefer to erect the wall section and then install the lower cripples and rough sill. When this procedure is followed, the lower ends of the cripple studs are toenailed to the sole plate.

Add studs or blocking at positions where partitions will intersect outside walls, Fig. 8-20. Also install any wall bracing that may be required for special installations. Remember that the inside surface of the frame is turned down.

Fig. 8-18. To speed up construction, carpenters use heavy duty power nailers. Note that header goes all the way to the top plate. (Paslode Co.)

Fig. 8-19. Attaching rough sill plate and cripple studs to the wall frame. Safety glasses should always be worn when a power nailer is being used.

Fig. 8-20. Stud has been added (arrow) where a partition will intersect the outside wall. (Senco Products Inc.)

In modern construction, wall sheathing is often applied to the frame before it is raised. Make certain that the framework is square before starting the application. Check diagonal measurements across the corners. They must be equal. To keep the frame square while the sheathing is being applied, fasten a diagonal brace across one corner. If you prefer, nail two edges of the frame temporarily to the floor.

ERECTING WALL SECTIONS

Most one-story wall sections can be raised by hand, Fig. 8-21. Larger structures will require the

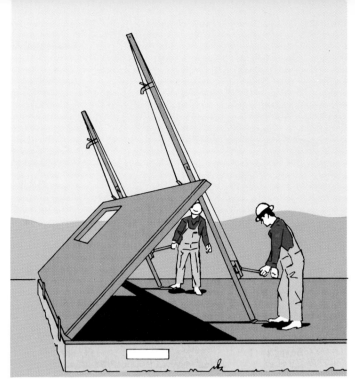

Fig. 8-22. Wall jacks are used to lift larger wall sections. (Proctor Products Co., Inc.)

use of a crane or other equipment, Fig. 8-22.

Before raising a section, be sure it is in the correct location. Have bracing at hand and ready to be attached. If the section is large, have extra help available. Make sure each worker knows what to do.

When raising sections to which sheathing has not been applied, it is good practice to install temporary diagonal bracing if regular bracing is not included. Some carpenters attach temporary blocking to the edge of the floor frame before raising wall sections. It prevents their sliding off the platform.

Fig. 8-21. Raising a section of wall. Headers for the window and door openings are made from laminated veneer lumber (LVL). (Gang-Nail Systems, Inc.)

Fig. 8-23. Waferboard sheathing was added to this wall section before it was raised. Section is being nailed in place through the sole plate. (Blandin Wood Products Co.)

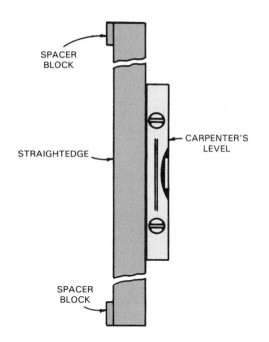

Fig. 8-24. Use a straightedge and level to plumb walls if there is warpage in the studs.

When plumbing a wall with a carpenter's level, hold the level so you can look straight in at the bubble. If the wall framing member or surface is warped, you should hold the level against a long straightedge that has a spacer lug at the top and bottom.

Immediately after the wall section is up, secure it with braces attached near the top and running to the subfloor at about a 45 deg. angle. Make final adjustments in the position of the sole plate. Be sure it is straight. Then nail it to the floor frame using 20d nails driven through the subfloor and into the joists, Fig. 8-23.

Loosen the braces one at a time and carefully plumb the corners and midpoint along the wall. This can be done with a plumb line, but on one-story construction, a carpenter's level is generally used. If the wall is slightly warped, use a straightedge, Fig. 8-24.

After one section of the wall is in place, proceed to other sections. No particular sequence needs to be followed. Most carpenters prefer to erect main side walls first and then tie in end walls and smaller projections. Procedures must be determined on each individual project. Design and construction methods help determine how to proceed.

PARTITIONS

When the outside wall frame is completed, partitions are built and erected. At this stage it is important to enclose the structure and make the roof watertight. Only bearing partitions (those that support the ceiling and/or roof) are usually installed. Erection of nonbearing partitions can be put off until later.

Roof trusses, often used in modern construction, are supported entirely by the outside walls. When they are used, inside partition work is seldom started until the roof is complete. Refer to Fig. 8-1 once more.

The centerlines of the partitions are established from a study of the plans and then marked on the floor with a chalk line. Plates are laid out; studs and headers are cut; partitions are assembled and erected in the same way as outside walls. See Fig. 8-25. Erect long partitions first, then cross partitions. Finally build and install short partitions that form closets, wardrobes, and alcoves.

The corners and intersections are constructed as described for outside walls. The size and amount of blocking, however, can be reduced, especially in nonbearing partitions. The chief concern is to provide nailing surfaces at inside and outside corners for wall-covering material. Refer again to Fig. 8-6.

NONBEARING PARTITIONS

Nonbearing partitions do not require headers. Many openings can be framed with single pieces of 2 x 4 lumber as shown in Fig. 8-26. Most carpenters, however, include trimmers around openings because they are more rigid. Furthermore, they provide additional framework for attaching casing and trim. Door openings in partitions (and outside walls) are framed with the sole plate included at the bottom of the opening. After the

Fig. 8-25. Raising a main bearing partition. A section of the double plate has been installed to stiffen the top plate.

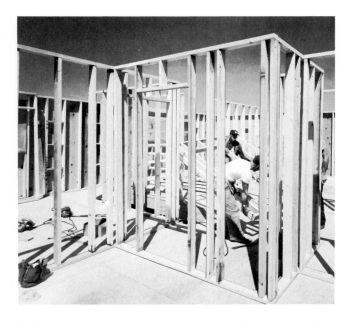

Fig. 8-26. Headers are not needed in nonbearing partitions.

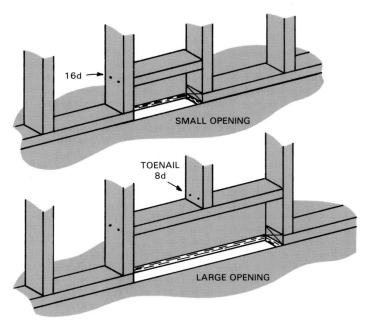

Fig. 8-28. How to install special framing for heat ducts.

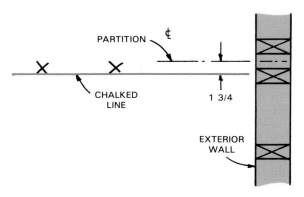

Fig. 8-27. Snap a chalk line on floor to mark position of partitions.

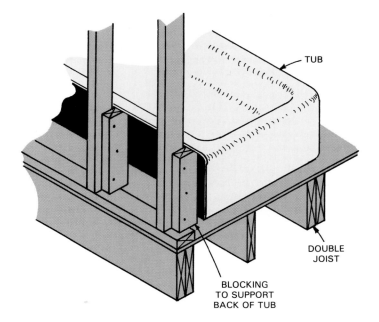

Fig. 8-29. Extra joists and blocking are needed to support tubs.

framework is erected, the sole plate is cut out with a handsaw. Rough door openings are generally made 2 1/2 in. wider than the finished door size.

The soundproofing of partitions between noisy areas and quiet areas is receiving increased emphasis in modern homes and may require a special method of framing. See Unit 13 for information on insulation.

Small alcoves, wardrobes, and partitions in closets are often framed with 2 x 2 material or by turning 2 x 4 stock sideways, thus saving space. This is usually satisfactory where the thinner constructions are short and intersect regular walls. Snap a chalkline across rough floor to mark position of partitions, Fig. 8-27.

During wall and partition framing, various important details can be added. Openings for the in-

stallation of heating ducts, Fig. 8-28, are easily cut and framed at this time. Bath tubs and wall mounted stools require extra support, Fig. 8-29.

Basic provisions for recessed and surface hung cabinets, tissue-roll holders, and similar items should be added to the framing at this stage. Architectural plans usually provide information concerning their size and location. Wall backing, Fig. 8-30, drapery brackets for towel bars, shower curtains, and wall mounted plumbing valves

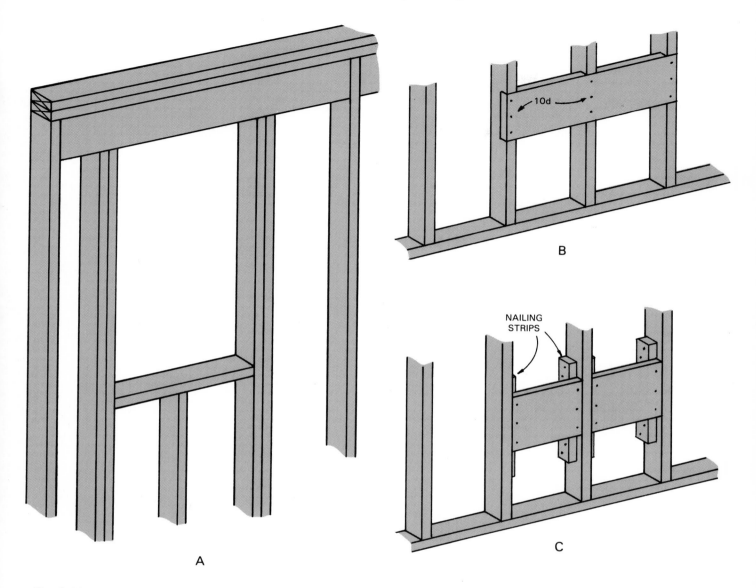

Fig. 8-30. Backing for mounting various fixtures and appliances. A—Extending header over windows provides a base for attaching drapery rod brackets. B—Backing let into studs. Never cut back more than 25 percent of stud width on bearing walls or partitions. C—Backing attached to nailing strips.

should also be added. Plumbing fixture rough-in drawings, Fig. 8-31, will be helpful in locating the backing. For most items, 1 in. thick backing material will provide adequate support.

Small items, not critical to the structure, can often increase efficiency and quality for work during the finishing stages. For example corner blocks, Fig. 8-32, will make it possible to nail baseboards some distance back from the end. This eliminates the possibility of splitting the wood.

PLUMBING IN WALLS

Where plumbing is run through walls, special construction may be required. Depending on the size of the drain and venting pipes, a partition may

The carpenter must continually study and plan the sequence of the job, so that neither the weather nor work of other tradespeople will cause slowdowns or bottlenecks.

have to be made wider. Usually a 6 in. frame is sufficient. Fig. 8-33 shows several methods of construction.

Lateral (horizontal) runs of pipe will require drilling of holes or notching of the studs. A metal strap can be attached to bridge the notch and strengthen the stud. The strap also protects the pipe from accidental damage should a nail be driven into the

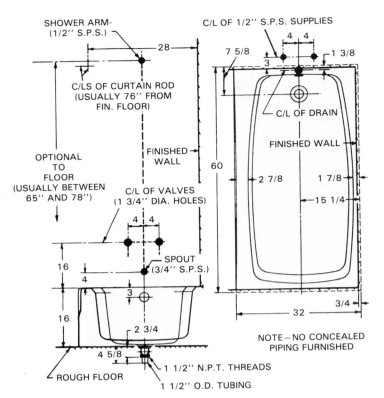

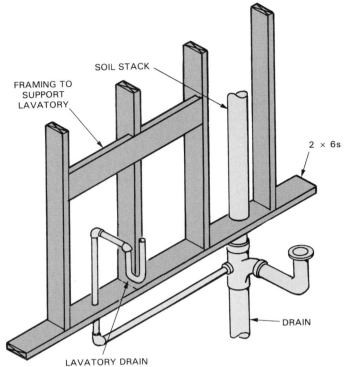

Fig. 8-31. Typical rough-in dimensions are helpful reference when locating special backing. (American Standard)

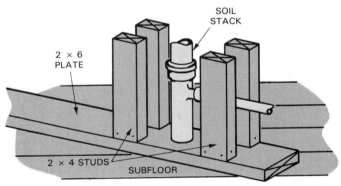

Fig. 8-33. A partition containing plumbing pipes may need to be constructed differently to provide room for the plumbing. Top. Wall constructed of 2 x 6 studs and plate. Note special framing for supporting lavatory. Bottom. 2 x 4 studs on 2 x 6 plate eliminate need to notch or bore studs for lateral runs.

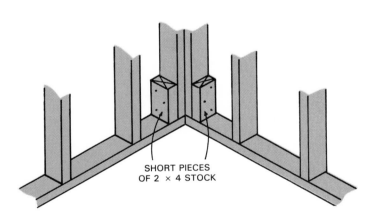

Fig. 8-32. Blocking in corners provides better nailing surface for attaching baseboards.

stud at this point. Similar protection should be provided for electrical wiring in walls. Sometimes a wooden block may be used to bridge notch cut for plumbing. See Fig. 8-34.

BRACING

Exterior walls usually need some type of bracing to resist lateral (sideway) stresses. Some applications of material, such as plywood, provide sufficient rigidity and bracing can be eliminated.

Always check the exact requirements of the local building code.

A standard installation of let-in corner bracing is shown in Fig. 8-35. Note how the bracing is applied when an opening interferes with the diagonal run from top plate to sole plate.

In modern construction, metal strap bracing is widely used. It is made of 18 or 20 gauge galvanized steel 2 in. wide. One type includes a 3/8 in. center rib, Fig. 8-36.

To install the ribbed type, snap a chalk line across the wall frame after it has been assembled on the floor deck. Use a portable saw and make multiple cuts to form the groove for the rib. The

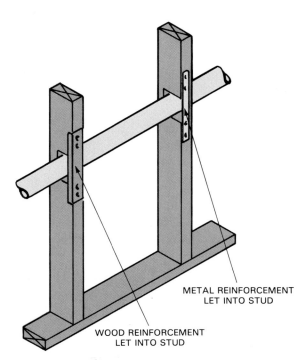

Fig. 8-34. Methods of reinforcing studs when notched or bored for plumbing. Similar methods are used to protect wiring from nails.

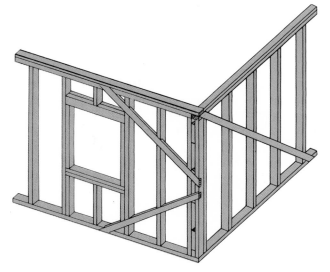

Fig. 8-35. Bracing a wall with 1 x 4 lumber. Material is cut away from the studs to receive (let in) the braces. Use two 8d nails at each intersection.

strap is usually installed after the frame has been erected and carefully plumbed. Drive two 8d nails through the rib and into each 2 x 4 stud and other framing members. For 2 x 6 studs, use one 16d nail. See Fig. 8-37.

DOUBLE PLATE

To add support under ceiling joists and rafters, the top plate is doubled. This also serves to further tie the wall frame together. Select long straight lumber. Install the double plate with 10d nails. Place two nails near the ends of each piece. The others are staggered 16 in. apart.

Joints should be located at least 4 ft. from those in the lower top plate. At corners and intersections, the joints are lapped as shown in Fig. 8-38.

SPECIAL FRAMING

In addition to the standard framing just described, the carpenter will often be required to build structures that include special design features. It is preferred these special designs be carefully engineered by the architect. Complete details can be included in the plans.

Occasionally, only the shape and size is included. The construction details must then be developed by the carpenter. A bay window is a

Fig. 8-36. Metal strap bracing. See how it is run from the floor trusses to the top plate.

good example. See Fig. 8-39. Here the carpenter must:
1. Visualize the construction details.
2. Lay out and construct the floor frame to carry the projection.
3. Build the wall and roof frame.
If the carpenter understands regular framing requirements and procedures, little difficulty is encountered in applying them to special problems.

The tri-level or split-level house presents some extra problems in wall framing. Generally, a platform type of construction is used. However, the floor joists for upper levels could be carried on ribbons cut into the studs.

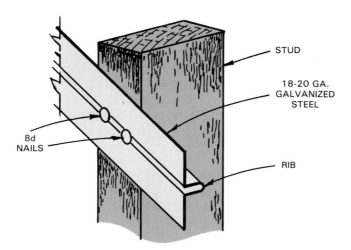

Fig. 8-37. Detail of metal strap bracing.

The plans will prescribe the type of construction. They should also include careful calculations of distances between floor levels.

When working with split-level designs, the good carpenter should prepare accurate story poles which show full-size layouts of vertical distances and actual sizes of the construction materials.

WALL SHEATHING

If the wall sections were not covered with sheathing before erection, this should be done before roof framing is started. Sheathing adds rigidity, strength, and some insulating qualities to the wall.

Today, plywood, composite board, fiberboard, and insulating panels are widely used for sheathing. Large sheets can be applied rapidly. Plywood and some composite board products provide enough lateral strength that diagonal bracing can be eliminated.

Fiberboard sheathing is made largely from wood fibers with added weather-proofing ingredients. It is commonly available in 4 x 8 ft. sheets. However, sheets as large as 8 x 14 ft. are manufactured. Regular fiberboard sheathing is also available in a 2 x 8 ft. size with tongue and groove or shiplap joints. It is applied horizontally and supplementary corner bracing is required. Regular fiberboard sheathing cannot be used as a nailing base for exterior wall finish materials. However, a special nail-base fiberboard is available.

Standard thicknesses of fiberboard sheathing are 1/2 and 25/32 in. Use 1 1/2 in. roofing nails for the 1/2 in. thickness and 1 3/4 in. for the 25/32 in. thickness. Diagonal bracing is recommended when using 1/2 in. fiberboard. Adequate bracing can be provided by installing 1/2 in. plywood at each corner as shown in Fig. 8-40.

Provide a 1/8 in. space between all edge joints when applying fiberboard sheathing.

When gypsum sheathing (usually 1/2 in. thick) is used, it is necessary to include corner bracing. This type of sheathing is usually installed with 1 3/4 in. nails spaced 4 in. around the edge and 8 in. on intermediate supports.

Plywood sheathing may be either of the interior or exterior type in a structural grade. It should be at least 5/16 in. thick for studs spaced 16 in. on centers and 3/8 in. thick for studs spaced 24 in. on centers. A 1/2 in. thickness is often used so that the exterior finish material can be fastened directly to the plywood. The sheets can be applied vertically or horizontally, Fig. 8-41. Provide 1/16 in. of space between panel edges and 1/8 in. of space between panel ends.

Plain wood boards or shiplap can also be used for wall sheathing. They should be at least 3/4 in. thick and not over 12 in. wide. End joints should fall over the centers of the studs. Diagonal bracing is required when the siding is installed horizontally. When using boards with side and end matching (tongue and groove) for sheathing, the joints need not be over a stud. When applied diagonally no corner braces are needed. Boards should be slanted in opposite directions on adjoining walls for greatest strength and stiffness. Also, use two 8d nails on each stud for 6 or 8 in. boards. Three nails should be used for wider boards.

Foamed plastic sheathing, Fig. 8-42, is made from polystyrene or polyurethane. The use of this type of sheathing has grown dramatically in recent years because of its high resistance to heat flow (R). Panels of this kind of material are not very

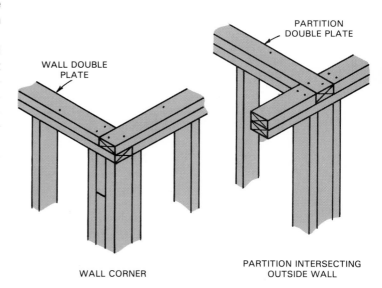

Fig. 8-38. Double plates are lap jointed for strength wherever they intersect.

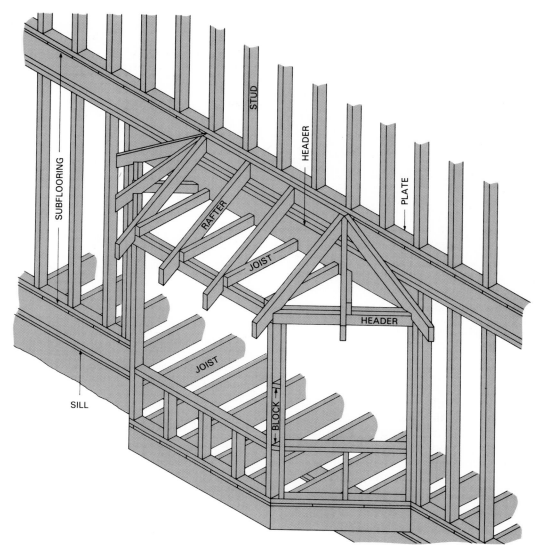

Fig. 8-39. Framing for a bay window. Subfloor is not shown so you can see how floor joists are extended over the sill. (National Forest Products Assoc.)

Fig. 8-40. Corner bracing is left out when plywood is used to sheath corners. (American Plywood Assoc.)

strong and must be handled with extra care. Corner bracing is required. Always follow the manufacturer's directions for application. Procedures may vary somewhat from one product to another.

CEILING FRAME

A ceiling frame is the assembly just below the roof that carries the ceiling surface. On other levels, the ceiling cover is carried on the floor joists.

Basic construction is similar to floor framing. The main difference is that lighter joists are used and headers are not included around the outside.

When trusses are used to form the roof frame, a ceiling frame is not required. The bottom chords of the trusses carry the ceiling surface.

The main framing members are also called joists. Like floor joists, their size is determined by the length of span and the spacing used. To coordinate

Fig. 8-41. Installing oriented strand board sheathing. Note use of metal strap bracing. (Georgia-Pacific Corp.)

SIZE IN.	SPACING IN.	GROUP A FT. IN.	GROUP B FT. IN.	GROUP C FT. IN.	GROUP D FT. IN.
2 × 4	12	9 - 5	9 - 0	8 - 7	4 - 1
	16	8 - 7	8 - 2	7 - 9	3 - 6
2 × 6	12	14 - 4	13 - 8	13 - 0	9 - 1
	16	13 - 0	12 - 5	11 - 10	7 - 9
2 × 8	12	19 - 6	18 - 8	17 - 9	14 - 3
	16	17 - 9	16 - 11	16 - 1	12 - 4
2 × 10	12	24 - 9	23 - 8	22 - 6	19 - 6
	16	22 - 6	21 - 6	20 - 5	16 - 10

Fig. 8-43. Spans for ceiling joists are figured for a normal dead load and a live load of 20 psf (pounds per sq. ft.). This permits the attic to be used for storage. Always check local codes. (National Building Code)

with walls and permit the use of a wide range of surface materials, a spacing of 16 in. O.C. is commonly used. Size and quality requirements must also be based on the type of ceiling finish (plaster or drywall) and what use will be made of the attic space, Fig. 8-43. The architectural plans will usually include specifications. These requirements should be checked with local building codes.

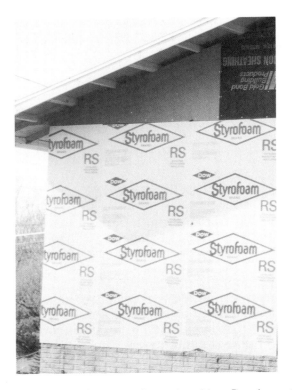

Fig. 8-42. Rigid polystyrene foam sheathing. Panels are 1 in. thick and 96 in. long. They have an R value of 5.50.

Ceiling joists usually run across the narrow dimensions of a structure. However, some may be placed to run in one direction and others at right angles as shown in Fig. 8-44. By running joists in different directions, the length of the span can often be reduced.

In large living rooms, the midpoint of the joists may need to be supported by a beam. This beam can be located below the joists or installed flush with the joists. In the latter installation, the joists may be carried on a ledger, as shown in Fig. 8-45. Joist hangers, like those described in the unit on floor framing, can also be used, Fig. 8-46. Sometimes a beam is installed above the joists in the attic area. It is tied to the joists with metal straps.

At their outer ends, the upper corners of the joists must be cut at an angle to match the slope of the roof. To lay out the pattern for this cut, you may use the framing square as illustrated in Fig. 8-47. When the amount of stock to be removed is small, the cuts can be made after the joists and rafters are in place. Use a hatchet or saw.

When ceiling joists run parallel to the edge of the roof, the outside member will likely interfere with the roof slope. This often occurs in low-pitched hip roofs. The ceiling frame in this area should be constructed with stub joists running perpendicular to the regular joists, Fig. 8-48.

Lay out the position of the ceiling joists along the top plate using a rod. When a double plate is used, the joists do not need to align with the studs in the wall. The layout, however, should put the joists alongside the roof rafters so that the joists can be nailed to them.

Ceiling joists are installed before the rafters. Toenail them to the plate using two 10d nails on each side.

Partitions or walls that run parallel to the joists must be fastened to the ceiling frame. A nailing strip or drywall clip to carry the ceiling material must be installed. Various size materials can be in-

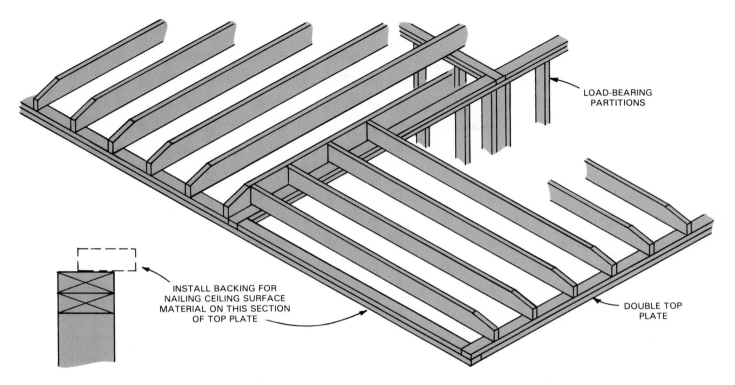

INSTALL BACKING FOR
NAILING CEILING SURFACE
MATERIAL ON THIS SECTION
OF TOP PLATE

LOAD-BEARING
PARTITIONS

DOUBLE TOP
PLATE

Fig. 8-44. Ceiling frame. Joists in foreground are turned at right angles to reduce the span.

Fig. 8-45. When a flush beam is required in a ceiling, a ledger strip can be used to support ceiling joists.
(Western Wood Products Assoc.)

stalled in a number of ways. The chief requirement is that they provide adequate support. Fig. 8-49 shows a typical method of making such an installation. Fig. 8-6 shows clip installation for a ceiling.

An access hole (also called a scuttle hole) must be included in the ceiling frame to provide an entrance to the attic area. Fire regulations and building codes usually list minimum size requirements. The building plans generally indicate size and show where it should be located. The opening is framed following the procedure used for openings in the floor. If the size of the opening is small (2 to 3 ft. square) doubling of joists and headers is not required.

STRONGBACKS

Long spans of ceiling joists may require a strongback. This is an L-shaped support which is attached across the tops of joists to strengthen them and maintain the space between them. It also evens up the bottom edges of the joists so the ceiling will not be wavy after the drywall is applied.

Fig. 8-46. Carpenter installing joist hangers on a flush beam. Joists were toenailed to beam during initial assembly.
(Southern Forest Products Assoc.)

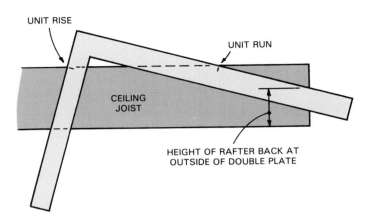

Fig. 8-47. Using the framing square to lay out the trim cut on the end of a ceiling joist to match the slope of the rafter.

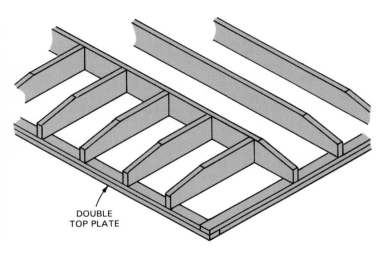

Fig. 8-48. Stub ceiling joists along the end wall. These are required for a low-pitched hip roof.

The strongback is built with two pieces of dimension lumber. One is usually a 2 x 4 and the other a 2 x 6 or 2 x 8.

To construct it, first mark off the proper spacing (16 or 24 in. O.C.) on a 2 x 4. Position the 2 x 4 across the ceiling joists and fasten it with two 16d nails at each joist. Apply pressure against the joists as needed to maintain proper spacing.

Select a straight 2 x 6 or 2 x 8 for the second member. Place it on edge against either side of the 2 x 4 just attached to the joists. Attach one end to the 2 x 4 with 16d nail. Work across the full length of the strongback aligning and nailing. Stepping on either the 2 x 4 or the member on edge will help align each joist. Nail the vertical member to each joist and to the 2 x 4, Fig. 8-50.

ESTIMATING MATERIALS

To estimate wall and ceiling framing materials, first determine the total lineal feet by adding

together the length of each wall and partition. The plans will include the dimensions of outside walls. These can be added together. Partitions, especially those that are short, may not be dimensioned and you will need to scale the drawing. It is a good idea to place a colored pencil check mark on each wall and partition as its length is added to the list.

For plates, multiply the total figure by three (one sole plate plus two top plates). Add about 10 percent for waste. Order this number of linear feet of lumber in random lengths or convert to the number of pieces of a specific length. Three pieces of 2 x 4 one foot long equals 2 bd. ft. Thus, the total lineal feet of walls and partitions can be quickly converted to board measure if required. For example:

Wall & partition length = 240
Total plate material = length × 3 +10%
= 240 × 3 + 10%
= 792 lineal ft.
or 57 pcs., 2″ × 4″ × 14′-0″

ESTIMATING STUDS

The total length of all walls and partitions is also used to estimate the number of studs required. When studs are spaced 16 in. O.C., multiply the total length by 3/4 and then add two more studs for each corner, intersection, and opening.

Using the preceding dimensions for the wall length, assume there are 12 corners, 10 intersections, and 20 openings. Find the number of studs

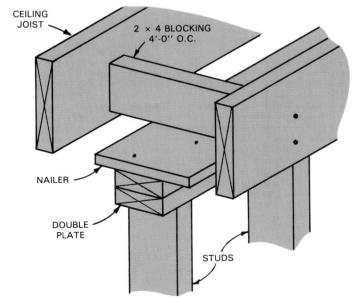

Fig. 8-49. Use special blocking to anchor partitions to the ceiling frame when they run parallel to the joists.

needed including outside walls and partitions:

Total studs = total length × 3/4 + 2 (corners + intersections + openings)

$$= \frac{240 \times 3}{4} + 2(12 + 10 + 20)$$

$$= \frac{\overset{60}{\cancel{240}} \times 3}{\cancel{4}} + 2(42)$$

$$= 60 \times 3 + 84 = 264$$

Many carpenters estimate the number of studs by simply counting one stud (spaced 16 in. O.C.) for each lineal foot of wall space. About 10 percent is added for waste. The over-run on spacing provides the extras needed for corners and openings. This method is rapid and fairly accurate. It will not allow for sufficient studs in a small house cut up into many rooms. On the other hand, too many studs will likely be figured for a large house with wide windows and open interiors.

Lumber for headers must be calculated by analyzing the requirements for each opening. Use the R.O. width plus the thickness of the trimmers.

Ceiling joists are estimated by about the same method used for floor joists. Since ceilings will be relatively free of openings, no extras or waste needs to be included. Because of this, the short method should not be applied to ceiling joists. Use the following formula and include the size of the joists required:

Number of ceiling joists = wall length × 3/4 + 1

ESTIMATING WALL SHEATHING

To estimate the amount of wall sheathing, first find the total perimeter of the structure. Multiply this figure by the wall height measured from the top of the foundation when the sheathing extends over the sill construction. The product will be the gross square footage of the wall surface. Now calculate the area of each major opening (windows and doors) and subtract this total from the original figure. Round the opening sizes downward to the nearest foot.

Net area = perimeter × height − wall openings

To the net area to be sheathed add allowances for waste and other extras when common boards or shiplap are used. See Unit 7 for more information. When using sheet materials for sheathing, there is only slight waste. Divide the net area by the square footage per sheet to secure the number

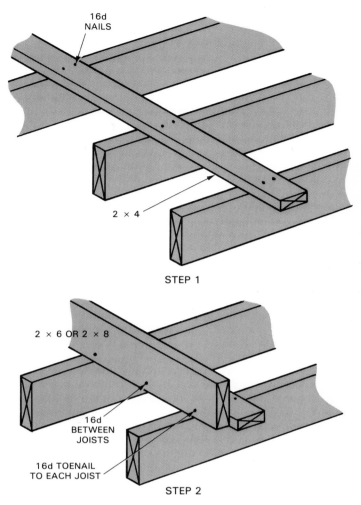

Fig. 8-50. Building a strongback. Step 1, nail the 2 x 4 to the joists. Step 2, turn the second 2 x 6 or 2 x 8 on edge and nail it to the 2 x 4 and to joists.

of pieces required. For example, if the net area to be sheathed is 1060 sq. ft. and the fiberboard sheets selected are 4' × 9', calculate quantity needed as follows:

Net area = 1060

Fiberboard sheet size = 4 × 9 = 36

Number of sheets = $\frac{1060}{36}$ = 29 + or 30

L = X × Z
L = 24 × Z
  = 24 × 3
  = 72

Most problems in estimating, although based on simple formulas, usually contain so many variables that a lot of good judgment must be applied to their solution. This judgment is acquired through experience.

TEST YOUR KNOWLEDGE — UNIT 8

1. What is a sole plate?
2. How many studs are required for a plain wall panel 8'-0" long if they are spaced 16 in. O.C.?
3. Trimmer studs stiffen the sides of an opening and carry the weight of the _____.
4. What is a story pole?
5. The first layout to be marked on the plates is the _____ _____ spacing.
6. A master stud pattern is laid out somewhat like a _____ _____.
7. The layout of the cripple studs on the rough sill can be marked directly from the_____ _____.
8. Most carpenters prefer to erect the _____ (side, end) walls first.
9. What is the difference between let-in bracing and metal strap bracing?
10. Joints formed along the doubled top plate should be at least _____ apart.
11. Regular fiberboard sheathing (can, cannot) be used as a nailing base for exterior wall finish materials.
12. The position of the ceiling joists along the double plate should be coordinated with the _____.
13. The first step in estimating the number of studs required is to figure the total length of all _____ and _____.
14. A strongback is needed for:
    a. Strengthening a long span of ceiling joists.
    b. Maintaining proper spacing between ceiling joists.
    c. Keeping joists even along their lower edges.
    d. None of the above.
    e. All of the above.

OUTSIDE ASSIGNMENTS

1. Secure a set of architectural plans for a one-story house. Study the details of construction, especially typical wall sections. Prepare a scale drawing of the framing required for the front walls. Be sure the rough openings are the correct size and in their proper location. Your drawing should look somewhat like the one in Fig. 8-2.
2. Working from the same set of plans, develop an estimated list of materials for the wall frame and sheathing. Include the number and length of studs; the number and size of lumber for headers; the material for plates, and type and amount of sheathing. Secure prices from your local supplier and figure the total cost of the materials.
3. Secure literature about fiberboard, foamed plastic, and gypsum sheathing. Secure this material from local lumber dealers or write directly to manufacturers. Also study books and other reference materials.

Prepare a report for the class based on the information you obtain. Include grades, manufacturing processes, characteristics, and application requirements. Discuss current prices and purchasing information. Be prepared to discuss specifications given in the literature. Relate these "specs" to your local code.

Long one-piece joists are made by joining short lengths with finger-joints (see arrows). The long lengths save construction time and permit flexibility in locating supporting partitions and beams. Available in odd and even lengths from 16 to 40 feet. (Winton Co.)

Left. Laminated veneer lumber (LVL) consists of select veneers bonded together with waterproof adhesives. Laminate lumber has all the capabilities of solid lumber without defects such as knots and warpage. It is made in nominal sizes from 2 x 4 in. to 2 x 14 in. and in lengths from 16 to 48 ft. Right. Ceiling beam is constructed from 2 x 12s spiked together like regular solid lumber. (Gang-Nail Systems, Inc.)

Left. Floor and ceiling structural members, called beams, are manufactured from laminated wood chords and oriented strand board webs. They are straighter, stronger, and lighter than solid lumber and will not shrink. The units are manufactured in nominal depths of 10 to 16 in. Right. Ceiling and roof frame constructed from manufactured units. Note how joist hanger is installed. (Gang-Nail Systems, Inc.)

GAMBREL

MANSARD ROOF

CONTINUOUS LOW SLOPE GABLE ROOF

DUTCH HIP ROOF

GABLE ROOF

HIP ROOF

COMBINATION FLAT
AND SHED ROOF

Distinctive differences in roof styles set these architectural
renderings apart from one another. Each will require slightly
different methods and measurement of the carpenter.
(Garlinghouse)

Unit 9

ROOF FRAMING

Roof framing provides a base to which the roofing materials will be attached. The frame must be strong and rigid. Besides this, the roof, if carefully designed and proportioned, can contribute a distinctive and decorative feature to the structure.

ROOF TYPES

Roof styles vary widely. Most of them can be grouped into the following types. See Fig. 9-1:
1. Gable roof. Two surfaces slope from the centerline (ridge) of the structure. This forms two triangular shaped ends called gables. Because of their simple design and low cost, gabled roofs are often used for homes.
2. Hip roof. All four sides slope from a central point or ridge. The angles created where two sides meet are called hips. An advantage of this type is the protective overhang formed over end walls as well as over side walls.
3. Gambrel roof. In this variation of the gable roof,

each slope is broken, usually near the center or ridge. This style is used on two-story construction. It permits more efficient use of the second floor level. Dormers are usually included. Refer to the sketch in Fig. 9-1 once more. The gambrel roof is a traditional style typical of colonial America and the period immediately following.
4. Flat roof. In this type, the roof is supported on joists which also carry ceiling material on the underside. It may have a slight pitch (slope) to provide drainage.
5. Shed roof. This simplest of pitched roofs is sometimes called a "lean-to" roof. The name comes from its frequent use on additions to a larger structure. It is often used in contemporary designs where the ceiling is attached directly to the roof frame.
6. Mansard roof. Like the hip roof, the mansard has four sloping sides. However, each of the four sides has a double slope. The lower, outside slope is nearly vertical. The upper slope is

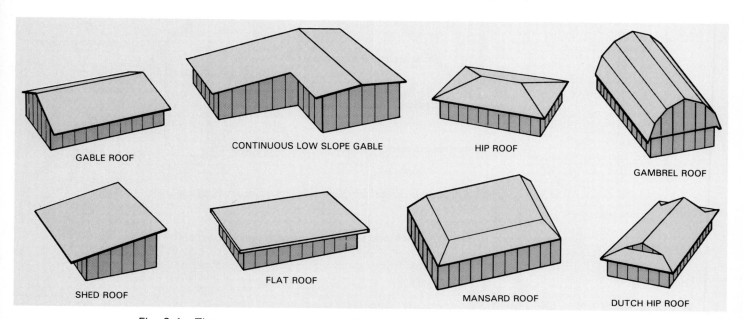

Fig. 9-1. These common types of roofs are to be found in residential construction.

slightly pitched. Like the gambrel roof, the main advantage is the additional space gained in the rooms on the upper level. The name comes from its originator, architect Francois Mansart (1598–1666). One of the photographs on the opening page of this unit shows a modern variation of the mansard.

ROOF SUPPORTS

Roofs, depending on the type of rafter design, are supported by one or all of the following systems:
1. Outside walls.
2. Ceiling joists (beams which hold the ceiling materials).
3. Interior bearing walls.

PARTS OF ROOF FRAME

The plan view of a roof, Fig. 9-2, combines several roof types. The kinds of rafters are identified.

The COMMON rafters are those that run at a right angle (plan view) from the wall plate to the ridge. A plain gable roof consists entirely of rafters of this kind.

HIP rafters also run from the plate to the ridge, but at a 45 deg. angle. They form the support where two slopes of a hip roof meet.

VALLEY rafters extend diagonally from the plate to the ridge in the hollow formed by the intersection of two roof sections.

Three kinds of jack rafters are:
1. HIP JACK. This is the same as the lower part of a common rafter but intersects a hip rafter in-

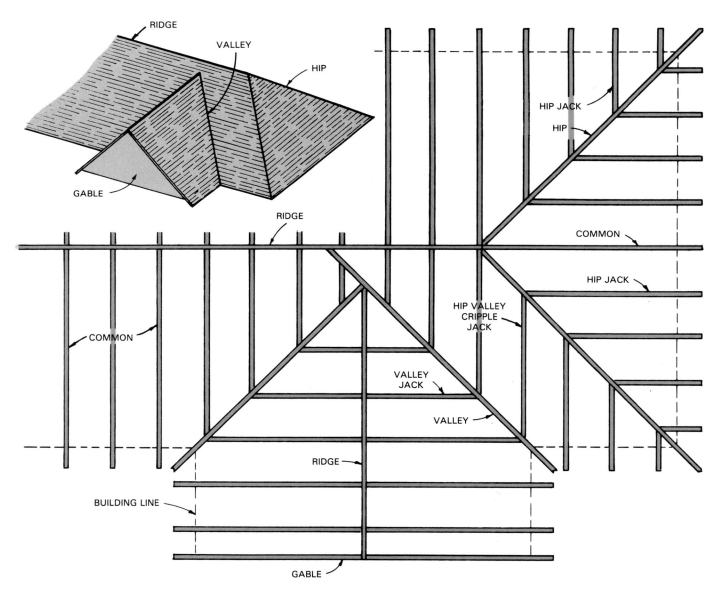

Fig. 9-2. Plan (looking down from top) view of roof frame. Note types of rafters and their names. Inset shows how finished roof will look.

stead of the ridge.

2. VALLEY JACK. This is the same as the upper end of a common rafter but intersects a valley rafter instead of the plate.

3. CRIPPLE JACK. Also called a cripple rafter, it intersects neither the plate nor the ridge and is terminated at each end by hip and valley rafters. The cripple jack rafter is also called a hip valley cripple jack or a valley cripple jack.

PARTS OF A RAFTER

Rafters are formed by laying out and making various cuts. Fig. 9-3 shows the cuts for a common rafter and the sections formed. The ridge cut allows the upper end to fit tightly against the ridge. The bird's mouth is formed by a seat cut and plumb (vertical) cut when the rafter extends beyond the plate. This extension is called the overhang or tail. When there is no overhang, the bottom of the rafter is ended with a seat cut and a plumb cut.

LAYOUT TERMS AND PRINCIPLES

Roof framing is a practical application of geometry. This is an area of mathematics that deals with the relationships of points, lines, and surfaces. It is based largely on the properties of the right triangle where:

1. The horizontal distance is the base.
2. The vertical distance is the altitude.
3. The length of the rafter is the hypotenuse.

If any two sides of a right triangle, Fig. 9-4, are known, the third side can be found mathematically. The formula used is: $H^2 = B^2 + A^2$. H is the hypotenuse; B is the base, and A is the altitude or height. The solution involves extracting the square root. This is a rather time consuming process. The answer could also be found through the application of trigonometric functions.

On-the-job, carpenters use the tables on the framing square or a direct layout method. Either method is rapid and practical for their work.

In rafter layout, the base of the right triangle is called the run. It is the distance from the outside of the plate to a point directly below the center of the ridge. The altitude or rise is the distance the rafter extends upward above the wall plate. Other layout terms and the relationship of the parts of the roof frame are illustrated in Fig. 9-5.

SLOPE AND PITCH

Slope indicates the incline of a roof as a ratio of the vertical rise to the horizontal run. It is properly expressed as X distance in 12.

For example, a roof that rises at the rate of 4 in. for each foot of run, is said to have a 4 in 12 slope. A triangular symbol above the roof line in the architectural plans gives this information. The slope of a roof is sometimes called the "cut of the roof."

Pitch indicates the incline of the roof as a ratio of the vertical rise to the span (twice the run). It is given as a fraction. For example, if the total roof rise is 4 ft. and the total span 24 ft., the pitch is 1/6 (4/24 = 1/6).

UNIT MEASUREMENTS

The framing square, also called a STEEL SQUARE or carpenter's square, is the basic layout

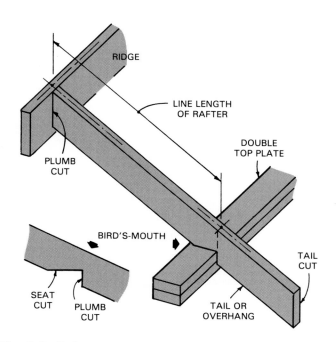

Fig. 9-3. Rafter parts. Various cuts and surfaces have special meaning.

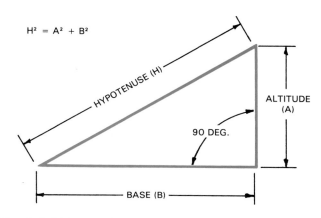

Fig. 9-4. Using math to find rafter length. If you square the rise and the run of a roof and add the answers together, you have the square of the rafter length.

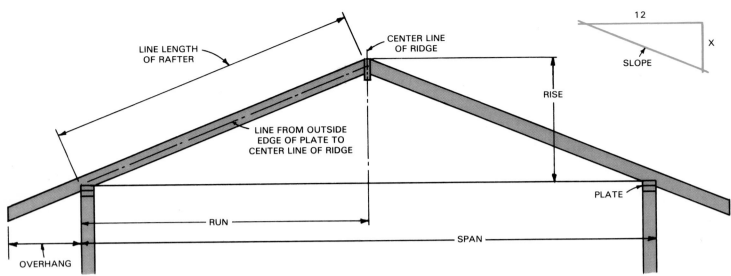

Fig. 9-5. These terms are basic in rafter layout.

tool in roof framing. The side with the manufacturer's name is called the face and the opposite side the back. The longer 24 in. arm is called the body or blade while the shorter one (16 in.) is called the tongue.

The framing square is not large enough to make the rafter layout at one setting. It is necessary, therefore, to use smaller divisions called UNITS. The foot (12 in.) is the standard unit for horizontal run. The unit for rise is always based on how many inches the roof rises in every foot of run. The unit run and rise is used to lay out plumb and level cuts, Fig. 9-6, required at the ridge, bird's-mouth, and tail of the rafter.

FRAMING PLANS

When working with simple designs, the carpenter can easily visualize the roof framing. Since the wall framing is already erected, the only additional information needed will be:
1. The slope of the roof.
2. The amount of overhang required.

These items are included in the house plans.

When the structure has a complicated roof design, the architect often includes a roof framing plan. If, however, a framing plan is not included, the carpenter should prepare one. This can be done by making a scaled drawing on tracing paper laid directly over the floor plans. Include ridges, overhang, and every rafter. The drawing may be made similar to the one in Fig. 9-2. However, you can use a single line to represent the framing members. In your drawing, maintain accurate spacing between rafters. Draw hips and valleys at a 45 deg. angle and make jack rafters parallel to the common rafters.

RAFTER SIZES

As in floor and ceiling framing, the cross section size of the rafter is determined by the spacing and span or length. Local building codes must be consulted for their specifications.

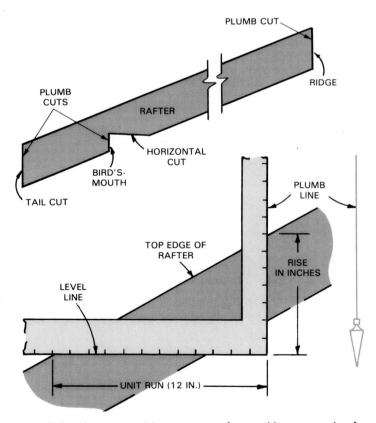

Fig. 9-6. How to position a square for marking one unit of measure on a rafter. Run is measured along the blade. Inches of rise per foot of run are measured off the tongue. Same unit is used to lay out horizontal and plumb cuts on the rafter.

A table listing rafter sizes for various loads and spacings is shown in Fig. 9-7. Stock for hips, valleys, and ridges is usually larger than that of the other framing members.

For purposes of estimating and ordering material, the rafter lengths can be determined with fair accuracy by making a scaled-down layout with the framing square. Use the back of the square where the outside edge of the blade and tongue are divided into inches and twelfths. Assume the inches to be feet and each division an inch.

First draw a triangle, using the unit run and unit rise specified by the slope of the roof. If the total run is more than 12 ft., lengthen the base and hypotenuse. Now lay the blade of the square along the base line until the point of total run on the square falls over the acute angle. Mark the point where the tongue crosses the sloping line, Fig. 9-8. Use either the blade or tongue of the square to measure the hypotenuse. This will give the rafter length. Be sure to add the overhang. Allow extra material, if needed, for cuts at the ridge and tail.

LAYING OUT COMMON RAFTERS

Rafters can be laid out by:
1. The step-off method.
2. The table on the framing square.

Either method will give the correct rafter length but most carpenters prefer the step-off method. They will then use the rafter tables to check their work.

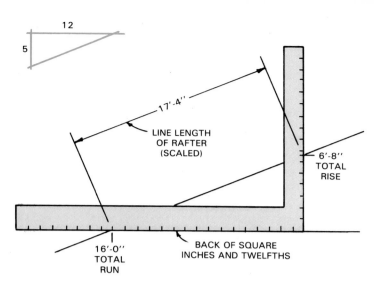

Fig. 9-8. Framing square can be used to estimate approximate length of rafters.

By either method, a pattern rafter is first carefully laid out, checked, and cut. This rafter is used to mark other rafters of the same size and kind.

For pattern layout, select a piece of lumber that is straight and true. Place it on a pair of sawhorses.

Usually the carpenter will stand on the crowned side that will be the top edge of the rafter. It is easier to hold and manipulate the framing square from that position. In order to describe rafter layouts that can be quickly understood, this position has been reversed. The rafter is shown in the position it will have when installed.

To lay out a rafter by the step-off method, place the framing square on the stock and align the figures with the top edge of the rafter—unit run (12 in.) on the blade and the unit rise on the tongue.

Patented clips are available that will maintain the square in this position, Fig. 9-9. A pair of

SIZE OF RAFTER (Inches)	SPACING OF RAFTER (Inches)	MAXIMUM ALLOWABLE SPAN (Feet and Inches Measured Along the Horizontal Projection)			
		Group I	Group II	Group III	Group IV
2 × 4	12	10-0	9-0	7-0	4-0
	16	9-0	7-6	6-0	3-6
	24	7-6	6-6	5-0	3-0
	32	6-6	5-6	4-6	2-6
2 × 6	12	17-6	15-0	12-6	9-0
	16	15-6	13-0	11-0	8-0
	24	12-6	11-0	9-0	6-6
	32	11-0	9-6	8-0	5-6
2 × 8	12	23-0	20-0	17-0	13-0
	16	20-0	18-0	15-0	11-6
	24	17-0	15-0	12-6	9-6
	32	14-6	13-0	11-0	8-6
2 × 10	12	28-6	26-6	22-0	17-6
	16	25-6	23-6	19-6	15-6
	24	21-0	19-6	16-0	12-6
	32	18-6	17-0	14-0	11-0

Fig. 9-7. Sample table showing maximum runs allowed for rafters sloped 4 to 12 or greater. Groups refer to species of wood. Secure this kind of information from local building codes.

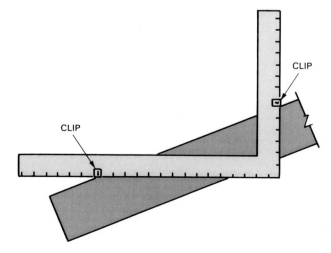

Fig. 9-9. Some carpenters use clips attached to the square to mark the rise and run. As each step is laid out, the clips automatically position the square against the rafter edge.

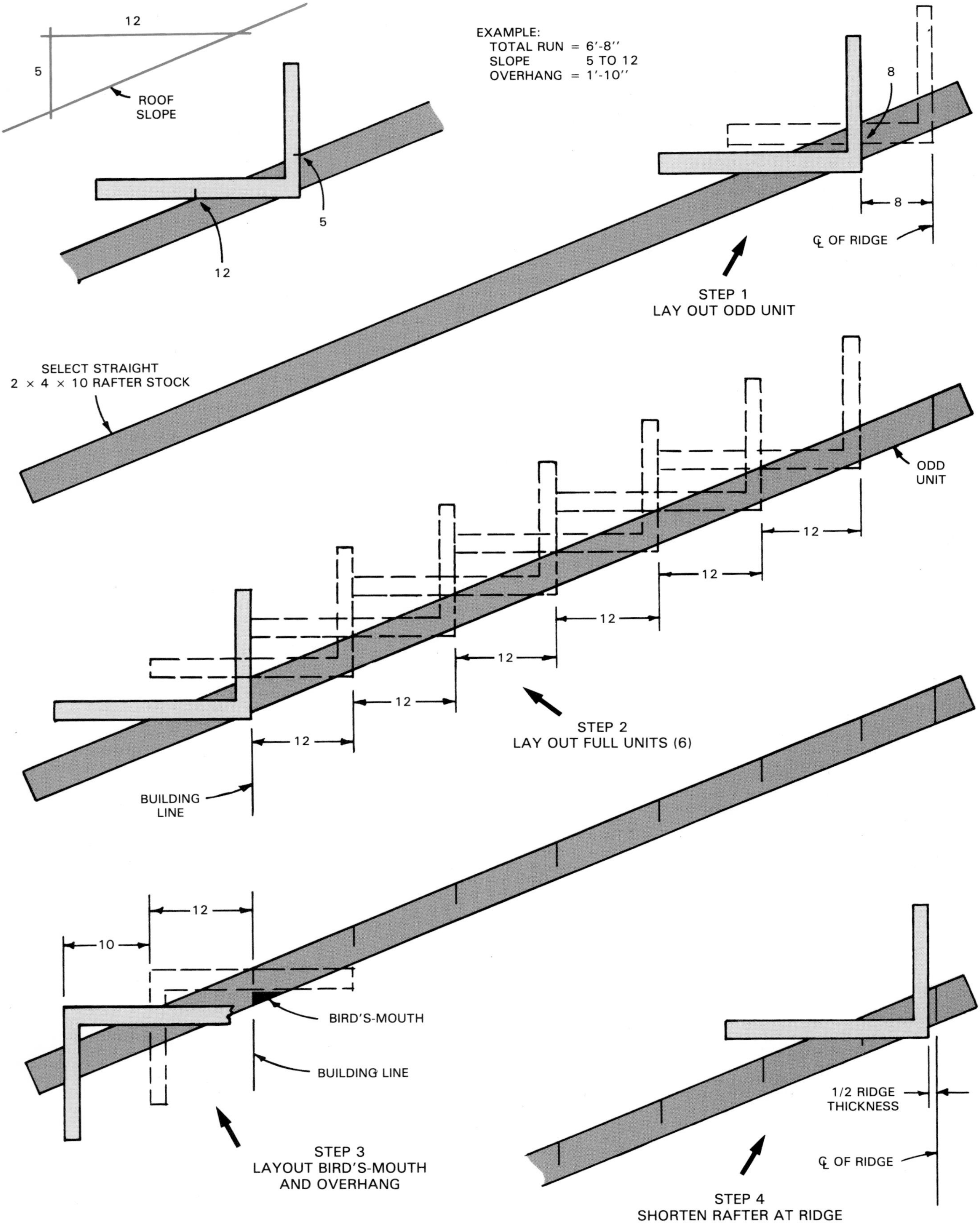
12
5
ROOF SLOPE
5
12
EXAMPLE:
TOTAL RUN = 6'-8''
SLOPE 5 TO 12
OVERHANG = 1'-10''
8
8
℄ OF RIDGE
STEP 1
LAY OUT ODD UNIT
SELECT STRAIGHT
2 × 4 × 10 RAFTER STOCK
ODD UNIT
12
12
12
12
12
12
STEP 2
LAY OUT FULL UNITS (6)
BUILDING LINE
12
10
BIRD'S-MOUTH
BUILDING LINE
STEP 3
LAYOUT BIRD'S-MOUTH
AND OVERHANG
1/2 RIDGE THICKNESS
℄ OF RIDGE
STEP 4
SHORTEN RAFTER AT RIDGE

Fig. 9-10. Use this procedure for laying out a common rafter by the step-off method.

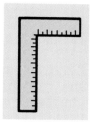

As you lay out rafters, try to visualize how each will appear when set in the completed roof frame. Forming the habit of visualizing the rafter in its proper place will help to eliminate errors.

handscrews, clamped to the blade and tongue, will also do the job.

To insure accuracy, the correct marks on the square must be positioned exactly over the edge of the stock each time a line is marked. Be sure to use a sharp pencil to make the layout lines.

Start at the top of the rafter. Hold the square in position and draw the ridge line along the edge of the tongue. Continue to hold the square in the same position and mark the length of the odd unit (8 in. used in the example). Now shift the square along the edge of the stock until the tongue is even with the 8 in. mark. Draw a line along the tongue and mark the 12 in. point on the blade for a full unit, Fig. 9-10.

Move the square to the 12 in. point just marked and repeat the marking procedure. Continue until the correct number of full units are laid out. (This number will be the same as the number of feet in the total run [six used in the example]).

Form the bird's-mouth by drawing a horizontal line (seat cut) to meet the building line so the sur-face will be about equal to the width of the plate. The size of the bird's-mouth may vary depending upon the design of the overhang. In Fig. 9-10, note that the square has been turned over to mark these cuts and also to lay out the overhang. This may or may not be necessary depending on the length of the rafter blank.

To lay out the overhang, start with the plumb cut of the bird's-mouth and mark full units first. Then add any odd unit that remains. The tail cut may be plumb, square, or a combination of plumb and level. Check the cornice details shown in the architectural plans for exact requirements.

The final step in the layout consists of shortening the rafter at the ridge. With the square in position, draw a new plumb line back from the ridge line half the thickness of the ridge, Fig. 9-10. Now make the ridge, bird's-mouth, and tail cuts you have laid out. Label the rafter as a pattern, indicating the roof section to which it belongs.

USING THE RAFTER TABLE

You can calculate the length of a common rafter using the table on the framing square. See Fig. 9-11. Under the full-scale number that corresponds with the unit rise secure the number in the first line. This is the line length of the rafter in inches for one foot of run. To find the length of the rafter from the building line to the center of the ridge, multiply the units of run by the figure from

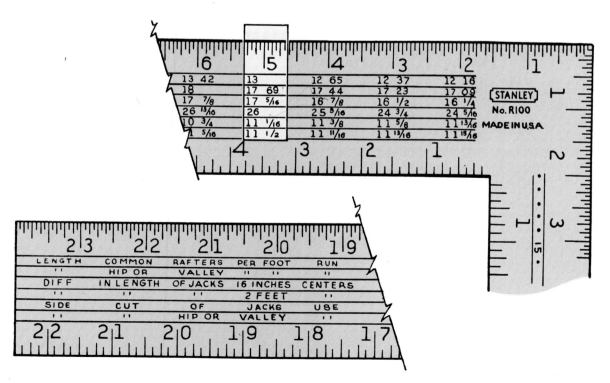

Fig. 9-11. The blade of the square carries tables for figuring length of rafters. For example, if unit rise is 5 in., you will find the rafter length for 12 in. of run is 13 in.

the table as follows:

Example No. 1 — Run = 6'-8'' Slope = 5 to 12

Run = 6'-8'' = 6 2/3 units
Table No. = 13
Rafter Length = 6 2/3 units × 13''
= 86 2/3''
= 7'-2 2/3''

Example No. 2 — Run = 10'-4'' Slope = 4 to 12

Run = 10'-4'' = 10 1/3 units
Table No. = 12.65
Rafter Length = 10 1/3 × 12.65''
= 130.72''
= 10'-10.72''
= 10'-10 3/4''

These calculations give the line length of the rafter, running from the center of the ridge to the outside of the plate. If used to make the pattern layout, the overhang will need to be added and the rafter will also need to be shortened half the ridge's thickness.

Using the rafter pattern, cut the number required. Some carpenters prefer to stand the completed rafters along the outside wall so they will be easy to reach during the assembly of the roof frame, Fig. 9-12.

ERECTING A GABLE ROOF

In conventional framing, it is considered good practice to lay out the rafter spacing along the wall plate at the same time the layout is made for the ceiling joists. When rafters are spaced 2 ft. O.C.

and ceiling joists are spaced 16 in. O.C., the layout is coordinated as shown in Fig. 9-13. The plate layout is important and the roof framing plan should be carefully followed.

Select straight pieces of ridge stock and lay out the rafter spacing by transferring the marking directly from the plate or a layout rod. Joints in the ridge should occur at the center of a rafter. Cut the pieces that will make up the ridge and lay them across the ceiling joists, close to where they will be assembled with the rafters.

In preparing to assemble a roof frame, be sure that rafters, ridge boards, and temporary bracing are readily at hand. Select straight rafters for the gable end and nail one in place at the plate. Install a rafter on the opposite side with a worker at the ridge supporting both rafters. Now place the ridge

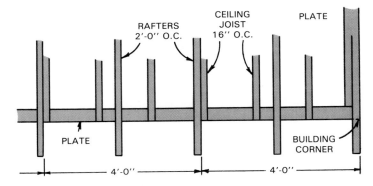

Fig. 9-13. Plan view of ceiling joist and rafter layout. A joist is nailed to every other rafter. Joists then act as a tie beam to keep walls from spreading.

Fig. 9-12. Some carpenters like to stand completed rafters against the outside wall where they are easy to reach from above. (Forest Products Lab.)

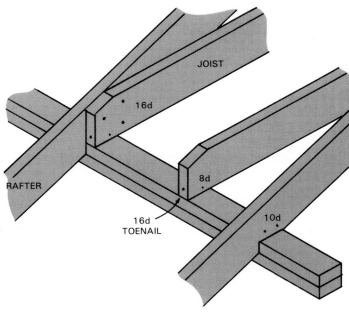

Fig. 9-14. Nailing pattern for joists and rafters at wall plate. Toenailing is done on opposite side too.

between the two rafters and nail it temporarily in place. Move about five rafter spaces from the end and install another pair of rafters. Plumb and brace the assembly and make any adjustments necessary in the nailing of the first rafters. Fig. 9-14 shows the assembly and nailing pattern at the plate. Special framing anchors, Fig. 9-15, are often used.

To make the initial assembly, some carpenters prefer to first attach the ridge to several rafters on one side. This assembly is then raised; the rafters are nailed to the plate. Then, several rafters are installed on the opposite side.

Install the rafters in between. First nail the rafter at the plate and then at the ridge, Fig. 9-16. As shown in Fig. 9-17, drive 16d nails through the ridge into the rafter. The rafters on the opposite side of the ridge are toenailed. Install only a few rafters on one side before placing matching rafters on the opposite side. This practice will make it easier to keep the ridge straight.

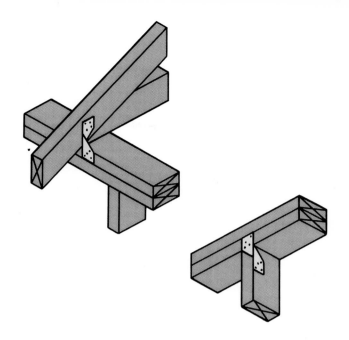

Fig. 9-15. Top. Special framing anchors strengthen rafter attachment to walls. Bottom. Same clip attaches plate to studding. (Panel Clip Co.)

Fig. 9-16. Installing common rafters. Be sure bird's-mouth is pushed tightly against the plate. (Southern Forest Products Assoc.)

Fig. 9-17. Nailing rafter to ridge board. Align edge of rafter with layout lines. (Southern Forest Products Assoc.)

Continue to add sections of ridge and assemble the rafters. Sight along the ridge to see that it is straight and level. Add bracing when required. Always install rafters with the crown (curve or warp) turned upward.

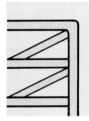

Use extra care when framing a roof to prevent a fall. Erect solid scaffolding wherever it will be helpful. Avoid working directly above another person.

GABLE END FRAME

Square a line across the end wall plate directly below the center of the gable. If a ventilator is to be installed, measure one-half of the opening size on each side of the center line, and mark for the first stud. Lay out the balance of the stud spacing.

Hold a stud upright at the first space and plumb it with a level. Mark across the edge of the stud at the underside of the rafter. Repeat the operation at the second space.

The distance between the two lengths, Fig. 9-18, will be the common difference and can be used to lay out the length of all the other studs. When the spacing is laid out from a centerline, as described, the studs can be cut in pairs.

The common difference of the stud lengths can also be secured with the framing square, Fig. 9-19. Set the square on the stud for the unit run and rise. Mark a line along the blade.

Now slide the square along this line until the number for the stud spacing (16 in. O.C., for example) aligns with the edge. Distance along the tongue of the square will be the common difference.

In modern residential construction, roof designs often include an extended rake (gable overhang). Typical framing, as illustrated in Fig. 9-20, requires the construction of the gable end frame before the roof frame is completed.

When constructing the gable ends for a brick or stone veneer building, the frame must be moved

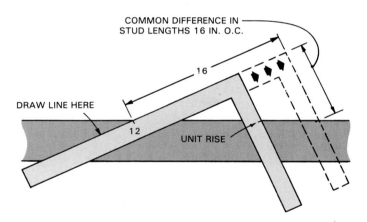

Fig. 9-19. Using framing square to find length and angle of gable studs. Left. Lining up run and rise on stud. Right. Relationship of the square method to the roof line.

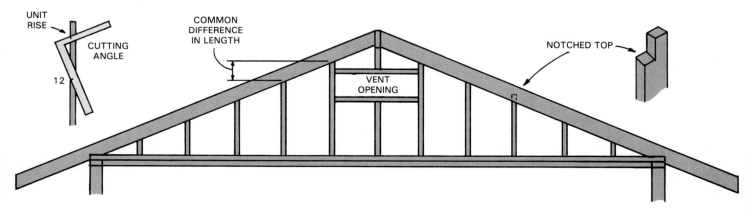

Fig. 9-18. Lay out studs for a gable end as shown.

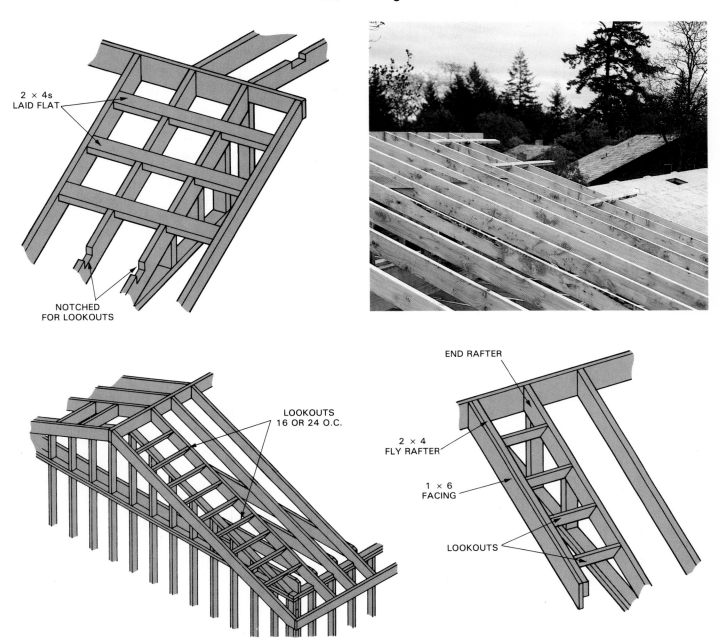

Fig. 9-20. Methods of framing gable overhang. Top. Lookouts laid flat over notched rafters. (Photo: American Plywood Assoc.) Bottom. Left: Plate atop framing studs supports lookouts. Be sure top of plate lines up with bottoms of rafters. Right: Small overhang with short lookouts supporting 2 × 4 fly rafter and face board.

outward to cover the finished wall. This projection can be formed by using lookouts and blocking attached to a ledger. (See Fig. 12-3 in the unit on exterior wall finish.)

When the top of the veneer is aligned on the sides and ends of the building, the ledger should be attached at the same level as the one used in the cornice construction. See Fig. 9-21.

Studs are mounted on this projection and attached to the roof frame in various ways. The architectural plans usually include details covering special construction features of this type.

HIP AND VALLEY RAFTERS

Hip roofs or intersecting gable roofs will have some or all of the following rafters: common, hip, valley, jack, and cripple.

First cut and frame the common rafters and ridge boards. The ridge of a hip roof is cut to the length of the building minus twice the run plus the thickness of the rafter stock. It should intersect with the common rafters as shown in Fig. 9-22.

From the corners of the building, lay out along the side walls a distance equal to half the span.

Through factory fabrication, truss rafters cut down the labor involved in constructing rafters on the site. This roof is somewhat like the roof plan shown in Fig. 9-22 with intersecting ridge lines. (American Plywood Assoc.)

These points will be the centerline of the first common rafters. All other rafters, both common and jack, are laid out from this position.

Two roof surfaces slanting upward from adjoining walls will meet on a sloping line called a hip. The rafter supporting this intersection is known as a hip rafter.

In a plan view, the hip rafter will be seen as a diagonal of a square, Fig. 9-23. Two common rafters form two sides of the square while the outside walls form the other two sides. The diagonal of this square is the total run of the hip rafter.

Since the unit run of the common rafter is 12 in., the unit run of the hip rafter will be the diagonal of a 12 in. square. When calculated accurately, this is 16.97. For actual application a rounding to 17 in. is close enough. To lay out a hip rafter, therefore, follow the same procedure used for a common rafter. Use the 17 in. mark on the blade of the square instead of the 12 in. mark, Fig. 9-24.

The odd unit must also be adjusted. Its length is found by measuring the diagonal of a square, the sides of which are equal to the length of the odd unit.

Valley rafters will also be the diagonal of a square. The sides are formed by ridges and common rafters. The layout is the same as described for hip rafters with 17 in. used for the unit run.

The length of hip and valley rafters can be determined from rafter tables just like common rafters. Using the same figures as were used in a previous example, the calculations are as follows:

Run = 6'-8''	Slope 5 to 12
Run = 6'-8''	= 6 2/3 units
Table No. 2	= 17.69
Hip or Valley Length	= 17.69 × 6 2/3
	= 117.93
	= 9'-9 15/16''
	= 9'-10''

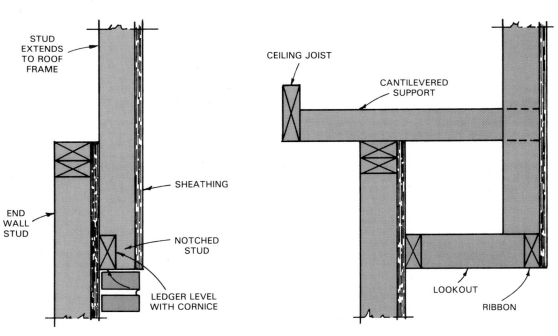

Fig. 9-21. Gable end framing for masonry veneered buildings. Left. Building out framing to cover brick or stone veneer. Right. Alternate framing extends the gable end further. Could be used over doors and other special features.

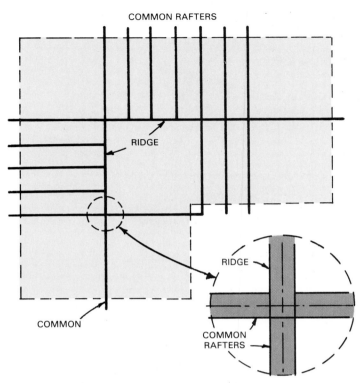

Fig. 9-22. First step in framing a hip or intersecting roof is to install the common rafters and ridges.

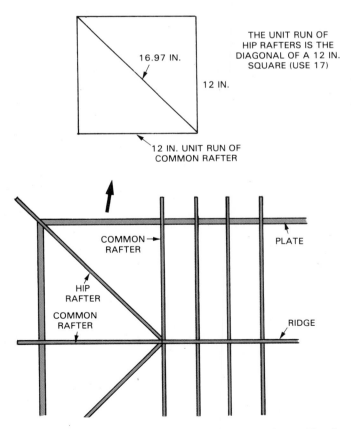

THE UNIT RUN OF HIP RAFTERS IS THE DIAGONAL OF A 12 IN. SQUARE (USE 17)

16.97 IN.

12 IN.

12 IN. UNIT RUN OF COMMON RAFTER

COMMON RAFTER

PLATE

HIP RAFTER

COMMON RAFTER

RIDGE

Fig. 9-23. Hip rafter is the diagonal of a square formed by the walls and two common rafters.

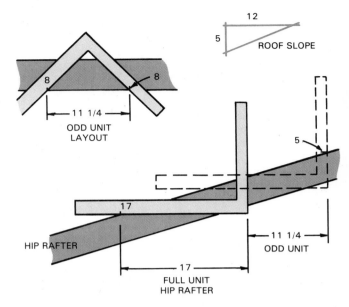

ROOF SLOPE

ODD UNIT LAYOUT

HIP RAFTER

ODD UNIT

FULL UNIT HIP RAFTER

Fig. 9-24. Starting layout of hip rafter. Slope and odd-unit size are the same as those used in common rafter layout.

Hip and valley rafters must be shortened at the ridge by a distance equal to half of the 45 deg. horizontal thickness of the ridge, Fig. 9-25. The side cuts are then laid out as shown in the illustration. Another method is to use the numbers from the sixth line of the rafter table, for example, 11 1/2. All the numbers in the table are based on or related to 12, so line up 12 on the blade and 11 1/2 on the tongue along the edge of the rafter as shown in Fig. 9-26. Draw the angle for each cut. Now draw plumb lines. Use 17 on the blade and the unit rise on the tongue. Tail cuts at the ends of rafters are laid out using the same angle.

A centerline along the top edge of a hip rafter is where the roof surfaces actually meet. The corners of the rafter will extend slightly above this line. Some adjustment must be made. The corners could be planed off. However, it is easier to make the seat cut of the bird's-mouth slightly deeper lowering the entire rafter, Fig. 9-27.

Using the framing square, align the number 17 and the number for the unit rise on the bottom edge of the rafter. Mark the position with a short line drawn along the body of the square. Measure back (toward the tail) half the thickness of the rafter and mark. Shift the square toward the tail of the rafter. Be sure to maintain the alignment of 17 in. and the unit rise. When the square aligns with the mark previously made, mark the seat of the bird's-mouth. Do not mark the plumb cut until you have read the next paragraph.

The plumb (vertical) cut of the bird's-mouth for valley rafters must be trimmed so it will fit into the corner formed by the intersecting walls. Although side cuts of approximately 45 deg. could be made,

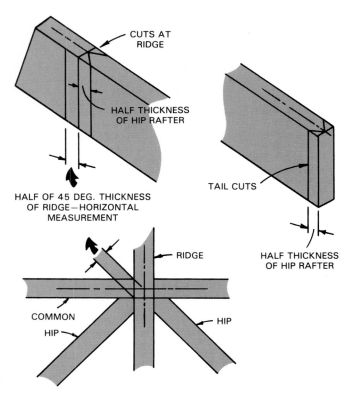

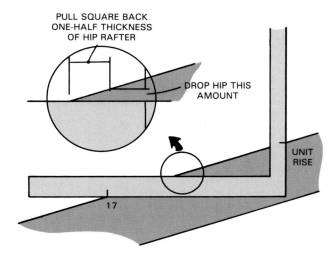

Fig. 9-27. Using square to find and mark distance to drop hip rafter. Bottom of rafter is up.

Fig. 9-25. Shortening hip rafter and making side cuts. With the exception of the tail cut, the same layout can be used for a valley rafter.

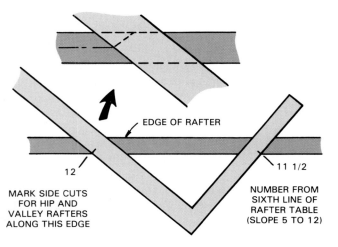

Fig. 9-26. Using the framing square to lay out side cuts on hip and valley rafters.

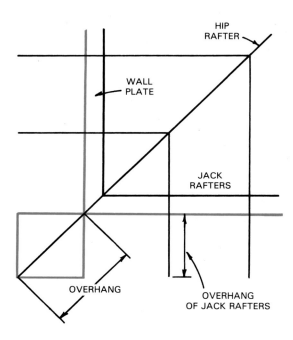

Fig. 9-28. Finding length of hip rafter overhang. You can construct the square full size on paper. Then measure the length of the diagonal. A rafter table on the square will also give this information.

it is more practical to move the plumb cut toward the tail of the rafter by a horizontal distance equal to the half 45 deg. thickness of the rafter.

The tail of the hip rafter is actually the diagonal of a square formed by extending the line of each of the walls the length of the jack rafters, Fig. 9-28. You can find its length by constructing on paper a full-size square. Then measure the diagonal. Mark the plumb cut using the run (17 in.) and the rise (in inches) on the square.

The tail must form a nailing surface for intersecting fascia boards. Refer again to Fig. 9-25 for directions on how to make the tail cut.

After hip and valley rafters are laid out and cut, they are installed on the roof. See Fig. 9-29.

JACK RAFTERS

Hip jack rafters are short rafters which run between the wall plate and a hip rafter. They run parallel to common rafters and are the same in

210

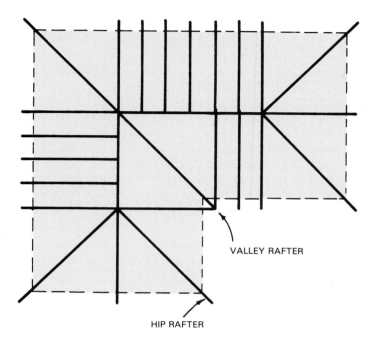

Fig. 9-29. Rafter plan showing hip and valley rafters in place.

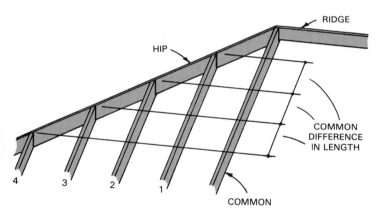

Fig. 9-30. When evenly spaced, hip jack rafters have a common and predictable difference in length from one to the next.

every respect except length from bird's-mouth to the hip rafter. When equally spaced along the plate the change in length from one to the next is always the same. This consistent change is called the common difference, Fig. 9-30. You ran into this term earlier in cutting the cripple studs for the gable end.

The common difference can be secured from the third and fourth line of the rafter table. For a roof slope of 5 to 12, with rafters speced 24 in. O.C., the figure from the table on the rafter square is 26 in. See Fig. 9-31.

The common difference can also be found with the layout method illustrated in Fig. 9-32. Hold the square along the edge of a smooth piece of lumber according to the unit run and rise of the roof. Draw a line along the blade and then slide the square along this line to a point equal to the rafter spacing. The distance thus laid out along the edge of the lumber will be the required difference in length.

To lay out jack rafters, select a piece of straight lumber. Lay out the bird's-mouth and overhang from the common rafter pattern. Next, lay out the line length of a common rafter. This is the distance from the plumb cut of the bird's-mouth to the center line of the ridge. For the first jack rafter down from the ridge, lay out the common difference in length.

Now take off half the 45 deg. thickness of the hip, Fig. 9-33. Square this line across the top of the rafter and mark the center point. Through this point, lay out the side cut as shown in the illustration. Another method is to use a square and the number from the fifth line of the rafter table, Fig. 9-34. Mark the plumb lines that will be followed when the cut is made.

For the next hip jack, move down the rafter the common difference and mark the cutting line. A sliding T-bevel is a good tool to use. Continue until jacks are laid out. Then use the T-bevel to mark the jack rafters required.

Fig. 9-31. Securing the common difference from the fourth line of the rafter table.

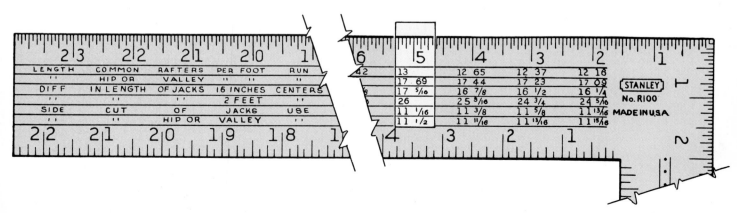

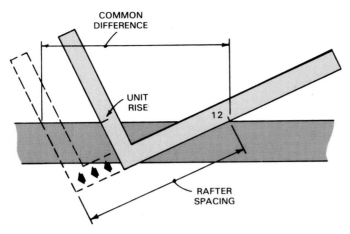

Fig. 9-32. How to use the framing square layout method of determining the common difference of jack rafters.

Each hip in the roof assembly requires one set of jack rafters made up of matching pairs. A pair consists of two rafters of the same length with the side cuts made in opposite directions.

VALLEY JACKS

Similar procedures are followed in laying out valley jack rafters. For these, however, it is usually best to start the layout at the building line, Fig. 9-35, and move toward the ridge. The longest valley jack will be the same as a common rafter except for the side (angle) cut at the bottom.

Use the common rafter pattern and extend the plumb cut of the bird's-mouth to the top edge. Lay out the side cut by marking (horizontally) half the thickness of the rafter. If you prefer, use the framing square and apply the numbers located in the fifth line of the rafter table.

The common difference for valley jack rafters is obtained by the same procedure used for hip jacks. Lay out this distance from the longest valley jack to the next. Continue along the pattern until all lengths are marked. Now use this pattern to cut all the valley rafters. They are cut in pairs in the same way as hip jacks.

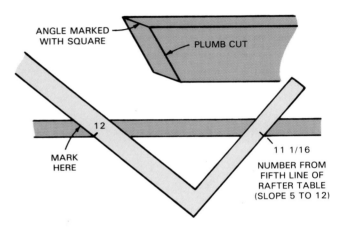

Fig. 9-34. Using a framing square to mark the angle cut on the top of the hip jack rafter.

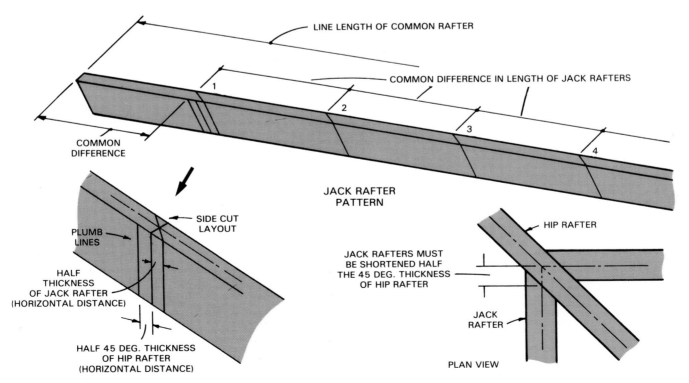

Fig. 9-33. Pattern layout for jack rafters.

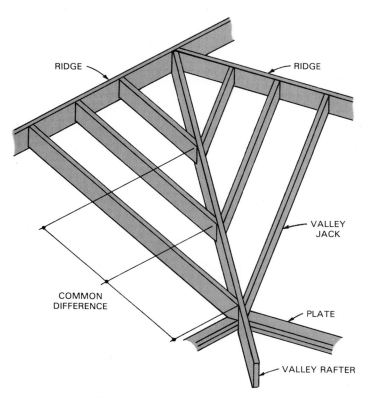

Fig. 9-35. Valley jack rafters. Finding the common difference will be the same as for the hip jacks.

Fig. 9-36. Side cuts for jack rafters can be accurately made with a radial arm saw.

When all jack rafters are laid out, they should be carefully cut. It requires a great deal of skill to make side cuts with a hand saw. Usually it is better to use a portable electric saw with an adjustable base and guide. Radial arm saws are designed to do accurate cutting and are well suited for this kind of work, Fig. 9-36.

The strength of a roof frame depends a great deal on the quality of the joints. Use special care on the side cuts of jack rafters so the joining surfaces will fit tightly together.

ERECTING JACK RAFTERS

When all jack rafters are cut, assemble them into the roof frame, Fig. 9-37. Nailing patterns will depend on the size of the various members. Use 10d nails. Space them so they will be near the heel of the side cut as they go from the jack into the hip or valley rafter.

Jack rafters should be erected in pairs to prevent the hip and valley rafters from being pushed out of line. It is good practice to first place a pair about halfway between the plate and ridge. Carefully

sight along the hip or valley to determine that it is straight and true.

Temporary bracing could also be used for this purpose. Be sure that outside walls running parallel to the ceiling joists are securely tied into the ceiling frame before hip jack rafters are installed. These rafters tend to push outward.

After rafters have been erected and securely nailed, check over the frame carefully. If some rafters are bowed sideways, they can be held straight with a strip of lumber located across the center of the span. Each rafter is sighted, moved as needed. A nail is driven through the strip to hold it in place. When the roof has been sheathed to this point, the spacer strip is removed.

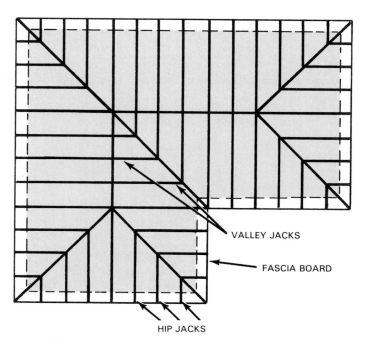

Fig. 9-37. Plan view of complete roof frame. All hip and valley jack rafters are assembled.

The last material to be attached to the roof frame before the decking is the fascia. This is the main trim member which is attached to the vertical ends of the rafters. It conceals the rafters, provides a finished appearance, and furnishes a surface to which guttering may be attached.

Fascia may be attached directly to the rafter ends. Some carpenters prefer to install 2 x 4s first to even up the ends of the rafters and provide a solid nailing surface for the 1 in. fascia board.

The upper edge of the fascia board should be cut at an angle to match the slope of the roof. Corners should be mitered and carefully fitted.

SPECIAL PROBLEMS

When framing intersecting roofs where the spans of the two sections are not equal, the ridges will not meet. To support the ridge of the narrow section, one of the valley rafters is continued to the main ridge. See Fig. 9-2 at the beginning of this unit. The length of this extended or supporting valley is found by the same method used in the layout of a hip rafter. It is shortened at the ridge just like the hip but only a single side cut is required. The other valley rafter is fastened to the supporting valley with a square, plumb cut.

A rafter framed between the two valley rafters is called a valley cripple jack. The angle of the side cut at the top is the reverse of the side cut at the lower end. The run of the valley cripple is one side of a square, Fig. 9-38. This run is equal to twice the distance from the centerline of the valley cripple jack to the intersection of the centerlines of the two valley rafters. Lay out the length of the cripple

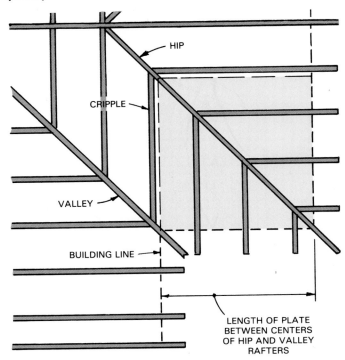

Fig. 9-39. The run of a hip-valley cripple jack is equal to the length of the wall plate.

by the same method used for a jack rafter. Shorten each end half the 45 deg. thickness of the valley rafter stock. Make the side cuts in the same way as for regular jack rafters.

Rafters running between hips and valleys are called hip-valley cripple jacks. They require side cuts on each end. Since hip and valley rafters are parallel to each other, all cripple rafters running between them, in a given roof section, will be the same length.

The run of a hip-valley cripple rafter will be equal to the side of a square, Fig. 9-39. The size of the square is determined by the length of the plate between the hip and valley rafter. Use this distance and lay out the cripple in the same manner used for a common rafter. Shorten each end by an amount equal to half the 45 deg. thickness of the hip and valley rafter stock. Now lay out and mark the side cuts. Follow the same steps used for hip and valley jacks. Side cuts, required on each end, form parallel planes.

ROOF OPENINGS

Some openings may be required in the roof for chimneys, skylights, and other structures. To frame large openings, follow about the same procedure used in floor framing.

To construct small-size openings, the entire framework is first completed. Then the opening is laid out and framed.

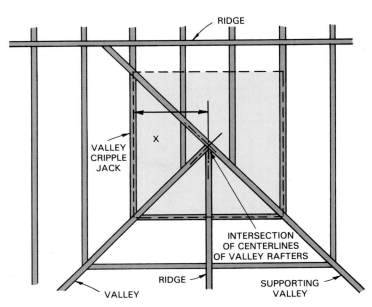

Fig. 9-38. The run of a valley cripple jack is twice the distance at X.

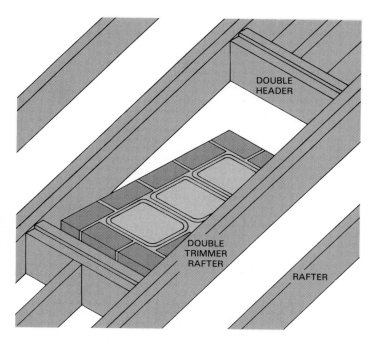

Fig. 9-40. Framing an opening in the roof for a chimney. Allow 2 in. clearance on each side and end. Note that the headers above and below the chimney are plumbed.

For a chimney opening, Fig. 9-40, use a plumb line to locate the opening on the rafters from openings already formed in the ceiling or floor frame.

Nail a temporary wooden strip across the top of the rafters to be cut. The supporting strip should be long enough to extend across two additional rafters on each side of the opening. This will support the ends of the cut rafters while the opening is being formed.

Now cut the rafters and nail in the headers. If the size of the opening is large, double the headers and add a trimmer rafter to each side.

ROOF ANCHORAGE

Rafters usually rest only on the outside walls of a structure. They lean against each other at the ridge, thus providing mutual support. This causes an outward thrust along the top plate that must be considered in the framing design.

Sidewalls are normally well secured by the ceiling joists which are also tied to some of the rafters. End walls, however, will be parallel to the joists. They need extra support, especially when located under a hip roof. Stub ceiling joists and metal straps are one method for reinforcing such walls. Framing anchors can be substituted for the metal straps especially when subflooring is included in the assembly. See Fig. 9-41.

COLLAR BEAMS

Collar beams tie together two rafters on opposite sides of a roof, Fig. 9-42. They do not support the roof but provide bracing and stiffening to hold the ridge and rafters together. In standard construction 1 x 6 boards are installed at every third or fourth pair of rafters.

PURLINS

Additional support must be provided when the rafter span exceeds the maximum allowed. A purlin, usually a 2 x 4, is attached to the underside of the rafters. This member is supported by bracing, also 2 x 4 stock, resting on a plate located over a supporting partition. Bracing under the purlin may be placed at any angle to transfer loads from the mid-point of the rafter to the support below. See Fig. 9-43.

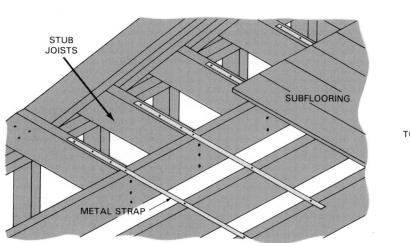

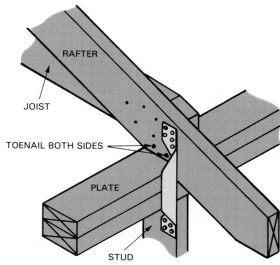

Fig. 9-41. Two methods of anchoring roof. Left. Metal straps and stub joists can be used to tie down hip roof to end walls. Right. Tie-down anchor can be used to fasten rafter, plate, and studding together.   (TECO)

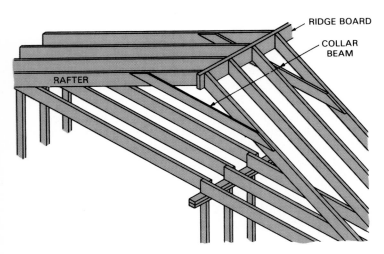

Fig. 9-42. Collar beams tie rafters and ridge together reinforcing the roof frame.

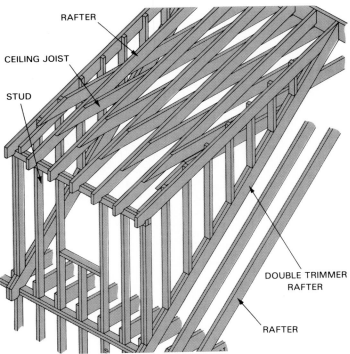

Fig. 9-44. Typical framing for shed dormer. Nailer strip is added along double trimmer to carry roof sheathing.

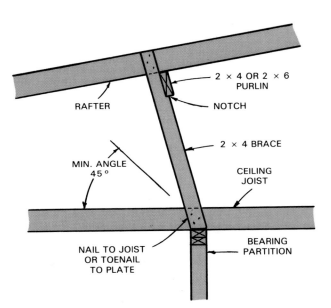

Fig. 9-43. A purlin is a plate which supports long runs of rafters at midpoint. Weight is transferred to a bearing partition through braces.

DORMERS

A dormer is a framed structure projecting above a sloping roof surface, and normally contains a vertical window unit. Although its chief purpose is to provide light, ventilation, and additional interior space, it should also enhance the exterior appearance of the structure.

The shed dormer's width is not restricted by its roof design. It is used where a large amount of additional interior space is required, Fig. 9-44.

In the simplest construction, the front wall is extended straight up from the main wall plate. Double trimmer rafters carry the side wall.

The rise of the roof is figured from the top of the dormer plate to the main roof ridge. Run will be the same as the main roof. Be sure to provide sufficient slope for the dormer roof.

Gable dormers, Fig. 9-45, are designed to provide openings for windows. They can be located at various positions between the plate and ridge of the main roof. See Fig. 9-46.

FLAT ROOFS

Flat roofs provide the long, low appearance desired in contemporary designs. Improvements in roofing surface materials and methods of application make this type of roof practical.

Methods and procedures used to frame flat roofs are about the same as those followed in constructing a floor. Most designs will require an overhang with the ends of the joists tied together by a header or band. Cantilevered rafters (extend beyond their supports) are tied to doubled roof joists. See Fig. 9-47. Corners can be formed as shown or carried on a longer diagonal joist that intersects the double joist.

Fig. 9-48 shows the similarity between a flat roof frame and a floor frame. The main members are called roof joists and support both the roof and ceiling. Because of this combined load, 2 x 10s or 2 x 12s spaced 16 in. O.C., are generally used.

Requirements will vary in different localities. Local building codes must be checked.

Fig. 9-46. View of gable dormers on a Cape Cod. Windows and exterior materials match the rest of the dwelling.

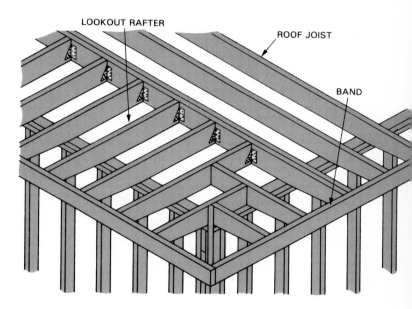

Fig. 9-47. Lookout rafters are cantilevered over outside wall to form an overhang.

Fig. 9-45. Framing a gable dormer. Top. Rafters are spaced 24 in. O.C. Joists will be added to carry ceiling. Bottom. Inside view shows 2 x 6 studs. Note that they extend through roof opening and rest on a sole plate.

GAMBREL ROOF

The gambrel roof is typical of an architectural style commonly called Dutch Colonial, Fig. 9-49. It is often used in two-story construction because it provides additional living space with a minimum of exterior wall framing. The upper roof surface usually forms about a 20 deg. angle with a hori-

zontal plane while the lower surface forms about a 70 deg. angle. See Fig. 9-50.

In residential construction, this type of roof is usually framed with a purlin located where the two roof surfaces meet. Rafters are notched to receive the purlin which is supported on partitions and/or tied to another purlin on the opposite side of the building with collar beams.

Procedures used to frame a gable roof can be applied to the gambrel roof. The rise and run of each surface is secured from the architectural plans. The two sets of rafters are laid out in the same way as previously described for a common rafter.

217

Rafters are installed over conventional wall framing. Rough openings (arrow) are for dormer alcoves instead of doors or windows. Floor joists extend beyond first floor wall plate to support rafters.

Upper roof level is framed the same as a hip roof with the lower end of the rafters extending over the plate. This extension provides support for the mansard rafters (arrow).

Framing is complete with sheathing applied to the upper roof section. Face frames for dormers were cut and assembled inside and then installed. Note how the lower mansard rafters run from the upper rafters to the 2 x 4 plate installed along the floor joists.

Sheathing being applied to lower roof. Rough fascia board (arrow) is installed along ends of floor joists. The rounded roofs of the dormers are constructed from closely spaced blocking and then sheathed with flexible composition board. Shingle bundles help hold underlayment in place on upper roof.

Framing and sheathing a mansard roof designed to provide an overhang for the first floor level.

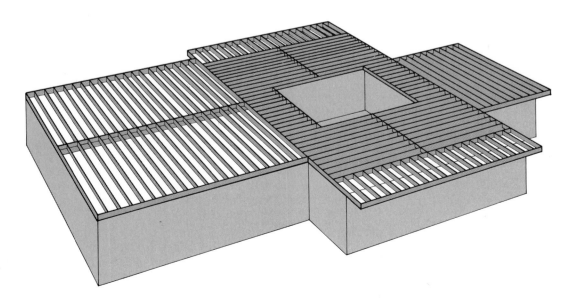

Fig. 9-48. Typical framing for a flat roof is similar to floor framing.

Roof Framing

Fig. 9-49. Gambrel roof provides extra living space with minimal exterior framing.

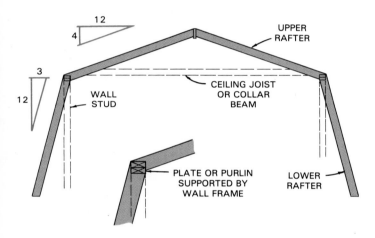

Fig. 9-50. Basic framing for a gambrel roof. Rafter patterns can be developed by making a full-size layout on the subfloor.

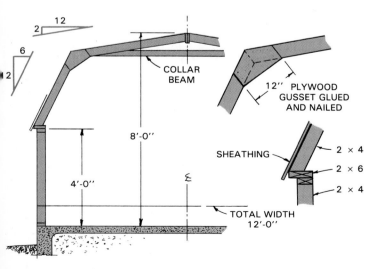

Fig. 9-51. Gambrel roof for a small building is easy to design and build.

It may help to make a full-size sectional drawing (if not included in the plans) at the intersection of the two slopes. It is then easier to visualize and proportion the end cuts of the rafters. Fig. 9-51 shows basic gambrel roof framing for a small building such as a garden house or tool shed. Note the use of gussets to join the rafter segments. Also note the simple framing used to form the roof overhang.

SPECIAL FRAMING

Fig. 9-52 shows upper floor framing for a 1 1/2-story structure. Generally there is some saving of material since walls and ceilings can be made a part of the roof frame. Knee walls are usually about 5 ft. high. Standard ceiling height is 7 1/2 ft.

Low-sloping roofs like the one shown in Fig. 9-53 usually require extra points of support. Strength derived from the triangular shapes of regular pitched roofs is greatly reduced. Thus, carefully prepared architectural plans are essential for this type of roof structure.

ROOF TRUSS CONSTRUCTION

A truss is a framework that is designed to carry a load between two or more supports. The principle used in its design is based on the rigidity of the triangle. Triangular shapes are built into the frame in such a way that the stresses of the various parts are parallel to the members making up the structure.

Roof trusses are frames that carry the roof and ceiling surfaces. They rest on the exterior walls and span the entire width of the structure. Since no load bearing partitions are required, more freedom in the planning and division of interior space is possible. They permit larger rooms without extra beams and supports. Another advantage is the opportunity to apply surface materials to outside walls, ceilings, and floors before partitions are constructed.

There are many types and shapes of roof trusses. One commonly used in residential construction is the W or Fink truss which is illustrated in Fig. 9-54.

If trusses are built on the job they should always be constructed according to designs developed from engineering data. There are several sources for such material. Usually not only detailed construction drawings are included but also specifications concerning materials and fasteners. A variety of roof truss designs are shown in the Technical Information section.

Roof trusses must be made of structurally sound lumber and assembled with carefully fitted joints.

219

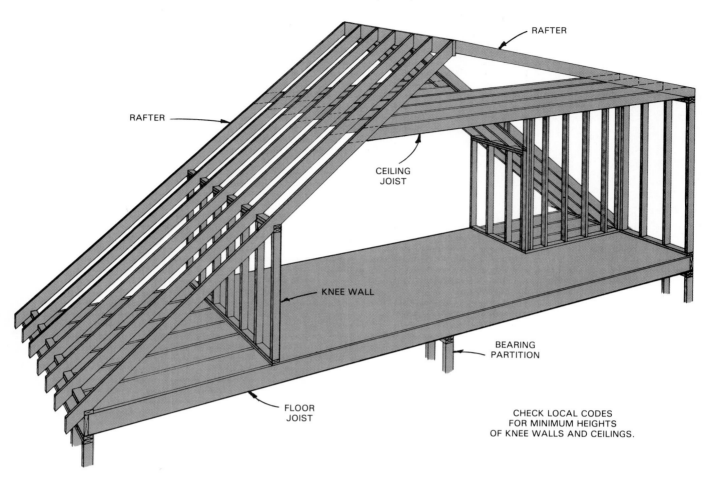

Fig. 9-52. Wall, ceiling, and roof framing for a 1 1/2 story structure. (National Forest Products Assoc.)

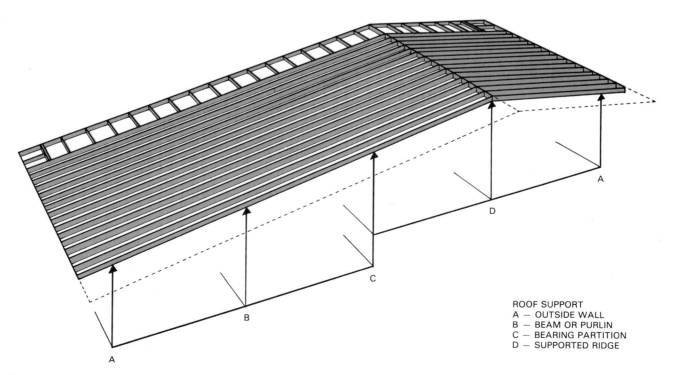

ROOF SUPPORT
A — OUTSIDE WALL
B — BEAM OR PURLIN
C — BEARING PARTITION
D — SUPPORTED RIDGE

Fig. 9-53. Modern low-slope roof for a split-level design. Arrows indicate support points provided by outside walls, bearing partitions, and purlins or beams.

Although the carpenter is seldom required to determine the sizes of truss members or the type of joints, she or he should understand their design well enough to appreciate the necessity of first-class work in their construction.

Trusses are pre-cut and assembled at ground level. Spacing of 24 in. O.C. is common. However, 16 in. O.C. and other spacing may be required in some designs.

When the truss is in position and loaded, there will be a slight sag. To compensate for this, the lower member (called the bottom chord) is raised slightly during fabrication of the unit. This adjustment is called camber and is measured at the mid-point of the span. A standard truss, 24 feet long, will usually require about 1/2 in. of camber.

In truss construction, it is essential that joint slippage be held to a minimum. Regular nailing patterns are usually not satisfactory. Special connectors must be used. Various kinds are available. All of them hold the joint securely and are easy to apply, Fig. 9-55. Plywood gussets are applied with glue and nails to both sides of the joint. Fig. 9-56 shows how.

When the number of trusses needed is small, they can be laid out and constructed on any clear floor area. First make a full-size layout on the floor, snapping chalk lines for long line lengths and using straightedges to draw shorter lines. Carefully follow the data provided.

Align the lumber with the layout to mark the size. Cut the pieces accurately. Work first with the top and bottom chords and then the web members. Cut enough material from straight lumber for a single truss and one extra of each piece to use as a pattern.

Align all members and check the fit. If ring connectors are used, bore the holes. Tack all joints together and then apply the truss plates or other types of connectors. Turn the truss over and fasten the opposite side. This truss will serve as a pattern. Check it over carefully and then tack it to the floor.

Using the pattern pieces, cut the amount of material needed to build the additional trusses required. Assemble the trusses, one at a time, by clamping the parts in position over the pattern truss as shown in Fig. 9-57. Be sure the joints are tied together. Use extra clamps on pieces that are warped. When all members are in place, apply the connectors, Fig. 9-58. Unclamp the truss, turn it over, and apply fasteners to the opposite side.

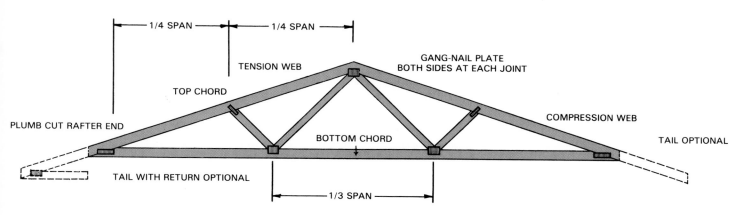

Fig. 9-54. Standard W or Fink truss is common in residential construction.

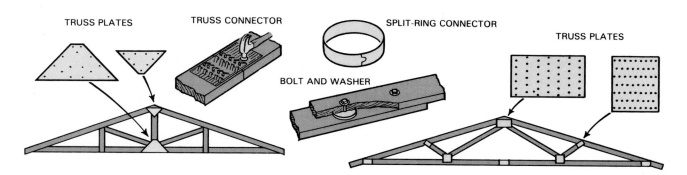

Fig. 9-55. Plates and connectors for roof trusses. Truss plates are made in many sizes, shapes, and types. Some are perforated for nails; some require no nails. Split-ring connectors fit into recesses bored in mating joints.

Fig. 9-56. Plywood gussets are applied with glue and nails to both sides of the roof truss during assembly.

Fig. 9-58. Clamps are especially useful when power nailers are used to fasten the metal plates. (Bostitch)

Fig. 9-57. Clamp truss being assembled to the pattern.

Fig. 9-59. Portable truss assembly unit adjusts to different sizes and designs.

When a large number of trusses must be built for a major building project, it is worthwhile to use a portable truss assembly unit, Fig. 9-59. This kind of equipment will insure accurate assemblies. Furthermore, it raises the assembly to a more comfortable working height.

Trusses for residential structures can normally be erected without special equipment. Each truss is simply placed upside down on the walls at the point of installation. Then the peak of the rafter is swung upward.

 *Use extra care when raising roof trusses. The first truss should be held with guy wires and all succeeding trusses carefully braced to prevent overturning.*

Fig. 9-60. Most roof trusses are prefabricated and delivered to the building site ready to install.

In modern construction, roof trusses are usually fabricated in a manufacturing plant and delivered to the job ready to install. See Fig. 9-60. These units are very satisfactory because accuracy and quality can be carefully controlled.

Roofs framed with trusses need not be limited to gable types. Today, a wide range of configurations can be produced. Designs are based on carefully prepared data covering load and lumber specifications. Modern computers apply this data to develop specific designs. Further efficiency results from the use of specialized methods, machines, and fasteners. Fig. 9-61 shows a variety of trusses used in a hip roof. Also, see Unit 20.

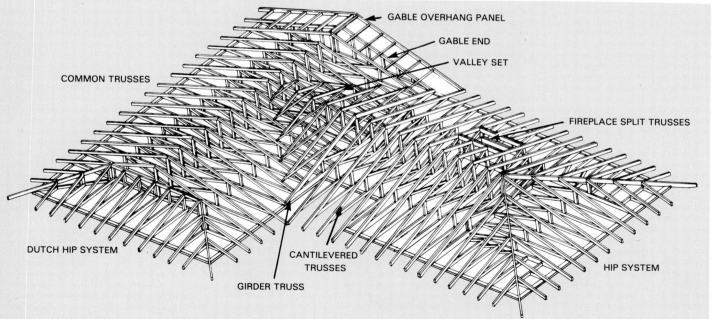

Fig. 9-61. Top. Intersecting roof uses many different styles: regular hip, Dutch hip, and gable end. Bottom. Many different prefabricated truss units were used for the frame. (TrusWal Systems Corp.)

Before installing roof trusses, always refer to the framing plans. The final erection and bracing of a roof system must be carried out according to plans and specifications if basic design requirements are to be met.

The theory of truss bracing is to apply sufficient support at right angles to the plane of the truss to hold each member in its correct position. The carpenter is responsible for truss handling and for proper temporary and permanent bracing, Fig. 9-62. Never leave the job at night until all appropriate bracing is in place. See Fig. 9-63.

When the roof frame is complete, check it carefully to see that all members are secure and that nailing patterns are adequate, Fig. 9-64.

ROOF SHEATHING

Sheathing provides a nailing base for the roof covering and adds strength and rigidity to the frame. Sheathing materials include plywood composites, oriented strand board, waferboard, particle board, shiplap, and common boards.

Other special materials are also available. For example, one product consists of panels formed with solid wood boards, bonded together with heavy kraft paper, Fig. 9-65.

Before starting the sheathing, erect the necessary scaffold that will make it easy and safe to install the boards or panels along the lower edge of the roof. This scaffold can also be used later to

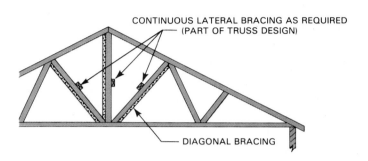

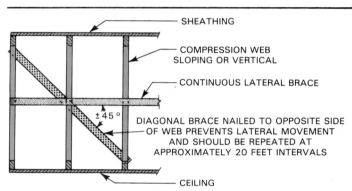

Fig. 9-63. Truss rafters may require additional lateral and diagonal bracing as shown.

Fig. 9-62. Truss erection usually requires machines or extra help to hoist them into place. Here a crane and a sling are being used. (TrusWal Systems Corp.)

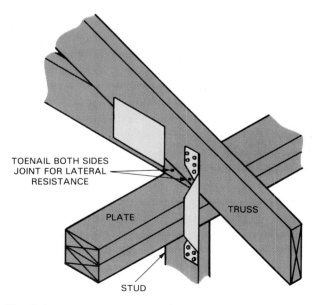

Fig. 9-64. Trusses can be attached to plates by toenailing with 10d nails. Metal tie-downs, however, are sometimes used where toenailing would damage gussets or where high winds require it. (TECO)

Fig. 9-66. Sheathing boards can be spaced when wood shingles, corrugated metal, or tile are the covering materials. (Paslode Co., Div. of Signode Corp.)

Fig. 9-65. Sheathing materials are manufactured for rapid application. Top. Panels are formed from solid boards bonded into sheets with heavy craft paper. (Western Wood Products Assoc.) Bottom. Plywood sheathing is being applied. To avoid waste at hip and valley intersections, some cutoffs can be turned over and used on adjoining slopes. (Georgia-Pacific Corp.)

build the cornice work after the roofing has been completed.

Shiplap and common boards must be applied solid if asphalt shingles or other composition materials are used for the finished roof decking. For wood shingles, metal sheets, or tile, board sheathing may be spaced according to the course arrangement, Fig. 9-66. They should be attached with two 8d nails at each rafter. Joints must be located over the center of the rafter. For greatest rigidity, use long boards, particularly at roof ends.

When end-matched boards are used, the joints may be made between rafters. Joints in the next board must not occur in the same rafter space. No boards should be used that are not long enough to be carried on at least two rafters.

Fit sheathing boards carefully at valleys and hips and nail them securely. This will insure a solid, smooth base for the installation of flashing materials. Around chimney openings, the boards should have a 1/2 in. clearance from masonry. Framing members must have a 2 in. clearance. Always nail material securely around openings.

STRUCTURAL PANELS

Structural panels are an ideal material for roof sheathing. They can be installed rapidly, hold nails well, resist swelling and shrinkage, and, because the panels are large, they add considerable rigidity to the roof frame. Plywood is laid with the face grain perpendicular to the rafters. End joints should be directly over the center of the rafter. Small pieces can be used but they should always cover at least two rafter spaces.

For wood or asphalt shingles with a rafter spacing of 16 in., 5/16 in. panels are usually recommended. For a 24 in. span, a 3/8 in. thickness should be used. Slate, tile, and mineral fiber shingles require 1/2 in. thicknesses for 16 in.

rafter spacing and 5/8 in. for 24 in. spacing. Panels should be nailed to rafters with 6d nails, spaced 6 in. apart on edges and 12 in. elsewhere. If wood shingles are used and the sheathing is less than 1/2 in. thick, 1 by 2 in. nailing strips, spaced according to shingle exposure, should be nailed to the sheathing. For a flat deck under built-up roofing, use 1/2 in. thickness.

When handling large sheets, use extra precaution. They may slide off a roof if they are not properly secured. A special rack, Fig. 9-67, will be helpful in moving sheets from the ground to the roof. It will also serve for storage until they are nailed in place. On large construction jobs a power panel elevator may be used to save time.

PANEL CLIPS

A patented clip is manufactured to strengthen roof sheathing panels between rafters. See Fig. 9-68. The clips are slipped onto the panels midway between the rafter or truss spans. Two clips should be used where supports are 48 in. O.C.

Clips are manufactured to fit five panel thicknesses: 3/8, 7/16, 1/2, 5/8, and 3/4 in. An average house requires 250 panel clips.

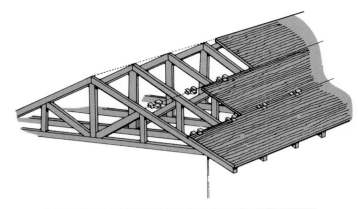

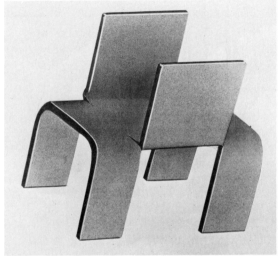

Fig. 9-68. Panel clips, sometimes called H clips, eliminate blocking on long truss or rafter spans. (Panel Clip Co.)

ESTIMATING MATERIALS

The number of rafters required for a plain gable roof is easy to figure. Simply multiply the length of the building by 3/4 for spacing 16 in. O.C. (3/5 for 20 in. and 1/2 for 24 in.) and add one more. Double this figure for the other side of the roof. To determine the length of the rafter, use the 12th scale on the framing square as previously described in this Unit. For example, estimate rafters for the following building:

Building Size 28 x 40. Roof slope 4 to 12.
Overhang 2'-0''. Rafter spacing 24 in. O.C.

Total Rafter Run   = 16'-0''
Total Rafter Length = 16'-11''
Nearest Std. Length = 18'-0''
Number of pieces   = 2 (Length of wall × 1/2 + 1)
                                = 2 (40 × 1/2 + 1)
                                = 2 × 21
                                = 42
Rafter Estimate: 42 pcs.   2 × 8 × 18'-0''

Fig. 9-67. Ladder rack will hold stack of roof sheathing where carpenter can reach it. (American Plywood Assoc.)

Use special care in handling sheet materials on a roof, especially if there is a wind. You may be thrown off balance or the sheet may be blown off the roof and strike someone.

When estimating a hip roof, it is not necessary to figure each jack rafter. The number of jack rafters required for one side of a hip is counted to obtain the number of pieces of common rafter stock. This will normally supply sufficient rafter material for the other side of the hip.

For a short method on a plain hip roof, proceed as if it were a gable roof. Add one extra common rafter for each hip. Also figure and add the hip rafters required.

With complicated roof frames, it is best to work from a complete framing plan. Apply the methods described for plain roofs to the various sections. Make colored check marks on the rafters as they are figured so you will not double up on some areas and skip others. In estimating material for the total roof frame, remember to include material for ridges, collar ties, and bracing.

To estimate the roof sheathing, first figure the total surface. Then apply the same procedures as used for subflooring and wall sheathing. Since the total area of the roof surface will also be needed to estimate shingles, building paper, or other roof surface materials, it is worth the extra time to figure the area accurately.

To figure sheathing for a plain gable roof, multiply the length of the ridge by the length of a common rafter and double the amount. Figure a plain hip roof as though it were a gable roof. However, instead of multiplying the length of a common rafter by the ridge, multiply it by the length of the building plus twice the overhang.

When working with complicated plans and intersecting roof lines, first determine the main roof areas. Multiply common rafter length times the length of the ridge times 2. Now add the triangles that make up the other sections (located over jack rafters). Remember that the area of a triangle equals half the base times the altitude. The altitude of most triangular roof areas will be the length of a common rafter located in or near the perimeter of the triangle. A plan view of the roof lines will be helpful since all horizontal lines (roof edges and ridges) will be seen true length and can be scaled.

Always add an extra percentage for waste when estimating sheathing requirements for roofs that are broken up by an unusually large number of valleys and hips.

Fig. 9-69. Models can be constructed for carpentry experience. Use a brad pusher to install small pieces. Model below was based on a modular layout. Roof trusses can be constructed in a jig.

MODEL CONSTRUCTION

Students of carpentry can often get worthwhile experiences through construction of scale model framing. Work of this kind requires much time. It is often best to construct only a part or section of a given structure. See Fig. 9-69.

A scale of 1 1/2'' = 1'-0'' will usually make it possible to apply regular framing procedures in the construction of a model that is not too large to handle and store. Cut framing members to their nominal (name) size. For example, a 2 x 4 cut to this scale would actually measure 1/4'' × 1/2'' while a 2 x 10 would measure 1/4'' × 1 1/4''.

Make all framing materials from clear white pine or sugar pine. Either has sufficient strength and is easy to work. Use small brads and fast-setting glue for assembly.

Materials other than wood can often be simulated from a wide range of items. For example, foundation work can be built of rigid foamed plastic (Styrofoam) and then brushed with a creamy mixture of Portland cement and water.

TEST YOUR KNOWLEDGE — UNIT 9

1. A type of sloping roof that simplifies the construction of an overhang for all outside walls is called a _____ roof.
2. The pitch of a roof is indicated by a fraction formed by placing the rise over the _____.
3. The tongue of a framing square is _____ in. long.
4. When laying out a rafter for a run that includes an odd unit, the _____ (full unit, odd unit) is laid out first.
5. The bird's-mouth is formed by a _____ cut and a plumb cut.
6. The final step in laying out a common rafter is to shorten it at the _____.
7. When assembling a roof frame, joints in the ridge should occur at the _____ of a rafter.
8. The part of a gable roof that extends beyond the end walls is called the _____.
9. When laying out a hip or valley rafter, use _____ in. for the unit run.
10. Figures used to make side cuts for hip and valley rafters are found in the _____ line of the rafter table on the framing square.
11. Hip jack rafters have the same tail and overhang as _____ rafters.
12. Jack rafters should be erected in _____ to keep the hip or valley rafters straight.
13. Horizontal ties between rafters on opposite sides of the ridge and usually located in the upper half of the frame are called _____.
14. The two general types of dormers are _____ and gable.
15. In residential construction, the gambrel roof is usually framed with a _____ located where the two surfaces of different slopes are joined together.
16. The adjustment in the lower chord of a roof truss to compensate for sag is called _____.
17. The sheathing on a roof frame provides a nailing base for shingles and also adds _____.
18. End joints in the sheathing boards can be made between rafters when _____ lumber is used.
19. The thickness of plywood required for a sheathing application will vary for different roofing materials and different _____.
20. To calculate the area of a plain gable roof, multiply the length of the ridge by the length of a _____ _____ and then double the product.

OUTSIDE ASSIGNMENTS

1. Secure a set of house plans where the design includes a hip roof and/or intersecting sections. It should not have a roof framing plan. Study the elevations and detail sections. Then prepare a roof framing plan.

   Overlay the floor plan with a sheet of tracing paper. Trace the well and draw all roof framing members to accurate scale. Be sure to include openings for chimneys and other items that would be helpful to the carpenter.
2. Prepare an estimate of the framing materials required for the roof in No. 1. Include the dimensions for all lumber needed. Refer to the detail drawings or specifications to find lumber size requirements. If this information is not included in the plans, secure it from the local building code.
3. Working from a set of architectural plans for a residential structure, make a layout for a common rafter in one of the roof sections. Use a good straight piece of stock. If dimension lumber is not available, a piece of 1 in. material may be used. Make the layout by the step-off method and cover all operations including the shortening at the ridge. When completed, put on a brief demonstration to the class, showing them the procedure you followed.
4. Study the various types of roof trusses. Learn their names and the basic design patterns. List the advantages and disadvantages of each. Find out where they are most commonly used. Prepare a display board with line drawings of about eight of the most practical types and label each with an appropriate caption.

Installing jack trusses in a hip roof system. These can be prefabricated on order.

Engineers subject roof truss designs to rigid testing. Pressure along top chords is carefully controlled and deflection is accurately measured.   (Forest Products Lab.)

SHAKERTOWN CORP.

REINKE SHAKES

GARLINGHOUSE

MANVILLE BUILDING MATERIALS CORP.

Roofing materials are selected for their durability, ease of application, and suitability for the climate conditions. In addition, modern roof materials must add to the beauty of the dwelling, especially where roofs are steeply pitched. Choices of material are broad, ranging from asphalt, to aluminum, pitch and gravel, wood, tile, and cement asbestos. Several of these materials are represented on this page.

Unit 10

ROOFING MATERIALS

Roofing materials protect the structure and its contents from the sun, rain, snow, wind, and dust. In addition to weather protection, a good roof should offer some measure of fire resistance and have a high durability factor. Due to the large amount of surface that is usually visible, especially in sloping roofs, the materials can contribute to the attractiveness of the building. Roofing materials can add color, texture, and pattern, Fig. 10-1.

Roof construction and finish consist of a number of operations. Most must follow a definite sequence.

All items that will project through the roof should be built or installed before roofing begins. These structures include chimneys, vent pipes, and special facilities for electrical and communications service. Performing any of this work after the finished roof is applied may damage the roof covering.

Fig. 10-1. Proper sequence of operations must be followed in installation of wood shingles so that they will shed water. Note that roofing felt, commonly called tar paper, is laid down before each course. (American Plywood Assoc.)

TYPES OF MATERIAL

Materials used for pitched (sloping) roofs include:
1. Asphalt, wood, metal, and mineral fiber shingles.
2. Slate and tile.
3. Sheet materials such as roll roofing, galvanized iron, aluminum, and copper.

For flat roofs and low-sloped roofs, a membrane system is used. It consists of a continuous watertight surface, usually obtained through built-up roofs or seamed metal sheets.

Built-up roofs are fabricated on the job. Roofing felts are laminated (stuck together) with asphalt or coal tar pitch. Then this surface is coated with crushed stone or gravel.

Metal roofs of this type are assembled from flat sheets. Seams are soldered or sealed with special compounds to insure watertightness.

When selecting roofing materials it is important to consider such factors as: initial cost, maintenance costs, durability, and appearance. The pitch of the roof limits the selection. Low-sloped roofs require a more watertight system than steep roofs, Fig. 10-2. Materials such as tile and slate require heavier roof frames.

Local building codes may prohibit the use of certain materials because of the fire hazard or because they will not resist the high winds or other elements found in a certain locality.

ROOFING TERMS

Slope and pitch have already been defined in Unit 9. Several other terms commonly used include:

● Square. Roofing materials are estimated and sold by the square. This is the amount of a given type of material needed to provide 100 sq. ft. of finished roof surface.

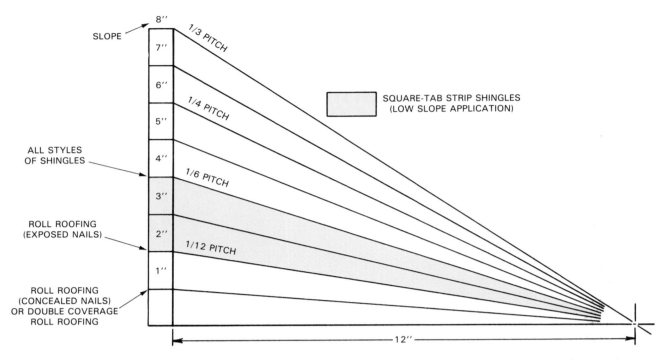

Fig. 10-2. Roofing manufacturers list minimum pitch and slope requirements for various asphalt roofing products. (Asphalt Roofing Manufacturers Assoc.)

● Coverage. This indicates the amount of weather protection provided by the overlapping of the shingles. Depending on the type of material and method of application, the shingles may furnish one (single coverage), two (double coverage), or even three (triple coverage) thicknesses of material on the roof.

● Exposure. The distance in inches between the edges of one course and the next higher course measured at right angles to the ridge. See Fig. 10-3.

● Head Lap. The distance in inches from the lower edge of an overlapping shingle or sheet, to the top edge of the shingle beneath, Fig. 10-3.

● Side Lap. The overlap length in inches for side-by-side elements of roofing. See Fig. 10-3.

● Shingle Butt. The lower, exposed edge of shingle.

PREPARING THE ROOF DECK

The roof sheathing should be smooth and securely attached to the frame. It must provide an adequate base to receive and hold the roofing nails and fasteners.

All types of shingles can be applied over solid sheathing. Spaced sheathing is sometimes used for wood shingles. When solid boards are used for sheathing, they should not be over 6 in. wide.

It is also important that the attic space be properly ventilated to minimize condensation of mois-

ture after the building is completed and ready for use. Moisture vapor from the lower stories may sometimes enter the attic. If it becomes chilled below its dew point it will condense on the under-side of the roof deck. This causes sheathing boards to warp and buckle.

To avoid this, louvered openings should be constructed either:
1. High up under the eaves in the gable ends.
2. Or at locations that insure adequate ventilation.

Louvers should provide 1/2 sq. in. of opening per square foot of attic space. Refer to Unit 13 for additional information.

Inspect the roof deck to see that nailing patterns are complete and that there are no nails sticking up. Joints should be smooth and free of sharp edges that might cut through roofing materials. Repair large knot holes over 1 in. diameter by covering with a piece of sheet metal. Clean the roof surface of chips or other scrap material.

ASPHALT ROOFING PRODUCTS

Asphalt roofing products are widely used in modern construction. They include three broad groups: saturated felts, roll roofing, and shingles.

Saturated felts are used under shingles for sheathing paper and for laminations in constructing a built-up roof. They are made of dry felt soaked with asphalt or coal tar.

Saturated felt is made in different weights, the

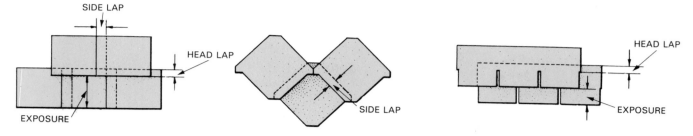

Fig. 10-3. Illustrating terms used in application of roofing materials.

most common being 15 lb. The weight indicates the amount necessary to cover 100 sq. ft. of roof surface with a single layer.

Roll roofing and shingles are outer roof covering. They must be weather-resistant. Their base material is organic felt and/or fiberglass. This base is saturated and then coated with a special asphalt that resists weathering. A surface of ceramic-coated mineral granules is then applied. The mineral granules shield the asphalt coating from the sun's rays, add color, and provide fire resistance.

Data on shingles and other groupings of asphalt products are given in Fig. 10-4. Additional products within each group differ in weight and size. For example, the three-tab square butt shingle is available in many qualities and colors and in weights from 215 to 245 lb. per square.

Asphalt shingles are the most common type of roofing material used today. They are manufactured as strip shingles, interlocking shingles, and large individual shingles. Dimensions of a standard three-tab strip shingle are shown in Fig. 10-5.

Many of the shingles are available with a strip of factory-applied, self-sealing adhesive. Heat from the sun will soften the adhesive and bond each shingle tab securely to the shingle below. This bond prevents tabs from being raised by heavy winds.

The self-sealing action usually takes place within a few days during warm weather. In winter, the sealing time is considerably longer, depending on the climate.

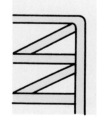

Safety considerations are very important in roofing work. Be sure to erect a secure scaffold that will support the worker at a waist-high level with the eaves. Study Unit 24 for more information and directions.

UNDERLAYMENT

An underlayment is a thin cover of asphalt-saturated felt or other material. It has a low vapor resistance. This underlayment:

1. Protects the sheathing from moisture until the shingles are laid.
2. Provides additional weather protection by preventing the entrance of wind-driven rain and snow.
3. Prevents direct contact between shingles and resinous areas in the sheathing.

Materials such as coated sheets or heavy felts, which might act as a vapor barrier, should not be used. They allow moisture and frost to gather between the covering and the roof deck. Although 15 lb. roofer's felt is commonly used for this purpose, requirements will vary depending on the kind of shingles and the roof slope.

Do not put down underlayment on a damp roof. Moisture may be trapped and damage the roof.

General application standards for underlayment suggest a 2 in. top lap at all horizontal joints and a 4 in. side lap at all end joints. See Fig. 10-6. It should be lapped at least 6 in. on each side of the centerline of hips and valleys.

DRIP EDGE

The roof edges along the eaves and rake should have a metal drip edge. Various shapes, formed from 26 ga. galvanized steel, are available. They extend back about 3 in. from the roof edge and are bent downward over the edge. This causes the water to drip free of underlying cornice construction. At the eaves the underlayment should be laid over the drip edge. At the rake, place the underlayment under the drip edge.

FLASHING AT EAVES

In cold climates it is highly recommended that an eaves flashing strip be installed over the underlayment and metal drip edge. This strip may be smooth or mineral-surfaced roll roofing. It should be wide enough to extend from roof's edge to about 12 in. inside the wall line.

This flashing will prevent leaks from water backed up by ice dams on the roof. The lower edge of this strip should be placed even with the drip edge.

Table I: Typical Asphalt Shingles

PRODUCT	Configuration	Per Square		Size		Exposure	Underwriters Laboratories Listing
		Approximate Shipping Weight	Shingles	Bundles	Width	Length	
Self-sealing random-tab strip shingle 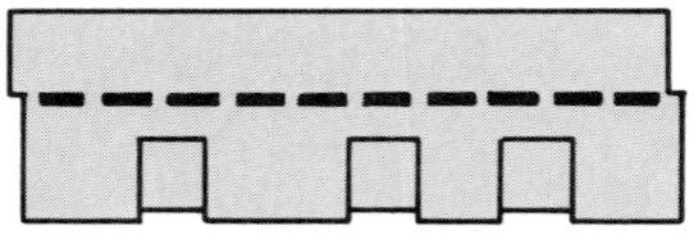 Multi-thickness	Various edge, surface texture and application treatments	285# to 390#	66 to 90	4 or 5	11½" to 14"	36" to 40"	4" to 6"	A or C - Many wind resistant
Self-sealing random-tab strip shingle 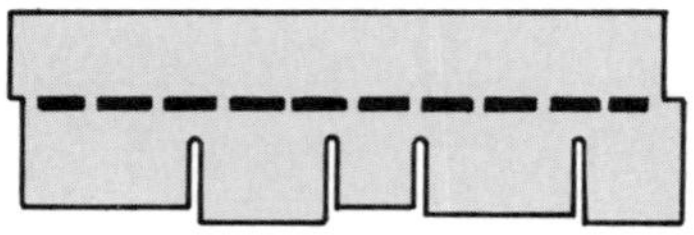 Single-thickness	Various edge, surface texture and application treatments	250# to 300#	66 to 80	3 or 4	12" to 13¼"	36" to 40"	5" to 5⅝"	A or C - Many wind resistant
Self-sealing square-tab strip shingle 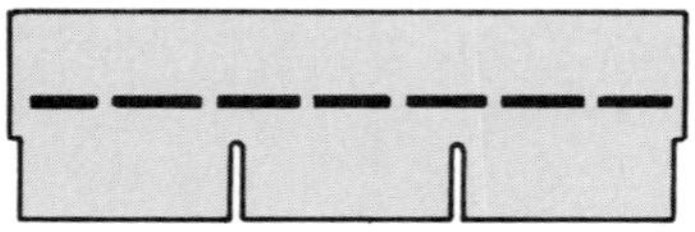 Three-tab	Two-tab or Four-tab	215# to 325#	66 to 80	3 or 4	12" to 13¼"	36" to 40"	5" to 5⅝"	A or C - All wind resistant
	Three-tab	215# to 300#	66 to 80	3 or 4	12" to 13¼"	36" to 40"	5" to 5⅝"	
Self-sealing square-tab strip shingle 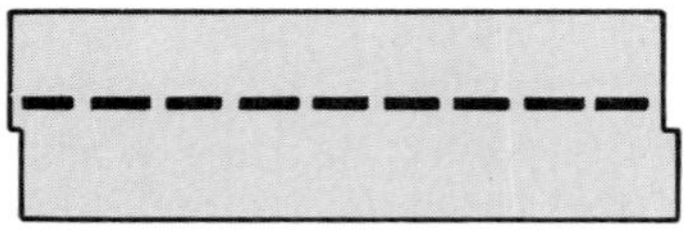 No-cutout	Various edge and surface texture treatments	215# to 290#	66 to 81	3 or 4	12" to 13¼"	36" to 40"	5" to 5⅝"	A or C - All wind resistant
Individual interlocking shingle 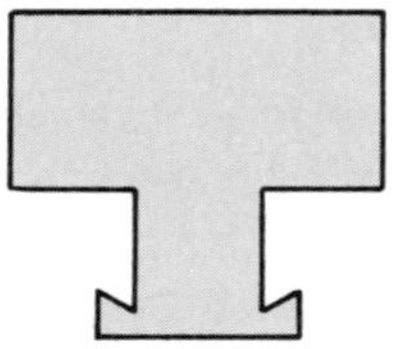 Basic design	Several design variations	180# to 250#	72 to 120	3 or 4	18" to 22¼"	20" to 22½"	—	C - Many wind resistant

Fig. 10-4. Follow this chart of specifications for installation data on common asphalt roofing products. (Asphalt Roofing Manufacturers Assoc.)

Table II: Typical Asphalt Rolls

PRODUCT	Approximate Shipping Weight		Squares Per Package	Length	Width	Side or End Lap	Top Lap	Exposure	Underwriters Laboratories Listing *
	Per Roll	Per Square							
Mineral surface roll	75# to 90#	75# to 90#	1	36' to 38'	36"	6"	2" to 4"	32" to 34"	C
	Available in some areas in 9/10 or 3/4 square rolls.								
Mineral surface roll (double coverage)	55# to 70#	110 # to 140 #	½	36'	36"	6"	19"	17"	C
Smooth surface roll	40# to 65#	40# to 65#	1	36'	36"	6"	2"	34"	None
Saturated felt (non-perforated)	60#	15# to 30#	2 to 4	72' to 144'	36"	4" to 6"	2" to 19"	17" to 34"	None

*UL rating at time of publication. Reference should be made to individual manufacturer's product at time of purchase.

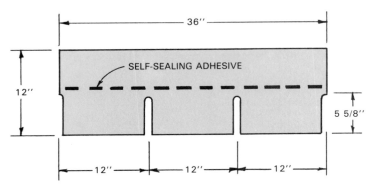

Fig. 10-5. Standard three-tab asphalt strip shingle. These are the most common dimensions.

OPEN VALLEY FLASHING

The installation of roofing materials is complicated by the intersection of other roofs, adjoining walls, and such projections as chimneys and soil stacks. Making these areas watertight requires a special construction that is called flashing. Materials used for flashing include: tin-coated metal, galvanized metal, copper, lead, aluminum, asphalt shingles, and roll roofing.

A valley is the surface where two sloping roofs meet. Water drainage is heavy at this point and leakage would create a serious problem. Flashing is one method of leakproofing valleys. For asphalt shingles, the recommended flashing material is 90 lb. mineral surfaced asphalt roll roofing, installed as illustrated in Fig. 10-7.

The first strip, at least 18 in. wide, is centered in the valley and laid with the mineral surface down. After nailing this strip, a second strip 36 in. wide is laid in place with the mineral side up.

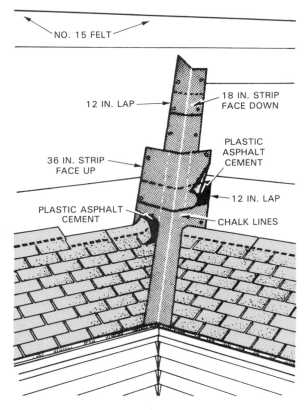

Fig. 10-7. Open valley flashing. Use two layers. Both layers should be 90 lb. roll roofing, mineral surfaced.

Joints are lapped at least 12 in. and are sealed with plastic asphalt cement. As each strip is laid, first nail one edge and then press the material firmly into the valley before the second edge is attached.

Before applying the shingles, snap a center chalk line in the valley and one on each side.

The outside lines will mark the width of the waterway. This should be 6 in. wide at the ridge, gradually widening. The lines should move away from the valley at the rate of 1/8 in. for every foot as they approach the eave. (A valley 8 ft. long would be 7 in. wide at the eave.)

When a course of shingles meets the valley, the chalk line serves as a guide in trimming the last unit. After the shingle is trimmed, cut off the upper corner at about a 45 deg. angle with the valley line. Cement the end of the shingle over the flashing.

WOVEN AND CLOSED-CUT VALLEYS

Some roofers prefer to use a woven or closed-cut valley design, especially on reroofing work. Strip shingles are the only type that can be used.

It is essential that a single unit be wide enough to straddle the valley with a minimum of 12 in. of

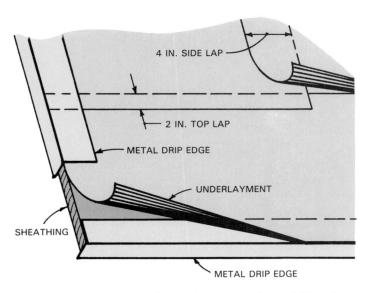

Fig. 10-6. Application of underlayment and metal drip edge. Underlayment goes under the drip edge on the rake; over it at the eaves.

material on either side. To provide this margin, it is necessary to cut some of the preceding shingle strips. Nails or other fasteners must stay at least 6 in. away from the valley centerline. See Fig. 10-8.

When reroofing, it is sometimes necessary to build up the trough of the valley to the average level of the existing roof surface. This can usually be accomplished with a beveled wood strip.

Before installing either type of valley apply a 36 in. width of 50 lb., or heavier, roll roofing across the valley. For a woven valley, lay the first course along the eave of one of the roof surfaces. Extend one strip across the valley a distance of at least 12 in. Now lay the first course along the intersecting roof and extend it across the valley over the previously applied shingle.

Succeeding courses are alternated; first along one roof area and then the other as shown in Fig. 10-8. One or both of the roof areas could be partially laid before laying the valley.

When laying shingles across the valley, be sure to press them firmly into the valley. Position nails at least 6 in. away from either side of the centerline. Use two nails at the end of each terminal strip as shown in the drawing.

To construct a closed-cut valley, install the roll roofing and then apply all the shingles on one roof surface. Carry each course across the valley and onto the adjoining roof. Follow the procedure described for the woven valley. When the roof surface is complete, apply the first course of shingles along the eaves of the intersecting roof. Where this course meets the valley, trim the shingle along

a line 2 in. back from the centerline of the valley. Trim off the upper corner of the shingle to prevent water from running back along the top edge. Embed the end of the shingle in a 3 in. wide strip of plastic asphalt cement. Succeeding courses are applied and completed as shown in Fig. 10-9.

Either the open, woven, or closed-cut valley method can be used to lay shingles between a main roof and a gable dormer. Lay the main roof area up to the point just above the lower end of the valley. Carefully flash the roof where it meets the dormer walls. Install the valley lining and extend it about 1/4 in. below the edge of the dormer roof section. Shingle units may now be applied.

Before laying the shingles on the main roof, it is good procedure to measure. If necessary, you can make slight adjustments so the courses will align with those of the dormer section.

FLASHING AT A WALL

Where the roof joins a vertical wall, it is best to install metal flashing shingles. They should be 10 in. long and 2 in. wider than the exposed face of the regular shingles. The 10 in. length is bent so that it will extend 5 in. over the roof and 5 in. up the wall as shown in Fig. 10-10.

As each course of shingles is laid, a metal flashing shingle is installed and nailed at the top edge as shown. Do not nail flashing to the wall as settling of the roof frame could damage the seal.

Wall siding is installed after the roof is completed and serves as cap flashing, Fig. 10-11.

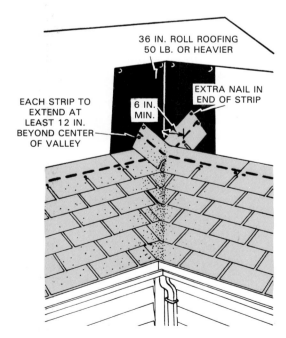

Fig. 10-8. Method of laying a woven valley. This is common during reroofing.

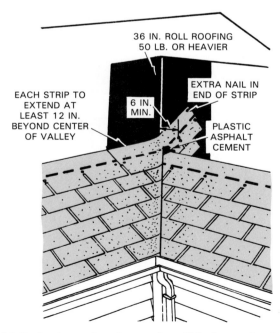

Fig. 10-9. Laying a closed-cut valley. Shingles on the right are cut along a line spaced 2 in. back from the valley centerline.

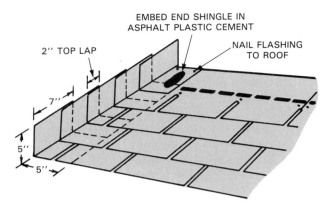

Fig. 10-10. Apply metal flashing shingles with each course. Follow these lap and bending directions.

Position the siding just above the roof surface. Allow enough clearance to paint the lower edges.

STRIP SHINGLES

On small roofs, strip shingles may be laid starting at either end. When the roof surface is over 30 ft. long, it is usually best to start at the center and work both ways. Start from a chalk line perpendicular to the eaves and ridge.

Asphalt shingles will vary slightly in length (plus or minus 1/4 in. in a 36 in. strip). There may be

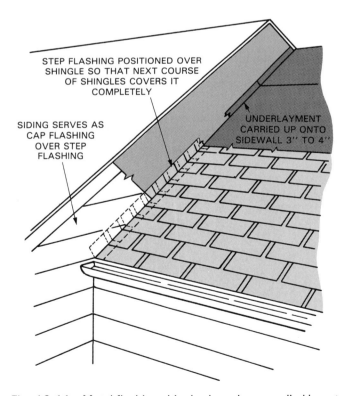

Fig. 10-11. Metal flashing shingles have been applied here to waterproof joint between sloping roof and vertical wall. Generally, this is called step flashing.

some variations in width. Thus, to achieve the proper placement so shingles will be accurately aligned horizontally and vertically, chalk lines should be used.

When laying shingles from the center of the roof toward the ends, snap a number of chalk lines between the eaves and ridge. They will serve as reference marks for starting each course. Space them according to the type of shingle and laying pattern.

These lines are used in the same way that the rake edge of the roof is used when the application is started at the roof end. The shingles do not need to be cut. Instead, full shingles are aligned with the chalk lines to form the desired pattern.

Chalk lines, parallel to the eaves and ridge, will help maintain straight horizontal lines along the butt edge of the shingle. Usually, only about every fifth or sixth course need be checked if the shingles are skillfully applied. Inexperienced workers may need to set up chalk lines for every second or third course.

When roofing materials are delivered to the building site, they should be handled with care and protected from damage. Try to avoid handling asphalt shingles in extreme heat or cold.

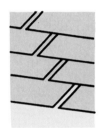

To get the best performance from any roofing material, always study the manufacturer's directions. Make the installation as directed.

FASTENING SHINGLES

Nails used to apply asphalt roofing must have a large head (3/8 in. to 7/16 in. dia.) and a sharp point. Fig. 10-12 shows standard nail designs and suggests lengths for nominal 1 in. sheathing. Most manufacturers recommend 12 ga. galvanized steel nail with barbed shanks. Aluminum nails are also

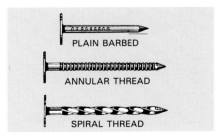

Fig. 10-12. Nails suited for installing asphalt shingles. They must be long enough to penetrate roofing materials and the decking. Usually, 1 1/4 in. lengths are enough for new roofs sheathed with 1 in. boards. Use 1 3/4 in. nails for reroofing.

used. The length should be sufficient to penetrate nearly the full thickness of the sheathing or 3/4 in. through wood boards.

The number of nails and correct placement are both vital factors in proper application of a roofing material. For three-tab square-butt shingles, use a minimum of four nails per strip as shown in the application diagrams. Align each shingle carefully and start the nailing from the end next to the one previously laid. Proceed across the shingle. This will prevent buckling. Drive nails straight so the edge of the head will not cut into the shingle. The nail head should be driven flush, not sunk into the surface. If, for some reason, the nail fails to hit

solid sheathing, drive another nail in a slightly different location.

In modern construction, pneumatic powered staplers are often used to install asphalt shingles, Fig. 10-13. Special staples with an extra wide crown should be used.

Always follow manufacturer's recommendations for staples and special power nailing equipment. In general, 16 ga. staples a minimum of 3/4 in. long should be used to attach asphalt strip shingles to new construction. See Fig. 10-14 for cross sections of well set and poorly set staples.

Staple gun pressure can be adjusted for proper staple application. If a staple must be removed from a shingle, repair the hole with asphalt plastic cement according to the manufacturer's directions. Generally speaking, staples should be placed 5/8 in. below adhesive strips, never on the adhesive itself. Follow recommendations in Fig. 10-15.

STARTER STRIP

The purpose of a starter strip is to back up the first course of shingles and fill in the space between the tabs. Use a strip of mineral surfaced roofing, 9 in. or wider, of a weight and color to match the shingles. Apply the strip so it overhangs the drip edge slightly. Secure it with nails spaced 3 to 4 in. above the edge. Space the nails so they will not be exposed at the cutouts between the tabs of the first course of shingles. Sometimes an inverted (upside down) row of shingles is used instead of the starter strip.

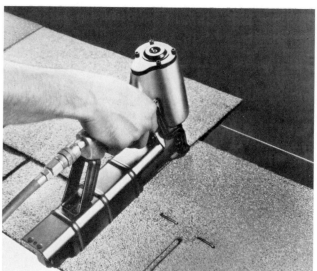

Fig. 10-13. Pneumatic stapler is being used to install three-tab shingles. Top. Proper method of handling stapler. (Paslode Co., Div. of Signode Corp.) Bottom. Proper nailing pattern using wide crown staples. (Senco Products, Inc.)

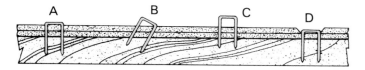

Fig. 10-14. Proper stapling of shingles. A—Crown should be parallel to and tight against shingle surface without cutting into it. B, C—Set deeper if you can see daylight under crown. D—If set too deep, staple will cut the shingle.
(Senco Products, Inc.)

THICKNESS OF WOOD DECK	MINIMUM STAPLE LEG LENGTH
3/8''	7/8''
1/2''	1''
5/8'' and thicker	1 1/4'' 1 1/2''

Fig. 10-15. Recommendations for staple length when they are applied to new construction. These specifications apply in most parts of the country. However, always check local building codes.

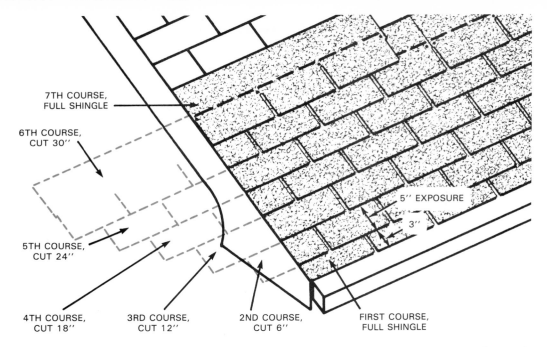

7TH COURSE,
FULL SHINGLE

6TH COURSE,
CUT 30"

5TH COURSE,
CUT 24"

4TH COURSE,
CUT 18"

3RD COURSE,
CUT 12"

2ND COURSE,
CUT 6"

FIRST COURSE,
FULL SHINGLE

5" EXPOSURE

3"

Fig. 10-16. These three-tab square-butt shingles are laid so the cutouts are centered over the tabs in the course directly below. This is called the 6 in. method. (Manville Bldg. Materials Corp.)

For self-sealing strip shingles, the starter strip is often formed by cutting off the tabs of the shingles being used. These units are then nailed in place, right side up, and provide adhesive under the tabs of the first course.

FIRST AND SUCCEEDING COURSES

The first course is started with a full shingle. Succeeding courses are then started with either full or cut strips, depending upon the type of shingle and the laying pattern.

Three-tab square-butt shingle strips are commonly laid so the cutouts are centered over the tab in the course directly below, thus the cutouts in every other course will be exactly aligned.

For this pattern, start the second course with a strip from which 6 in. has been cut. The third course is started with a strip with a full tab removed and the fourth with half a strip. Continue as shown in Fig. 10-16.

Reduce the length of shingle in the same sequence for subsequent courses. A pair of tin snips can be used to cut the shingles.

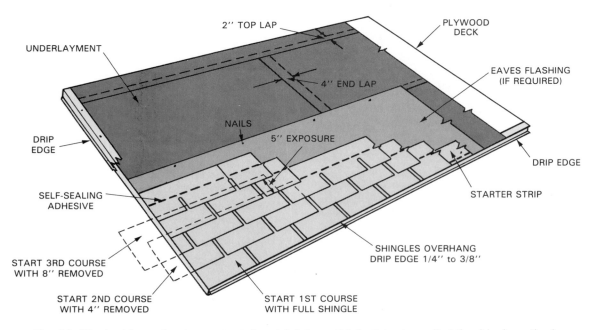

Fig. 10-17. In this application, cutouts break joints on thirds. It is also called the 4 inch method. (Bird and Son Inc.)

240

The diagram in Fig. 10-17 shows the procedure to follow for a pattern where the cutouts break joints on thirds. The second course is started with a strip shortened by 4 in., the third by 8 in. The fourth course is started with a full strip.

Using an approved nailing pattern for three-tab shingles is very important in securing the best appearance and full weather protection. Manufacturers recommend that four nails be used as shown in Fig. 10-18. When shingles are applied with an exposure of 5 in., nails should be placed 5/8 in. above tops of cutouts. Locate one nail above each cutout and one nail in 1 in. from each end. Nails should not be placed in or above the factory applied adhesive strip.

CHIMNEY FLASHING

Flashing around the chimney must allow for some movement from settling or shrinkage of the building frame. To provide for this movement, the flashing is divided into two parts:
1. The base flashing that is secured to the roof deck.
2. The cap flashing (also called counter flashing) that is secured to the chimney.

Before base flashing is applied, lay the shingles up to the front face of the chimney. Then lay out and cut, from 90 lb. mineral surfaced roofing, the front section of flashing as shown in Fig. 10-19. A similar section can be cut for the back if there is no saddle. Such a structure is not required when the chimney is small and located high on the roof near the ridge.

Cement the front base flashing into place. Then cut and apply the side pieces as shown. The base flashing at the back of the chimney is applied last. All the sections are cemented together as they are applied.

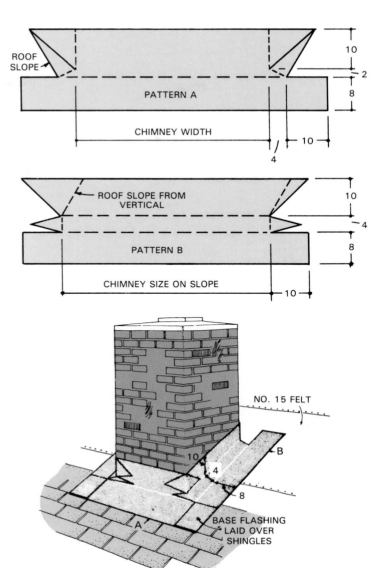

Fig. 10-19. Base flashing seals chimney opening against leaks. Top. Pattern layouts. Bottom. Flashing cut and attached with cement.

Fig. 10-18. Approved nailing pattern for three-tab square-butt shingles. (Asphalt Roofing Manufacturers Assoc.)

Sheet metal is often used for base flashing. It should be applied by the step method previously described in the section on wall flashing.

Cap flashing consists of metal sheets set into the mortar joints of the chimney when it is constructed. Bend the metal strips down over the base flashing. The metal should be set into the mortar joints to a depth of 1 1/2 in. Cap flashing on the front of the chimney can be one continuous piece while on the sides it must be stepped up in sections to align with the roof slope, Fig. 10-20.

CHIMNEY SADDLE

Large chimneys on sloping roofs generally require an auxiliary roof deck on the high side. This

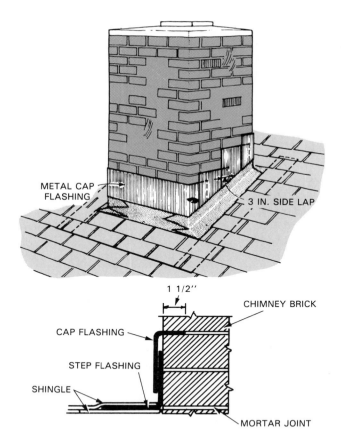

METAL CAP
FLASHING

3 IN. SIDE LAP

1 1/2"

CHIMNEY BRICK

CAP FLASHING

STEP FLASHING

SHINGLE

MORTAR JOINT

Fig. 10-20. Metal cap flashing is set into the mortar joints as the chimney is built. It must go over the top of the base flashing as shown in the cross section.

structure is called a saddle. Its purpose is to divert the flow of water and to prevent ice and snow from building up behind the chimney.

Fig. 10-21 shows a chimney saddle and suggests a framing design. The frame is nailed to the roof deck and then sheathed. A small saddle could be constructed from triangular pieces of 3/4 in. exterior plywood.

Saddles are usually covered with corrosion-resistant sheet metal. However, mineral surfaced roll roofing could be used. Valleys formed by the saddle and main roof should be carefully sealed in the manner described for regular roof valleys.

VENT STACK FLASHING

Pipes projecting through the roof must also be carefully flashed. Prefabricated flanges are available for this purpose. Asphalt products can also be used successfully for the flashing.

The roofing must be laid up to where the stack projects. Cut and fill shingles around the stack, Fig. 10-22. Then carefully cement a flange in place. Lay the roof shingles over the top as shown. The flange must be large enough to extend at least 4 in. below, 8 in. above, and 6 in. on each side.

When laying asphalt shingles, it is a good idea to wear soft-soled shoes that will not damage the surface and edge of the shingles. Asphalt products are easy to damage when worked at high temperatures. Try to avoid laying these materials on extremely hot days.

HIPS AND RIDGES

Special hip and ridge shingles are usually available from the manufacturer. The special shingles can be easily made, however. Cut pieces 9 in. by 12 in. from either square-butt shingle strips or mineral surfaced roll roofing that matches the color of the shingles.

After the shingles are cut, bend them lengthwise in the centerline. In cold weather, the shingle should be warmed before bending to prevent cracks and breaks. Begin at the bottom of the hips or at one end of the ridge. Lap the units to provide a 5 in. exposure as illustrated in Fig. 10-23. Secure with one

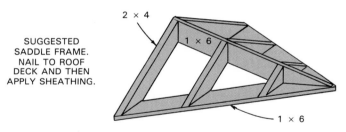

2 × 4

1 × 6

SUGGESTED SADDLE FRAME. NAIL TO ROOF DECK AND THEN APPLY SHEATHING.

1 × 6

SLOPE EQUAL TO MAIN ROOF

Fig. 10-21. Chimney saddle. Small saddles need not be framed. They can be formed from triangular pieces of 3/4 in. exterior plywood.

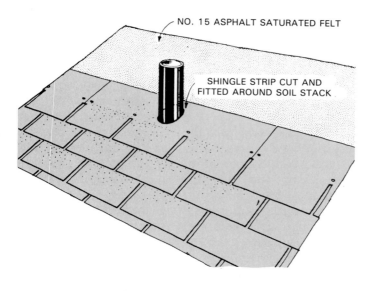

NO. 15 ASPHALT SATURATED FELT

SHINGLE STRIP CUT AND FITTED AROUND SOIL STACK

nail on each side, 5 1/2 in. back from the exposed end and 1 in. from the edge.

Metal ridge roll is not recommended for asphalt shingles. Corrosion may discolor the roof.

WIND PROTECTION

Shingles that are provided with factory applied adhesive under each tab are available for use in localities where high winds are frequent. After installation, only a few warm days are needed to thoroughly seal the tabs to the course below. This will prevent them from being blown up by strong winds. This precaution is especially important on low-sloping roofs where it is easier for the wind to get under the shingles.

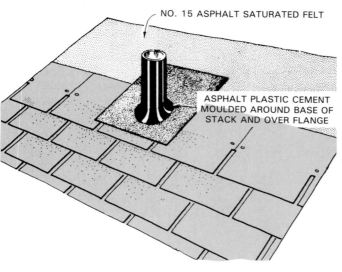

NO. 15 ASPHALT SATURATED FELT

ASPHALT PLASTIC CEMENT MOULDED AROUND BASE OF STACK AND OVER FLANGE

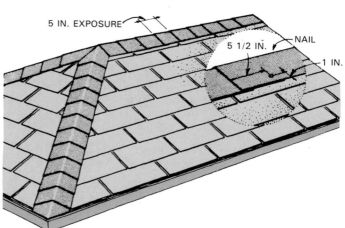

5 IN. EXPOSURE

5 1/2 IN.

NAIL

1 IN.

Fig. 10-23. Nail hip and ridge shingles 5 1/2 in. back from edge. Use one nail on each side.

Self-sealing shingles are satisfactory for roofs with slopes up to about 60 degrees. For very steep slopes, like those used in mansard roofs, Fig. 10-24, special application procedures must be followed. You may have to seal them in place with quick-setting asphalt cement. Follow recommendations provided by the roofing manufacturer.

If regular shingles are used, the tabs can be cemented. Apply a spot of special tab cement about 1 in. square with a putty knife or caulking gun and then press the tab down. Avoid lifting the tab any more than necessary while applying the cement.

A variety of interlocking shingles are designed to provide resistance against strong winds. They are used for both new construction and reroofing. Details of the interlocking devices and methods of application vary considerably. Always study and follow the manufacturer's directions when installing all types of shingles.

SHINGLE COURSES LAID OVER UPPER PORTION OF FLANGE

Fig. 10-22. Steps for vent stack flashing. Top. Lay shingles up to stack and fit last course around the stack. Middle. Install flange. Bottom. Apply shingles over the upper side of flange.

Fig. 10-24. Mansard roofs may require special application procedures if asphalt shingles are applied in windy areas. Wood shingles are often used in these conditions. Check local codes. (Shakertown Corp.)

INDIVIDUAL ASPHALT SHINGLES

Roof surfaces may be laid with an individual asphalt shingle. There are several sizes and designs available. One commonly used is 12 in. wide and 16 in. long. Several patterns can be used in its application. See Fig. 10-25. Follow the same procedure that was described for strip shingles. Horizontal and vertical chalk lines should be used to insure accurate alignment.

LOW-SLOPE ROOFS

When applying asphalt shingles to slopes less than 4 in 12, certain additional procedures should be followed. Slopes as low as 2 in 12 can be made watertight and windtight if the installation includes:

1. A double-thickness felt underlayment. Lap each course over the preceding one 19 in., starting with a 19 in. strip.
2. In areas where the January daily average temperature is 25 °F or colder, cement the two felt layers together from the eaves up the roof to 24 in. inside the interior wall line of the building. See Fig. 10-26.
3. Shingles provided with factory applied adhesive and manufactured to conform to the Underwriters Laboratories Standard for Class "C" Wind Resistant shingles. "Free" tab

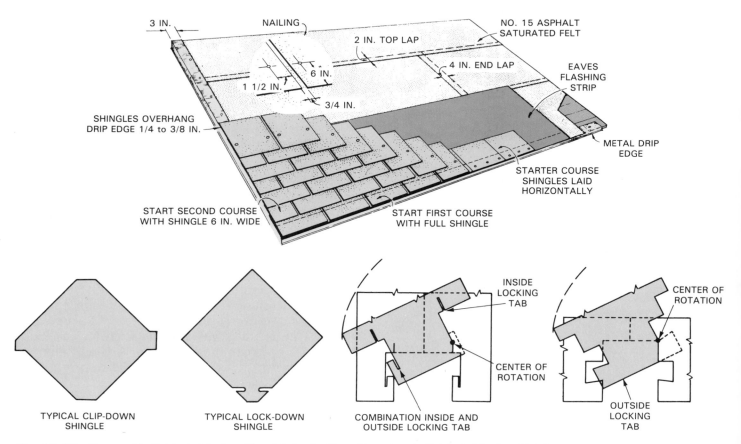

Fig. 10-25. Types of individual shingles. Top. Method of installing giant individual shingles. This is called the American method. Bottom left. Two types of hex shingles are intended primarily for application over old roofing. Slope must be 4 in. per ft. or greater. Bottom right. Interlocking devices on individual shingles provide increased wind resistance. (Asphalt Roofing Manufacturers Assoc.)

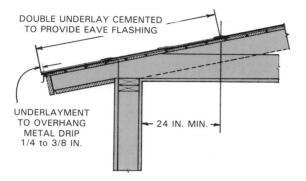

Fig. 10-26. Plies of underlayment are cemented together to form watertight eave flashing for low-slope roofs.

square-butt strips can be used if you cement all the tabs.
4. Follow special application methods shown in Fig. 10-27.

ROLL ROOFING

Asphalt roll roofing is manufactured in a variety of weights, surfaces, and colors. It is used as a main roof covering or as a flashing material. For best results install it at temperatures of 45 °F or above.

In residential construction, a double coverage roll roofing provides good protection. The roofing can be used on slopes as low as 1 in. per foot. The 36 in. width consists of a granular surfaced area 17 in. wide and a smooth surface, called a selvage, that is 19 in. wide.

Although double coverage roll roofing can be applied parallel to the rake, it is usually applied parallel to the eaves, as shown in Fig. 10-28. You can make the starter strip by cutting off the granular surfaced portion. Use two rows of nails to install the starter strip. One should be spaced 4 3/4

in. below the upper edge and the other 1 in. above the lower edge.

Cover the entire starter strip with asphalt cement and overlay a full-width sheet. Attach the sheet with a row of nails spaced 4 3/4 in. from the upper edge and a second row spaced 8 1/2 in. below the first row. The nail interval should be about 12 inches.

Position each succeeding course so that it overlaps the full 19 in. selvage area. Nail the sheet in place and then carefully turn the sheet back to apply the cement. Spread the cement to within about 1/4 in. of the granular surface. Press the overlaying sheet firmly into the cement using a stiff broom or roller. Avoid excessive use of cement. Be sure to check the manufacturer's recommendations.

REROOFING

When reroofing, a choice must be made between removing the old roofing or leaving it in place. It is usually not necessary to remove old wood shingles, old asphalt shingles, or old roll roofing before putting on a new asphalt roof provided that:
1. The strength of the existing deck and framing is adequate to support the weight of workers and additional new roofing, as well as snow and wind loads.
2. The existing deck is sound and will provide good anchorage for the nails used in applying new roofing.

When putting on new roofing over old wood shingles, all loose or protruding nails should be removed and the shingles renailed in new locations. Renail loose, warped, and split shingles. Replace missing shingles. Cut back shingles at eaves and rakes far enough to allow the application of 4 in. to 6 in. nominal 1 in. thick strips.

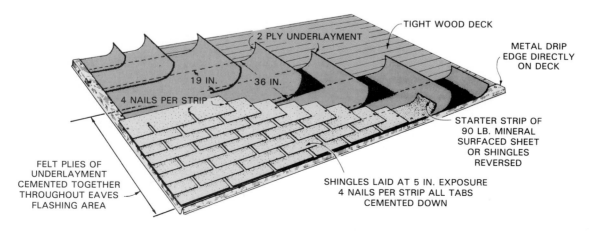

Fig. 10-27. Use special application methods for shingling low-slope roofs.

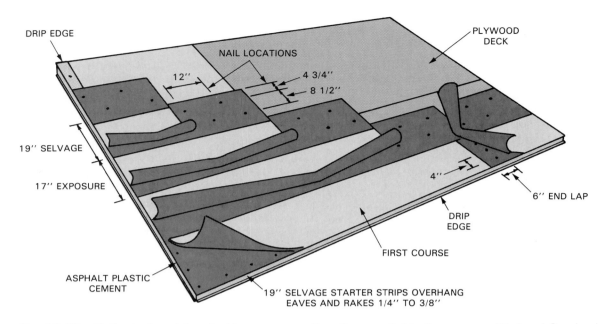

Fig. 10-28. Method of applying double-coverage roll roofing parallel to the eaves. (Bird and Son Inc.)

These strips should be nailed in place with their outside edges projecting beyond the roof deck the same distance as the old wood shingles.

To provide a smooth surface for asphalt roofing, it is often advisable that a "backer board" be applied over the wood shingles. As an alternative, beveled wood "feathering strips" can be used along the butts of each course of old shingles.

When the old roof consists of square-butt asphalt shingles with a 5 in. exposure, new self-sealing strip shingles can be applied as shown in Fig. 10-29. This application pattern will insure a smooth, even appearance.

The joint between a vertical wall and roof surface should be sealed in a reroofing application. First apply a strip of smooth roll roofing about 8 in. wide. Nail each edge firmly, spacing nails about 4 in. O.C. As the shingles are applied, asphaltic plastic cement is spread on the strip and the shingles are thoroughly bedded. To insure a tight joint, use a caulking gun to apply a final bead of cement between the edges of the shingles and the siding.

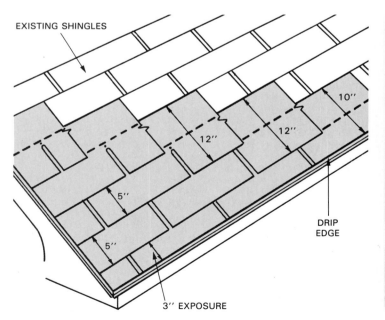

Fig. 10-29. Applying strip shingles over old asphalt shingle roof. Left. Note measurements for 5 in. exposure. (Asphalt Roofing Manufacturers Assoc.) Right. Stopper on pneumatic stapler makes spacing easier. (Paslode Corp.)

Fig. 10-30. Use of a flat-bladed shovel is common for removal of old asphalt shingles. (Asphalt Roofing Manufacturers Assoc.)

Fig. 10-31. Flat and low pitched roofs must be covered with a watertight membrane.

When old shingles are to be removed before applying a new roof, it is common to use a flat-bladed shovel as shown in Fig. 10-30. Both asphalt and wood shingles can be removed in this manner. A shovel can also be used to remove the old underlayment.

BUILT-UP ROOFING

A flat roof or a roof with very little slope, Fig. 10-31, must be covered with a watertight membrane. Most flat roofs are covered with built-up roofing which, when properly installed, is very durable. Companies that manufacture the components (mainly saturated felt and asphalt) provide detailed specifications for making the installation.

On a wood deck, a heavy layer of saturated felt is first nailed down with galvanized nails, Fig. 10-32. Nails must have a large head or be driven through tin caps. Each succeeding layer is then mopped in place with hot asphalt. When the felts are all in place, they are coated with hot asphalt and covered with slag, gravel, crushed stone, or marble chips. These materials provide a weathering surface and also improve the appearance. The mineral covering is usually applied at a rate of 300 to 400 lb. per 100 sq. ft.

Built-up roofs for residential structures normally have three or four plies (layers of asphalt-saturated felts). The asphalts used between each layer and to bed the surface coating are products of the petroleum industry. They are graded on the basis of their melting point. Actually, they begin to flow, very slowly, at a lower temperature. This results in a self-healing property that is essential for flat roofs where water is likely to stand. A special low-

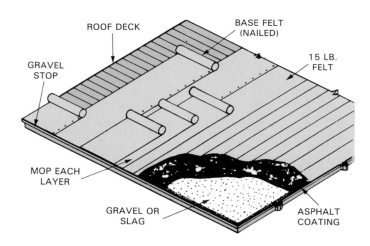

Fig. 10-32. The steps shown here are generally followed in constructing a built-up roof. Proper materials are indicated layer by layer.

temperature asphalt known as "dead-flat" asphalt is used for this type of roof, Fig. 10-33.

For sloping roofs, an asphalt with a high melting point is used. It is generally classified as "steep" asphalt. In hot climates, only high-temperature asphalts can be used.

A gravel stop, usually fabricated from galvanized sheet metal, is attached to the roof deck to serve as a trim member. It also keeps the mineral surface and asphalt in place, Fig. 10-34. Gravel stops are installed after the base felt has been laid. Joints

Fig. 10-33. Mopping felts into place is done with a mop and asphalt heated to 450 °F. Gravel stop was installed after the base felt was laid.

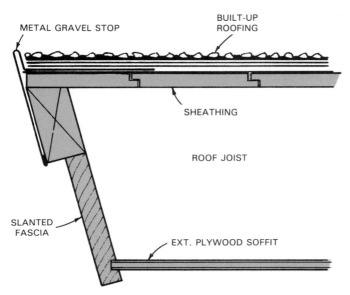

Fig. 10-34. Section through edge of flat roof overhang shows metal gravel stop installation.

between sections of gravel stop are bedded in a special mastic that permits expansion and contraction in the metal.

Flashings around chimneys, vents, or where the roof joins a wall must be constructed with special care. Leaks are most likely to occur at these locations. The best flashing materials include lead jackets, sheet copper, asbestos fabrics, and special flashing cement. Basic construction where a roof section joins a wall is shown in Fig. 10-35. The cant strip provides support for the felt layers as they curve from a horizontal to vertical position.

Bare spots on a built-up roof should be repaired. First clean the area. Apply a heavy coating of hot asphalt and then spread more gravel or slag.

Felts which have fallen apart should be cut away and replaced with new felt. The new felt should be mopped in place, allowing at least one

additional layer of felt to extend not less than 6 in. beyond the other layers.

WOOD SHINGLES

Wood shingles have been used for many years in residential construction. A disadvantage is that wood shingles, unless treated, have little resistance to fire. Building codes often prohibit their use. Since wood weathers to a soft, mellow color after exposure, wood shingles provide an appearance that is desired by many homeowners. When properly installed, they also provide a very durable roof.

Wood shingles are made from western red cedar, redwood, and cypress. All are highly decay resistant. They are taper sawed and graded No. 1, No. 2, and No. 3, plus a utility grade. The best grade is cut in such a way that the annular rings are perpendicular to the surface. Butt ends vary in thickness from 1/2 to about 3/4 in. as shown in Fig. 10-36. Wood shingles are manufactured in random widths and in lengths of 16, 18, and 24 inches. They are packaged in bundles. Four bundles contain enough shingles to cover one square (100 sq. ft.) when a standard application is made.

The exposure of wood shingles depends on the slope of the roof. When the slope is 5 in 12 or greater, standard exposures of 5, 5 1/2, and 7 1/2 inches are used for 16, 18, and 24 inch sizes respectively. On roofs with lower slopes, the exposure should be reduced to 3 3/4, 4 1/4, and 5 3/4 in. This will provide a minimum of four

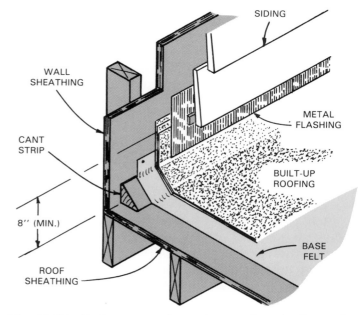

Fig. 10-35. Basic construction at intersection of a flat roof and wall. Cant strip provides gently curving base for layers of roofing felt.

GRADE	Length	Thickness (at Butt)	No. of Courses Per Bundle	Bdls/Cartons Per Square		Description
No. 1 BLUE LABEL	16" (Fivex) 18" (Perfections) 24" (Royals)	.40" .45" .50"	20/20 18/18 13/14	4 bdls. 4 bdls. 4 bdls.		The premium grade of shingles for roofs and sidewalls. These top-grade shingles are 100% heartwood. 100% clear and 100% edge-grain.
No. 2 RED LABEL	16" (Fivex) 18" (Perfections) 24" (Royals)	.40" .45" .50"	20/20 18/18 13/14	4 bdls. 4 bdls. 4 bdls.		A good grade for many applications. Not less than 10" clear on 16" shingles, 11" clear on 18" shingles and 16" clear on 24" shingles. Flat grain and limited sapwood are permitted in this grade.
No. 3 BLACK LABEL	16" (Fivex) 18" (Perfections) 24" (Royals)	.40" .45" .50"	20/20 18/18 13/14	4 bdls. 4 bdls. 4 bdls.		A utility grade for economy applications and secondary buildings. Not less than 6" clear on 16" and 18" shingles, 10" clear on 24" shingles.
No. 4 UNDER-COURSING	16" (Fivex) 18" (Perfections)	.40" .45"	14/14 or 20/20 14/14 or 18/18	2 bdls. 2 bdls. 2 bdls. 2 bdls.		A utility grade for undercoursing on double-coursed sidewall applications or for interior accent walls.
No. 1 or No. 2 REBUTTED-REJOINTED	16" (Fivex) 18" (Perfections) 24" (Royals)	.40" .45" .50"	33/33 28/28 13/14	1 carton 1 carton 4 bdls.		Same specifications as above for No. 1 and No. 2 grades but machine trimmed for parallel edges with butts sawn at right angles. For sidewall application where tightly fitting joints are desired. Also available with smooth sanded face.

Fig. 10-36. Wood shingles are made in several grades and to certain specifications.
(Red Cedar Shingle and Handsplit Shake Bureau)

layers of shingles over the entire roof area. In any type of construction there should be a minimum of three layers at any given point to insure complete protection against heavy wind-driven rain.

SHEATHING

Solid sheathing for wood shingles may consist of matched or unmatched 1 in. boards, shiplap, or plywood. Open or spaced sheathing, Fig. 10-37, is sometimes used because it costs less and permits shingles to dry out quickly. One reason for using solid sheathing is to gain the added insulation and resistance to infiltration that such a deck offers.

One method of applying spaced roof sheathing is to space 1 by 3 in., 1 by 4 in., or 1 by 6 in. boards the same distance apart as the anticipated shingle exposure. Each course of shingles is nailed to a separate board.

Fig. 10-37. Either open or spaced sheathing can be used for wood shingles. Sheathing has been laid solid over an open cornice.
(Western Wood Products)

Another method uses 1 by 6 in. lumber as sheathing boards with two courses of shingles nailed to each one.

UNDERLAYMENT

Normally an underlayment is not used for wood shingles when they are applied on either spaced or solid sheathing. If it is desirable to use roofing paper to prevent air infiltration, the roof may be covered with rosin-sized building paper or "dry" unsaturated felts. Saturated building paper is usually not recommended because of the condensation trouble it may cause.

UNDERLAYMENT (FIRE-RESISTANT)

Recent tests by qualified agencies have shown that flame-spread and burn-through rates for wood shingles and shakes can be reduced. This is achieved by using an underlayment of asbestos felt reinforced with glass fibers (weight 12 lb. per square). For wood shakes this special felt is also used between the courses of shingles.

Flame penetration time can also be increased by using 1/2 in. Type X gypsum board under solid or spaced sheathing. For more information see Uniform Building Code Standard No. 32-14.

FLASHING

In areas where outside temperatures drop to 0°F or colder and there is a possibility of ice forming along the eaves, an eaves flashing strip is recommended. Follow the same procedure for making the installation as described for asphalt shingles.

It is important to use good materials for valleys and flashings. Materials used for this purpose include tin plate, lead-clad iron, galvanized iron, lead, copper, and aluminum sheets.

If galvanized iron (mild steel coated with a layer of zinc) is selected, 24 or 26 ga. metal should be used. Tin, or galvanized sheets with less than 2 oz. of zinc per sq. ft., should be painted on both sides with white lead and oil paint and allowed to dry before being used.

When making bends, care should be taken not to crack the zinc coating. On roofs of 1/2 pitch or steeper, the valley sheets should extend up on both sides of the center of the valley for a distance of at least 7 in. On roofs of less pitch, wider valley sheets should be used. Minimum extension should be at least 10 in. on both sides, Fig. 10-38. The open portion of the valley is usually about 4 in. wide and should gradually increase in width toward the lower end.

Tight flashing around chimneys is also essential to a good roofing job. One method of installing

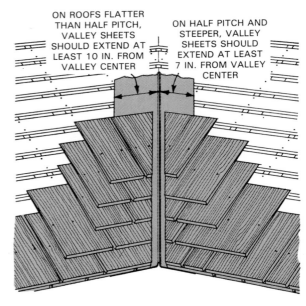

Fig. 10-38. Valley flashing for wood shingles is similar to flashing for asphalt shingles.
(Red Cedar Shingle and Handsplit Shake Bureau)

base and cap or counter flashing around a chimney is shown in Fig. 10-39.

In new construction, the cap flashing is laid in the joints when the chimney is built. If the chimney is laid up without flashing, the mortar joints must be chiseled out and the flashing forced in. It may be held in place by nails driven into the mortar. Finally, the joints must be filled or pointed with good mortar.

NAILS

Only rust resistant nails should be used with wood shingles. Hot-dipped, zinc-coated nails, which have the strength of steel and the corrosion resistance of zinc, are recommended. Check Fig. 10-40 on sizes of nails for various jobs.

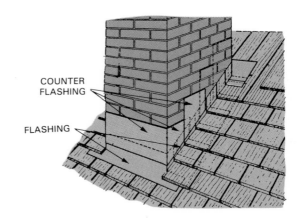

Fig. 10-39. Flashing should be tucked into mortar joints.

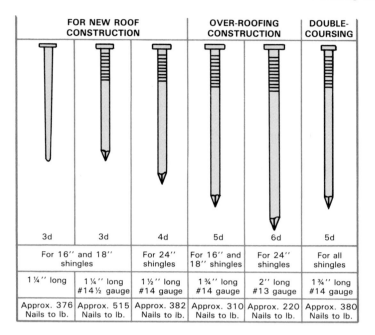

FOR NEW ROOF CONSTRUCTION			OVER-ROOFING CONSTRUCTION		DOUBLE-COURSING
3d	3d	4d	5d	6d	5d
For 16" and 18" shingles		For 24" shingles	For 16" and 18" shingles	For 24" shingles	For all shingles
1¼" long	1¼" long #14½ gauge	1½" long #14 gauge	1¾" long #14 gauge	2" long #13 gauge	1¾" long #14 gauge
Approx. 376 Nails to lb.	Approx. 515 Nails to lb.	Approx. 382 Nails to lb.	Approx. 310 Nails to lb.	Approx. 220 Nails to lb.	Approx. 380 Nails to lb.

Fig. 10-40. Use only rust resistant nails of the size recommended for your particular application.

Most carpenters prefer to use a shingler's or lather's hatchet, Fig. 10-41, to lay wood shingles. This has a blade for splitting and trimming. Some have a gauge for spacing the weather exposure.

APPLYING SHINGLES

The first course of shingles at the eaves should be doubled or tripled. All shingles, when laid on the roof, should be spaced 1/4 in. apart to provide for expansion when they become rain soaked.

Use only two nails to attach each shingle. Of considerable importance in nailing is the proper placing of these two nails. They should be near the butt line of the shingles in the next course that is to be applied over the course being nailed, but should never be driven below this line so they will be exposed to the weather. Driving the nails 1 to 1 1/2 in. above the butt line is good practice. Two in. above is an allowable maximum. Nails should be placed not more than 3/4 in. from the edge of the shingle at each side. When nailed in this manner, the shingles will lie flat and give good service.

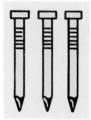

Use care when nailing wood shingles. Drive the nail just flush with the surface. The wood in shingles is soft. It can be easily crushed and damaged under the nail heads.

The second layer of shingles in the first course should be nailed over the first layer so the joints in each course are at least 1 1/2 in. apart. See Fig. 10-42. A good shingler will use care in breaking the joints in successive courses, so they do not match up in three successive courses. Joints in adjacent courses should be at least 1 1/2 in. apart.

It is good practice to use a board as a straightedge to line up rows of shingles. Tack the board temporarily in place to hold the shingles until they are nailed. Two shinglers often work together; one distributes and lays the shingles along the straightedge while another nails them in place. As shingling progresses, check the alignment every five or six courses with a chalk line. Measure down from the ridge occasionally to be sure shingle courses are parallel to the ridge.

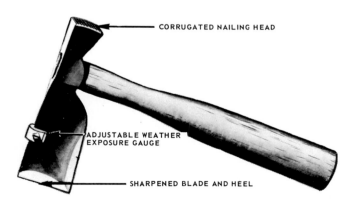

CORRUGATED NAILING HEAD

ADJUSTABLE WEATHER EXPOSURE GAUGE

SHARPENED BLADE AND HEEL

Fig. 10-41. Shingler's hatchet is especially suited for laying wood shingles.
(Red Cedar Shingle and Handsplit Shake Bureau)

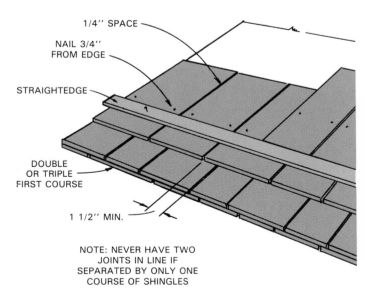

1/4" SPACE

NAIL 3/4" FROM EDGE

STRAIGHTEDGE

DOUBLE OR TRIPLE FIRST COURSE

1 1/2" MIN.

NOTE: NEVER HAVE TWO JOINTS IN LINE IF SEPARATED BY ONLY ONE COURSE OF SHINGLES

Fig. 10-42. A wooden straightedge should be used as a guide in laying wood shingles. Stagger joints and use two nails per shingle.

On a roof section where one end terminates at a valley, shingles for the valley should be carefully cut to the proper miter at the butts. Use wide shingles. Nail the shingles in place along the valley first, Fig. 10-43.

Dripping from gables may be prevented by using a piece of 6 in. bevel siding along the edge and parallel to the end rafter, Fig. 10-44.

SHINGLED HIPS AND RIDGES

Good, tight ridges and hips are required to avoid roof leakage. In the best type of hip construction, Fig. 10-45, nails are not exposed to the weather. Shingles of approximately the same width as the roof exposure are sorted out. Two lines are then marked on the shingles on the roof the correct distance back from the center line of the ridge on each side. On small houses, hip caps may be made narrower.

Factory assembled hip and ridge units are available, Fig. 10-46. Weather exposure should be the same as that used for the regular shingles. Be sure

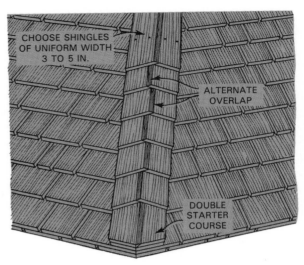

Fig. 10-45. How to install wood cap shingles at hips and ridges.

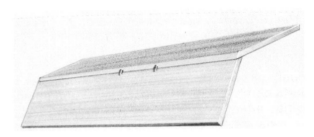

Fig. 10-46. Hip and ridge units can be purchased prefabricated. (Red Cedar Shingle and Handsplit Shake Bureau)

to use longer nails that will penetrate well into the sheathing.

SPECIAL EFFECTS

By staggering or building up wood shingles, usually in random patterns, shadow lines and texture can be emphasized. This is sometimes a feature used in contemporary as well as traditional architecture. Several applications are shown in Fig. 10-47. The ocean wave effect is secured by placing a pair of shingles butt-to-butt under the regular course and at right angles to the butt line. These cross shingles should be about 6 in. wide.

The Dutch weave effect is made by doubling or super-imposing extra shingles, completely at random. This effect can be emphasized by using two shingles instead of one. It is generally referred to as a pyramid pattern. Note that joints are always broken by at least 1 1/2 in.

REROOFING WITH WOOD SHINGLES

Wood shingles may be applied to old as well as new roofs, Fig. 10-48. If the old roofing is in rea-

Fig. 10-43. Mark, cut, and lay wood shingles along a valley first. Then complete each course to the roof's edge.

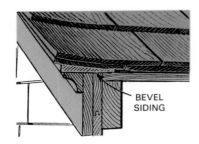

Fig. 10-44. Length of beveled siding tilts shingles inward to prevent dripping off the edge.

Fig. 10-47. Special patterns for wood shingles are used on contemporary as well as traditional houses. Left. Ocean wave. Center. Dutch weave. Right. Pyramid.

sonably good shape, it need not be removed.

Before applying new shingles, all warped, split, and decayed shingles should be nailed tightly or replaced. To finish the edges of the roof, the exposed portion of the first two rows of old shingles along the eaves should be cut off with a sharp hatchet. Nail a 1 in. wood strip in this space. Place the outer edge flush with the eave line. Edges along the gable ends should be treated in a similar manner.

The level of the valleys should be raised by applying wood strips. New flashings should be installed over the strips. Remove old hip and ridge caps to provide a solid base for new shingles.

New shingles should be spaced 1/4 in. apart to allow for expansion in wet weather. Let them project 1/2 to 3/4 in. beyond the edge of the eaves.

The procedure described for new roofs should be followed. However, longer nails are required. For 16 in. and 18 in. shingles, rust-resistant or zinc clad 5d box nails or special over-roofing nails 1 3/4 in. long, 14 ga., should be used. A 6d, 13 ga., rust-resistant nail is desirable for 24 in. shingles.

Usually, no particular attention needs to be given to how the nails penetrate the old roof beneath. It does not matter whether they strike the sheathing strips or not because, with the larger nails that are used, complete penetration is obtained through the old shingles. Enough nails to anchor all the shingles of the new roof will strike sheathing or nailing strips.

New flashings should be placed around chimneys. Do not remove the old flashing. Use high grade non-drying mastics liberally to get a watertight seal between the brick and the metal.

In reroofing houses covered with composition material, whether in the form of roll roofing or imitation shingles, it is usually best to strip off the old material. Otherwise, moisture may condense on the roof deck below. Rapid decay of sheathing could follow.

WOOD SHAKES

Often called the aristocrat of roofing materials, wood shakes provide the most pleasing surface texture, Fig. 10-49. They are highly durable and, if properly installed, may outlast the structure itself.

Generally, wood shakes are available as hand-split and resawn, taper split, and straight split, Fig. 10-50. Like regular wood shingles, they are available in random widths. Various lengths and thicknesses are standardized as listed in Fig. 10-51.

Do not apply shakes to roofs that have too little slope for good drainage. The recommended minimum is 4 in 12. Maximum weather exposure is 13 in. for 32 in. shakes, 10 in. for 24 in. shakes, and 8 1/2 in. for 18 in. shakes.

Start the application by placing a 36 in. strip of 30 lb. roofing felt along the eaves. The beginning or starter course is doubled, just like regular wood shingles. After each course of shakes, apply an 18 in. strip of 30 lb. felt. This must cover the top portion of the shakes and extend onto the sheathing. The bottom edge of the felt is placed above the butt a distance equal to twice the exposure.

For example, if 24 in. shakes are being laid at a

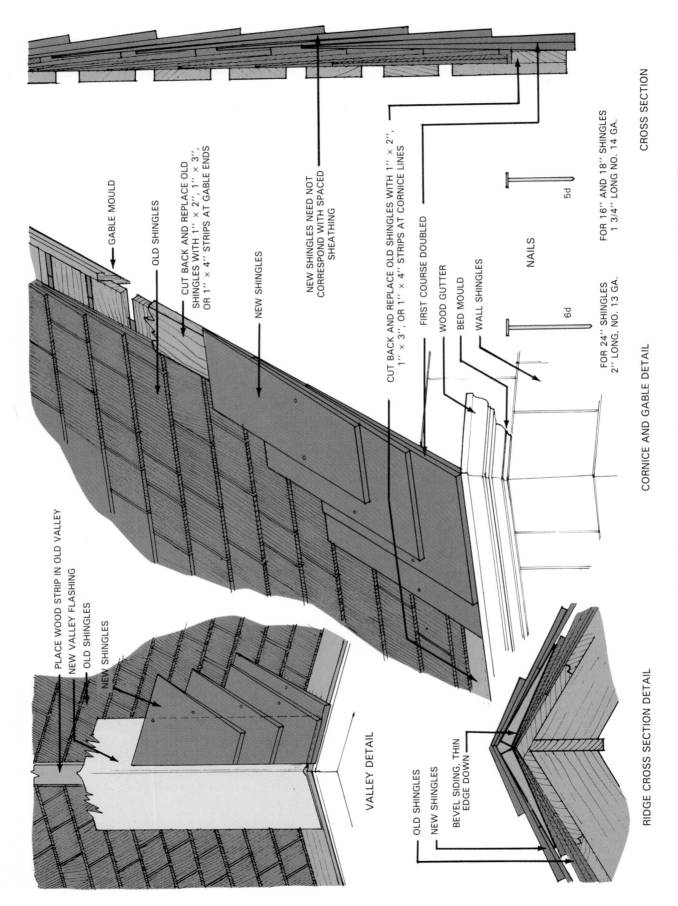

GABLE MOULD

OLD SHINGLES

CUT BACK AND REPLACE OLD SHINGLES WITH 1" × 2", 1" × 3", OR 1" × 4" STRIPS AT GABLE ENDS

NEW SHINGLES

NEW SHINGLES NEED NOT CORRESPOND WITH SPACED SHEATHING

CUT BACK AND REPLACE OLD SHINGLES WITH 1" × 2", 1" × 3", OR 1" × 4" STRIPS AT CORNICE LINES

FIRST COURSE DOUBLED

WOOD GUTTER

BED MOULD

WALL SHINGLES

NAILS

5d

6d

FOR 16" AND 18" SHINGLES 1 3/4" LONG NO. 14 GA.

FOR 24" SHINGLES 2" LONG, NO. 13 GA.

CROSS SECTION

CORNICE AND GABLE DETAIL

PLACE WOOD STRIP IN OLD VALLEY

NEW VALLEY FLASHING

OLD SHINGLES

NEW SHINGLES

VALLEY DETAIL

OLD SHINGLES

NEW SHINGLES

BEVEL SIDING, THIN EDGE DOWN

RIDGE CROSS SECTION DETAIL

Fig. 10-48. Detail drawings for reroofing method with wood shingles. Note how valleys and ridges are built up.

Fig. 10-49. Wood shakes provide an attractive and durable roof.

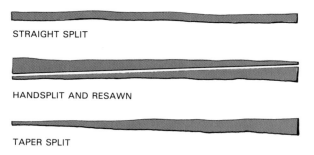

STRAIGHT SPLIT

HANDSPLIT AND RESAWN

TAPER SPLIT

Fig. 10-50. There are three basic types of wood shakes.

10 in. exposure, place the roofing felt 20 in. above the butts of the shake, Fig. 10-52.

Individual shakes should be spaced from 1/4 to 3/8 in. apart to allow for expansion. These joints should be offset at least 1 1/2 in. from course to course.

Proper nailing is important. Use rust-resistant nails, preferably the hot-dipped zinc-coated type. The 6d size, which is 2 in. long, normally is adequate. Longer nails should be used, if necessary, because of unusual shake thickness and/or weather exposure. Nails should be long enough for adequate penetration into the sheathing boards.

Two nails should be used for each shake. Drive them at least one inch from each edge and about one or two inches above the butt line of the following course. Do not drive nailheads into the shakes so that wood fibers are crushed, Fig. 10-53.

Valleys are laid as recommended for regular wood shingles. Underlay all valleys with 30 lb. roofing felt. Metal valley sheets must be at least 20 in. or wider.

Chimneys or other structures that project through the roof should be flashed and counterflashed on all edges. Flashing should extend at least 6 in. under the shakes and should be covered as shown in Fig. 10-54.

For the final course at the ridge line, try to select a uniform size of shakes and trim off the ends so they meet evenly. Carefully apply a strip of 30 lb. felt along all ridges and hips. Install shakes that have a uniform width of about 6 in. Nail them in place following the procedure described for regular wood shingles.

Prefabricated hip-and-ridge units are available. Their use will save time and provide uniformity. See Fig. 10-55.

Shakes are also prefabricated in 8 ft. panels. Shingles are bonded to a backing of 5/16 in. exterior sheathing plywood. Underlayment specifications are similar to those for other shakes. Shake panels can be applied over solid sheathing or furring strips on roofs with a 4 in 12 or steeper slope. See Fig. 10-56.

GRADE	Length and Thickness	18″ Pack**		Description
		♯ Courses Per Bdl.	♯ Bdls. Per Sq.	
No. 1 HANDSPLIT & RESAWN	15″ Starter-Finish	9/9	5	These shakes have split faces and sawn backs. Cedar logs are first cut into desired lengths. Blanks or boards of proper thickness are split and then run diagonally through a bandsaw to produce two tapered shakes from each blank.
	18″ x 1/2″ Mediums	9/9	5	
	18″ x 3/4″ Heavies	9/9	5	
	24″ x 3/8″	9/9	5	
	24″ x 1/2″ Mediums	9/9	5	
	24″ x 3/4″ Heavies	9/9	5	
No. 1 TAPERSAWN	24″ x 5/8″	9/9	5	These shakes are sawn both sides.
	18″ x 5/8″	9/9	5	
No. 1 TAPERSPLIT	24″ x 1/2″	9/9	5	Produced largely by hand, using a sharp-bladed steel froe and a wooden mallet. The natural shingle-like taper is achieved by reversing the block, end-for-end, with each split.
No. 1 STRAIGHT-SPLIT		20″ Pack		
	18″ x 3/8″ True-Edge*	14 Straight	4	Produced in the same manner as tapersplit shakes except that by splitting from the same end of the block, the shakes acquire the same thickness throughout.
	18″ x 3/8″	19 Straight	5	
	24″ x 3/8″	16 Straight	5	

NOTE: * Exclusively sidewall product, with parallel edges.
** Pack used for majority of shakes.

Fig. 10-51. Wood shakes are manufactured in four grades with various specifications and sizes.
(Red Cedar Shingle and Handsplit Shake Bureau)

Fig. 10-52. Wood shakes are laid with an 18 in. strip of 30 lb. asphalt saturated felt between each course.
(Red Cedar Shingle and Handsplit Shake Bureau)

MINERAL FIBER SHINGLES

Mineral fiber shingles are manufactured from asbestos fiber and Portland cement. They are formed in molds under high pressure and provide a finished product that is immune to rot and decay, unharmed by exposure to salt air, unaffected by ice and snow, and fireproof because they contain nothing that will burn.

Mineral fiber shingles are available in a variety of colors, textures, and shapes, Fig. 10-57. Finishes include plastic coatings and embedded mineral granules. Shingles are sold by the square, and are equally well adapted for use on new buildings or over old roofs. On old roofs, the new shingles can usually be applied right over the old shingles.

Fig. 10-53. Putting down wood shakes. Butts can be aligned or laid random as shown here. Be sure to lap joints.
(American Plywood Assoc.)

Fig. 10-54. Base flashing for a vertical projection such as a chimney or wall should extend at least 6 in. under wood shakes.
(Red Cedar Shingle and Handsplit Shake Bureau)

Fig. 10-55. Preformed hip-and-ridge units will reduce the labor of attaching the ridge row.

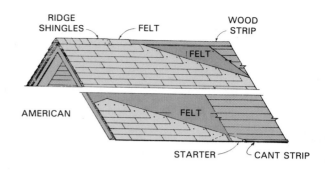

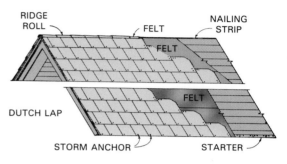

Fig. 10-57. Mineral fiber shingles are made in several shapes, textures, and colors. Minimum slopes are 4 in 12 for American. 5 in 12 for Dutch Lap.

SAFETY

Great care must be used in working with mineral fiber shingles. Asbestos fibers are harmful to the body if inhaled or swallowed. OSHA requires proper dust control and use of personal protection. Tools used for cutting should be equipped with dust collectors. Dust and debris caused by the removal of the product from a roof should always be vacuumed up. Use of brooms should be avoided since they introduce the dust into the air. If a broom must be used, wet down the area first.

A respirator, gloves, hat, and other protective clothing should be worn when handling asbestos products. Place waste in a sealed bag before disposing of it.

Because the mineral fiber material is rigid and hard, nail holes are prepunched during the manufacturing process. Eave starter strips, hip and ridge shingles, and ridge rolls are prefabricated. Some designs, such as the Dutch lap, have additional holes at the exposed corner for a storm anchor. It holds the corner to the shingle below it.

Fig. 10-56. Applying shingle panels to a roof. Panels are prefabricated in 8 ft. lengths with 11 1/4 in. exposure. (Shakertown Corp., Cedar Panel Div.)

Shingles should be installed with galvanized needlepoint nails. For new roofs, use 1 1/4 in. nails. When reroofing over old shingles, use 2 in. nails.

Special equipment is required to cut the material. Dealers handling mineral fiber products will usually have shingle cutters on hand for customers. They cut the shingles quickly, accurately, and neatly. Dealers also have a punch for forming nail holes where extra holes are required.

If a cutter is not available, asbestos shingles can be cut by hand. (Observe all safety precautions.) Use an old chisel, a drift punch, or the blade of the hatchet. Score the shingle with the tool being drawn along a straightedge. After scoring, place the shingle over a solid piece of wood and break along the scored line, Fig. 10-58.

Irregular cuts or round holes are made by punching holes along the line of the cut and breaking out the piece which is to be discarded. There is a punch on the shingle cutter for punching additional holes. They may be drilled or punched with a drift punch or other suitable pointed tool. A drift punch is recommended because it will punch a clean hole without splitting the material.

STORING SHINGLES

While still packed in bundles, whether in the yard or on the job, mineral fiber shingles should be kept dry. Moisture trapped between bundled shingles may cause discoloration due to efflorescence.

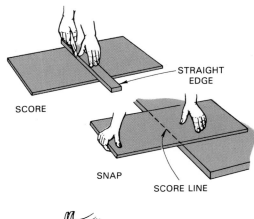

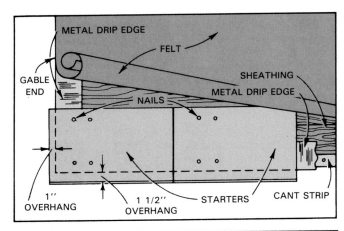

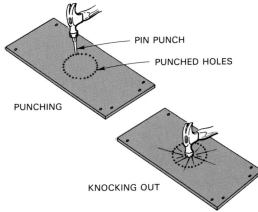

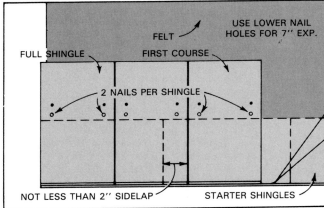

Fig. 10-58. Mineral fiber shingles can be formed by hand. Top. Score and snap as you would drywall. Carbide tipped knife blade works best. Bottom. How to make interior cuts. (Mineral Fiber Products Bureau)

Fig. 10-59. Diagrams for installing standard rectangular shingles. Top. Requirements for starter course. Use shingles turned sideways. Bottom. First course with shingles placed over felt. (Supradur Mfg. Corp.)

This ordinarily is known as "blooming." If it is necessary to use outdoor storage, stack the shingles on planks and use roofing felt, waterproof paper, or a tarpaulin for cover.

APPLICATION OF MINERAL FIBER SHINGLES

It is important that the roof deck be in good condition to receive the shingles. Deck lumber should be well seasoned, dry, and of uniform thickness. Tongue-and-groove 1 × 6 boards or plywood of adequate thickness is recommended. Nailheads must be driven down. High spots or rough edges must be removed. Install a wood cant strip along the eaves, flush with the lower edge, to give proper pitch for the first course of shingles. Also install metal drip edges like those used for asphalt shingles.

Fig. 10-59 shows the application of standard rectangular shingles. Underlayment felt is rolled back and the starter course (regular shingles turned sideways) is nailed in place. Turn starter shingles upside down so the exposed overhang will match the roof surface.

Start the first course with full shingles as shown. Then start the second course with a half shingle. The third course is applied in the same manner as the first course.

Attach a furring strip along each side of hips and ridges to provide a nailing base for the hip and ridge covering. The roof shingles are butted against these furring strips. Therefore, they must be the same thickness as the shingles. Since furring strips must be covered by the ridge units, Fig. 10-60, they should not be over 2 in. wide.

On new work or on reroofing jobs, where the old materials have been removed, the roof boards should be covered with one course of waterproof sheathing paper. Horizontal laps need not extend beyond 2 in. End laps should be 6 in. Hips, ridges, and valleys are covered with a 12 in. lap so double thickness will be assured at these points.

FLASHING FOR MINERAL FIBER SHINGLES

Use care in applying valley flashings. First, a corrosion resistant metal, such as copper or stain-

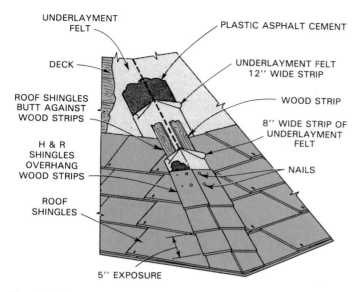

Fig. 10-60. Hip and ridge treatment with mineral fiber shingles. Use wood strips as a nailing surface and butt courses against them. (Mineral Fiber Products Bureau)

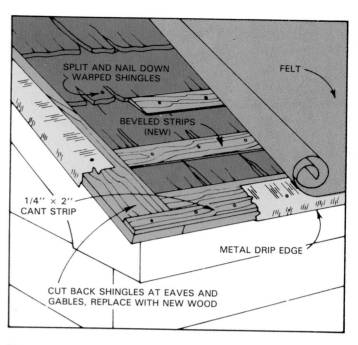

Fig. 10-61. Old roof is being prepared for reroofing with mineral fiber shingles. (Supradur Mfg. Corp.)

less steel, should be laid when a mineral fiber roof is being put on. Do not use aluminum.

Second, to insure a leakproof roof, the metal valley should extend out onto the roof deck well beyond the edge of the overlapping shingles on both sides.

Finally, shingles should be bedded in a layer of asphalt cement. The asphalt bed should extend for a distance of 6 in. back from the edge where the shingles meet the valley.

Open valleys are usually used. The edges of the metal should be turned back 1/2 in., forming a hem or water seal. The valley metal should be attached to the deck with cleats at the hem. *Shingle nails should not be driven through metal valley linings.*

REROOFING OVER OLD SHINGLES

Nail down old shingles which are badly curled or warped. Generally, underlayment material is not required under asbestos shingles when they are laid over old wood or asphalt shingles. However, missing or badly decayed wood shingles should be replaced and the surface leveled.

If the top edges or corners of the new shingles do not rest on the butts of the old shingles, tilting will result. Wood strips, beveled and of the same thickness as the butts of the old wood shingles, can be used for leveling an old roof deck, as shown in Fig. 10-61.

Sometimes at the edge of the roof, the wood shingles and sheathing are found to be in bad condition. In such cases, cut away the old shingles and lay a new board 4 to 6 in. wide by 7/8 in. thick

along the edges. This provides a solid base to which the new shingles may be nailed.

Build up old valleys with wood strips to bring the surface flush with the butts of the old wood shingles. Lay waterproof felt over hips, ridges, and at valleys.

TILE ROOFING

The most commonly used roofing tile are manufactured products consisting of molded, hard-burned shale, or mixtures of shale and clay. Tiles are also made from metal and concrete.

When well made, clay tile is hard, fairly dense, and durable. A variety of shapes and textures are sold. Most roofing tile of clay is unglazed, although glazed tile is sometimes used. Typical tile roof application details are shown in Fig. 10-62.

Tile may be used over an old roof provided:
1. The old covering is in reasonably good condition.
2. The roof framing is heavy enough to stand the additional weight.

Additional roof framing or bracing, if required, should be added before laying tile. Fig. 10-63 shows a roof surface formed from tiles made of concrete.

GALVANIZED SHEET METAL ROOFING

Only galvanized sheets that are heavily coated with zinc (2.0 oz. per sq. ft.) are recommended for

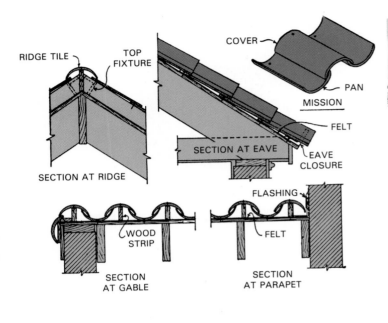

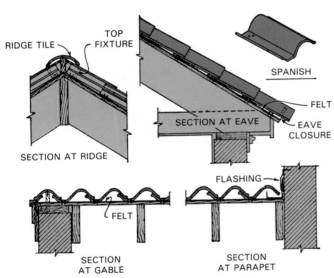

Fig. 10-62. Typical application methods for tile roofs. Top. Two-piece pan and cover commonly known as Mission tile. Bottom. Spanish tile.

Fig. 10-63. Roof surface of Spanish tile. This kind of roof is durable and will withstand hurricane force winds and rain. (Gory Associated Industries, Inc.)

roof has a pitch of 1/4 or more, 4 in. end laps are usually satisfactory.

To make a tight roof, sheets should be lapped 1 1/2 corrugations at either side, Fig. 10-64. The wind is likely to drive rainwater over single-corrugation lap joints. When using roofing 27 1/2 in. wide with 2 1/2 in. corrugations and 1 1/2 corrugation lap, each sheet covers a net width of 24 in. on the roof.

When 27 1/2 in. roofing is not available, sheets of 26 in. in width may be used. In laying the narrower sheets, every other one should be turned upside down. So laid, each alternate sheet laps over the two intermediate sheets. The 26 in. roofing with 2 1/2 in. corrugations cover a net width of 22 1/4 in. of roof.

SHEATHING AND NAILS

If 26 ga. sheets are used, supports may be 24 in. apart. If 28 ga. sheets are used, supports should be not more than 12 in. apart. The heavier gauge has no particular advantage except its added strength. A zinc coating for durability is more important than strength for this type of roofing.

For best results, galvanized sheets should be fastened with lead-headed nails or galvanized nails and lead washers. Nails properly located are driven only into tops of the corrugations. To avoid corrosion, use nails specified by the manufacturer.

permanent-type construction. The sheets with lighter coatings of zinc are less durable and are likely to require painting every few years. On temporary buildings, and in cases where the most economical construction is required, lighter metal can be used. It will give satisfactory results if protected by paint.

SLOPE AND LAPS

Galvanized sheets may be laid on slopes as low as a 3 in. rise to the foot (1/8 pitch). If more than one sheet is required to reach the top of the roof, the ends should lap no less than 8 in. When the

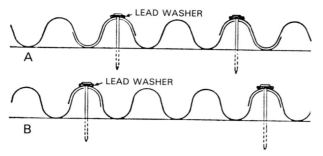

Fig. 10-64. Guide to application of corrugated sheet metal roofing. A—Sheets properly laid with one-and-one-half corrugation lap. B—Single corrugation lap is not recommended.

ALUMINUM ROOFING

Corrugated aluminum roofing, if properly applied, usually makes a long-lasting roof. Seacoast exposure tests reported by the Bureau of Standards indicate this material is capable of resisting corrosion in such localities unless subjected to direct contact with salt-laden spray. Where this is likely to happen, aluminum roofing is not recommended.

Aluminum alloy sheets available for roofing usually have a corrugation spacing of 1 1/4 or 2 1/2 in. Recommendations for the installation of sheet metal regarding side lap and end lap are applicable to the laying of aluminum sheets.

An important precaution to observe in laying aluminum roofing is to make sure that contact with other kinds of metal is avoided. Where this is not possible, both metals should be given a heavy coating of asphalt paint wherever the surfaces are in contact.

As aluminum is soft and the sheets used for roofing are relatively thin, they should be laid on tight sheathing or on decks with openings no more than 6 in. wide. Aluminum roofing should be nailed with no less than 90 nails to a square or about one nail for each square foot. It is recommended that aluminum alloy nails be used and that nonmetallic washers be used between nail heads and the roofing.

If desired, the sheathing may be covered with water-resistant building paper or asphalt impregnated felt. Paper that absorbs and holds water should never be used.

To avoid corrosion, aluminum sheets should be stored so that air will have free access to all sides. Otherwise, a white deposit will form. This deposit creates pinholes very quickly.

TERNE METAL ROOFING

Terne metal roofing is made of copper-bearing steel, heat-treated to provide the best balance between malleability (easy to form) and toughness. It is hot dip-coated with Terne metal, an alloy of 80 percent lead and 20 percent tin. The high weather resistance factor (notable in this type of roofing) is due primarily to the lead. Tin is included because the alloy makes a better bond with steel.

Grades are expressed as the total weight of the coating on a given area. This area, by old trade custom, is the total area contained in a box of 112 sheets that are 20 in. × 28 in. in size, and amounts to 436 square feet. The best grade of Terne coating is 40 lb. It provides a roof surface that will last for many years.

A wide variety of sheet sizes are available, as well as 50 ft. seamless rolls in various widths. This permits its use for many different types of roofs and methods of application. It is used extensively for flashing around both roof and wall openings.

For best appearance and longest wear, Terne metal roofs must be painted. Use a linseed oil-based iron oxide primer for a base coat. Almost any exterior paint and color can be used over this base.

ALUMINUM SHAKES

Aluminum shakes are manufactured of an aluminum-magnesium alloy with a nominal 0.019 in. thickness. They are available in brown, red, dark gray, white, and natural aluminum. See Fig. 10-65. For installation, see manufacturer's instructions.

GUTTERS

Gutters or eaves troughs collect rainwater from the edge of the roof and carry it to downspouts. Downspouts direct water away from the foundation or into a drainage system.

Fig. 10-65. Aluminum shingles should be installed over solid sheathing. Use 1 1/4'' aluminum nails or the best zinc-coated nails. The three nails needed here will remain exposed. (Reinke Shakes)

The term "gutter" refers to a separate unit that is attached to the eave. The term "eaves trough" usually applies to a waterway built into the roof surface over the cornice.

Eaves troughs must be carefully designed and built since any leakage will penetrate the structure. Because of this and the extra cost of construction, they are seldom used in modern residential work.

WOOD GUTTERS

Wood gutters of fir or red cedar are used in many parts of the country. When properly installed and maintained, the life of wood gutters is usually equal to that of the main structure.

The installation of modern wood gutters is usually made before the shingles are applied. One type is attached to the fascia board after the roof is sheathed, Fig. 10-66.

Some designs are coordinated with the fascia board and are installed with this trim unit before the roof sheathing is complete. In this case, the sheathing overhangs the fascia and gutter, thus reducing the possibility of leakage. When wood gutters are specified, the architectural plans usually include installation details.

Cutting, fitting, and drilling is done on the ground before the various units are set in place. Most wood guttering is primed or prepainted. Always use galvanized or other types of weatherproof nails and/or brass wood screws.

Gutter ends may be sealed with blocks, returned, and mitered, or butted against an extended rake frieze board. In general, the gutter is treated like cornice molding and should present a smooth, trim appearance.

The roof surface should extend over the inside edge of the gutter with the front top edge at approximately the height of a line extended from the top of the sheathing. For correct appearance, wood gutters are set nearly level. They will drain satisfactorily if kept clean and if adequate downspouts are provided.

METAL GUTTERS

A wide variety of metal gutters are available to control roof drainage. Manufacturers have perfected gutter and downspout systems that include various component parts. Whole systems can be quickly assembled and installed on the building site. Materials consist of galvanized iron and aluminum. Many systems are available in either a primed or prefinished condition to match a wide range of colors. One development consists of a guttering system made from molded vinyl plastic.

Gutter systems include inside and outside mitered corners, joint connectors, pipes, brackets, and other

Fig. 10-66. Properly installed and treated, wood gutters will last as long as the rest of the structure.   (Weyerhaeuser Co.)

items. All are carefully engineered and fabricated. Parts slip together easily and are generally held with soft pop-rivets or sheet metal screws. Fig. 10-67 shows standard parts of a typical gutter and downspout system and how they are assembled.

Fig. 10-68 shows a gutter system that provides both gutter and fascia in a single unit. This unit also includes a channel to support the outside edge of a prefabricated soffit, of either panel or coil stock. The system saves time and material and results in a clean-line appearance. In new construction, it can be mounted directly on the rafter tails, as shown, or it can be attached to an existing fascia in remodeling work.

For best results, gutters and eaves troughs must be sized to suit the roof areas from which they receive water. For roof areas up to 750 sq. ft., a 4 in. wide trough is suitable. For areas between 750 and 1400 sq. ft., 5 in. troughs should be used. For larger areas, a 6 in. trough is recommended. Quality of gutters, like that of flashing, should correspond to the durability of the roof covering. If galvanized steel guttering is used, it should have a heavy zinc coating.

The size of downspouts or conductor pipes required also depends on the roof area. For roofs up to 1000 sq. ft., downspouts of 3 in. diameter have sufficient capacity if properly spaced. For larger roofs, 4 in. downspouts should be used.

Gutters having the proper slope stay clean. Metal gutters are usually sloped 1 in. for every 12 to 16 ft. of length.

ESTIMATING MATERIAL

To estimate roofing materials, first calculate the total surface area to be covered. In new construction, the figures used to estimate the sheathing can

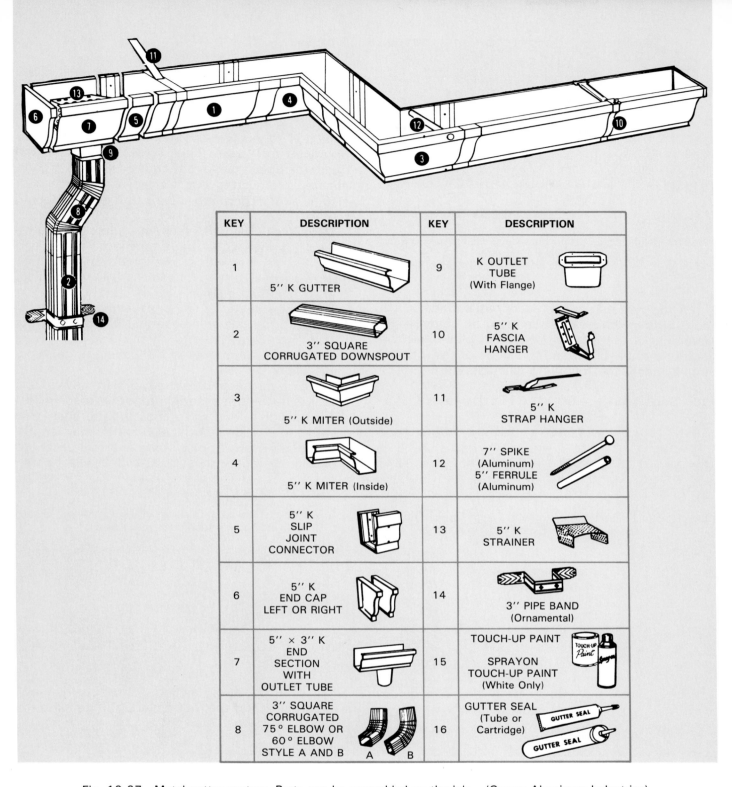

KEY	DESCRIPTION	KEY	DESCRIPTION
1	5'' K GUTTER	9	K OUTLET TUBE (With Flange)
2	3'' SQUARE CORRUGATED DOWNSPOUT	10	5'' K FASCIA HANGER
3	5'' K MITER (Outside)	11	5'' K STRAP HANGER
4	5'' K MITER (Inside)	12	7'' SPIKE (Aluminum) 5'' FERRULE (Aluminum)
5	5'' K SLIP JOINT CONNECTOR	13	5'' K STRAINER
6	5'' K END CAP LEFT OR RIGHT	14	3'' PIPE BAND (Ornamental)
7	5'' × 3'' K END SECTION WITH OUTLET TUBE	15	TOUCH-UP PAINT SPRAYON TOUCH-UP PAINT (White Only)
8	3'' SQUARE CORRUGATED 75° ELBOW OR 60° ELBOW STYLE A AND B	16	GUTTER SEAL (Tube or Cartridge)

Fig. 10-67. Metal gutter system. Parts can be assembled on the job.   (Crown Aluminum Industries)

also be used to estimate the underlayment and finished roofing materials. When these figures are not available, they can be calculated by the same methods used for roof sheathing, described in Unit 9.

Another method sometimes used to estimate roof area is to determine the total ground area of the structure. Include all eave and cornice overhang. Convert the ground area to roof area by adding a percentage determined by the roof slope, as follows:

Slope 3 in 12, add 3% of area.
Slope 4 in 12, add 5 1/2% of area.
Slope 5 in 12, add 8 1/2% of area.
Slope 6 in 12, add 12% of area.
Slope 8 in 12, add 20% of area.

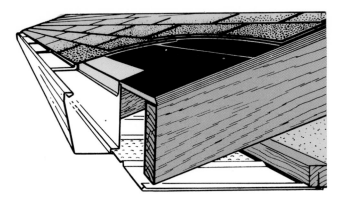

Fig. 10-68. Modern gutter system includes complete cornice trim. (Omni Products)

Divide the total square feet of roof surface by 100 to find the number of squares to be covered. For example, if the total ground area, plus overhang, is found to be 1560 and the slope of the roof is 4 in 12, apply the following calculations:

$$\text{Roof Area} = 1560 + (1560 \times 5\ 1/2\%)$$

$$= 1560 + (1560 \times .055)$$

$$= 1560 + 85.80 \text{ or } 86 = 1646$$

Number of squares = 16.46 or 16 1/2

After the number of squares is established, additional amounts must be added. For asphalt shingles, it is generally recommended that 10 percent be added for waste. This, however, may be too much for a plain gable roof and too little for a complicated intersecting roof. Certain allowances must also be added for reduced exposure on low sloping roofs.

Usually the 10 percent waste figure can be reduced if allowance is made for hips, valleys, and other extras. For wood or asphalt shingles, one square is usually added for each 100 lineal (running) feet of hips and valleys.

Quantities of starter strips, eaves flashing, valley flashing, and ridge shingles must be added to the total shingle requirements. All of these are figured on linear measurements of the eaves, ridge, hips, and valleys.

For a complicated structure, a plan view of the roof will be helpful in adding together these materials. All ridges and eave lines will be seen true length and can simply be scaled to find their length. Hips and valleys will not be seen as their true length and a small amount must be added. Using the per-

Fig. 10-69. Frame the roof through folded carpenter's rule held at arm's length. Adjust until sides parallel roof slope. Center the end of rule to get proper reading at "reading point." (Asphalt Roofing Manufacturers Assoc.)

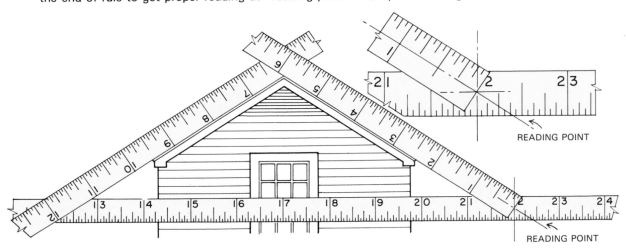

Fig. 10-70. Reading point conversions. Locate reading point on chart and read downward for pitch and slope. (Asphalt Roofing Manufacturers Assoc.)

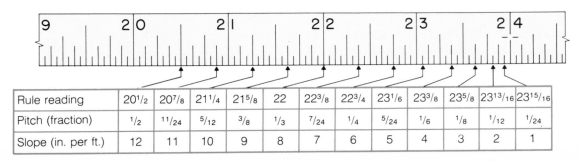

Rule reading	20½	20⅞	21¼	21⅝	22	22⅜	22¾	23⅙	23⅜	23⅝	23¹³/₁₆	23¹⁵/₁₆
Pitch (fraction)	½	¹¹/₂₄	⁵/₁₂	⅜	⅓	⁷/₂₄	¼	⁵/₂₄	⅙	⅛	¹/₁₂	¹/₂₄
Slope (in. per ft.)	12	11	10	9	8	7	6	5	4	3	2	1

centage listed for converting ground area to roof area will usually provide sufficient accuracy.

On a reroofing job there is a simple method for determining the roof pitch when it is not known. You can estimate it from the ground with the help of a folding carpenter's rule.

Stand away from the building some distance and fold the rule into a triangle. Hold the folded rule at arms length and frame the roof through it. Adjust the triangle until the two sloping sides line up with the roof as in Fig. 10-69. Be sure the base of the triangle is level.

Read off the dimension on the base of the rule which is marked the "reading point" (Fig. 10-69). Now, refer to the chart in Fig. 10-70 and locate the proper pitch and slope.

TEST YOUR KNOWLEDGE — UNIT 10

1. The selection of roofing materials is influenced by such factors as cost, durability, appearance, application methods, and _____ of the roof.
2. In shingle application, the distance between the edges of one course and the next, measured at a right angle to the ridge, is called_____.
3. Asphalt saturated felt is available in a roll that is _____ wide.
4. Three-tab square-butt shingles are _____ in. long.
5. The most commonly used underlayment for asphalt shingles is _____lb. roofer's felt.
6. The waterway of a valley should diverge (grow wider) as it approaches the eaves at a rate of _____ in. per foot.
7. The minimum number of nails recommended for the application of each three-tab square-butt shingle unit is _____.
8. Indicate the correct answer(s). When using a power staple gun for attaching asphalt shingles:
   a. Install staples, generally speaking, 5/8 in. below adhesive strips.
   b. Adjust staple gun pressure for proper staple application.
   c. Be sure staples are sunk well into shingle surface.
   d. If a staple must be removed, repair hole with asphalt cement.
   e. Gun should have enough pressure to drive the staple flush with the top of shingle.
9. Chimney flashing consists of two parts: the cap or counter flashing and the _____ flashing.
10. Asphalt shingles can be used on roofs with a slope as low as _____ if special application procedures are followed.
11. Where is a chimney saddle used and why?

12. List the mineral materials which may form the top coat of a built-up roof.
13. A formed metal strip, called a _____ _____ is attached to the edge of built-up roofs.
14. In the best grade of wood shingles, the annual rings run _____ (parallel, perpendicular) to the surface.
15. When laying wood shingles, they should be spaced _____ in. apart (horizontally) to provide for expansion when they become rain soaked.
16. The vertical joints between wood shingles in adjacent courses should be spaced at least _____ in. apart.
17. A strip of 30 lb. roofing felt is placed between each course when applying _____ _____.
18. Mineral fiber roofing products are made from asbestos and _____ _____.
19. When installing corrugated steel roofing, the joints should be lapped _____ (1, 1 1/2, 2) corrugations.
20. Terne metal roofing consists of sheets of copper-bearing steel coated with an alloy of _____ and _____.
21. The vertical pipes of a gutter system are called _____ or _____.

OUTSIDE ASSIGNMENTS

1. Study the kinds and qualities of asphalt shingles used in your locality. Secure manufacturer's literature from a local builders supply center. Prepare a report including information about kinds, grades, and costs. Also include information about materials for underlayment, valley flashing, hip and ridge finish, and fasteners.
2. Report on application procedures used to install a roof on a residence in your community. Visit the building site and observe the methods used. Note the type of sheathing, special preparation of the roof deck, how valley flashing is applied, use of drip edges, type of starter courses, procedures used to align shingle courses, and how ridges are finished. *Always be sure to obtain permission from the foreman or head carpenter when visiting a building site.*
3. Construct a full-size visual aid showing the application of wood or asphalt shingles. Use a piece of 3/4 in. plywood about 4 ft. square to represent a lower corner of a roof deck. Apply underlayment (if required), drip edges, starter strips, and then carefully lay the shingles according to an approved pattern. By making only a partial coverage of the various layers, all of the application steps and materials used can be easily studied and observed.

Window treatment adds comfort and beauty to a residence. This multi-level home has matching windows on two levels. Each consists of a fixed (picture) window with two casement units on each side. Frame and sash are wood encased in vinyl. (Andersen Corp.)

Main entrance in traditional styling. Door facings are of 0.024 in. steel with thermal break. Cavity is filled with expanded polystyrene foam. (Pease Industries Inc.)

Garage door in modern style. Roll-up style is most common design for modern construction. (Stanley Door Systems)

Unit 11

WINDOWS AND EXTERIOR DOORS

Windows and doors are an important part of a structure. The carpenter should:
1. Have a basic understanding of the various types, sizes, and standards of construction.
2. Be able to recognize good quality in materials, fittings, weather stripping, and finish.
3. Appreciate the importance of careful installation of the various units.
4. Be an expert in installation.

Placement of the outside doors and the windows starts when the sheathing has been installed, Fig. 11-1. Careful installation will assure a close fit so that air infiltration is kept at a minimum.

MANUFACTURE

Today windows and doors are built in large millwork plants. They arrive at the building site as completed units ready to be installed in the openings of the structure.

Some windows used in residences are made from aluminum and steel. Most of them, however, are made from wood. Since wood does not transmit heat readily, there is less tendency for window frames made of wood to become cold and condense moisture vapor.

Wood will decay under certain conditions and

Fig. 11-1. Residential structure has been framed, sheathed, and roofed. It is ready for windows and exterior doors. (Weyerhaeuser Co.)

must be treated with preservatives. Exposed surfaces must be kept painted.

Metal is stronger than wood and thus permits the use of smaller frame members around the glass. Aluminum has the added advantage of a protective film of oxide which eliminates the need for paint.

Ponderosa pine is commonly used in the fabrication of windows. It is carefully selected and kiln dried to a moisture content of 6 to 12 percent.

MANUFACTURING STANDARDS

The control of quality in the manufacture of windows is based on standards established by the National Woodwork Manufacturers Association. These standards cover every aspect of material and fabrication. Included are such details as the projection of the drip cap and the slope of the sill.

For example, NWMA industry standard I.S. 2-74 states that the weather stripping for a double-hung window shall be effective to the point that it will prevent air leakage (infiltration) in excess of 0.50 cfm (cubic feet per minute) per linear foot of sash crack when tested at a static air pressure of 1.56

psf (pounds per square foot). The latter is equivalent to the pressure subjected by a 25 mph wind. Standards also cover the methods and procedures used in preservative treatments.

TYPES OF WINDOWS

In general, windows can be grouped under one or a combination of three basic headings:
1. Sliding.
2. Swinging.
3. Fixed.

Each of these includes a variety of designs or method of operations. Sliding windows include the double-hung and horizontal sliding. Swinging windows that are hinged on a vertical line are called casement windows while those hinged on a horizontal line can be either awning or hopper windows.

DOUBLE-HUNG

A double-hung window consists of two sash that slide up and down in the window frame. These are held in any vertical position by a friction fit against the frame or by springs and various balancing devices. Double-hung windows are widely used because of their economy, simplicity of operation, and adaptability to many architectural designs.

Fig. 11-2 shows an outside view of a double-hung window unit. Screen and storm sash are installed on the outside of the window.

HORIZONTAL SLIDING

Horizontal sliding windows have two or more sash. At least one of them moves horizontally within the window frame. The most common design consists of two sash, both of which are movable. See Fig. 11-3. When three sash are used, the center one is usually fixed.

CASEMENT

A casement window has a sash that is hinged on the side and swings outward. Installations usually consist of two or more units, separated by mullions. Sash are operated by a cranking mechanism or a push-bar mounted on the frame, Fig. 11-4. Latches are used to close and hold the sash tightly against the weather stripping.

The swing sash of a casement window permits full opening of the window. This provides good ventilation. Frequently, fixed units are combined with operating units where row of windows is desirable.

Crank operators make it easy to open and close windows located above kitchen cabinets or other built-in fixtures. Screen and storm sash are attached to the inside of standard casement windows.

Fig. 11-2. Eight light double-hung window is in a traditional style but has an air infiltration rate of less than one third of the industry standard of 0.5 cfm. (Rolscreen Co.)

Fig. 11-3. Horizontal sliding window is sometimes called "glide-by" unit. (Andersen Corp.)

Fig. 11-4. Casement windows are hinged on the side and are controlled by a crank or push bar.

AWNING

Awning windows, Fig. 11-5, have one or more sash that are hinged at the top and swing out at the bottom. They are often combined with fixed units to provide ventilation. Several operating sash can be stacked vertically in such a way that they close on themselves or on rails that separate the units.

Most awning windows have a so-called projected action where sliding friction hinges cause the top rail to move down as the bottom of the sash swings out. Crank and push bar operators are similar to those on casement windows. Screens and storm sash are mounted on the inside. Awning windows are often installed side by side to form a "ribbon" effect. Such an installation provides privacy for bedroom areas and also permits greater flexibility in furniture arrangements along outside walls.

Consideration of outside clearance must be given to both casement and awning windows. When open, they may interfere with movement on porches, patios, or walkways that are located adjacent to outside walls.

HOPPER

The hopper window has a sash that is hinged along the bottom and swings inward, Fig. 11-6. It is operated by a locking handle located in the top rail of the sash. Hopper windows are easy to wash and maintain. They often interfere with drapes, curtains, and the use of inside space near the window.

Fig. 11-5. Awning windows have one or more sash that are hinged at the top. (Andersen Corp.)

Fig. 11-6. Hopper window is hinged at bottom and swings inward.   (Andersen Corp.)

MULTIPLE-USE WINDOW

The multiple-use window is a single outswinging sash designed so it can be installed in either a horizontal or vertical position. Fig. 11-7 shows two units installed to operate similar to a casement window. These windows are simple in design and do not require complicated hardware.

JALOUSIES

A jalousie window is a series of horizontal glass slats held at each end by a movable metal frame. The metal frames are attached to each other by levers. The slats tilt together in about the same man-

ner as a venetian blind. Jalousie windows provide excellent ventilation. Weather-tightness values are low. Their use in northern climates is usually limited to porches and breezeways.

FIXED WINDOWS

The fixed window unit can be used in combination with any of the movable or ventilating units. Its main purpose is to provide daylight and a view of the outdoors. When used in this type of installation the glass is set in a fixed sash mounted in a frame that will match the regular ventilating win-

Fig. 11-7. Multi-use windows have been installed vertically.   (Andersen Corp.)

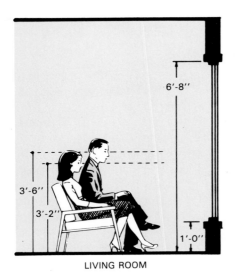

LIVING ROOM

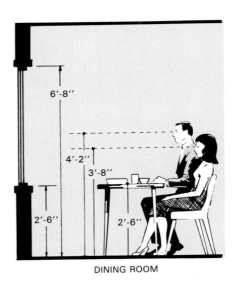

DINING ROOM

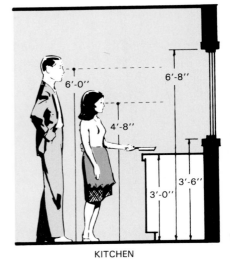

KITCHEN

Fig. 11-8. Architectural drawings should take into account the standards for window heights. Horizontal framework should be avoided at eye levels shown.

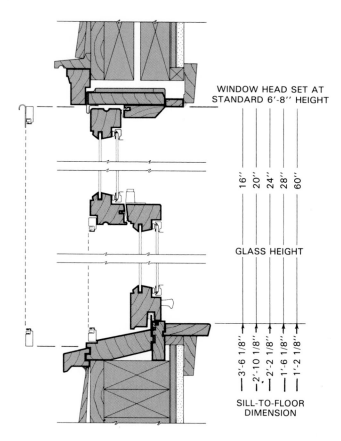

WINDOW HEAD SET AT
STANDARD 6'-8'' HEIGHT

16'' 20'' 24'' 28'' 60''

GLASS HEIGHT

3'-6 1/8'' 2'-10 1/8'' 2'-2 1/8'' 1'-6 1/8'' 1'-2 1/8''

SILL-TO-FLOOR
DIMENSION

Fig. 11-9. Manufacturer's product literature gives window head and sill heights. (Rolscreen Co.)

FLOAT GLASS		GLAZING QUALITY	NOMINAL THICKNESS	
			in	mm
Clear	SS	B	3/32	2.5
	DS	B, Select	1/8	3
			5/32	4
			3/16	5
		Mirror	1/4	6
		Select		
Heavy Duty Clear		Select	5/16	8
			3/8	10
			1/2	12
			5/8	15
			3/4	19
			7/8	22

Fig. 11-10. Chart lists standard thicknesses of glass used in residential construction. Metric thicknesses are not exact equivalents of Conventional measure. However, they do represent sizes currently being produced in other countries. (Libby-Owens-Ford Co., Glass Div.)

dows. Large sheets of plate glass are often separated from other windows to form ''window-walls.'' They are usually set in a special frame formed in the wall opening.

WINDOW HEIGHTS

One of the important functions of a window is to provide a view of the outdoors. The architect will be aware of the various considerations and drawings should reflect the dimensions shown in Fig. 11-8. Kitchens used by handicapped in wheelchairs will have different window requirements, of course.

In residential construction the standard height from the bottom side of the window head to the finished floor is 6'-8''. When this dimension is used, the heights of window and door openings will be the same. If inside and outside trim must align, 1/2 to 3/4 in. must be added to this height for thresholds and door clearances. Window manufacturers usually provide exact dimensions for their standard units. See Fig. 11-9.

WINDOW GLASS

Sheet glass used in regular windows is produced by floating. In this process, melted glass flows onto the flat surface of molten tin contained in a vat more

than 150 ft. long. As it flows over the tin, a ribbon of glass is formed that has smooth, parallel surfaces. The glass cools, becoming a rigid sheet which is then carried through an annealing oven on smooth rollers. Finally, the continuous sheet of glass is inspected and cut to usable sizes. Fig. 11-10 lists standard thicknesses produced. The letters SS and DS stand for single-strength and double-strength.

ENERGY EFFICIENT WINDOWS

Glass areas of a dwelling account for much of the heat loss in winter and heat gain in summer. Glass conducts heat more readily than most other building materials.

The resistance of any material to passage of heat is measured in R values. When a building material has a low R value, it means that the material has little resistance to heat passage.

A single pane of glass has an R value of about 0.88. A second pane of glass with a 1/2 in. air space will increase the R value to about 2.00. Storm sashes have long been used to improve R value of window space. Normally it was attached to the outside of the window frame and was removed and stored in the spring. ''Triple track'' storms did away with the removal and storage problem.

Another variation of the storm sash is the storm panel. See Fig. 11-11. Panes of glass are mounted in metal frames. The frame can be attached to either the inside or outside of the window sash. It is generally used on horizontal sliding sash and on casement or other hinged sash. When a higher R value is required, the storm panel can be equipped with sealed double glazing as shown.

271

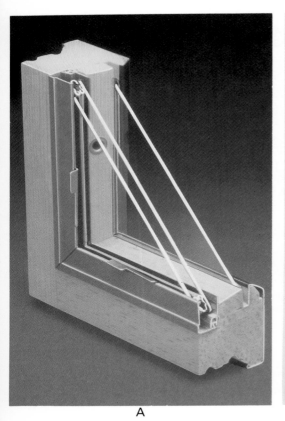

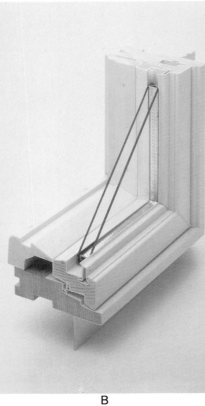

Fig. 11-11. Cutaway views of energy efficient windows with removable interior panels. A—Triple glazing with 3/4 in. and 1/4 in. air spaces (R 3.23). (Rolscreen Co.) B—Sealed double glazed unit installed in a modern window sash. Flange projecting from the bottom of the window is used to mount and fasten frame in structural opening. Standard insulating glass, as shown, has an R rating of 2.0. High performance insulating glass has an R 3.3 rating. (Andersen Corp.) C—Double glazed unit has 3/16 in. air space for an R 2.43 insulating factor.

DOUBLE AND TRIPLE SEALED GLAZING

For movable sash, two or three layers of 1/8 in. glass are fused together with a 3/16 in. air space between layers. Double or triple layers of plate glass are used in large fixed units. Special seals are used to trap the air between panes. Air spaces are generally from 1/4 to 1/2 in. The air is dehydrated (moisture removed) before the space is sealed. Fig. 11-11 (B) shows sealed double glazing in a standard casement unit.

Double and triple glazing offers the following advantages:
1. Lower heat loss in cold weather.
2. Down drafts along window surface are reduced.
3. Heat penetration in summer months is reduced.
4. Sweating and fogging of windows in cold weather is reduced or eliminated.
5. Less outside noise is transmitted through the window.

LOW-EMISSIVITY GLAZING

Emissivity is the relative ability of a material to absorb or re-readiate heat. Research in the area of glazing technology has resulted in a totally new method of raising the R value of double glazed windows. Commonly referred to as ''low-e'' or ''high performance'' windows, they have a clear outer pane, an air space, and a special coating on the air-gap side of the inner pane.

The special factory-applied coating consists of an atoms-thin layer of metal oxide. It reflects infrared (heat wave) radiation but lets regular light waves pass through.

During winter months, warm surfaces within a room (wall, floor, furniture) radiate heat waves. When these waves strike the low-emissivity surface of the window, they are reflected back into the room—thus reducing heat loss. During summer months, heat waves from walks, drives, and other outside surfaces are prevented from entering. This lowers air-conditioner loads.

Fig. 11-12 shows a low-emissivity glass panel being attached to a window sash. The narrow slat blind, which is suspended between the two panes of glass, is also coated with a low-emissivity finish. The total R value for this kind of window and blind is 4.35.

Fig. 11-12. Removable interior insulating panel is easy to install. Inside surface of the panel has a low-emissivity coating which reflects radiant heat waves. (Rolscreen Co.)

SCREENS

Ventilating windows require screens to keep out insects. The mesh should have a minimum of 252 openings per square inch, and be made of such non-corrosive materials as aluminum, bronze, plastic, or stainless steel. Manufacturers have perfected many unique methods of mounting and storing screen panels. They are attached to the inside of the window frame on horizontal sliding and outward swinging windows. Most modern screens use a light metal frame as shown in Fig. 11-13.

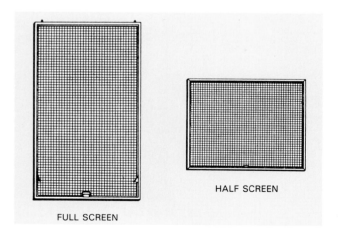

HALF SCREEN

FULL SCREEN

Fig. 11-13. Window screens. Half screens are used on double-hung windows and fit under the top sash and between the jambs.

MUNTINS

Years ago window glass was available only in small sheets. By using rabbeted strips called muntins, small panes of glass could be used to fill large openings.

Today, even though large sheets of glass are available, muntins are still used for special effects in traditional architecture. They are, however, usually applied as an overlay and do not actually separate or support small panes of glass. Made of wood or plastic and in various patterns—they snap in and out of the sash for easier painting and cleaning, Fig. 11-14.

PARTS OF WINDOWS

Because much of the actual construction of a window is hidden, sectional views are used to show the parts and how they fit together. It is standard practice to use sections through the top, side, and bottom of a window. See Fig. 11-15. These drawings also include wall framing members and surface materials. See Fig. 11-16.

A section view of a typical mullion shows how window units fit together. A mullion is formed by the window jambs when two units are joined together as shown in Fig. 11-17.

A drip cap, shown in the head section of Fig. 11-16 is designed to carry rain water out over the window casing. When the window is protected by a wide cornice, this element is seldom included.

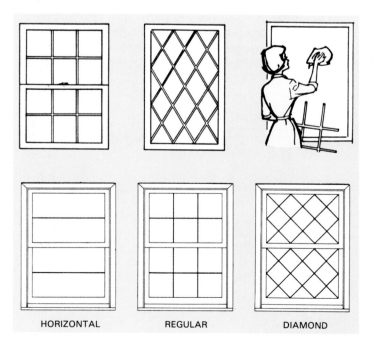

HORIZONTAL   REGULAR   DIAMOND

Fig. 11-14. Top. Muntin assemblies detach easily for cleaning and painting. Below. Standard muntin patterns.

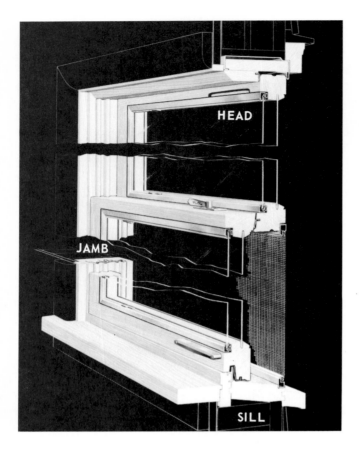

Fig. 11-15. Standard sections used to show details of all types of windows. (Rolscreen Co.)

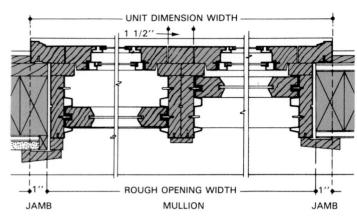

Fig. 11-17. Section drawing of a mullion. This is where two window jambs are joined. (Rock Island Millwork)

WINDOWS IN PLANS AND ELEVATIONS

The carpenter studies the plans and elevations of the working drawings to become informed about the type of windows and their location. It is common practice to locate the horizontal position of windows and exterior doors by including a dimension line to the center of the opening as shown in Fig. 11-18.

In masonry construction the dimension is given to the edge of the opening. The latter method is sometimes used in frame construction.

Elevations will show the type of windows, Fig. 11-19, and may include glass size and heights. The position of the hinge line (point of dotted line) will indicate the type of swinging window. Sliding windows require a note to indicate they are not fixed units. Supporting mullions will be included in the plans and also in elevations. See Fig. 11-20.

WINDOW SIZES

Besides the type and position of the window, the carpenter should know the size of each unit or combination of units. Window sizes may include several

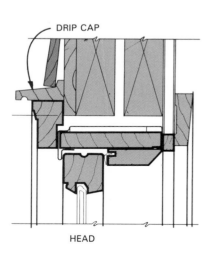

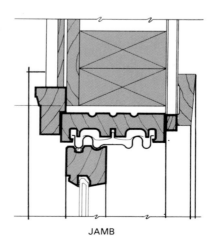

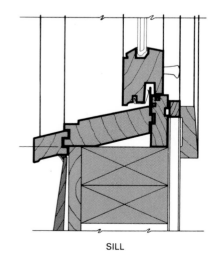

Fig. 11-16. Typical detail drawing shows a section through a window. Such details are included with the house plans. They help the carpenter to see how windows should be installed.

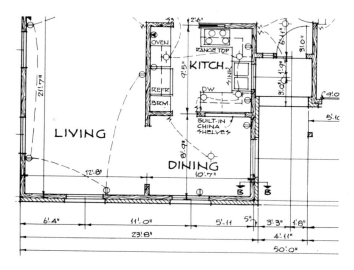

Fig. 11-18. Floor plans show location of windows and doors.

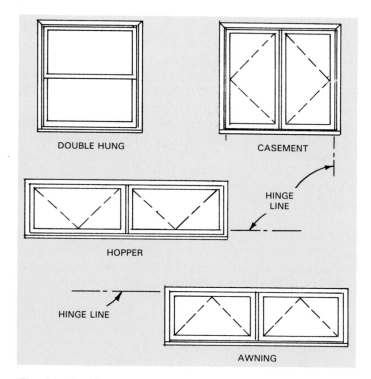

Fig. 11-19. How window types are shown in elevation views. Horizontal sliding windows are noted to define them from fixed units.

or all of the following:
1. Glass size.
2. Sash size.
3. Rough frame opening.
4. Masonry or unit opening.

Fig. 11-21 shows the position of these measurements and approximately how they are figured from the glass size. They will vary slightly from one manufacturer to another.

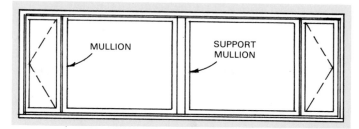

Fig. 11-20. Elevation of windows will show supporting and nonsupporting mullions.

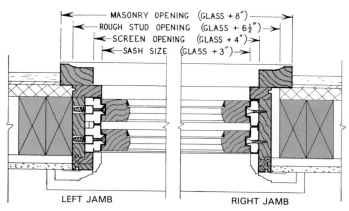

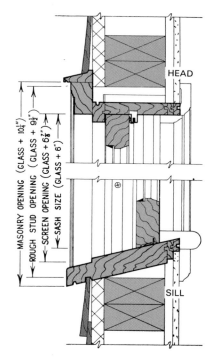

Fig. 11-21. Window sizes and location of measurements. The rough opening is larger than the overall size of the frame to permit alignment and leveling when unit is installed.

A complete set of architectural drawings should provide detailed information about window sizes. This information is usually listed in a table called a window and door schedule. Refer to Fig. 5-19, page 106. It includes, among other things, the manufacturer's numbers and rough opening sizes for each unit or combination. An identifying letter is located at each opening on the plan and a corresponding letter is then used in the schedule to specify the required window unit and the necessary information.

When this information is not included in the architectural plans, the carpenter will need to study manufacturer's catalogs and other descriptive literature. Sizes for basic units are often given in diagrams as shown in Fig. 11-22. The size of the rough opening is of major importance to the carpenter. However, she or he will also need other dimensions—for example, the height of the R.O. above the floor.

When window and door sizes are given, whether they consist of rough openings, sash size, or other items, the horizontal dimension is listed first and the height second. For example, the unit No. 222W shown in Fig. 11-22 requires a R.O. of 4'-8 1/2'' × 2'-7''.

Rough openings and other sizes are readily secured from manufacturer's data. An example of sizes for casement combinations is shown in Fig. 11-23. For other special combinations, apply the amounts specified in the details included at the bottom of Fig. 11-23. Note that the extra allowance for a support mullion is 2 in.

DETAILED DRAWINGS

Sectional drawings are helpful to the carpenter since they show each part of the window and how the unit is placed in a wall structure. Fig. 11-24 shows detail drawings for a typical double-hung window. Similar drawings are available for other types of windows from other manufacturers. The architect will often include selected detail views. Whether included in the architectural drawings or made available through manufacturers' catalogs, detailed drawings such as these will be essential in building the rough frame of the wall structure and in installing the window units. See Fig. 11-25

JAMB EXTENSIONS

The thickness of window units may be adjusted to various walls. For example, a frame wall with 3/4 in. sheathing and a standard interior surface of lathe and plaster may be 5 1/8 in. compared with a single-layer drywall construction of about 4 3/4 in. Adjustments in the window frame may be made by applying a jamb extension, Fig. 11-26. Some manufacturers build their frames to a basic size, such as 4 5/8 in. Then they equip the unit with an extension as specified by the builder or architect.

STORY POLE

A door and window story pole (refer again to Fig. 8-13, page 179) is helpful when installing doors and windows. It will help insure the alignment of doors and window heads, Fig. 11-27. It can be marked with the additional dimensions given in Fig. 11-28. To use the pole, hold it against the trimmers and transfer the marks.

The construction details of the window head will normally be the same throughout a building. The sill height may vary. Additional positions can be superimposed over other layouts if each one is carefully labeled or indexed so the correct distance will always be applied to the proper opening.

Residential doors are normally 6'-8'' high. The tops of windows are usually held at this same height. See Fig. 11-29. An extra 1/2 in. is added to allow for thresholds under entrance doors and clearance under interior doors.

INSTALLING WINDOWS

If rough openings are plumb, level, and the correct size, it is easy to install modern windows.

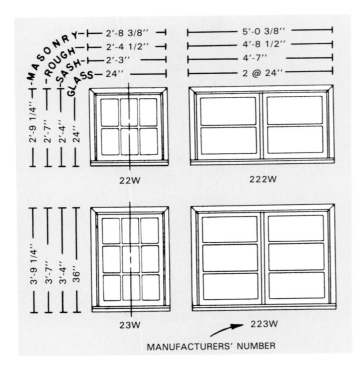

Fig. 11-22. This is typical of illustrations used by window manufacturers to give sizes of their standard units. (Rolscreen Co.)

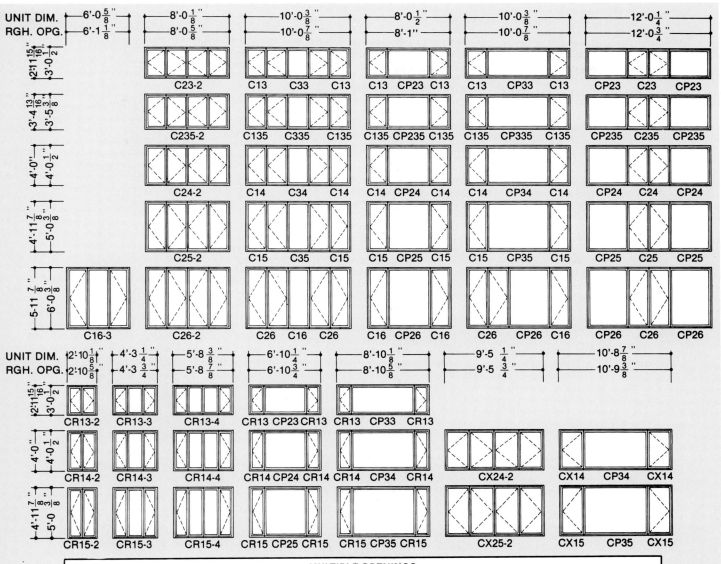

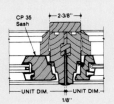  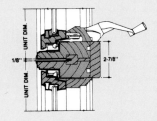

MULTIPLE OPENINGS

A number of suggested combinations are shown above using the narrow (no support) mullion.

Additional combinations using support mullion and transom joining can be arrived at by using the dimensions in the formulas listed below.

CASEMENT NARROW MULLION

Joining basic casement units to form multiple units or picture window combinations without vertical support between units.

scale: 1½″ = 1′0″

Overall Unit Dimension Width — The sum of individual unit dimensions, plus ⅛″ for each unit joining.

Overall Rough Opening Width — Add ½″ to Overall Unit Dimension Width.

CASEMENT SUPPORT MULLION

Joining basic casement units using a 2 x 4 vertical support between units.

Overall Unit Dimension Width — The sum of individual unit dimensions, plus 2″ for each unit joining.

Overall Rough Opening Width — Add ½″ to Overall Unit Dimension Width.

CASEMENT TRANSOM

Joining basic casement units by stacking units to form combinations.

Overall Unit Dimension Height — The sum of individual unit dimension heights, plus ⅛″ for each unit joining.

Overall Rough Opening Height — Add ½″ to Overall Unit Dimension Height.

Fig. 11-23. Manufacturer's catalogs and brochures picture the variety of window units they supply. This page shows sizes of casement and picture window units available. (Andersen Corp.)

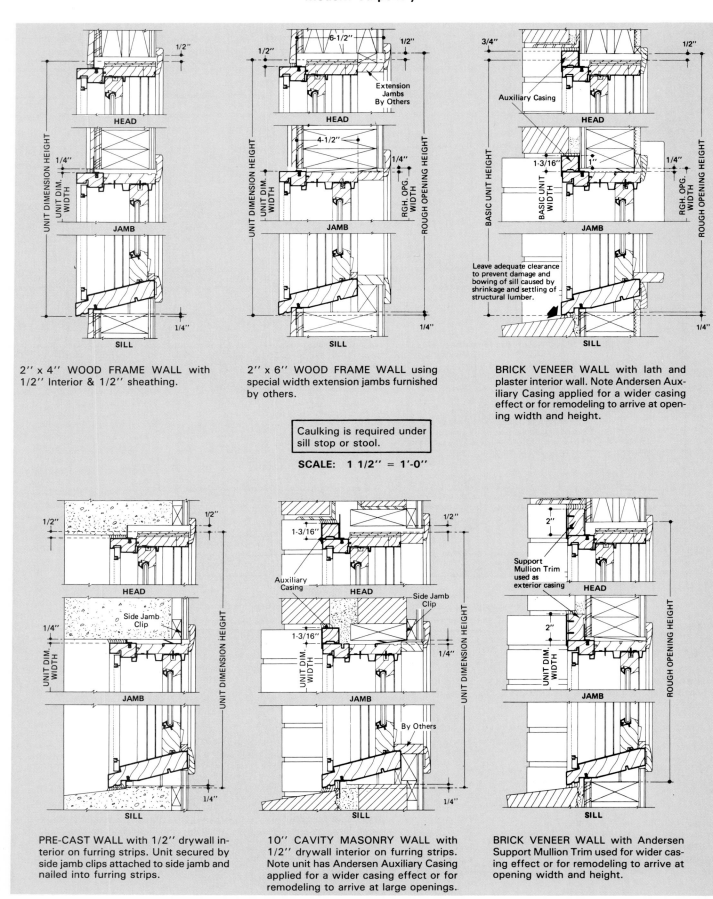

HEAD

JAMB

SILL

2'' x 4'' WOOD FRAME WALL with 1/2'' Interior & 1/2'' sheathing.

2'' x 6'' WOOD FRAME WALL using special width extension jambs furnished by others.

BRICK VENEER WALL with lath and plaster interior wall. Note Andersen Auxiliary Casing applied for a wider casing effect or for remodeling to arrive at opening width and height.

Caulking is required under sill stop or stool.

SCALE: 1 1/2'' = 1'-0''

PRE-CAST WALL with 1/2'' drywall interior on furring strips. Unit secured by side jamb clips attached to side jamb and nailed into furring strips.

10'' CAVITY MASONRY WALL with 1/2'' drywall interior on furring strips. Note unit has Andersen Auxiliary Casing applied for a wider casing effect or for remodeling to arrive at large openings.

BRICK VENEER WALL with Andersen Support Mullion Trim used for wider casing effect or for remodeling to arrive at opening width and height.

Fig. 11-24. Detail drawings show a standard double-hung window installed in different wall structures. (Andersen Corp.)

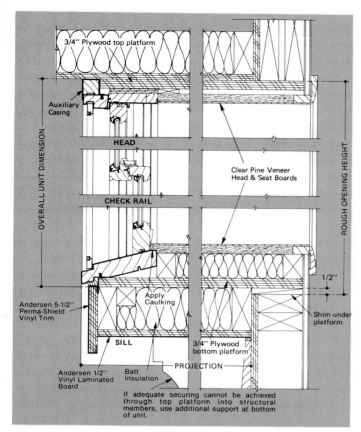

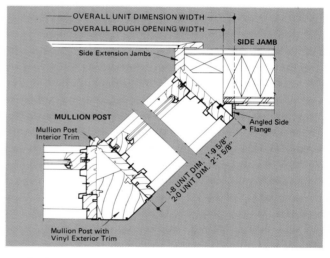

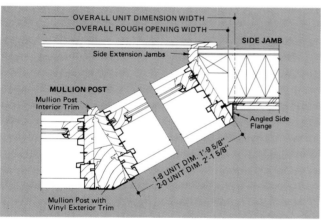

Fig. 11-25. Bay window construction is more complicated than standard windows. Details for installation are extremely important for the carpenter so that she or he will do a good job. (Andersen Corp.)

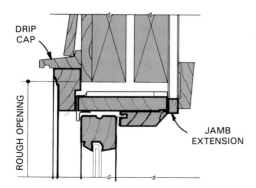

Fig. 11-26. Jamb extensions are intended to be applied to a standard window frame to adjust for various wall thicknesses.

Manufacturers furnish directions that apply specifically to their various products. The carpenter should follow them carefully.

When windows are received on the job they should be stored in a clean, dry area. If they are not fully packaged, some type of cover should be used to prevent damage from dust and dirt. You should allow wood windows to adjust to the humidity of the locality before they are installed.

Check carton labels and move window units (still packaged) into the various rooms and areas where they will be installed.

Unpack the window and check for shipping damage. Do not remove any diagonal braces or

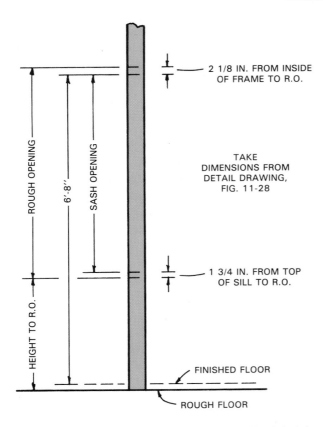

2 1/8 IN. FROM INSIDE
OF FRAME TO R.O.

TAKE
DIMENSIONS FROM
DETAIL DRAWING,
FIG. 11-28

1 3/4 IN. FROM TOP
OF SILL TO R.O.

FINISHED FLOOR

ROUGH FLOOR

Fig. 11-27. Use the story pole once more to check height of certain parts of the window frame.

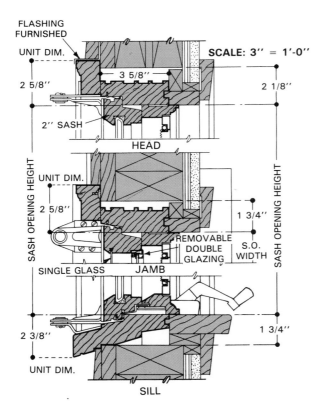

Fig. 11-28. Details of a window may also carry information on installation heights of certain window features.

Fig. 11-29. Usually, tops of doors and window frames are at same level. (Andersen Corp.)

spacer strips until after the installation is complete.

Check the rough opening to make certain it is the correct size. Most windows require at least 1/2 in. clearance on each side and 3/4 in. above the head for plumbing and leveling. When board sheathing is used, a strip of building paper should be tacked around the rough opening to reduce air infiltration.

If the window has not been primed at the factory, this may be done before installation. Weather stripping and special channels should not be painted. Follow the manufacturer's recommendations. Although outside casings are usually attached, some windows, especially those with a metal frame, are set in the opening. The outside trim is then installed.

Most window units and multiple unit combinations are installed from the outside. The window, if stored inside, can be easily moved to the outside through the rough opening in single story construction. Secure sufficient help to carry windows and handle them carefully, Fig. 11-30. Place the window in the opening, Fig. 11-31, and secure it temporarily.

Fig. 11-30. Windows must be handled carefully. Get help when installing larger units.

Fig. 11-31. Place the window in the opening and fasten it temporarily. Use extra props in windy weather.

Fig. 11-32. Adjust shims and wedges until the frame is level. (Andersen Corp.)

Fig. 11-33. Permanently fasten the window frame by nailing through the casing into the frame of the building.

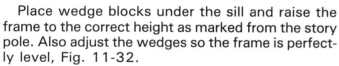

Workers' hands should be clean when handling windows. The National Woodwork Manufacturers Association recommends wearing clean canvas gloves.

Place wedge blocks under the sill and raise the frame to the correct height as marked from the story pole. Also adjust the wedges so the frame is perfectly level, Fig. 11-32.

There may be a tendency for the sill to sag on wide windows or multiple units. Place wedge blocks not only at the ends but at several places in the center. Nail through the lower end of the side casing to secure the bottom of the frame.

Plumb the side jambs with the level and check the corners with a framing square. Usually the sash should be closed and locked in place. Drive nails temporarily into the top of the side casing. Now check over the entire window to see if it is square and level. Open the ventilating sash to see that they operate smoothly. A sag in the head or bow in the jambs should be straightened with a spacer strip.

Finally, nail the window permanently in place with aluminum or galvanized casing nails. Space the nails about 16 in. O.C. and be certain they are long enough to penetrate well into the building frame, Fig. 11-33. Window casing is made of soft wood and will dent easily. Therefore, a nail set should be

used for the last driving strokes to sink the nail head below the surface.

At this point, many builders prefer to cover the inside of the window with a sheet of polyethylene film to protect it during the application of inside wall surface materials.

INSTALLING FIXED UNITS

The need for ventilating windows will likely be reduced as more buildings are air conditioned. Fixed units (do not open) are less expensive than those that are movable.

When the fixed glass panel is of medium size, it is usually mounted in a sash and frame, Fig. 11-34. Often it is combined with matching ventilating units. The installation of such a unit is the same as for regular windows, Fig. 11-35. They are, of course, larger and heavier and extra precautions should be observed in handling and making the installation.

Large insulating units are made of 1/4 in. glass which results in considerable weight, Fig. 11-36. They are seldom installed in window frames at the factory although they are sometimes mounted in a

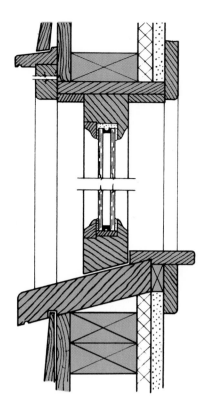

Fig. 11-34. Detail of 1 in. double glazed window in a conventional sash and frame.

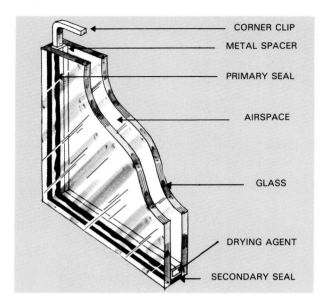

Fig. 11-36. Cutaway view of modern double glazed window unit. Two layers of 1/4 in. glass are separated by a dry air space which is hermetically sealed.
(Libby-Owens-Ford Co., Glass Div.)

CORNER CLIP
METAL SPACER
PRIMARY SEAL
AIRSPACE
GLASS
DRYING AGENT
SECONDARY SEAL

Fig. 11-35. Picture window units are heavy and must be handled carefully to avoid damage to the unit or injury to carpenters. (Andersen Corp.)

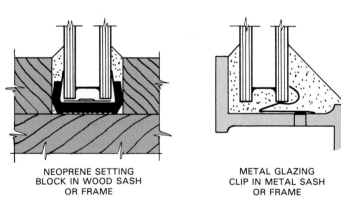

NEOPRENE SETTING
BLOCK IN WOOD SASH
OR FRAME

METAL GLAZING
CLIP IN METAL SASH
OR FRAME

Fig. 11-37. Setting blocks and clips. The use of metal clips is limited to welded glass units.

Openings must be square, free of twists, and rugged enough to bear the weight of the glass unit. Use only high grade wood materials that are dry and free from warp. Special setting blocks and clips, Fig. 11-37, may be used to hold the glass in position with clearance on all edges. For wood frames or sash, use two neoprene setting blocks (at least 4 in. long) located at quarter points along the lower edge. The width of the blocks should be equal to, or greater than, the thickness of the glass unit.

The edge should be completely surrounded with a high-quality, nonhardening glazing sealant. There must be no direct contact between the glass unit and frame. Fig. 11-38 shows clearances recommended by one manufacturer.

sash. Usually, large glass units are glazed (set in opening with glazing sealant) as a separate operation after the frame and/or sash have been installed.

Windows and Exterior Doors

GLASS THICKNESS		DIMENSIONAL TOLERANCE				MINIMUM CLEARANCE				BITE C	
		UP TO 48" 1220mm		OVER 48" 1220mm		FACE A		EDGE B			
in	mm	in	mm	in	mm	in	mm	in	mm	in	mm
1/2	12	±1/16	±1.6	+1/8 −1/16	+3.2 −1.6	1/8	3.2	1/8	3.2	1/2	12.7
5/8	15	+1/8 −1/16	+3.2 −1.6	+3/16 −1/16	+4.8 −1.6						
23/32	18										
3/4¹	19										
3/4²	19	±1/16	±1.6	+1/8 −1/16	+3.2 −1.6	3/16	4.8	1/4	6.4	1/2	12.7
7/8	22	+1/8 −1/16	+3.2 −1.6	+3/16 −1/16	+4.8 −1.6						
31/32	24										
1	25										

¹ 1/4" (6 mm) Air Space
² 1/2" (12 mm) Air Space

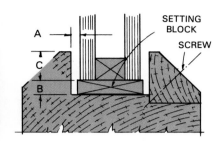

Fig. 11-38. Recommended clearances for sealed insulating glass. The "C" dimension is commonly referred to as bite.

1—Select a good grade of softwood lumber like ponderosa pine. For a medium sized frame use a 1 1/2 in. thickness. Make the frame slightly larger than the insulating glass. See recommended clearances listed in Fig. 11-38. After assembly, apply a coating of water-repellent preservative.

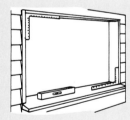

2—Install frame in opening. Place wedge blocks under sill and at several points around perimeter of frame. When sill is level and frame perfectly square, nail the frame securely to structural members of the building.

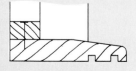

3—Prepare inside stop members according to recommended sizes (Fig. 11-38). Use miter joints in corners and nail down stops. Also prepare outside stops.

4—Using recommended grade of glazing sealant, apply thick bed to stops, and top, sides, and bottom of frame. Apply enough material to fill space between frame and glass unit when installation is made.

5—Install neoprene setting blocks on bottom of frame or edge of glass unit as shown. Blocks (use only two) should be moved in from the corners a distance equal to one-quarter of frame width.

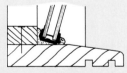

6—Place bottom edge of insulating glass unit on setting blocks as shown. Then carefully press unit into position against stops until proper thickness of glazing sealant is secured around entire perimeter. Small gauge blocks may be helpful.

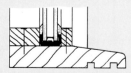

7—Nail on outside stop. Additional sealant may be required to fill joint. Remove excess sealant from both sides of unit and clean surfaces with an approved solvent.

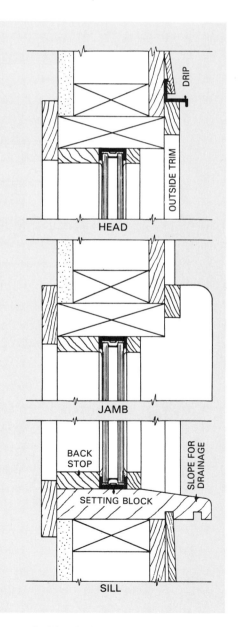

Fig. 11-39. Details and method for constructing a frame and installing sealed insulating glass.

All glazing systems should be designed and installed to insure that the seal between the layers of glass (organic type) are not exposed to water for long periods of time. The inclusion of weep holes as recommended by SIGMA (Sealed Insulating Glass Manufacturers Assoc.) should be included.

Insulating glass units cannot be altered in any way on the building site. It is essential, then, that sash and frames be carefully designed and that specified dimensions are followed. Wooden window frames should be treated with a wood preservative.

Fig. 11-39 gives step-by-step procedures for building a simple wood frame and installing insulating glass. Some standard thicknesses of insulating glass for fixed window units are listed in Fig. 11-40.

UNIT CONSTRUCTION							
SINGLE GLASS THICKNESS		OVERALL UNIT THICKNESS		AIR SPACE		APPROXIMATE WEIGHT	
in	mm	in	mm	in	mm	lb/ft²	kg/m²
1/8	3	1/2	12	1/4	6	3.27	16
		3/4	19	1/2	12		
3/16	5	5/8	15	1/4	6	4.90	24
		7/8	22	1/2	12		
		1	25	5/8	15		
1/4	6	3/4	19	1/4	6	6.54	32
		1	25	1/2	12		

Fig. 11-40. Glass manufacturers publish charts of specifications for the glass they produce. (Libby-Owens-Ford Co., Glass Div.)

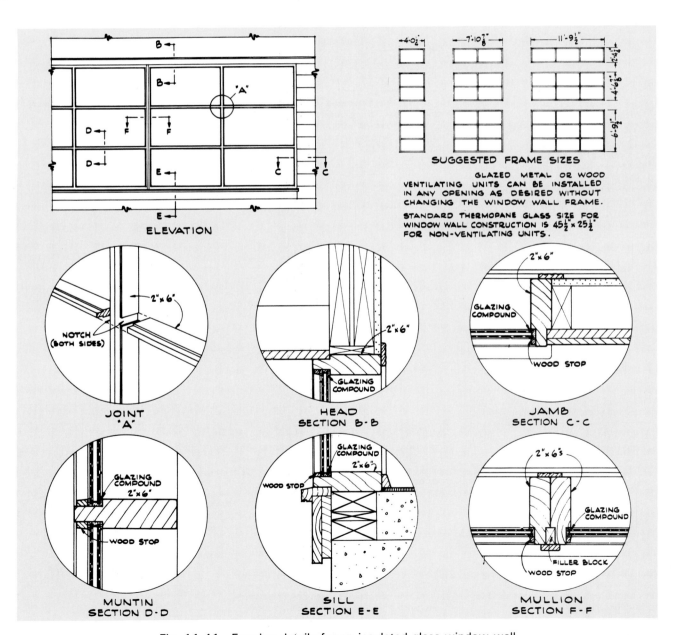

Fig. 11-41. Framing details for an insulated glass window wall.

Details for the construction of a window wall are shown in Fig. 11-41. The photo, Fig. 11-42, is an example of how window walls can create pleasing effects by bringing the out-of-doors inside.

GLASS BLOCKS

Glass blocks, Fig. 11-43, have good insulating properties. Used in outside walls they provide light, help prevent drafts, dampen disturbing noises, cut off unpleasant views, and insure privacy where it is desired. Inside the home, partitions and screens of glass blocks add a pleasant touch to rooms they divide.

Glass blocks are made of two formed pieces of glass fused together to leave an insulating air space between. They come in several different patterns and are usually available in three nominal sizes: 6 x 6 in., 8 x 8 in., and 12 x 12 in. See Fig. 11-44. All are 3 7/8 in. thick. Special shapes are available for turning corners and for building curved panels. The blocks come in both light-diffusing and light-directing types.

Installing glass blocks is not difficult. Use regular masonry tools. Even though the carpenter will seldom make the actual installation, she or he will be required to build the framework. Therefore, knowledge of the design requirements is helpful.

INSTALLING SMALL GLASS BLOCK PANELS

Details for installing glass block panels of 25 sq. ft. and less are given in Fig. 11-45. In such panels, the height should not exceed 7 ft. nor the width 5 ft.

Panels may be supported by a "mortar key" at jambs in masonry, or by wood members in frame construction. No wall anchors and no wall ties are required in the joints. Expansion space is required at the head only.

INSTALLING LARGE GLASS BLOCK PANELS

When panel areas exceed 25 sq. ft., additional requirements must be met. See Fig. 11-45. Expansion strips (strips of resilient material) are used to partially fill expansion spaces at jambs and heads of larger panel openings.

Panels should never be larger than 10 ft. wide or 10 ft. high. Provide support at jambs with wall anchors or wood members. A portion of each anchor is embedded in masonry and in the glass block mortar joint. Anchors should be crimped within expansion spaces and spaced on 24 in. centers to rest in the same joints as wall ties. Anchors, which are corrosion resistant, are 2 ft. long and 1 3/4 in. wide.

Wall ties should be installed on 24 in. centers in horizontal mortar joints of larger panels and lap not

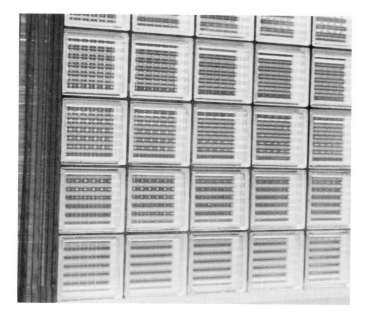

Fig. 11-43. Glass block windows provide a high level of security and privacy for basements and ground floor levels.

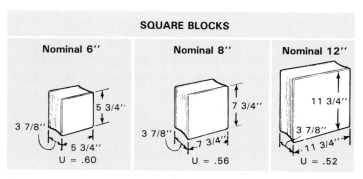

Fig. 11-44. Nominal and actual sizes of commonly used glass block.

Fig. 11-42. View from a window wall tends to create the feeling of bringing the outdoors inside.

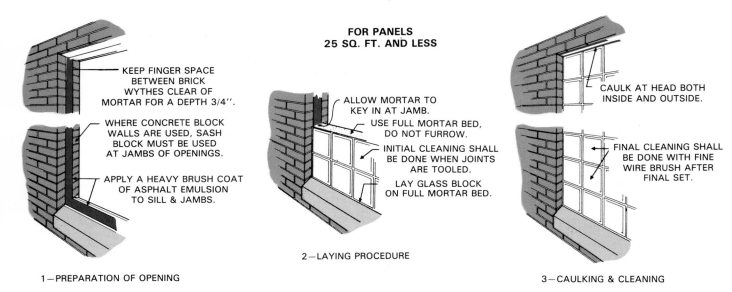

**FOR PANELS
25 SQ. FT. AND LESS**

KEEP FINGER SPACE BETWEEN BRICK WYTHES CLEAR OF MORTAR FOR A DEPTH 3/4".

WHERE CONCRETE BLOCK WALLS ARE USED, SASH BLOCK MUST BE USED AT JAMBS OF OPENINGS.

APPLY A HEAVY BRUSH COAT OF ASPHALT EMULSION TO SILL & JAMBS.

ALLOW MORTAR TO KEY IN AT JAMB.

USE FULL MORTAR BED, DO NOT FURROW.

INITIAL CLEANING SHALL BE DONE WHEN JOINTS ARE TOOLED.

LAY GLASS BLOCK ON FULL MORTAR BED.

CAULK AT HEAD BOTH INSIDE AND OUTSIDE.

FINAL CLEANING SHALL BE DONE WITH FINE WIRE BRUSH AFTER FINAL SET.

1—PREPARATION OF OPENING

2—LAYING PROCEDURE

3—CAULKING & CLEANING

TYPICAL INSTALLATION DETAILS

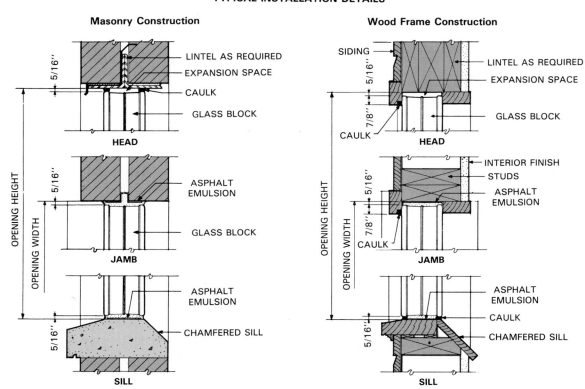

Masonry Construction

Wood Frame Construction

LINTEL AS REQUIRED
EXPANSION SPACE
CAULK
GLASS BLOCK
ASPHALT EMULSION
GLASS BLOCK
ASPHALT EMULSION
CHAMFERED SILL

SIDING
LINTEL AS REQUIRED
EXPANSION SPACE
GLASS BLOCK
INTERIOR FINISH
STUDS
ASPHALT EMULSION
CAULK
ASPHALT EMULSION
CAULK
CHAMFERED SILL

Fig. 11-45. Construction details for installing glass block. (PPG Industries, Inc.)

less than 6 in. whenever it is necessary to use more than one length of tie. Do not bridge expansion spaces. Ties are also corrosion resistant. They are 8 ft. long and 2 in. wide.

OPENINGS FOR GLASS BLOCKS

To determine heights or widths of openings required for panels, multiply the number of units by the nominal block size and then add 3/8 in. For example: what size opening would be required for a panel consisting of 8 in. blocks that was four units wide by five units high?

Width = 4 × 8 + 3/8 = 32 3/8 = 2'-8 3/8''
Height = 5 × 8 + 3/8 = 40 3/8 = 3'-4 3/8''

REPLACING WINDOWS

Older style windows waste tremendous amounts of heating and cooling energy. Today, many of these windows are being replaced with modern units that have excellent weather stripping and double or triple glazing. Manufacturers provide a wide range of sizes and also a variety of special trim members that are helpful in making the replacement. They also provide detailed instructions for installing their products.

First considerations in window replacement include the selection of type and size of the new units. To determine the size, it is best to remove the inside trim and measure the rough opening. If the trim members are to be reused, pry them off carefully. Remove nails by pulling them through the trim from the back side.

After the new window units are delivered and you have checked for size, remove the old units. Begin by removing the inside stops and lift out the lower sash. Weights or counterbalances may need to be disconnected. Next remove the parting stops and lift out the upper sash.

With the window sash removed, pry off the outside casing. Make saw cuts through the sill and frame, collapse the members and remove them from the opening as shown in Fig. 11-46.

Clean the rough opening to remove dirt, plaster, putty, and nails. Carefully measure the opening to determine the thickness of material needed to bring the rough opening to required size. Install the strips as shown in Fig. 11-47. Constantly check with a square and level. Use shims where necessary.

Check the rough opening for plumb and level. Insert the new unit, Fig. 11-48. With the window resting on the rough sill, make adjustments with shims as described for new construction on page 281. The window unit can be attached to the rough opening by nailing through special flanges, metal

Fig. 11-47. Installing wood furring to bring the rough opening to the correct size.

Fig. 11-48. Placing the new window unit in the rough opening. Get help if the unit is large. (Andersen Corp.)

Fig. 11-46. After removing the sash and trim members, cut through sill and frame and remove them. (Andersen Corp.)

clips, or exterior casing. Check the manufacturer's directions for their recommendations on specific units.

With the window unit carefully secured in the opening, install outside and inside trim. See Fig. 11-49. Manufacturers furnish a variety of trim pieces especially made for this purpose. Since the new window unit may not be exactly the same size as the one removed, standard casing may require various widths of filler strips.

All of these exterior trim members should be bedded in a high quality construction mastic to seal against air infiltration. If the top of the window is not well protected by roof overhang, metal flashing should be installed as shown in Fig. 12-18.

Fig. 11-49. Installing outside trim furnished by manufacturer.

Finally, completely seal the window by caulking the joints between the house siding and side trim, head trim, and sill. On the inside, carefully pack all openings between the wall and frame with insulation. Apply the trim members.

SKYLIGHTS

Modern skylights can make a dramatic difference in the appearance and feeling of a room, Fig. 11-50. Skylights in bathrooms provide good light and privacy. They may be the best way to secure natural light in stairwells and hallways. Skylights can convert attic space into living space at a fraction of the cost of dormers with regular windows.

Fig. 11-51 shows a skylight with sloped double glazing. The glazing consists of tempered glass with a 1/2 in. air space. The sash is hinged as shown and can be operated from below with the aid of an extension pole. Fixed units are also available.

Rough openings are framed in about the same way as described for chimneys. See page 215. A detail drawing of a typical installation is shown in Fig. 11-52. Flashing flanges, included in the frame structure, should be installed according to the manufacturer's directions. Skylights of this type should not be installed in roofs with less than a 3 in 12 slope.

In standard frame construction a shaft is required to connect the ceiling opening with the roof. Fig. 11-53 shows several types of shafts. It is very important that the shaft be constructed as airtight as possible. Insulation should be attached to the sides of the shaft to the same thickness as the ceiling insulation.

EXTERIOR DOOR FRAMES

Outside door frames are installed at the same time as the windows. Follow similar procedures. Secondary and service entrances usually have frames and trim members to match the windows. Main entrances, however, often contain additional elements that add an important decorative architectural feature. See Figs. 11-54 and 11-55.

Exterior doors in residential construction are nearly always 6'-8'' high, although 7'-0'' sizes are available. Main entrances usually are equipped with a single door that is 3'-0'' wide. Narrower (2'-8''

Fig. 11-50. A skylight provides natural overhead lighting for modern kitchens. (Andersen Corp.)

Fig. 11-51. Ventilating skylight. This unit is hinged at the top and opens up to 8 in. at the bottom. (Rolscreen Co.)

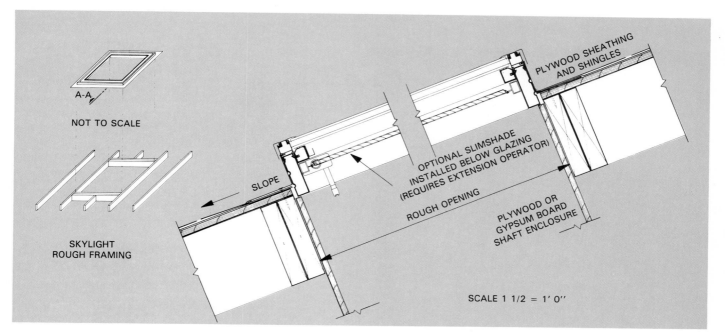

Fig. 11-52. Cross-section detail for installation of a skylight. (Rolscreen Co.)

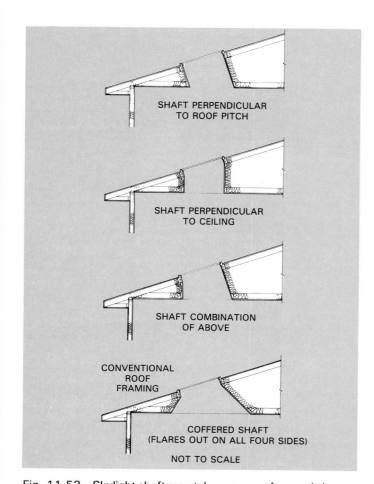

SHAFT PERPENDICULAR TO ROOF PITCH

SHAFT PERPENDICULAR TO CEILING

SHAFT COMBINATION OF ABOVE

CONVENTIONAL ROOF FRAMING

COFFERED SHAFT (FLARES OUT ON ALL FOUR SIDES)

NOT TO SCALE

Fig. 11-53. Skylight shaft may take any one of several shapes.

Fig. 11-54. Wood paneled door with insert and sidelights of leaded glass. Panels are full 1 5/32 in. thick for strength and heat loss reduction. (C-E Morgan)

Outside door frames, like windows, have heads, jambs, and sills. The head and jambs are made of 5/4 in. stock since they must carry not only the main door but also screen and storm doors. Fig. 11-56 shows an elevation view of door frames. The doors are in place.

Door frames are manufactured at a millwork plant and arrive at the building site either assembled and ready to install or disassembled (knocked down or K.D.). Sometimes K.D. units are assembled by the dealer or distributor. It is relatively easy to assemble door frames "on the job" when the joints are accurately machined and the parts are carefully packaged and marked.

While details of a door frame may vary, the

and 2'-6'') sizes are used for rear and service doors. FHA Minimum Property Standards specify a minimum exterior door width of 2'-6''.

Fig. 11-55. Main entry door. Traditional style is combined with 0.024 in. steel face with a thermal break. Cavity is insulated with expanded polystyrene foam. (Pease Industries, Inc.)

Fig. 11-56. Frames and doors for main entrances as they appear on elevation drawings.

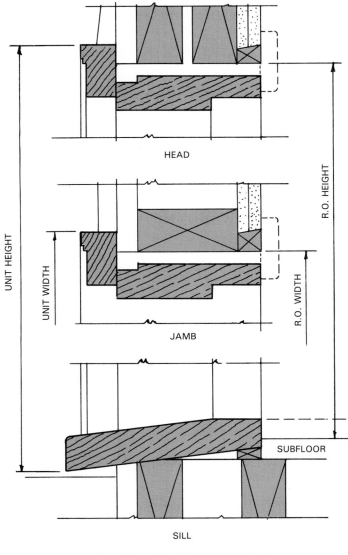

HEAD

JAMB

SILL

CALCULATIONS FOR DETERMINING FRAME
AND ROUGH OPENING SIZES:

UNIT HEIGHT = DOOR + 4 1/2 IN.
UNIT WIDTH = DOOR + 4 IN.
R.O. HEIGHT = DOOR + 2 1/2 IN.
R.O. WIDTH = DOOR + 2 1/2 IN.

Fig. 11-57. Always check drawings for exterior door frame details and sizes. Make sure rough openings are correct size.

general construction is the same, Fig. 11-57. The head and jambs are rabbeted, usually 1/2 in. deep, to receive the door.

In residential construction, outside doors swing inward and the rabbet must be located on the inside. Stock door frames are designed for standard wall framing. However, they can also be adapted to stone or brick veneer construction as shown in Fig. 11-58.

Stock frames can be fitted with extension strips, Fig. 11-59, that convert the frames to fit greater wall thicknesses. Extension strips are also used on window frames.

Door sill design varies considerably. However, the top is always level with the finished floor. Sills may be made of wood, metal, stone, or concrete. The outside stoop at the entrance is placed just below the door sill or may be lowered by a standard riser height (7 1/2 in.), Fig. 11-60.

Positioning a door sill so it will be level with the finished floor requires cutting away a section of the

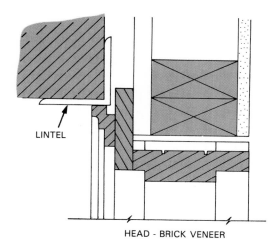

Fig. 11-58. Detail on how to install head of door frame in brick veneer construction.

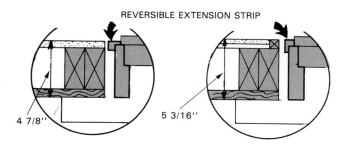

Fig. 11-59. Stock frames can be made wider. Reversible extension strips above convert 4 5/8 in. jambs to either 4 7/8 in. or 5 3/16 in. (C-E Morgan)

Fig. 11-60. Height of door sill may vary. Left. Raised slightly above porch, stoop, or sidewalk. Right. Elevated a standard riser height above outside surface.

rough floor. Part of the top edge of the floor joist must also be cut away. This is done at the time the frame is installed.

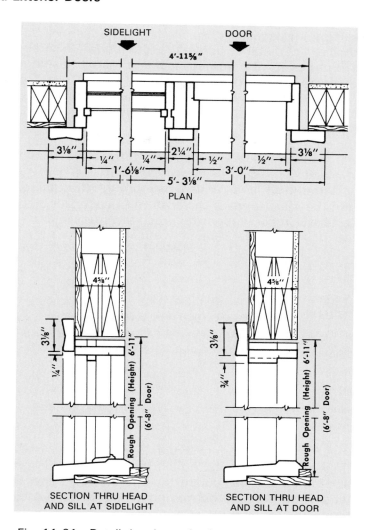

Fig. 11-61. Detail drawings of a front entrance door frame with sidelight. (C-E Morgan)

The framing of the rough opening (R.O.), however, comes earlier before the door frame is delivered to the job. The carpenter must check the working drawings carefully. The size of the rough opening will usually be included in the door and window schedule. The height of the opening is shown in detail sections. For standard construction, the R.O. can be calculated by adding about 2 1/2 in. to the door width and height.

When this information is not included in the working drawings, you should consult the manufacturer's literature. This will contain not only R.O. requirements but also detail drawings. The latter are especially important for front entrances that consist of more complicated structures and for fixed sidelight window units as shown in Fig. 11-61.

INSTALLING DOOR FRAMES

Check the size of the rough opening to make sure that proper clearances have been provided. Cut out sill area, if necessary, so the top of the sill will be the correct distance above the rough floor. In some structures it may be necessary to install flashing over the bottom of the opening.

Place the frame in the opening, center it horizontally, and secure it with a temporary brace. Using blocking and wedges, level the sill and bring it to the correct height. Be sure the sill is well supported. For masonry walls and slab floors, the sill is usually placed on a bed of mortar.

With the sill level, drive a nail through the casing into the wall frame at the bottom of each side. Insert blocking or wedges between the studs and the top of the jambs. Adjust wedges until frame is plumb. Use a level and straightedge as shown in Fig. 11-62.

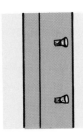

When setting door and window frames, never drive any of the nails completely into the wood until all nails are in place and a final check has been made to make sure that no adjustments are necessary.

Place additional wedges between the jambs and stud frame in the approximate location of the lock strike plate and hinges. Adjust the wedges until the side jambs are well supported and straight. Then, secure the wedges by driving a nail through the jamb, wedge, and into the stud.

Finally, nail the casing in place with nails spaced 16 in. O.C. Follow the same precautions suggested for window frame installation.

After the installation is complete, a piece of 1/4 or 3/8 in. plywood should be lightly tacked over the sill to protect it during further construction work. At this time, many builders prefer to hang a temporary combination door in order that the interior of the structure can be secured, thus providing a place to store tools and materials.

Setting the threshold and hanging the door is a part of the interior finishing operation and will be described in Unit 17. Exterior door types, designs, and sizes are described in the same unit.

When a prehung door unit is installed, the door should be removed from its hinges and carefully stored.

INSTALLING PREHUNG DOOR UNITS

A variety of prehung exterior door units are available. They include single doors, double doors, and doors with sidelights. Millwork plants provide detailed instructions for installing their products.

First check the rough opening. Make sure the size is correct and that it is plumb, square, and level.

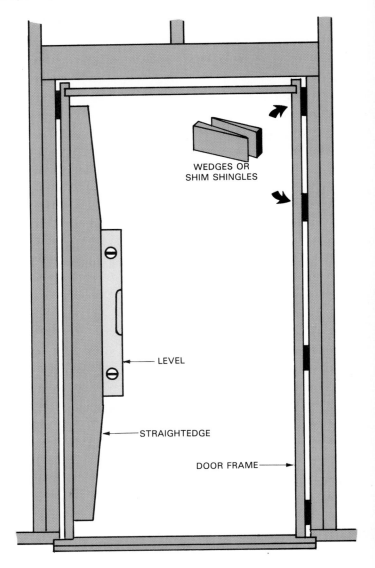

Fig. 11-62. How to plumb the door jambs.

Apply a double bead of caulking compound to the bottom of the opening, and set the unit in place, Fig. 11-63. Spacer shims located between the frame and door should not be removed until the frame is firmly attached to the rough opening.

Insert shims between the side jambs and opening. They should be located at the top, bottom, and midpoint. Drive 16d finish nails through the jambs, shims, and into the structural frame members. Manufacturers usually recommend that at least two of the screws in the top hinge be replaced with 2 1/4 in. screws. Finally, adjust the threshold so it makes smooth contact with the bottom edge of the door.

SLIDING GLASS DOORS

To accommodate outdoor living, a terrace or patio door is often included in residential designs. The French or casement type door, once used for this

Fig. 11-63. Prehung door is installed in a rough frame opening. (C-E Morgan)

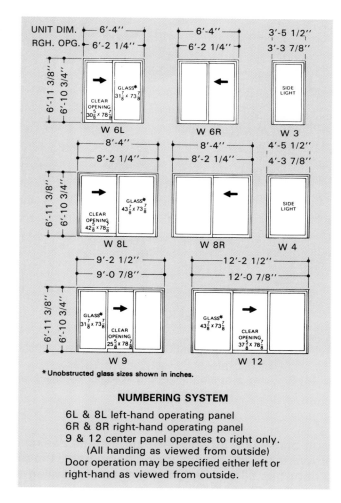

NUMBERING SYSTEM

6L & 8L left-hand operating panel
6R & 8R right-hand operating panel
9 & 12 center panel operates to right only.
(All handing as viewed from outside)
Door operation may be specified either left or right-hand as viewed from outside.

Fig. 11-64. Sliding glass door units come in standard sizes and shapes. (Andersen Corp.)

purpose, has been largely replaced with the modern sliding glass door.

This type of door, riding on nylon or stainless steel rollers, is easy to operate. When equipped with quality weather stripping and insulating glass, it restricts heat loss and condensation to a level satisfactory even in cold climates. It is available as a factory assembled frame and door unit. Parts and installation details are similar to those for sliding windows.

Sliding glass door units contain at least one fixed and one operating panel. Some may have three or four panels. The type of unit is commonly designated by the number and arrangement of the panels as viewed from the outside. The letters ''L'' or ''R'' mean an operating panel. This indicates which way the panel opens. See Fig. 11-64. Some manufacturers use the letter ''X'' to indicate operating panels and ''O'' for fixed panels. Sliding glass door units must be glazed with tempered (safety) glass.

Construction details will vary from one manufacturer to another, as will unit sizes and rough opening requirements. See Fig. 11-65. If detailed drawings or R.O. sizes are not included in the architectural plans, then the carpenter should secure the information from the manufacturer's literature. Fig. 11-66 identifies typical parts of a sliding door installation at a sill section.

The installation of sliding door frames is similar to the procedure described for regular outside doors. Before setting the frame in place a bead of sealing compound should be laid across the opening to insure a weathertight joint. If heavy glass doors are to slide properly the sill must be level and straight.

Plumb side jambs and install wedges in the same way as you would install regular door frames, Fig. 11-67. After careful checking, complete the installation of the frame by driving weatherproof nails through the side and head casings into structural frame members.

At this point, many carpenters prefer just to check the fit between the metal sill cover and doors rather than installing them. These items are carefully stored away. The opening is enclosed temporarily with plywood or polyethylene film attached to a frame. Then, during the finishing stages of construction, after inside and outside wall surfaces are completed, the sill cover, threshold, doors, and hardware are installed, Fig. 11-68.

Manufacturers of sliding glass doors provide detailed instructions for installing their particular product. This material should be read and studied carefully before proceeding with the work.

After installing large glass units in buildings under construction, it is considered good practice to place a large ''X'' on the glass. Use masking tape or washable paint. This will alert workers so they will not walk into it or damage it with tools and materials.

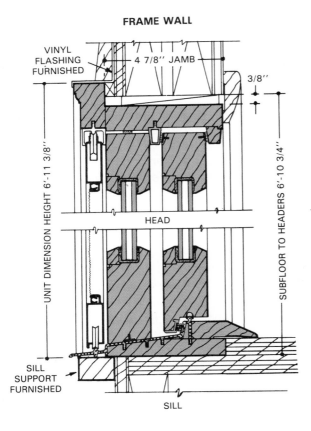

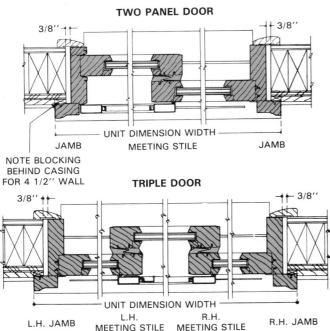

Fig. 11-65. Construction details like the above will be different from one manufacturer to the next. Be sure to secure this information before installation.

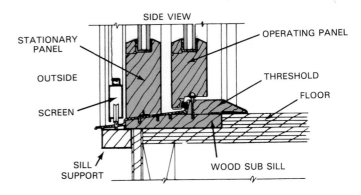

Fig. 11-66. Manufacturer's literature often labels main parts of sliding glass doors at sill section.

Fig. 11-67. Use a straightedge and wedges to check and straighten side jambs.

GARAGE DOORS

Basically, there are three types of garage doors; hinged or swinging, swing-up, and roll-up. As a result of refinement and perfection in the design of hardware and counterbalancing equipment, the roll-up door has all but replaced hinged and swing-up doors. See Fig. 11-69.

Wooden garage doors are constructed much like exterior passage doors. See page 425. Flush type doors with foamed plastic cores are available for installations that require high levels of sound and thermal insulation. Garage doors are also manufactured from steel, aluminum, and fiberglass. Many designs are available to match contemporary or traditional architecture. Stock sizes are listed in Fig. 11-70.

Fig. 11-68. Sliding glass doors are installed during the finishing stages of construction. (Andersen Corp.)

Fig. 11-69. Roll-up garage doors are hinged. Note how sections move upward and tilt to horizontal position. Rollers, attached to each section, run in a special steel track to support door. (Stanley Door Systems)

WIDTH	HEIGHT
8'0''	7'0'' or 6'6''
9'0''	7'0'' or 6'6''
10'0''	7'0'' or 6'6''
16'0''	7'0'' or 6'6''
18'0''	7'0'' or 6'6''

Fig. 11-70. Stock sizes of garage doors.

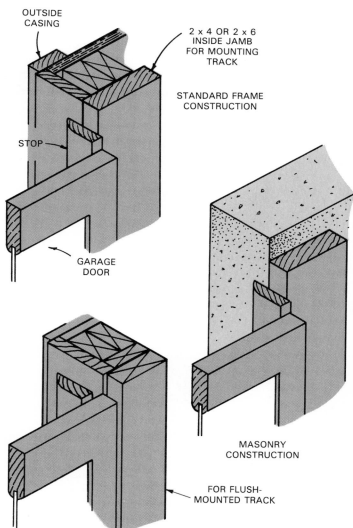

Fig. 11-71. Typical jamb construction for garage door frames when roll-up door is used.

GARAGE DOOR FRAMES

Frames for garage doors include side jambs and a head similar to exterior passage doors. No rabbet is required. The frame is usually included in the millwork order along with windows and doors so the outside trim will match. The size of the frame opening is usually the same size as the door. However, the manufacturer's specifications and details should be checked before placing the order.

Fig. 11-71 shows a typical jamb section for wood frame or masonry construction. Note the thickness (2 in. nominal) of the heavy inside frame to which the track and hardware will be mounted. The width of this member should be at least 4 in. wide with no projecting bolt or lag screw heads.

The rough opening width for frame construction will normally be about 3 in. greater than the door size. The height of the rough opening should be the door height plus about 1 1/2 in. as measured from the finished floor.

Give careful consideration to the inside headroom height. A minimum clearance between the top of the door and the ceiling must be provided for hardware, counterbalancing mechanisms, and the door itself when open. On some special low-headroom designs, this distance may be as little as 6 in. Always be sure to check the manufacturer's requirements for each door.

When installing garage door frames, follow the general procedure used for regular door frames. Be certain the jambs are plumb and the head is level.

INSTALLATION

The following are the major steps for installing a sectional roll-up door:

1. Tack stops temporarily in place.
2. Assemble the door sections in the opening and attach hinges.
3. Attach rollers to door.
4. Place track on rollers and attach track to jamb.
5. Mount horizontal track sections.
6. Raise and prop door in open position.
7. Attach counterbalancing mechanism.
8. Open and close door and make necessary adjustments.
9. Reset stops for smooth, tight fit.

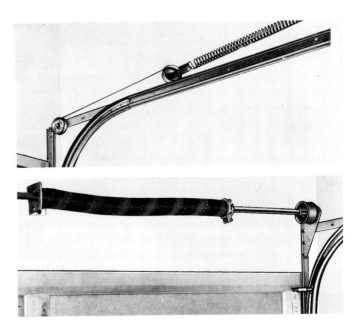

Fig. 11-73. Spring counterbalances for garage doors. Top. Extension spring. Bottom. Torsion spring.

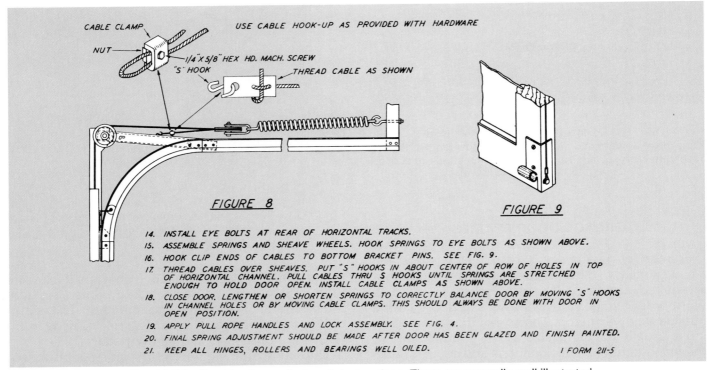

USE CABLE HOOK-UP AS PROVIDED WITH HARDWARE

CABLE CLAMP

NUT

1/4" x 5/8" HEX HD. MACH. SCREW

"S" HOOK

THREAD CABLE AS SHOWN

FIGURE 8

FIGURE 9

14. INSTALL EYE BOLTS AT REAR OF HORIZONTAL TRACKS.
15. ASSEMBLE SPRINGS AND SHEAVE WHEELS. HOOK SPRINGS TO EYE BOLTS AS SHOWN ABOVE.
16. HOOK CLIP ENDS OF CABLES TO BOTTOM BRACKET PINS. SEE FIG. 9.
17. THREAD CABLES OVER SHEAVES. PUT "S" HOOKS IN ABOUT CENTER OF ROW OF HOLES IN TOP OF HORIZONTAL CHANNEL. PULL CABLES THRU S HOOKS UNTIL SPRINGS ARE STRETCHED ENOUGH TO HOLD DOOR OPEN. INSTALL CABLE CLAMPS AS SHOWN ABOVE.
18. CLOSE DOOR. LENGTHEN OR SHORTEN SPRINGS TO CORRECTLY BALANCE DOOR BY MOVING "S" HOOKS IN CHANNEL HOLES OR BY MOVING CABLE CLAMPS. THIS SHOULD ALWAYS BE DONE WITH DOOR IN OPEN POSITION.
19. APPLY PULL ROPE HANDLES AND LOCK ASSEMBLY. SEE FIG. 4.
20. FINAL SPRING ADJUSTMENT SHOULD BE MADE AFTER DOOR HAS BEEN GLAZED AND FINISH PAINTED.
21. KEEP ALL HINGES, ROLLERS AND BEARINGS WELL OILED.

I FORM 211-5

Fig. 11-72. Sample of manufacturer's instructions. These are generally well illustrated and easy to understand.

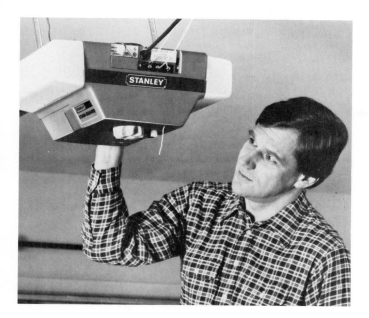

Fig. 11-74. Installing an automatic garage door opener. In addition to radio control, it features a smoke and heat alarm, and an intrusion (burglar) warning system.
(Stanley Door Systems)

Manufacturers always supply detailed directions and procedures for the installation of their products. The carpenter should follow these printed materials carefully. They are usually well illustrated and easy to understand. See Fig. 11-72.

HARDWARE AND COUNTERBALANCES

Garage door hardware must be well designed so the door will operate easily. Track, hinges, and bolts should be made of galvanized steel. Gauge weight must be heavy enough to last the life of the door.

To offset the weight of the door, various counterbalancing devices are used. Two of the most common types are shown in Fig. 11-73. The extension spring is commonly used on residential doors, either single or double. The torsion spring and its mechanism is somewhat more expensive but usually provides a smoother and more consistent action. It is especially recommended for wide doors and those that are heavier in construction.

Today, many residential plans include a two-car garage with a 16 ft. or wider door. The operation of the large door has led to the wide acceptance of electric powered door openers, Fig. 11-74.

TEST YOUR KNOWLEDGE — UNIT 11

1. Today, windows and doors are built in large _____ plants.
2. The kind of wood most often used to manufacture windows is _____ _____.
3. A type of window that is hinged on the side and swings outward is called a_____.

4. The height of a standard residential door is _____.
5. A single pane of glass in a window has an R value of about _____. Adding another pane of glass with 1/2 in. of air space increases the R value to _____.
6. The side section of a window is called the _____.
7. When there is little or no roof (cornice) overhang to protect the window, a _____ should be installed above the head casing.
8. If the R.O. for a window is listed as 3'-6'' x 3'-5'', the height of the rough opening would be _____.
9. To adjust for various wall thicknesses, _____ _____are applied to standard window frames.
10. After a window unit has been temporarily set in the rough opening, the next step is to _____ the _____ at the correct height.
11. Large insulating window units are made from polished glass that is _____ in. thick.
12. The thickness of standard glass block is _____ in.
13. When figuring the size of the opening for glass block panels, multiply the number of units by the _____ (actual, nominal) block size.
14. In brick veneer construction, the masonry over door and window heads is carried by a metal support called a _____.
15. Outside wood casing for windows and doors is attached with nails spaced _____ in. O.C.
16. The most popular type of garage doors is the _____ type.
17. A two-car garage door is usually _____ ft. wide.
18. The two common types of spring counterbalances for garage doors are extension and _____.

OUTSIDE ASSIGNMENTS

1. Prepare a written report on the manufacture of glass. Include such headings as: historical development, early production methods, modern processes, float method, drawing method, and grinding and polishing plate glass. Study encyclopedias, reference books, and booklets from glass manufacturers.
2. The rising cost of energy has resulted in special emphasis being placed on window design and construction. Some manufacturers are now producing triple glazed window units for homes located in northern climates. Secure information about these units that includes: R values, prices, special installation directions, and predicted fuel savings.

Exteriors of homes can be finished off with a variety of different materials. Top. Hardboard siding is embossed and toned to simulate random planked barnboards and shakes. (Abitibi-Price Corp.) Bottom left. Natural cedar shakes. (Shakertown Corp.) Bottom right. Stucco, an exterior wall finish widely used in the Southwest.

Unit 12

EXTERIOR WALL FINISH

The term "exterior finish" includes all exterior materials of a structure. It generally refers to the roofing materials, cornice trim boards, wall coverings, and trim members around doors and windows. The installation of special architectural woodwork at entrances, or the application of a ceiling to a porch or breezeway area would also be included under this broad heading.

Previous units have described the application of the finished roof and the installation of the trim around windows and outside doors. This unit will cover the construction and finish of cornice work, and the materials and methods used to provide a suitable outside wall covering.

CORNICE DESIGNS

The cornice, or eave, is formed by the roof overhang. It provides a finished connection between the wall and the edge of the roof. Architectural style will determine, to a large extent, the design requirements. Fig. 12-1 shows architectural details of cornice construction.

Diagrams of several closed or "boxed" cornice designs are illustrated in Fig. 12-2. An open cornice is sometimes used, exposing the rafters and underside of the roof sheathing. Wide overhangs are used often in modern buildings. These provide shade for large window areas, protect the walls, and add to attractiveness of the structure.

The rake is the part of a roof that overhangs a gable. It is usually enclosed with carefully fitted trim members.

PARTS OF CORNICE AND RAKE SECTION

Fig. 12-3 shows structural and trim parts of a boxed cornice. The fascia board is the main trim member along the edge of the roof. A ledger strip is nailed to the wall and carries the lookouts which are attached to each rafter.

A nailing strip is sometimes attached to the back of the fascia, between each rafter. Nailing strips, along with the lookouts and ledger, provide a frame to which the plancier or soffit material can be applied. Soffits can be plywood, hardboard, solid stock, mineral fiber board, or metal.

The trim used for a boxed rake section, Fig. 12-4, is supported by the projecting roof boards. In addition, lookouts or nailers are fastened to the side wall and the roof sheathing. As in cornice construction, these serve as a nailing base for the soffit and fascia.

When the rake projects a considerable distance, the sheathing does not provide adequate support. In such cases, the roof framing should be extended. See Fig. 9-20.

Parts of the cornice and rake structure exposed to view are generally called EXTERIOR TRIM. In a typical construction, these parts are cut on the job. The properties needed in material used for exterior trim include: good painting and weathering characteristics, easy working qualities, and maximum freedom from warp. Decay resistance is also desirable where materials may absorb moisture.

The cedars, cypress, and redwood have high decay resistance. Less durable species may be treated to make them decay resistant. Coat end joints or miters of members subjected to heavy moisture. White lead paste or special caulking compounds are commonly used for this purpose.

CORNICE AND RAKE CONSTRUCTION

In most construction, the fascia boards are installed on the rafter ends at the time the roof is sheathed. It is important that they be straight, true, and level with well fitted joints.

Before attaching fascia, some builders prefer to first nail on a 2 x 4 ribbon to align the tails of the rafters. The fascia is then attached to this ribbon. Corners of fascia boards should be mitered. End

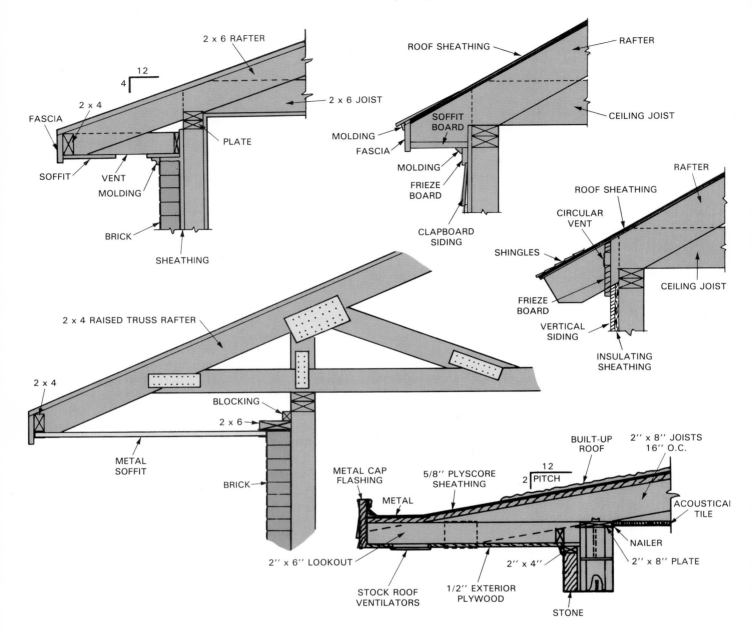

Fig. 12-1. Typical cornice details. These will be found in architectural plans.

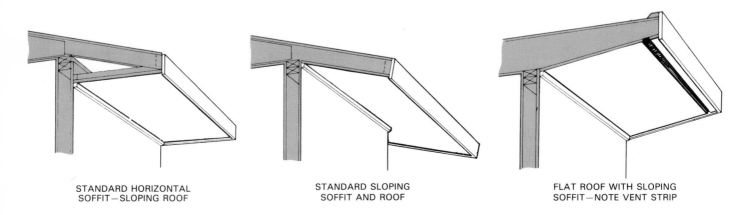

STANDARD HORIZONTAL
SOFFIT—SLOPING ROOF

STANDARD SLOPING
SOFFIT AND ROOF

FLAT ROOF WITH SLOPING
SOFFIT—NOTE VENT STRIP

Fig. 12-2. Three different cornice designs. Most cornices are boxed today.

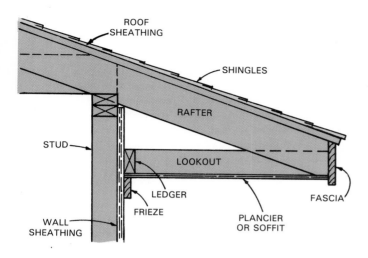

Fig. 12-3. Parts of a typical boxed cornice. Soffit materials are often prefabricated and usually are hardboard, plywood, or metal.

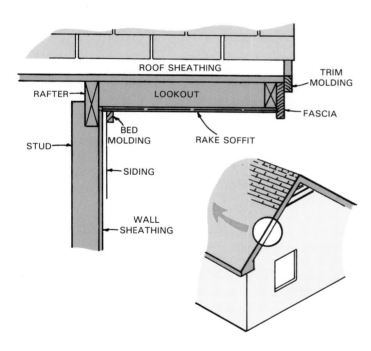

Fig. 12-4. Boxed rake section. Lookouts provide a nailing surface and support for soffit.

Lookouts are usually made from 2 x 4 stock. Locate them at each rafter or every other rafter depending on the kind of soffit material used. Cut the lookouts. Toenail one end to the ledger and nail the other end to the overhang of the rafter. When using thin material for the soffit, attach a nailing strip along the inside of the fascia to provide a nailing surface. Sometimes the back of the fascia is grooved to receive the soffit material.

After this frame is complete, apply the soffit. First cut the material to size and then secure it with rust resistant nails or screws. When regular casing or finish nails are used, they must be countersunk and the holes filled with putty. Do this after the prime coat of paint.

To install a sloping soffit refer to Fig. 12-6.

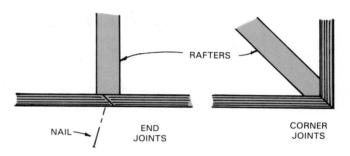

Fig. 12-5. Plan view of joints for fascia boards. Miter corners and make matching angled cuts on ends. Note how end joints are nailed.

Fig. 12-6. Installing a sloping soffit. The 3/8 in. plywood sheet is attached to the underside of the rafter. Frieze board may be attached first to help hold one edge while nailing the soffit. (American Plywood Assoc.)

joints should meet at a 45 deg. angle, as illustrated in Fig. 12-5.

To frame a cornice with a horizontal soffit, first install a ledger strip along the wall if lookouts are to be used. With a carpenter's level, locate points on the wall level with the bottom edge of the rafter. Study the architectural drawing details to secure dimensions. Snap a chalk line between these points and nail on the ledger.

Soffits should be provided with screened vents so there will be a flow of air through the enclosed cornice section into the attic space.

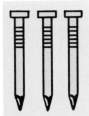

Always use rust resistant nails for outside finish work. They may be made of aluminum or galvanized (or cadmium plated) steel.

The rake should be constructed to match the cornice using the same general procedures. After the main trim members are installed, moldings are often set in corners to cover irregularities. In modern construction however, their use is minimal in order to maintain a smooth, trim appearance, Fig. 12-7.

PREFABRICATED CORNICE MATERIALS

Cornice construction is time consuming. Therefore, many builders prefer to purchase prefabricated materials. Various systems are available that provide a neat, trim appearance. One consists of 3/8 in. laminated wood-fiber panels. These are factory primed and available in a variety of standard widths (12 to 48 in.) and in lengths up to 12 ft. Panels can be equipped with factory-applied screened vents. See Fig. 12-8.

When installing large sections of wood fiber panels, fit each panel with some clearance for expansion. Space 4d rust resistant nails about 6 in. apart along edges and intermediate supports. Start nailing at the end butted against a previously placed panel. First nail to the main supports and then along the edges. Drive the nails carefully so the underside of the head is just flush with the panel surface.

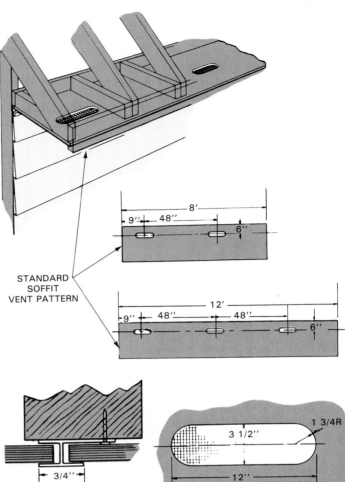

STANDARD SOFFIT VENT PATTERN

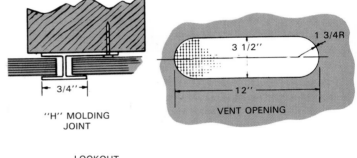

"H" MOLDING JOINT

VENT OPENING

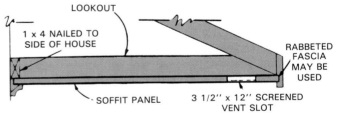

LOOKOUT

1 x 4 NAILED TO SIDE OF HOUSE

RABBETED FASCIA MAY BE USED

SOFFIT PANEL

3 1/2" x 12" SCREENED VENT SLOT

Fig. 12-8. Details of prefabricated soffit system. Soffits are manufactured material—usually hardboard or metal.

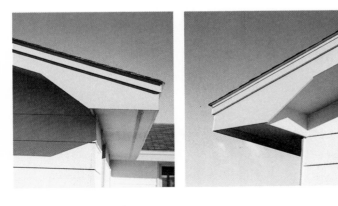

Fig. 12-7. Two views of a completed cornice and rake. Note continuous vent system in soffit. (Dickinson Homes, Inc.)

Lookouts may be left out of a soffit system where special supports are attached to the upper surface of the panels. These supports, in one system, are made of 20 ga. steel channels with prongs that make it easy to attach them to the back of the panels. The supports provide rigidity so the panels can be secured in place only at the front and back edges.

To install the channels, place the soffit face down on a solid surface and position the channel (usual-

Fig. 12-9. Use a section of 2 x 4 when attaching patented stiffener to concealed side of soffit panel.

ly located 24 in. O.C.). Insert a piece of 2 x 4 stock in the channel and drive the prongs into the soffit panel as illustrated in Fig. 12-9.

With channels attached to the structure, lift units into place, Fig. 12-10. They are held with a special molding. This molding is attached to the inside of the fascia along a nailer attached to the wall. Special H-shaped clips, shown in Fig. 12-8, are used to join the ends of panels. A view of the completed installa-

tion is shown in Fig. 12-11.

Porch and carport ceilings can be covered with factory-primed panels similar to those used for cornice soffits. Standard 4 x 8 ft. and larger units are designed for either 16 in. or 24 in. O.C. framing. Always leave a 1/8 in. space along all edges for expansion.

Start nailing in the center of the panels and move toward the outside. Space nails about 6 in. apart. Edges of panels can be secured and joints covered with special H-strips as shown in Fig. 12-12.

Fig. 12-11. View of completed soffit. Panels are preprimed. Regular spacing of vents assures adequate attic ventilation.

Fig. 12-10. Panel is lifted into place after channels are attached.

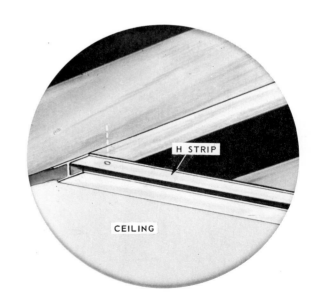

Fig. 12-12. Factory primed porch ceiling panels with H strips to conceal and protect joints.

Soffit systems made of prefinished metal panels and attachment strips are common. They consist of three basic units:
1. Wall hanger strips (also called frieze strips).
2. Soffit panels.
3. Fascia covers.

See Fig. 12-13. Panels include a vented area and are available in a variety of lengths.

HANGING METAL SOFFITS

To install a metal panel system first snap a chalk line on the side wall, level with the bottom edge of the fascia board. Use this line as a guide for nailing the wall hanger strip into place. Insert the panels, one at a time, into the wall strip. Nail the outer end to the bottom edge of the fascia board. See Fig. 12-14.

After all soffit panels are in place, cut the fascia cover to length and install as shown in Fig. 12-15. The bottom edge of the cover is hooked over the end of the soffit panels. It is then nailed into place through prepunched slots located along the top edge. Always study and follow the manufacturer's directions when making an installation of this type.

Fig. 12-14. To install metal soffit, slip inner end of panel into wall hanger strip. Nail outer end of panel to bottom edge of fascia board. These panels are 9 in. wide.

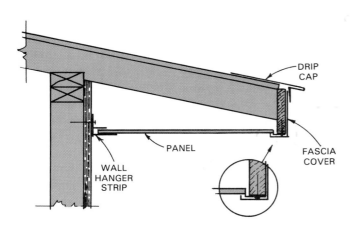

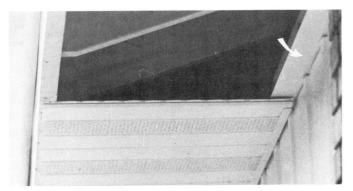

Fig. 12-13. Detail of a prefinished metal soffit system. Top. Drawing of system used on wood or aluminum sided wall. Bottom. Metal soffit system used with brick veneered wall. Note 2 x 6 nailer attached to wall (arrow).

Fig. 12-15. Install fascia cover after all soffit panels are attached. The cover is placed under the roof drip edge.

WALL FINISH

After the cornice and rake section of the roof are covered you can apply siding to the walls. When the structure includes a gable roof, the wall surface material is usually applied to the gable end before the lower section is covered. This permits scaffolding to be attached directly to the wall while siding the gable end.

All exterior trim members, if not factory-primed, should be given a primer coat of paint as soon as possible after installation.

HORIZONTAL SIDING

One of the most common materials used for the exterior finish of American homes is wood siding. The deep shadow cast on the wall by the butt edge, especially evident in bevel siding, emphasizes the horizontal lines preferred by many homeowners.

Siding is usually applied over sheathing. However, in mild climates or on buildings such as summer cottages, it may be applied directly to the studs. Where sheathing is omitted, or where the type of sheathing does not provide sufficient strength to resist a racking load, the wall framing should be braced as described in Unit 8.

End views of a number of types of horizontal siding are shown in Fig. 12-16. Bevel siding, most commonly used, is available in various widths. It is made by sawing plain surfaced boards at a diagonal to produce two wedge-shaped pieces. The siding is about 3/16 in. thick at the thin edge and 1/2 to 3/4 in. thick on the other edge, depending on the width of the piece.

Wide bevel siding often has shiplapped or rabbeted joints. The siding lies flat against the studding instead of touching it only near the joints as ordinary bevel siding does. This reduces the ap-

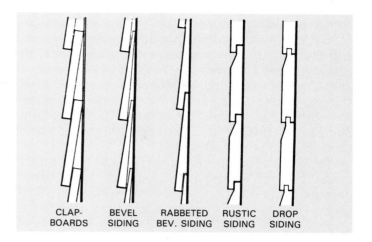

Fig. 12-16. Edge view of five types of horizontal siding.

parent thickness of the siding by 1/4 in. but permits the use of extra nails in wide siding and reduces the chance of warping. It is also economical, since the rabbeted joint requires less lumber than the lap joint used with plain bevel siding.

The rabbet, however, must be deep enough so that, when the siding is applied, the width of the boards can be adjusted upward or downward to meet window sill, head casing, and eave lines. The table in Fig. 12-17 lists the more common sizes of

SIDING	NOMINAL SIZE	DRESSED DIMENSIONS		
	WIDTH	STANDARD THICKNESS		STANDARD FACE WIDTH
	INCHES	INCHES		INCHES
		Butt	Tip	
Bevel	4	** 7/16	3/16	3 1/2
	5	10/16	3/16	4 1/2
	6	. . . .	. . . .	5 1/2
Wide Beveled	8	** 7/16	3/16	7 1/4
	10	9/16	3/16	9 1/4
	12	11/16	3/16	11 1/4
Rustic and Drop . . . . (Shiplapped)	4	9/16		3 1/8
	5	3/4		4 1/8
	6	. . . .		5 1/16
	8	. . . .		6 7/8
Rustic and Drop . . . . (Dressed and Matched)	4	9/16		3 1/4
	5	3/4		4 1/4
	6	. . . .		5 3/16
	8	. . . .		7

In patterned siding, 11/16, 3/4, 1, 1 1/4, and 1 1/2 inches thick, board measure, the tongue shall be 1/4 in. wide in tongue-and-groove lumber, and the lap 3/8 in. wide in shiplapped lumber, with the overall widths 1/4 in. and 3/8 in. wider, respectively, than the face widths shown above.
**Minimum thicknesses.

Fig. 12-17. Table lists sizes of horizontal wood siding. Nominal sizes are used in figuring footage of lumber. They are based on rough-sawn dimensions. These boards, being green, shrink slightly as they dry. Machining reduces the sizes even more.

horizontal siding and the actual size measurements.

Rustic and drop sidings are usually 3/4 in. thick, and 6 in. wide. They are made in a wide variety of patterns. Drop siding usually has tongue-and-groove joints, while rustic siding has shiplap type joints.

Drop siding is heavier, has more structural strength, and tighter joints than bevel siding. Because of this, it is often used on garages and other buildings not sheathed.

Wood used for exterior siding should be a select grade free of knots, pitch pockets, and other defects. Edge-grain is less likely to warp than a flat-grain. The moisture content at the time of application should be what it will reach in service. This is about 12 percent, except for the southwestern states, where the moisture content should average about 9 percent.

Siding should be handled carefully when it is delivered to the building site. The wood from which it is made is usually quite soft. The surface can be easily damaged. Try to store siding inside the structure or keep it covered with a weatherproof material until it is applied.

WALL SHEATHING AND FLASHING

Horizontal siding can be applied over various sheathing materials. When the sheathing consists of solid wood, plywood, or nail-base fiberboard, the siding is nailed directly to the material at about 24 in. intervals. End joints in the siding may occur between framing members. Gypsum board and regular fiberboard sheathing cannot be used as a nailing base and the siding should be attached by nailing through the sheathing and into the frame.

Certain special application methods may require wood strips to form a nailing base. Additional information about wall sheathing can be found in Unit 8.

Be sure the structural frame and sheathing are dry before applying the siding material. If excessive moisture is present during the application, later drying and shrinkage will likely cause the siding to buckle.

Sheathing paper is applied directly to the wall frame to resist the infiltration of air and moisture when sheathing is not used. It is also used to cover common board and shiplap sheathing. Sheathing paper is not necessary over plywood, fiberboard, or treated gypsum sheathing.

Sheathing paper, also called building paper, may be an asphalt-saturated felt which has a low vapor resistance. Materials such as coated felts or laminated waterproof papers that have a high moisture vapor resistance will act as a vapor barrier and should not be used. See Unit 13.

Before the application of siding, flashing should be installed where it is required around openings. Metal flashing is usually included over the drip caps of doors and windows, Fig. 12-18. In areas not subjected to wind driven rain, head flashing may be omitted when the vertical height between the top of the finished trim of the opening and the soffit of the cornice is equal to or less than one-fourth of the width of the overhang.

For structures with unsheathed walls, jambs of doors and windows should be flashed with a 6 in. wide strip consisting of either metal, 3 oz. copper-coated paper, or a 6 mil polyethylene film.

INSTALLATION PROCEDURES

Wood siding is precision-manufactured to standard sizes. It is easily cut and fitted. Plain beveled siding is lapped so it will shed water and provide a windproof and dustproof covering. A minimum lap of 1 in. is used for 6 in. widths, while 8 and 10 in. siding should lap about 1 1/2 in.

To install horizontal siding, first prepare a story pole:
1. Lay out the distance from the soffit to about 1

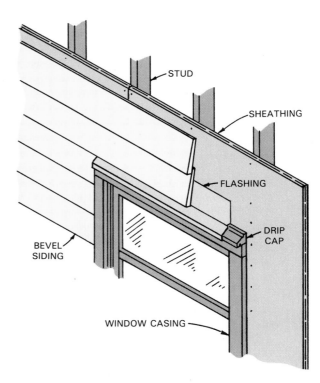

Fig. 12-18. Apply metal flashing over the drip caps above windows and doors that are not protected by roof overhang.

in. below the top of the foundation, Fig. 12-19.

2. Divide this distance into spaces equal to the width of the siding minus the lap.

3. Adjust the lap allowance (maintain minimum requirements) so the spaces are equal.

4. When possible, adjust the spacing so single pieces of siding will run continuously above and below windows or other wall openings without notching, Fig. 12-20.

5. When the layout is complete, mark the position of the top of each siding board on the story pole.

Now hold the story pole in position at each inside and outside corner of the structure and transfer the layout to the wall, Fig. 12-21. Also make the layout along window and door casings.

Some carpenters prefer to set nails at these layout points, since lines can be quickly attached to them and used to align the siding stock. They can also be used to hold the chalk line if guidelines are laid out by this method.

When snapping a chalk line stretched over a long distance, it is a good idea to hold it against the surface at the midpoint; then snap it on each side.

After the layout has been made carefully check it. Start the application of bevel siding by first nailing a strip along the foundation line equal to the thin edge of the siding, Fig. 12-22. This will provide the

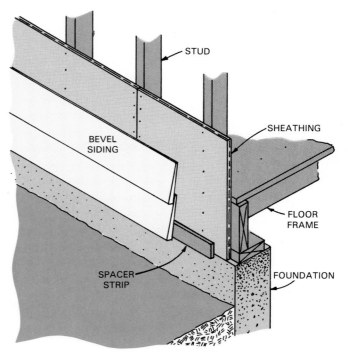

Fig. 12-21. Transfer story pole layout to all inside and outside corners as well as to door and window casings.

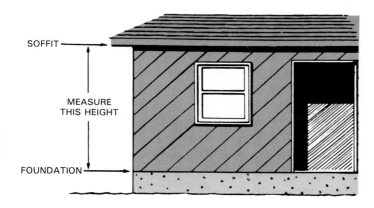

Fig. 12-19. Begin layout of story pole by taking the total measurement from beneath the soffit to about 1 in. below the top of the foundation.

Fig. 12-20. Try to adjust courses of siding to come out even, if possible with tops and bottoms of windows. Avoid notching siding if possible.

Fig. 12-22. Install a spacer strip under the first course of bevel siding to give it the proper tilt to match succeeding courses.

proper tilt for the first course. Now apply the first piece. Allow the butt edge to extend below the strip to form a drip edge.

Inside corners can be formed with a square length of wood, or metal corners can be used as shown in Fig. 12-23. Although outside corners could be lapped or mitered, in modern construction, metal corners are used almost universally. They can be installed quickly and provide a neat, trim appearance.

Corner boards, Fig. 12-24, are sometimes used for siding installation. They may be formed with two pieces of lumber. Thickness depends on siding. Attach the assembly to the structure. Corner boards may be plain or molded depending on the architectural treatment required. After the corner boards are in place, fit the siding tightly against them.

Cut and fit horizontal wood siding tightly against window and door casings, corner boards, and adjoining boards. For quality work, the carpenter first makes the cuts with a fine-tooth saw and then smoothes the ends with a few strokes of a block plane. Square butt joints are used between adjacent pieces of siding and should be staggered as widely as possible from one course to the next.

Wood siding can be given a coat of water-repellent preservative before it is installed, or the water repellent can be brushed on after the installation. Preservatives are sold by lumber dealers and paint stores. They contain waxes, resins, and oils. In addition to this treatment, joints in siding may be bedded in a special caulking compound to make them watertight.

NAILING

To fasten siding in place, zinc-coated steel, aluminum, or other noncorrosive nails are recom-

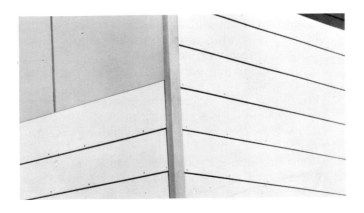

Fig. 12-24. Outside corners can also be finished with corner boards. (Armstrong World Industries)

Fig. 12-25. Horizontal siding should be face nailed over studs. A pneumatic nailer loaded with special, noncorrosive nails will make the job go faster. (Senco Products, Inc.)

mended. Avoid plain steel-wire nails, especially the large-headed ones designed for flush driving. They produce unsightly rust spots on most paints. Even small-headed plain steel nails, countersunk and puttied, are likely to rust.

Horizontal siding should be face-nailed to each stud, Fig. 12-25. For 1/2 in. siding over wood or plywood sheathing, use 6d nails. Over fiberboard or gypsum sheathing, use 8d nails. For 3/4 in. siding over wood or plywood sheathing, use 7d nails; and over fiberboard or gypsum sheathing, use 9d nails.

For narrow siding, the nail is generally placed about 1/2 in. above the butt edge. In this location the fastener passes through the upper edge of the lower course.

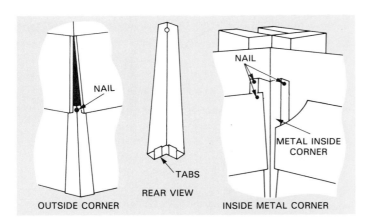

Fig. 12-23. Metal corners are almost always used for horizontal siding.

When applying wide bevel siding, the nail should be driven through the butt edge *just above* the lap so that it misses the thin edge of the piece of siding underneath. *This permits expansion and contraction of the siding boards with seasonal changes in moisture content.* It eliminates the tendency for the siding to cup or split when both edges are nailed. Since the amount of swelling and shrinking is proportional to the width of the material, move the nail slightly above the lap when boards are extra wide.

If there is a possibility the material will split when nailing end joints, holes should be drilled for the nails. Fig. 12-26 shows some wood siding nailing patterns.

Although the usual procedure for the installation of horizontal siding is to proceed upward along the wall, some carpenters prefer to start at the top and work down, Fig. 12-27. This is adaptable for multi-story or split level structures where scaffolds are attached to wall. Another advantage: the siding is less likely to be damaged after it is applied. When following such procedure, chalk lines are set at the butt edge of the siding instead of the top edge.

Fig. 12-27. Some carpenters prefer to install siding from the top down so they can attach scaffolding directly to the wall.

PAINTING AND MAINTENANCE

Wood siding is subject to decay and weathering. Neither will occur if simple precautions are taken. Decay is the disintegration of wood caused by the growth of fungi. These fungi grow in wood when the moisture content is too high.

If the structure is built on a foundation which has been carried well above the ground and the construction is such that water runs off instead of into the walls, decay should never be a problem.

A priming coat of paint should be put on as soon as possible after the siding has been applied. If an unexpected rain should wet unprimed wood siding, the first coat of paint should not be applied until the wood has dried.

ESTIMATING AMOUNT OF SIDING

To determine how much siding is needed, it is necessary to increase the footage to make up for the difference between nominal and finished sizes. More must also be added for the cutting of joints and the overlap in beveled siding. The table in Fig. 12-28 provides a factor. The net square footage of the wall surface to be covered should be multiplied by this factor. The following example shows the steps:

1 x 10 Bevel siding with 1 1/2 in. lap
Wall height = 8 ft.
Wall perimeter = 160 ft.
Door and window area = 240 sq. ft.

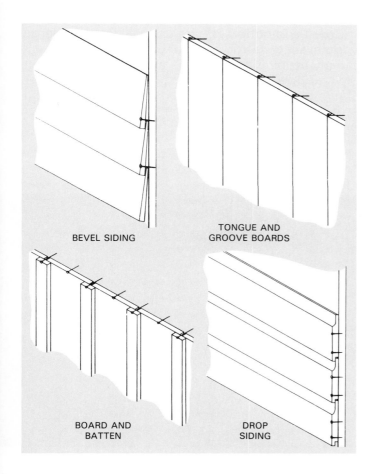

BEVEL SIDING

TONGUE AND GROOVE BOARDS

BOARD AND BATTEN

DROP SIDING

Fig. 12-26. Different types of wood siding require different nailing patterns.

TYPE	SIZE (INCHES)	LAP (INCHES)	MULTIPLY NET WALL SURFACE BY
Bevel Siding	1 x 4	3/4	1.45
	*1 x 5	7/8	1.38
	1 x 6	1	1.33
	1 x 8	1 1/4	1.33
	1 x 10	1 1/2	1.29
	1 x 12	1 1/2	1.23
Rustic and Drop Siding (Shiplapped)	1 x 4	————	1.28
	*1 x 5		1.21
	1 x 6		1.19
	1 x 8		1.16
Rustic and Drop Siding (Dressed and Matched)	1 x 4	————	1.23
	*1 x 5		1.18
	1 x 6		1.16
	1 x 8		1.14

*Unusual Sizes.

Fig. 12-28. When estimating horizontal wood siding, multiply the net wall surface to be covered by the factor in the last column.

Fig. 12-29. Vertical wood boards make a durable and beautiful siding material. (American Plywood Assoc.)

Total area to be covered = (8 × 160) − 240
= 1280 − 240
= 1040 sq. ft.

Siding needed = 1040 + (1040 × 1.29)
= 1342 sq. or bd. ft.

Area for gable ends can be calculated by multiplying the height above the eaves by the width and dividing by two. Considerable waste occurs in covering triangular areas and at least 10 percent should be added to this calculation. When the structure includes many corners due to projections and recesses in the wall line, an additional allowance of at least 5 points should be added to the factors shown in Fig. 12-28.

VERTICAL SIDING

Vertical siding is commonly used to set off entrances or gable ends. It is also often used for the main wall areas, Fig. 12-29. Vertical siding may be plain-surfaced matched boards, pattern matched boards, or square-edge boards covered at the joint with a batten strip.

Matched vertical siding, made from solid lumber, should be no more than 8 in. wide. It should be installed with two 8d nails not more than 4 ft. apart. Backer blocks should be placed horizontally between studs to provide a good nailing base. The bottom of the boards are usually undercut to form a water drip.

Board and batten applications are designed around wide square-edged boards, spaced about 1/2 in. apart. They are fastened at each bearing (blocking) with one or two 8d nails. Use 10d nails to attach the battens, Fig. 12-30. Locate nails in center of batten so the shanks will pass between the boards and into the bearing.

Board and batten effects are possible with large vertical sheets of plywood or composition material. Simply attach vertical strips over the joints and at several positions between the joints. Fig. 12-31 shows an application of this type with solid wood strips being applied to the surface of exterior plywood sheets.

Fig. 12-30. Battens over rough boards have a rustic appearance. They also provide for expansion and contraction. (Western Wood Products Assoc.)

Fig. 12-31. Board and batten effect is created by nailing battens over exterior plywood sheets applied vertically.

WOOD SHINGLES

Wood shingles, Fig. 12-32, are sometimes used for wall covering, and a large selection of types is available. Some are especially designed for sidewall application with a grooved surface and factory applied paint or stain.

Shingles are very durable and can be applied in various ways to provide a variety of architectural effects. Handsplit shingles are occasionally used, however they are more expensive and difficult to install.

Most shingles are made in random widths. No. 1 grade shingles vary from 3 in. to 14 in. wide. Only a small number of the narrow width are permitted. Shingles of a uniform width, known as dimension shingles, are also available.

For side wall application, follow these recommendations for maximum exposure:
1. For 16 in. shingles, 7 1/2 in.
2. For 18 in. shingles, 8 1/2 in.
3. For 24 in. shingles, 11 1/2 in.

Shingles on side walls are frequently laid in what is called "double-coursing." This is done by using a lower grade shingle under the shingle exposed to the weather. The exposed shingle butt extends about 1/2 in. below the butt of the under course.

When butt nailing is used, a greater weather exposure is possible. Frequently as much as 12 in. for 16 in. shingles, 14 in. for 18 in. shingles, and 16 in. for 24 in. shingles is satisfactory. Fig. 12-33 lists the sizes of a standard side wall shingle with grooved surface. Approximate coverage for various lengths and exposures is included.

In mild climates, sheathing boards can be spaced the distance of the shingle exposure. A high grade shingle provides a satisfactory wall. Roofing felt should always be used with such construction. Place it either between the shingles and sheathing or between the sheathing and the studding. Spaced sheathing is also satisfactory on implement sheds, garages, and other structures where protection from the elements is the principal consideration.

To obtain the best effect and to avoid unnecessary cutting of shingles, butt-lines should be even with the upper lines of window openings. Likewise, they should line up with the lower lines of such openings.

It is better to tack a temporary strip to the wall to use as a guide for placing the butts of the shingles squarely, rather than to attempt to shingle to a chalk line.

SINGLE COURSING OF SIDE WALLS

The single-coursing method for side wall application is similar to roof application. The major

Fig. 12-32. Wood shingles or shakes are an attractive and durable siding material. (Shakertown Corp.)

Grade	Length	Thickness (at Butt)	No. of Courses Per Bdl/Carton	Bdls/Cartons Per Square	Shipping Weight	Description
No. 1	16" (Fivex) 18" (Perfections) 24" (Royals)	.40" .45" .50"	33/33 28/28 13/14	1 carton 1 carton 4 bdls.	60* lbs. 60* lbs. 192 lbs.	Same specifications as rebutted-rejointed shingles, except that shingle face has been given grain-like grooves. Natural color, or variety of factory-applied colors. Also in 4-ft. and 8-ft. panels.

NOTE: * 70 lbs. when factory finished.

LENGTH AND THICKNESS	Approximate coverage of one square (4 bundles) of shingles based on following weather exposures																									
	3½"	4"	4½"	5"	5½"	6"	6½"	7"	7½"	8"	8½"	9"	9½"	10"	10½"	11"	11½"	12"	12½"	13"	13½"	14"	14½"	15"	15½"	16"
16" x 5/2"	70	80	90	100*	110	120	130	140	150I	160	170	180	190	200	210	220	230	240†	....	....	....	....	....	....	....	....
18" x 5/2¼"	....	72½	81½	90½	100*	109	118	127	136	145½	154½I	163½	172½	181½	191	200	209	218	227	236	245½	254½†	....	....	....	....
24" x 4/2"	....	....	....	....	....	80	86½	93	100*	106½	113	120	126½	133	140	146½	153I	160	166½	173	180	186½	193	200	206½	213†

NOTES: * Maximum exposure recommended for roofs.   I Maximum exposure recommended for single-coursing on sidewalls.   † Maximum exposure recommended for double-coursing on sidewalls.

Fig. 12-33. Chart of sizes and coverage for side wall shingles. Shingle thickness is based on number of butts required to equal a given measurement.

difference is in the exposures employed. In roof construction, maximum permissible exposures are slightly less than one-third of the shingle length. This produces a three-ply covering.

Vertical surfaces of side walls present less weather-resistance problems than do roofs. Accordingly, a two-ply covering of shingles is usually adequate.

In single-coursed side walls, weather exposure of shingles should never be greater than half the length of the shingle, minus 1/2 in. Thus, two layers of wood will be found at every point in the wall.

For example, when 16 in. shingles are used, the maximum exposure should be 1/2 in. less than 8 in. or 7 1/2 in.

Single-course side walls, Fig. 12-34, should have concealed nailing. This means that the nails must be driven about 1 in. above the butt line of the succeeding course, so the shingles of this course will adequately cover them. Two nails should be driven in each shingle up to 8 in. in width, and each nail placed about 3/4 in. from the edge of the shingle. On shingles wider than 8 in. a third nail should be driven in the center of the shingle, at the same distance above the butt line as the other nails. Use rust resistant 3d nails, 1 1/2 in. long.

ESTIMATING QUANTITIES

In estimating the quantity of shingles required for side walls, areas to be shingled should be calculated in square feet. Deduct window and door areas. Consult the table, Fig. 12-33. The coverage of one square (4 bundles) at the exposure to be used should be divided into the wall area to be covered. The figure arrived at will be the number of squares needed. Add about 5 percent to allow for waste in cutting and fitting around openings, and for the double starter course.

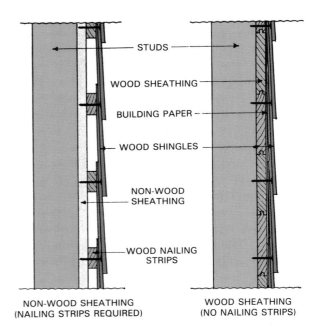

Fig. 12-34. Single-course method of applying shingles to sidewalls. Solid backing and nailing base is provided by wood sheathing or nailing strips over nonwood sheathing.

DOUBLE COURSING OF SIDE WALLS

In double coursing, a low-cost shingle is generally used for the bottom layer. This is covered with a No. 1 grade shingle. Many types of shingles are available for the outer course. Prestained shingles, which are available in attractive colors, are particularly suitable. Although wide exposures usually require the use of long shingles, this effect is obtained in double coursing by the application of doubled layers of regular 16 in. or 18 in. shingles. The maximum exposure to the weather of 16 in.

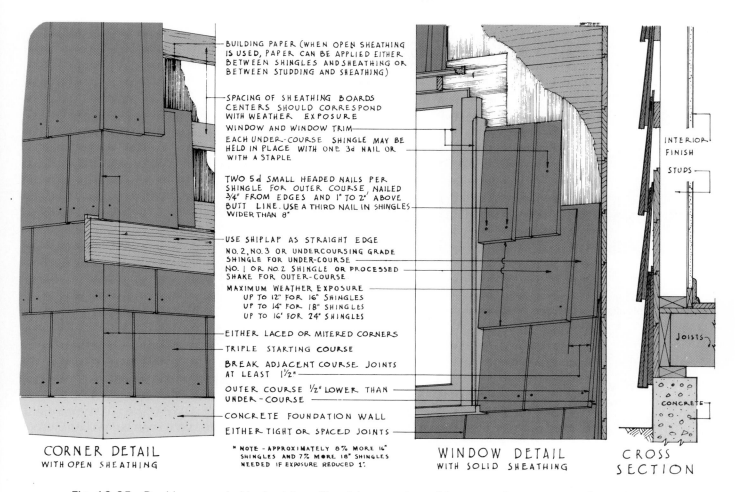

BUILDING PAPER (WHEN OPEN SHEATHING IS USED, PAPER CAN BE APPLIED EITHER BETWEEN SHINGLES AND SHEATHING OR BETWEEN STUDDING AND SHEATHING)

SPACING OF SHEATHING BOARDS CENTERS SHOULD CORRESPOND WITH WEATHER EXPOSURE

WINDOW AND WINDOW TRIM

EACH UNDER-COURSE SHINGLE MAY BE HELD IN PLACE WITH ONE 3d NAIL OR WITH A STAPLE

TWO 5d SMALL HEADED NAILS PER SHINGLE FOR OUTER COURSE, NAILED 3/4" FROM EDGES AND 1" TO 2" ABOVE BUTT LINE. USE A THIRD NAIL IN SHINGLES WIDER THAN 8"

USE SHIPLAP AS STRAIGHT EDGE

NO.2, NO.3 OR UNDERCOURSING GRADE SHINGLE FOR UNDER-COURSE

NO.1 OR NO.2 SHINGLE OR PROCESSED SHAKE FOR OUTER-COURSE

MAXIMUM WEATHER EXPOSURE
UP TO 12" FOR 16" SHINGLES
UP TO 14" FOR 18" SHINGLES
UP TO 16" FOR 24" SHINGLES

EITHER LACED OR MITERED CORNERS

TRIPLE STARTING COURSE

BREAK ADJACENT COURSE JOINTS AT LEAST 1½"

OUTER COURSE ½" LOWER THAN UNDER-COURSE

CONCRETE FOUNDATION WALL

EITHER TIGHT OR SPACED JOINTS

INTERIOR FINISH

STUDS

JOISTS

CONCRETE

CORNER DETAIL
WITH OPEN SHEATHING

* NOTE - APPROXIMATELY 8% MORE 16" SHINGLES AND 7% MORE 18" SHINGLES NEEDED IF EXPOSURE REDUCED 1".

WINDOW DETAIL
WITH SOLID SHEATHING

CROSS SECTION

Fig. 12-35. Double-coursed shingle siding. Sheathing may be solid or spaced on-center to the nailing line.

shingles double coursed is 12 in. For 18 in. shingles it is 14 in.

The application of shingles on a double-course side wall is illustrated in Fig. 12-35. Most procedures used for regular siding can be followed. *When the application is made over composition or spaced sheathing, mark the position of the nailing strips on the story pole when it is laid out.*

SHINGLE AND SHAKE PANELS

Shingles and shakes for side wall application are available in panel form. The panels consist of individual shingles (usually Western red cedar) permanently bonded to a backing. Standard size panels are 8 ft., Fig. 12-36. These are available in various textures, either unstained or factory-finished in a variety of colors. Special metal or mitered wood corners are also manufactured.

Shingle panels are applied by following the same basic precautions and procedures described for regular shingles, Fig. 12-37. Installation time however, is greatly reduced. Additional on-site labor

Fig. 12-36. Applying panelized shakes. Panels are 8 ft. long and consist of two 7 in. courses bonded to a plywood or veneer core base. (Shakertown Corp.)

SIDING PANEL APPLICATIONS

DIRECT TO STUDS (OVER FELT)
RECOMMENDED FOR SIDEWALLS
& MANSARDS 60° & STEEPER.

OVER SHEATHING WHERE LOCAL
CODES REQUIRE & FOR "A"
FRAMES (MINIMUM 12/12 PITCH).

STUDS 16" OR 24" O.C.

30-LB. FELT

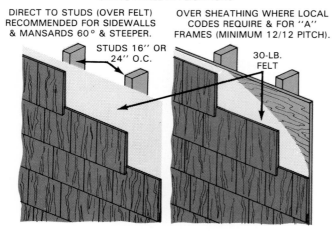

Fig. 12-37. Proper application of panelized siding. Panels are self-aligning. (Shakertown Corp.)

When you are making an application of a specialized or prefabricated product such as shingle panels, be sure to follow the recommendations furnished by the manufacturer.

RE-SIDING WITH WOOD SHINGLES

Shingles or shakes can be applied over old siding or other wall coverings that are sound and will hold nailing strips. First, apply building paper over the old wall. Next, attach nailing strips as previously described. Usually it is necessary to add new molding strips around the edge of window and door casings to trim the edge of the shingles. Fig. 12-39 illustrates how nailing strips are applied over an old stucco surface. Fig. 12-40 shows shingle panels being applied over stucco. No furring strips are necessary.

MINERAL FIBER SIDING

This product is produced from the same materials as mineral fiber roof shingles, (see Unit 10). Siding of this type is usually prefinished with modern long lasting factory baked coatings. It is available in a variety of textures and colors. Both textured and smooth surfaced siding may be obtained with either straight or wavy exposed butt lines.

Mineral fiber siding usually comes in units (called shingles) 24 in. wide and 12 in. deep. When applied, the shingles are lapped 1 1/2 in. at the head, leaving an exposed area of 10 1/2 by 24 in. Some units, however, are as large as 48 in. wide. The siding is sold in squares; a square being sufficient to cover 100 sq. ft. of wall surface. Ordinarily there are three bundles per square.

is also saved when factory-primed or factory-finished units are used. When applying the latter, the installation should be made with nails of matching color, supplied by the manufacturer. Fig. 12-38 shows a completed residential structure where shingle panels were used for the outside wall covering.

Fig. 12-38. View of shingle-paneled house. Surfaces may be factory finished, stained, painted, or left natural. (Shakertown Corp.)

Fig. 12-39. Re-siding a stucco wall with double-coursed wood shingles. Nailers are spaced to correspond to the nailing lines.

Fig. 12-40. Using shingle panels for a re-siding over stucco. No nailing strips are used. (Shakertown Corp.)

Fig. 12-41. Mineral fiber shingle application. Use a double layer of felt around corners. (Mineral Fiber Products Bureau)

Follow the same general procedures and methods for storing, handling, and cutting as are given in Unit 10 for mineral fiber shingles. The layout of a wall surface with a story pole and treatment of corners, Fig. 12-41, is basically the same as for other siding materials.

Composition sheathing will not hold nails. Nailing strips should be placed to overlay the top of the lower course by about 3/4 in. The ends of the nailing strips must be located over studs. Vertical joints between shingle units must not occur over nailing strip joints. The mineral fiber siding unit is applied with the bottom edge overlapping the wood strip 1/4 in. to provide a drip edge.

If the sheathing consists of lumber, plywood, or other materials that will serve as an adequate nailing base, the application can be made as shown in Fig. 12-42. The shingle backer produces an

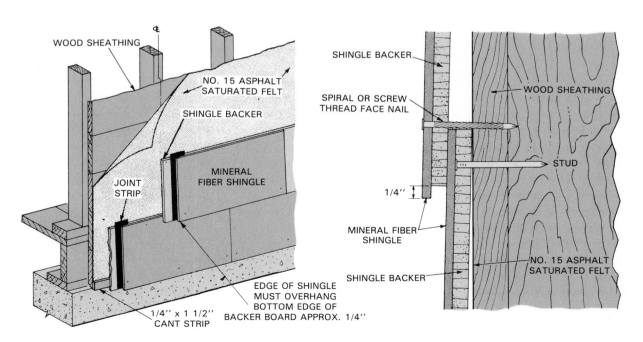

Fig. 12-42. Typical application of siding units over wood or nail-base sheathing. (Mineral Fiber Products Bureau)

attractive shadow line along the lower edge of each course of siding. Be sure the end joints of the backer and shingle units do not occur at the same location.

When the shingle backer method is employed, first nail the backer in place with 1 1/4 in. galvanized nails spaced about 3 in. from the top edge. If this is carefully positioned along the chalkline layout, the siding units can be quickly lined up along the top edge as shown. Joint flashing strips are not required when the backer board consists of a material that is water repellent or nonstaining.

Before making an installation of mineral fiber products, secure full information from manufacturers.

PLYWOOD SIDING

The use of plywood as an exterior wall covering permits a wide range of application methods and decorative treatments, Fig. 12-43. It may be used alongside other materials to:

1. Provide a vertical treatment to gable ends.
2. Provide emphasis as fill-in panels above and below windows.
3. Establish a continuous decorative band at various levels along an entire wall.

All plywood siding must be made from exterior type plywood. Douglas fir is the most commonly used species. However, cedar and redwood are also available. Panels come in either a sanded condition or with factory applied sealer or stain. Plywood siding with a special long-lasting surface or coating is also available. Information on grading standards may be obtained from Unit 1.

Panel sizes are 48 in. wide by 8, 9, and 10 ft. long. A 3/8 in. thickness is normally used for direct-to-stud applications. A 5/16 in. thickness may be used over an approved sheathing. Thicker panels are required when the texture treatment consists of deep cuts.

Application of large sheets is generally made with the panels in a vertical position. This eliminates the need for horizontal joints.

Fig. 12-44 shows the installation of a vertical panel over a sheathed wall. For unsheathed walls, Fig. 12-45, the thickness of the plywood should be not less than 3/8 in. for 16 in. stud spacing, 1/2 in. for 20 in. stud spacing, and 5/8 in. for 24 in. stud spacing. Vertical joints must occur over studs and horizontal joints must be over solid blocking.

Fig. 12-43. Examples of modern plywood siding styles. Top. Rough-sawn surface. Grooves are on 4 in. or 8 in. centers. Bottom. Reverse board and batten. Surface may be brushed, rough-sawn, coarse sanded, or otherwise textured. Both styles are manufactured in fir, cedar, redwood, southern pine, and other wood species. Edges are shiplapped. (American Plywood Assoc.)

Fig. 12-44. Textured plywood siding panel is being applied over insulation board sheathing.

Standard application requirements are given in Fig. 12-46.

Plywood lapped sidings may look the same as regular beveled siding. Heavy shadow lines are secured by using spacer strips at the lapped edges.

Fig. 12-45. Panel installation directly over studs. Panel is 5/8 in. thick.

at vertical joints. Nails should penetrate studs or wood sheathing at least 1 in.

If plywood lap siding is wider than 12 in. a wood taper strip should be used at all studs with nailing at alternate studs. Outside corners should butt against corner molding or they should be covered.

Fig. 12-48 shows several ways to handle joints between plywood panels. All edges of plywood siding—whether butted, V-shaped, lapped, covered, or exposed—should be sealed with a heavy application of high grade exterior primer, aluminum paint, or heavy lead and oil paint. Special caulking compounds are also recommended.

Because large sheets of plywood and hardboard siding provide tight, draft-free wall construction, it is important to have an effective vapor barrier. This should be between the insulation and the warm surface of the wall.

HARDBOARD SIDING

Improved techniques in manufacture have produced hardboard siding materials that are durable, easy to apply, and adaptable to various architectural effects. Installation methods are similar to those described for plywood sidings. Hardboard sidings may expand more than plywood. Special precautions should be observed in the application.

Application requirements are given in Fig. 12-47. A bevel of at least 30 deg. is recommended. The lap should be at least 1 1/2 in.

Vertical joints should be butted over a shingle and centered over a stud unless wood sheathing at least 3/4 in. thick is used. Nail siding to each of the studs along the bottom edge and not more than 4 in. O.C.

Stud spacing: 16'' o.c. for 3/8''; 24'' o.c. for 1/2 or 5/8'' plywood (1).

Caulk butt joints unless battened, ship-lapped, or backed with bldg. paper.

Omit diag. bracing and sheathing paper with rough sawn plywood.

Insulation as required.

Rough sawn panel siding (2).

Apply battens with 8d noncorrosive casing nails, 12'' o.c. and staggered.

For best results, paint plywood edges before installation.

Notes:
1. May use 3/8'' panel siding over 24'' o.c. supports; 5/16'' over 16'' o.c.
2. Nail 6'' o.c. at panel edges and 12'' o.c. at intermediate supports. Use galvanized, aluminum, or other noncorrosive casing or siding nails—6d for 3/8'' and 1/2'' panels; 8d for panels 5/8'' and thicker. Nail 3/8'' in from edges.

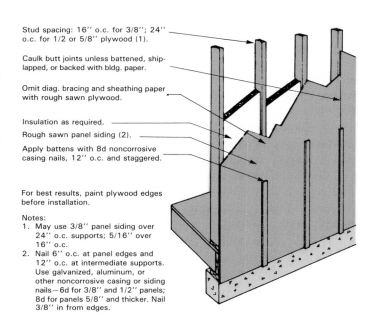

Fig. 12-46. Standard requirements for application of vertical siding. (American Plywood Assoc.)

Studs 16'' o.c. for 3/8'' lap siding applied directly to frame (1). 5/16'' rough sawn lap siding may be used over sheathing on same spacing.

Use shingle wedge under vertical joints for lap siding.

Add insulation as required.

Rough sawn plywood lap siding (2).

Plywood sheathing.

Use diag. bracing and sheathing paper if lap or bevel siding installed directly over studs.

Notes:
1. Use same nail schedule to install lap siding over studs or over sheathing. With sheathing, siding need not join over studs.
2. Use one nail per stud along bottom panel edges and 4'' o.c. at vertical joints. Nail 8'' o.c. at intermediate studs where siding is wider than 12''. Use galvanized, aluminum, or noncorrosive casing or siding nails. 6d for 3/8'' lap and 8d for thicker siding.

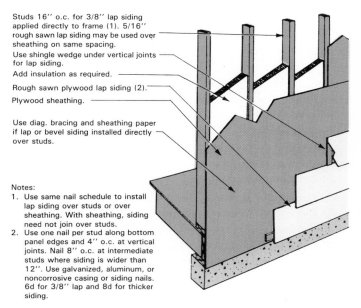

Fig. 12-47. Application requirements for lapped plywood siding. Local codes should be checked.

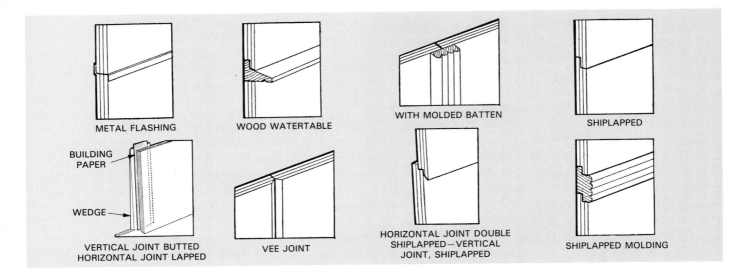

| METAL FLASHING | WOOD WATERTABLE | WITH MOLDED BATTEN | SHIPLAPPED |

BUILDING PAPER

WEDGE

| VERTICAL JOINT BUTTED HORIZONTAL JOINT LAPPED | VEE JOINT | HORIZONTAL JOINT DOUBLE SHIPLAPPED—VERTICAL JOINT, SHIPLAPPED | SHIPLAPPED MOLDING |

Fig. 12-48. Joint details for plywood siding. All edges should be sealed with paint or special caulking compound. (American Plywood Assoc.)

Manufacturers usually recommend leaving a 1/8 in. space where hardboard siding butts against adjacent pieces or trim members.

Hardboard siding panels are available in standard widths of 4 ft. and lengths of 8, 9, and 10 ft. Lap siding units are usually 12 in. wide by 16 ft. long. However, narrower widths can be purchased. The most common thickness is 7/16 in. Like plywood, hardboard sidings are furnished in a wide range of textures and surface treatments. See Fig. 12-49. Most panels have a factory-applied primer coat.

Prefinished units with matching batten strips and trim members are available.

When applying hardboard siding, follow standard installation procedures described for regular siding materials. Studs should not be spaced greater than 16 in. O.C. A firm and adequate nailing base is essential. Use a fine-tooth handsaw or power saw equipped with a combination blade to cut the panels. Nails must be galvanized or otherwise weatherproofed. Wood trim and corner boards should be at least 1 1/8 in. thick. Space nails at least 1/2 in. in from edges and ends. Use an approved caulking compound at joints.

SIDING SYSTEMS

Many manufacturers produce siding products that include special designs or patented devices which simplify the methods of application.

Fig. 12-50 shows the reverse side of a lap siding system. Application details are illustrated in Fig. 12-51.

Vertical siding is available in most systems with special matching trim and colored nails. On some installations, batten strips are used to cover hardboard batten backers. If desired, special snap-on batten strips can be furnished to cover hardboard batten backers, thus eliminating any exposed nailing. In addition to covering all vertical joints, intermediate batten strips may be installed to fulfill design requirements.

Another system is based on a patented aluminum fastener that not only eliminates face nailing but provides venting to minimize moisture traps behind the siding. A metal starter strip is attached to the bot-

Fig. 12-49. Home sided in shingle panels manufactured from hardboard. Panels are made in 8 ft. and 16 ft. lengths. (Masonite Corp.)

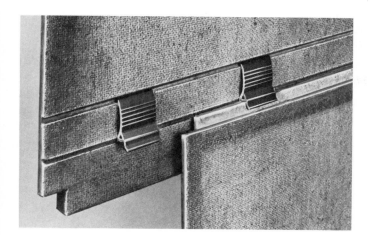

Fig. 12-50. Clip system holds siding courses together. All nails are concealed by the overlapping course. (Abitibi-Price Corp.)

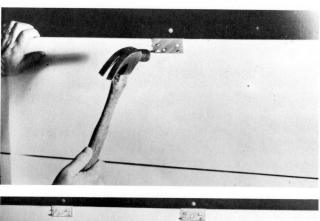

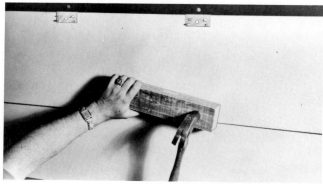

Fig. 12-52. Primed siding being installed with patented aluminum fastener. Top. Nail fastener and top edge of siding to stud. Bottom. Drive bottom edge of next course into fastener. (The Upson Co.)

tom of the sheathing or sill plate to carry the first course. The top edge of the siding is secured by fasteners, nailed in place at 16 in. intervals, Fig. 12-52. After the top edge is attached, the bottom edge is driven into the prongs of the fastener in the lower course.

Fig. 12-53 shows a completed installation of hardboard panelized shakes.

ALUMINUM SIDING

Aluminum siding offers low maintenance costs. It is factory finished with baked-on enamel, and provides an appearance that closely resembles painted wood siding. It is designed for use on new or existing construction. Aluminum can be applied over wood, stucco, concrete block, and other surfaces that are structurally sound. Basic specifications

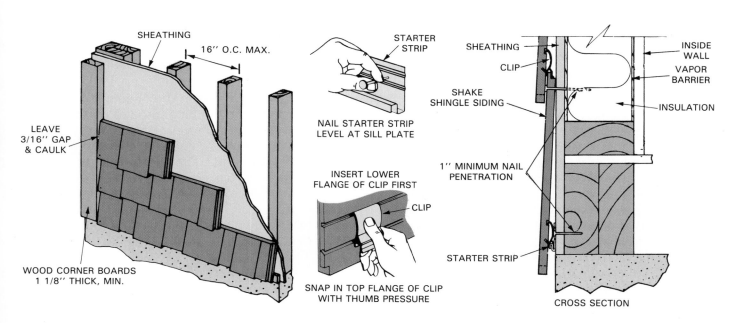

Fig. 12-51. How hardboard siding fastener system works. (Abitibi-Price Corp.)

319

Fig. 12-53. Hardboard shake application. No nails are visible.

Fig. 12-54. Installing an aluminum siding panel. Lower edge interlocks with previously applied course. (Alcoa Building Products)

Fig. 12-55. Special strips are available to cover top nailing surface of aluminum siding. Here it is being applied under windows.

(alloy and gauge) for aluminum siding are established by FHA and the Aluminum Siding Association.

A variety of horizontal and vertical panel styles in both smooth and textured designs are produced with varying shadow lines and size of face exposed to the weather. An insulated panel is also produced. It has an impregnated fiberboard material laminated to the back surface.

Manufacturers supply directions for the installation of their products. These should be carefully followed. Fig. 12-54 shows a horizontal siding unit being fastened in place. Panels are fabricated with prepunched nail and vent holes and special interlocking design. Standard strips, Fig. 12-55, are easily attached around windows and doors to provide a weathertight seal. Special corners and trim members are often formed on the job site as shown in Fig. 12-56.

As a precaution against faulty electrical wiring or appliances that might energize aluminum siding and create an electrical hazard, grounding should be included. The Aluminum Siding Association recommends that a No. 8, or larger, wire be connected to any convenient point on the siding and to the cold water service or the electrical service ground. Connectors should be UL (Underwriters Laboratories) approved.

VINYL SIDING

Recent developments in the chemical industries have resulted in economical production of a rigid polyvinyl chloride compound that is tough and durable. This material, commonly called vinyl, is extruded into siding units, either horizontal or vertical, and accessories, Fig. 12-57. Panel thickness of about 1/20 in. is available in various widths up to 8 in.

Like aluminum, vinyl siding is usually installed with a backer board or insulation board behind each sheet. This backer adds rigidity and strength as well as insulation. Panels are designed with interlocking joints that are moistureproof. Since the siding must be allowed to expand and contract slightly with temperature changes, the nail holes are slotted to permit movement.

Fig. 12-56. Skilled siding worker forms custom aluminum trim unit on portable bending brake on the job.

If platform framing is used, shrinkage of joists may cause distortion or cracks in the stucco.

The base for stucco consists of wood sheathing, sheathing paper, and metal lath, Fig. 12-58. The metal lath should be heavily galvanized and spaced at least 1/4 in. away from the sheathing so the base coat (called scratch coat) can be easily forced through, thoroughly embedding the lath. Metal or wood molding with a groove that "keys" the stucco is applied at edges and around openings. Galvanized furring nails, metal furring strips, and self-furring wire mesh, Fig. 12-59, are available. Nails should penetrate the sheathing at least 3/4 in. When fiberboard or gypsum sheathing is used, nailing with adequate penetration should occur over studs.

When installing vinyl siding, be sure to read and follow the directions furnished by the manufacturer.

STUCCO

When properly applied, stucco makes a satisfactory exterior wall finish. The finish coat may be tinted by adding coloring or the surface may be painted with a suitable material. Where stucco is used on houses more than one story high, the use of balloon framing for the outside walls is desirable.

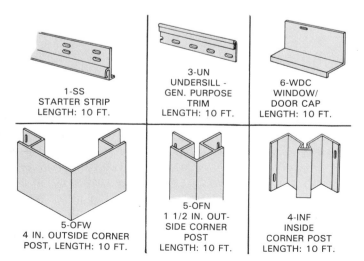

Fig. 12-57. Various accessories are made for use with vinyl siding. (Bird and Son, Inc.)

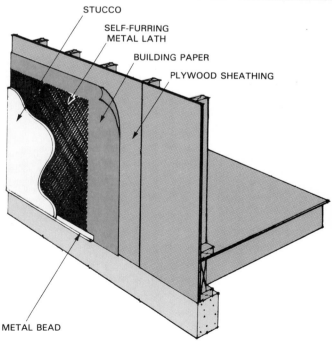

STUCCO
SELF-FURRING METAL LATH
BUILDING PAPER
PLYWOOD SHEATHING
METAL BEAD

Fig. 12-58. Top. Construction worker applies base coat of stucco. Bottom. Construction details for stucco finish. Building or sheathing paper is recommended over both interior and exterior plywood sheathing. (American Plywood Assoc.)

321

Fig. 12-59. A stapler is being used to install self-furring wire mesh for a stucco finish. (Bostitch)

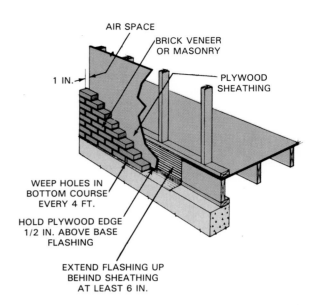

Fig. 12-60. General construction details for brick veneer siding. Plywood sheathing has been used. (American Plywood Assoc.)

BRICK OR STONE VENEER

A veneer wall is usually not referred to as a masonry wall but as a frame wall with masonry being used in place of the siding. In this type of wall, the masonry (which may be brick, concrete units, or stone) supports its own weight only.

The foundation must be wide enough to provide a base for the masonry, Fig. 12-60. A base flashing of noncorrosive metal should extend from the outside face of the wall, over the top of the ledge, and at least 6 in. up behind the sheathing, as shown. When the construction consists of plywood sheathing, and an air space of 1 in. is included, sheathing paper is not required.

Corrosion resistant metal ties are used to secure the veneer to the framework. These are usually spaced 32 in. apart horizontally and 15 in. vertically. Where other than wood sheathing is used, the ties should be secured to the studs. Weep holes (small openings in the bottom course) permit the escape of any water or moisture that may penetrate the wall. They are spaced about 4 ft. apart.

Select a type of brick suitable for exposure to the weather. Such brick will be hard and low in water absorption. Sandstone and limestone are most commonly used for stone veneer. These materials vary widely in quality. Be sure to select materials known locally to be durable.

BLINDS AND SHUTTERS

Some architectural designs may require the installation of blinds and shutters at the sides of window units. These consist of frame assemblies with solid panels or louvers, Fig. 12-61. In the early days of our country, they served an important function since they could be closed over the window. This protected the glass. Closing and locking the shutters also provided some security to the inhabitants.

Today, shutters and blinds are used mainly for decoration. They tend to extend the width of the window unit, emphasizing the horizontal lines of the structure. Hinging devices are seldom employed and they are usually attached to the exterior wall with screws or other fasteners so they can be easily removed for painting and maintenance. Stock sizes include various heights to fit standard window units. Widths range from 14 in. to 20 in. with 2 in. increments (steps).

TEST YOUR KNOWLEDGE — UNIT 12

1. In cornice construction the strip nailed to the wall to support the lookouts and soffit is called a _____.
2. Some prefabricated soffit systems utilize steel channels to provide rigidity to the soffit material, thus eliminating the need for _____.
3. Of the various types of horizontal wood siding, _____ siding usually has the most strength and tightest joints.
4. The best grade of solid wood siding is made from _____ (edge-grain, flat-grain) material.
5. Head flashing can usually be omitted over an opening if the vertical distance to the soffit is less than _____ the width of the overhang.
6. Plain beveled siding in a 10 in. width should be lapped about _____ in.
7. Corner boards are installed _____ (before, after) the horizontal siding units are fastened in place.
8. When plain square-edged boards are used for

322

Fig. 12-61. Shutters add a pleasing decorative effect and improve the appearance of window units. (Andersen Corp.)

vertical siding the joint is usually covered with a _____ strip.

9. Standard wood shingle lengths for side wall coverage are 16 in., 18 in., and _____ inches.

10. In single-coursed side walls, the weather exposure of wood shingles should not be greater than _____ their length, minus 1/2 in.

11. The most commonly used mineral fiber siding shingle is _____ in. wide.

12. When using plywood siding over an unsheathed wall, the thickness should not be less than _____ in. when studs are spaced 16 in. O.C.

13. Large sheets of plywood siding usually result in tight construction, therefore it is highly important that a vapor barrier be included on the _____ (warm, cold) side of the wall.

14. The most common thickness for hardboard siding material is _____ in.

15. As a protection against faulty electrical wiring or appliances, aluminum siding should be _____.

16. Small openings located in the bottom course of veneer construction that permits moisture to escape are called _____.

OUTSIDE ASSIGNMENTS

1. Study the cornice work on your home or some other residence in your neighborhood. Prepare a detail drawing (use a scale of 1'' = 1'-0'') showing a typical cross section. Since most of the construction will likely be hidden, you will need to develop your own structural design and select sizes for some of the parts. Study the details shown in various architectural plans and those in Fig. 12-1 of this unit. Make your drawing similar to these and be sure to include the sizes of all materials.

2. Visit a builders' supply store or lumber yard. Study the various types of siding carried in stock. Secure descriptive folders about various prefinished products and special siding systems. Also secure approximate costs. Write a report that includes a general description of the materials and summarizes the installation procedures. List some advantages and disadvantages of the various types.

3. Visit a building site in your neighborhood where the exterior wall finish is being applied. Be sure to secure permission from the builder or head carpenter. Observe the methods and materials being used. Make notes about: type of framing and stud spacing; type of sheathing; type of flashing; type, size, and quality of siding material; layout methods; how units are cut and fitted; nails and nailing patterns; and special joint treatment. Carefully organize your notes and make an oral report to your class.

Cutaway of modern home constructed and insulated to conserve energy. Rafters are the raised truss type which allow 12 in. or more of ceiling insulation to extend over the top of the wall plate. Walls carry 6 in. of insulation, also. Other energy conservation features include: double glazing of windows, sill sealer between foundation and sill, insulated floor over ventilated crawl space, vapor barrier over ground in crawl space, insulated heating ducts in attic and crawl space, insulated masonry walls in basement, vapor barrier under basement floor and slab-on-grade, perimeter insulation under slab-on-grade floor, insulated exterior door and storm door, cornice and roof vents for better attic ventilation. Deciduous trees provide shade for southern and western exposures during the summer but not in winter.

Unit 13

THERMAL AND SOUND INSULATION

Insulation requirements of homes and other buildings have changed radically since 1973 and the first energy crisis. As energy costs have increased so has the amount of insulation being placed in walls, floors, and ceilings. At the same time, the insulating qualities of doors and windows have been upgraded to hold down energy costs.

Likewise, duct work is likely to be insulated if it is not placed in the heated space itself, Fig. 13-1.

Insulation is necessary in buildings where temperature of the buildings must be controlled. In cold climates, the major concern is to retain heat in the building. In warmer regions, insulation is needed to keep heat from entering the building.

Fig. 13-1. Interior view of modern residence ready to receive insulation. To make room for 18 in. (R-60) of insulation in ceiling, wiring has been raised well above ceiling joists. Insulated ducts are already installed in the attic. Strip of 6 mil polyethylene was attached to interior walls before raising and will provide integrated ceiling vapor barrier. No pipes were installed in exterior walls. (Owens-Corning Fiberglas Corp.)

A wide range of insulation materials is available to fill the requirements of modern construction. The materials are engineered for efficient installation and come in convenient packages that are easy to handle and store.

Acoustical treatments and sound control employ many of the same materials used for thermal insulation and are, therefore, described in this unit.

BUILDING SEQUENCE

While carpenters are completing the exterior finish of the structure, other tradespeople work on the inside. Duct work is installed in the floors, walls, and ceiling for heating and air conditioning. The electrician strings rough wiring, Fig. 13-2, and installs other types of conduit in the framework. She or he also sets the metal boxes for convenience outlets, switches, and lighting fixtures that must be installed at this time.

The plumber installs the drains, vent stacks, and the pipes that provide water service and carry away waste water. See Fig. 13-3. Built-in plumbing fixtures (bath tubs and shower drains) are installed and then carefully covered with building paper or other materials to protect them during the inside finish operations.

During the installation of the heating and plumbing ''rough-in,'' the tradesperson often needs to cut through structural framework. The carpenter should check this work. Framing members should be reinforced wherever necessary.

HOW HEAT IS TRANSMITTED

Although carpenters are seldom required to design buildings or figure heat losses, they should have some knowledge and understanding of the theory and factors that are involved. Thus, they will more fully appreciate how important it is to carefully select and install thermal insulation materials.

Heat seeks a balance with surrounding areas. When the inside temperature is controlled within a

Fig. 13-2. Before interior of house is finished, other construction trades must make their installations. Here electrician is drilling joists to receive electrical wiring.

Fig. 13-3. Plumber must install pipes before insulation is placed.

given comfort range there will be some flow of heat. It will move from the inside to the outside in winter and from the outside to the inside during hot summer weather.

Heat is transferred through walls, floors, ceilings, windows, and doors at a rate directly related to:
1. The difference in temperature.
2. Resistance to heat flow provided by intervening materials.

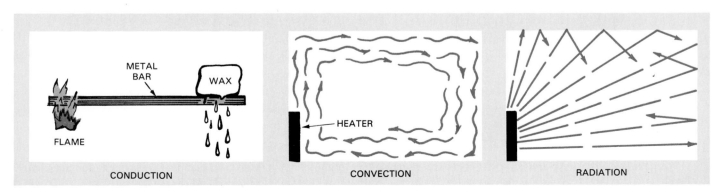

Fig. 13-4. Heat is transferred by conduction, convection, and radiation.

The transfer of heat takes place by one or more of three methods: conduction, convection, and radiation, Fig. 13-4.

CONDUCTION

Conduction is transmission of heat from one molecule to another within a material or from one material to another when they are held in direct contact. Dense materials such as metal or stone conduct heat more rapidly than porous materials such as wood and fiber products. Any material will conduct some heat when a temperature difference exists between its surfaces.

CONVECTION

Convection is the transfer of heat by another agent, usually air. In large spaces, the molecules of air can carry heat from warm surfaces to cold surfaces. When air is heated it becomes lighter and rises. Thus, a flow of air (called convection currents) is created within the space. Air is a good insulator when confined to a small space or cavity where convection flow is limited or absent. In walls and ceilings, air spaces restrict convection currents and will reduce the flow of heat.

RADIATION

Heat can be transmitted by wave motion in about the same manner as light. This process is called radiation because it represents radiant energy.

Heat obtained from the sun is radiant heat. The waves do not heat the space through which they move, but when they come in contact with a colder surface, a part of the energy is absorbed while some may be reflected.

Effective resistance to radiation comes about through reflection. Shiny surfaces, such as aluminum foil, are often used to provide this type of insulation.

Actually, heat transmission through walls, ceilings, and floors will be a result of all three of the methods. In addition to this, some heat is lost through cracks around doors, windows, and other openings in the structure.

THERMAL INSULATION

All building materials resist the flow of heat, mainly conduction, to some degree, depending on their porosity or density. As previously stated, air is an excellent insulator when confined to the tiny spaces or cells inside a porous material. Dense material such as masonry or glass contain few if any air spaces and are poor insulators.

Fibrous materials are generally good insulators not only because of the porosity in the fibers themselves but also because of the thin film of air that surrounds each individual fiber.

Commercial insulation materials are made of glass fibers, glass foam, mineral fibers, organic fibers, and foamed plastic. A good insulation material should be fireproof, vermin proof, moistureproof, and resistant to any physical change that would reduce its effectiveness against heat flow.

Selection is based on initial cost, effectiveness, durability, and the adaptation of its form to that of the construction and installation methods.

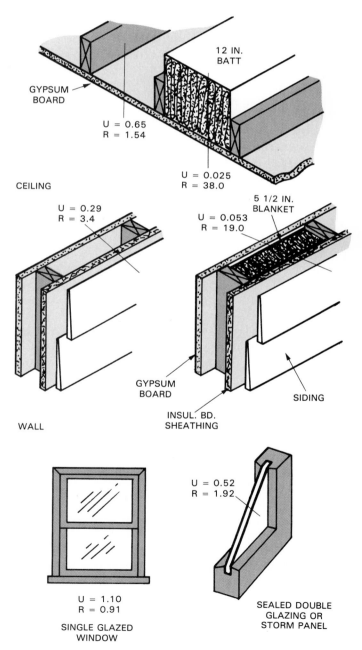

Fig. 13-5. Insulation reduces flow of heat. Trapped air, such as is provided by double-glazed windows, is a good insulator.

HEAT LOSS COEFFICIENTS

The thermal properties of common building materials and insulation materials are known or can be accurately measured. Heat transmission (the amount of heat flow) through any combination of these materials can be calculated. First, it is necessary to know and understand certain terms.

Btu: The abbreviation for British thermal unit. It is the amount of heat needed to raise the temperature of 1 lb. of water 1 °F.

k: The amount of heat, in Btu's, transferred in one hour through 1 sq. ft. of a given material that is 1 in. thick and has a temperature difference between its surfaces of 1 °F. It is also called the coefficient of thermal conductivity.

C: The conductance of a material, regardless of its thickness. It is the amount of heat (Btu's) that will flow through the material in one hour per sq. ft. of surface with 1 degree of temperature difference. For example, the C-value for an average hollow concrete block is 0.53.

R: Represents resistance which is the reciprocal (opposite) of conductivity or conductance. A good insulation material will have a high R-value.

$$R = \frac{1}{k} \text{ or } \frac{1}{C}$$

U: Represents the total heat transmission in Btu per sq. ft. per hour with 1 degree temperature difference for a structure (wall, ceiling, floor) which may consist of several materials or spaces. A standard frame wall with composition sheathing, gypsum lath, and plaster, with a 1 in. blanket insulation will have a U-value of about 0.11. To calculate the U-value where the R-values are known, apply the following formula:

$$U = \frac{1}{R_1 + R_2 + R_3}$$

Fig. 13-5 shows how insulation reduces the U-value for a conventional frame wall. Note that a 5 1/2 in. thick blanket reduces the U-value from 0.29 to 0.053. This is about an 81 percent reduction. Actually the U-value for the total wall structure will be slightly higher because the wood studs have a lower R-value than the blanket insulation.

A chart of wall structures with various types and amounts of insulation is shown in Fig. 13-6. The R-value provides a convenient measure to compare heat loss in specified materials and structural designs. However, to determine the total heat loss (or gain) through a wall, ceiling, or floor, R-values must be converted to U-values. The total of these U-values (Btu's per sq. ft., per hour with 1 deg.

temperature difference) will be needed to calculate the size of heating and cooling equipment. Directions for calculating U-values are presented in the Technical Information Section.

R-values for commonly used insulation and building materials are listed in Fig. 13-7. The original source of most data on this subject is the American Society of Heating, Refrigerating, and Air Condition-

Uninsulated 2 × 4 stud wall	
Air films ....R =	0.9
3/4″ wood exterior siding	1.0
1/2″ insulation board.....................	1.2
Air space	1.2
Vapor barrier	0
1/2″ gypsum board.....................	0.5
	4.8 total R

2 × 4 stud wall with batt insulation	
Air films ....R =	0.9
3/4″ wood exterior siding	1.0
1/2″ insulation board.....................	1.2
3 1/2″ batt or blanket insulation	11.0
Vapor barrier...........................	0
1/2″ gypsum board.....................	0.5
	14.6 total R

2 × 4 stud wall with rigid board	
Air films ....R =	0.9
3/4″ wood exterior siding	1.0
1″ polystyrene rigid board	5.0
3 1/2″ insulation blanket	11.0
Vapor barrier...........................	0
1/2″ gypsum board.....................	0.5
	18.4 total R

Improved insulated 2 × 4 stud wall	
Air films ....R =	0.9
3/4″ wood exterior siding	1.0
3/4″ insulation board.....................	2.0
3 5/8″ batt insulation	13.0
Vapor barrier...........................	0
5/8″ urethane insulation board	5.0
1/2″ gypsum board.....................	0.5
	22.4 total R

2 × 6 insulated stud wall	
Air films ....R =	0.9
3/4″ wood exterior siding	1.0
25/32″ insulation board...................	1.9
5 1/2″ insulating blanket	19.0
Vapor barrier...........................	0
1/2″ gypsum board.....................	0.5
	23.3 total R

Improved 2 × 6 insulated stud wall	
Air films ....R =	0.9
3/4″ wood exterior siding	1.0
3/4″ insulation board...................	2.0
5 1/2″ batt insulation	19.0
Vapor barrier...........................	0
5 8″ urethane insulation	5.0
5/8″ gypsum board.....................	0.6
	28.5 total R

Fig. 13-6. Types of wall construction and their R-values. Materials are listed in order from the outside in. Air films refer to the inside and outside film of stagnant air which forms on any surface. It makes a small contribution to the R-value. (Iowa Energy Policy Council)

Thermal and Sound Insulation

MATERIAL	KIND	INSULATION VALUE
Masonry	Concrete, sand, and gravel, 1 in.	R-0.08
	Concrete blocks (three core)	
	Sand and gravel aggregate, 4 in.	R-0.71
	Sand and gravel aggregate, 8 in.	R-1.11
	Lightweight aggregate, 4 in.	R-1.50
	Lightweight aggregate, 8 in.	R-2.00
	Brick	
	Face, 4 in.	R-0.44
	Common, 4 in.	R-0.80
	Stone, lime, sand, 1 in.	R-0.08
	Stucco, 1 in.	R-0.20
Wood	Fir, pine, other softwoods, 3/4 in.	R-0.94
	Fir, pine, other softwoods, 1 1/2 in.	R-1.89
	Fir, pine, other softwoods, 3 1/2 in.	R-4.35
	Maple, oak, other hardwoods, 1 in.	R-0.91
Manufactured Wood Products	Plywood, softwood, 1/4 in.	R-0.31
	Plywood, softwood, 1/2 in.	R-0.62
	Plywood, softwood, 5/8 in.	R-0.78
	Plywood, softwood, 3/4 in.	R-0.93
	Hardboard, tempered, 1/4 in.	R-0.25
	Hardboard, underlayment, 1/4 in.	R-0.31
	Particleboard, underlayment, 5/8 in.	R-0.82
	Mineral fiber, 1/4 in.	R-0.21
	Gypsum board, 1/2 in.	R-0.45
	Gypsum board, 5/8 in.	R-0.56
	Insulation board sheathing, 1/2 in.	R-1.32
	Insulation board sheathing, 25/32 in.	R-2.06
Siding and Roofing	Building paper, permeable felt, 15 lb.	R-0.06
	Wood bevel siding, 1/2 in.	R-0.81
	Wood bevel siding, 3/4 in.	R-1.05
	Aluminum, hollow-back siding	R-0.61
	Wood siding shingles, 7 1/2 in. exp.	R-0.87
	Wood roofing shingles, standard	R-0.94
	Asphalt roofing shingles	R-0.44
Insulation	Cellular or foam glass, 1 in.	R-2.50
	Glass fiber, batt, 1 in.	R-3.13
	Expanded perlite, 1 in.	R-2.78
	Expanded polystyrene bead board, 1 in.	R-3.85
	Expanded polystyrene extruded smooth, 1 in.	R-5.00
	Expanded polyurethane, 1 in.	R-7.00
	Mineral fiber with binder, 1 in.	R-3.45
Inside Finish	Cement plaster, sand aggregate, 1 in.	R-0.20
	Gypsum plaster, light wt. aggregate, 1/2 in.	R-0.32
	Hardwood finished floor, 3/4 in.	R-0.68
	Vinyl floor, 1/8 in.	R-0.05
	Carpet and fibrous pad	R-2.08
Glass	(see Unit 11 Windows and Exterior Doors)	

Fig. 13-7. R-values for commonly used construction and insulation materials. Additional resistance values can be obtained from the ASHRAE Handbook of Fundamentals. It is published by the American Society of Heating, Refrigerating, and Air Conditioning Engineers.

ing Engineers (ASHRAE). R-values can be converted to U-values by calculating the reciprocal (dividing the value into 1).

HOW MUCH INSULATION

Insulation is required in any building where a temperature above or below outdoor temperature must be maintained. The amount of insulation recommended has changed in recent years.

Comfort, health, and economy are the three considerations for thermal insulation. Comfortable and healthy indoor temperatures depend not only on temperature of the air but also on the temperature of the surfaces of walls, ceilings, and floors. It is possible to heat the air in a room to the correct comfort level and still have cold room surfaces. The human body will lose heat by radiation (or conduction if there is direct contact) to these colder surfaces.

Rising costs of energy, coupled with the prospect of fuel shortages has caused the recommended levels of insulation to be raised much higher than they were a few years ago. Today, the amount of insulation for a given structure must be based not only on comfort standards but also on factors such as insulation costs (material and labor), probable fuel costs in the future, and local climatic conditions. Fig. 13-8 shows a chart that has been devised as a guide. In warmer climates smaller amounts of insulation are needed—not for protection against cold but to keep out heat during summer months.

Fig. 13-9. One system of getting sufficient wall thickness for R-19 or better. Doubled wall has studs spaced 2 ft. O.C. Insulation is "woven" back and forth in a continuous band 6 in. thick. (Owens-Corning Fiberglas Corp.)

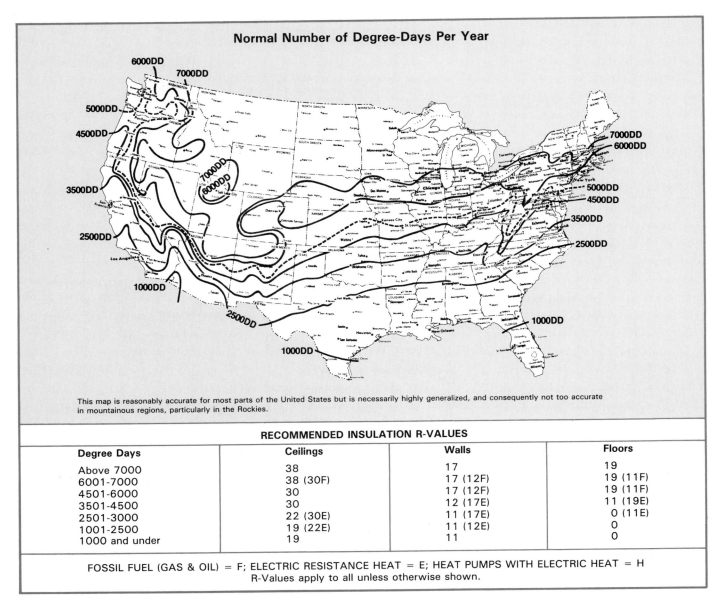

Normal Number of Degree-Days Per Year

This map is reasonably accurate for most parts of the United States but is necessarily highly generalized, and consequently not too accurate in mountainous regions, particularly in the Rockies.

RECOMMENDED INSULATION R-VALUES

Degree Days	Ceilings	Walls	Floors
Above 7000	38	17	19
6001-7000	38 (30F)	17 (12F)	19 (11F)
4501-6000	30	17 (12F)	19 (11F)
3501-4500	30	12 (17E)	11 (19E)
2501-3000	22 (30E)	11 (17E)	0 (11E)
1001-2500	19 (22E)	11 (12E)	0
1000 and under	19	11	0

FOSSIL FUEL (GAS & OIL) = F; ELECTRIC RESISTANCE HEAT = E; HEAT PUMPS WITH ELECTRIC HEAT = H
R-Values apply to all unless otherwise shown.

Fig. 13-8. Map outlines normal degree days (coldness) of different parts of the United States. R-values listed below map are recommended to meet FHA Minimum Property Standards for thermal resistance.

The harshness of climate is measured in degree days. The higher the number of degree days the colder the climate and the more insulation needed.

A degree day is the product of one day and the number of degrees F the mean temperature is below 65°F. Figures are usually quoted for a full year and are used by the heating engineer to determine the design and size of the heating system.

For example: High 60°F, Low 30°F

$$Degree\ Day = 65 - \frac{60 + 30}{2}$$
$$= 65 - \frac{90}{2}$$
$$= 65 - 45$$
$$Degree\ Day = 20$$

R-18 plus values in residential wall construction can be obtained by:
1. Using 2 x 6 studs which provide a thicker wall cavity for insulation.
2. Using 2 x 4 studs sheathed with thick, rigid insulation panels made from foamed plastic.
3. Using a double 2 x 4 frame, Fig. 13-9.

It is important to understand that heat transmission decreases as insulation thicknesses are increased, but not in a direct relationship. This can be noted through a study of U and R-values for various materials and structures. For example, in a frame wall with a U-value of 0.24, the addition of 1 in. of insulation will reduce the heat loss by 46

percent to U-0.13. A second inch of insulation will reduce the loss by 16 percent to U-0.09. And a third inch will reduce the loss by 10 percent to U-0.065. Additional thicknesses will continue to lower the U-value but at a still lower percentage. At some point it is useless to add more insulation.

Windows provide another example of how U-values decrease. The heat loss through a sash with a single pane of glass will be reduced from U-1.10 to about U-0.52 when a second pane is added. This provides a reduction of about 50 percent. A third pane (triple-glazing) will reduce the heat loss to U-0.35 and a fourth pane ("quad" glazing) will result in a U-value of about 0.27. Windows with three and four layers of glass are very expensive. In northern climates, this extra expense is usually justified by lower heating costs and added comfort.

TYPES OF INSULATION

Insulation is made in many forms, Fig. 13-10. It may be grouped into four broad classifications:
1. Flexible.
2. Loose fill.
3. Rigid.
4. Reflective.

Flexible insulation is manufactured in two types: Blanket or quilt, and batt. Blanket insulation is generally furnished in rolls or strips of convenient length and in various widths suited to standard stud and joist spacing. It comes in thicknesses of 3/4 in. to 12 in., Fig. 13-11.

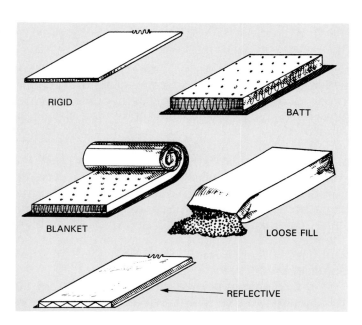

Fig. 13-10. Insulation is made in these four basic forms.

Fig. 13-11. Blanket insulation may be kraft faced, foil faced, or unfaced. This fiberglass blanket is 3 1/2 in. thick (R-11). (Owens-Corning Fiberglas Corp.)

The body of the blanket is made of loosely felted mats of mineral or vegetable fibers, such as rock, slag, glass fiber, wood fiber, and cotton. Organic fiber mats are usually treated chemically to make them resistant to fire, decay, insects, and vermin. Blanket insulation is often enclosed in paper with tabs on the side for attachment, Fig. 13-12. The covering sheet on one side may be treated to serve as a vapor barrier. In some cases the covering sheet is surfaced with aluminum foil or other reflective insulation. Fig. 13-13 shows an unfaced blanket that is easy to install between wall studs. It is held in place by friction. After all the insulation is in place, a vapor barrier is applied over the entire surface of the wall frame.

Batt insulation is made of the same fibrous material as blankets. Thickness can be greater in this form and may range from 3 1/2 in. to 12 in. They are generally available in widths of 15 and 23 in. and in 24 and 48 in. lengths. Batts are available with a single flanged cover or with both sides covered as shown in Fig. 13-14.

Loose fill insulation is composed of various materials used in bulk form and supplied in bags or bales, Fig. 13-15. It may be poured or blown. Loose insulation is commonly used to fill spaces between studs or to build up any desired thickness on a flat surface.

Loose fill insulation is made from such materials as rock, glass, slag wool, wood fibers, shredded

Fig. 13-13. Unfaced blanket fits snugly between studs and is held by friction.

Fig. 13-14. Thicker batts (12 in.) are designed to give high insulation values for ceilings and attics.

Fig. 13-12. Paper faced and flanged insulation. Flange is used to attach blanket to joists and studs.

redwood bark, granulated cork, ground or macerated wood pulp products, vermiculite, perlite, powdered gypsum, sawdust, and wood shavings. One of the chief advantages of this type is that when insulating an older structure, only a few boards need to be removed in order to blow the material into the walls.

Rigid insulation, as ordinarily used in residential construction, is often made by reducing wood, cane, or other fiber to a pulp and then assembling the pulp into lightweight or low-density boards that combine strength with heat and acoustical insulating properties.

It is available in a wide range of sizes, from tile 8 in. square, to sheets 4 ft. wide and 10 ft. or more long. Insulating boards are usually 1/2 in. to 1 in. in thickness. Boards of greater thickness are made by

Fig. 13-15. Loose insulation. Coverage and R values are listed on the bag.

laminating together boards of standard thickness.

Insulating boards are used for many purposes including:
1. Roof and wall sheathing.
2. Subflooring.
3. Interior surface of walls and ceilings.
4. Base for plaster.
5. Insulation strips for foundation walls and slab floors.

Fig. 13-16. Foil faced blanket insulation being applied to wall. Foils are fire resistant and must be used where insulation will remain exposed. (Owens-Corning Fiberglas Corp.)

Although expensive, foamed glass or cork board makes an excellent rigid form of insulation. A development widely used today is foamed plastic (polystyrene and polyurethane). Because it resists water, it is especially adaptable to masonry work.

Insulating boards should not be confused with ordinary wallboard. The wallboard is more tightly compressed and has less insulating value. Insulating sheathing board ordinarily comes in two thicknesses, 1/2 and 25/32 in. It is made in 2 x 8 ft. sheets for horizontal application, and in 4 x 8 ft. sheets or longer for vertical application.

REFLECTIVE INSULATION

Reflective insulation is usually a metal foil or foil-surfaced material. It differs from other insulating materials in that the number of reflecting surfaces, not the thickness of the material, determines its insulating value. In order to be effective, the metal foil must be exposed to an air space, preferably 3/4 in. or more in depth, Fig. 13-16.

Aluminum foil is available in sheets or corrugations supported on paper. It is often mounted on the back of gypsum lath. One effective form of reflective insulation has multiple spaced sheets.

OTHER TYPES OF INSULATION

There are available on the market today many insulations which do not fit the classifications covered. Some examples include the confetti-like material mixed with adhesive and sprayed on the surface to be insulated; multiple layers of corrugated paper; and lightweight aggregates like vermiculite and perlite used in plaster to reduce heat transmission.

Lightweight aggregates made from blast furnace slag, burned clay products, and cinders are commonly used in concrete and concrete blocks. They improve the insulation qualities of these materials.

WHERE TO INSULATE

Heated areas, especially in cold climates, should be surrounded with insulation by placing it in the walls, ceiling, and floors, Fig. 13-17. It is best to have it as close to the heated space as possible. For example, if an attic is unused, the insulation should be placed in the attic floor rather than in the roof structure.

If attic space or certain portions of the attic must be heated, walls and ceilings should be insulated. If the insulation is placed between the rafters, be sure to allow space between the insulation and the sheathing for free air circulation. The floors of rooms above unheated garages or porches require insulation if maximum comfort is to be maintained.

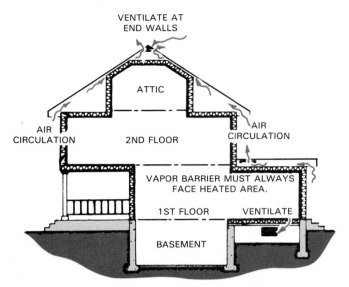

Fig. 13-17. Where to insulate residential structures.

When a basement is to be used as a living or recreation area, it will be necessary to insulate the walls, Fig. 13-18. This is highly recommended as an energy conservation measure, too. Not only does

Fig. 13-18. Insulating a basement wall with R-11 unfaced fiberglass. Concrete blocks have been furred out to provide support for drywall or paneling and to provide space for blanket insulation. Later, the insulation will be covered with a 6 mil polyethylene vapor barrier. (Owens-Corning Fiberglas Corp.)

it save heat and provide comfort, it gives the basement better acoustical qualities as well.

Whether applied to walls or not, insulation should be installed over the band joists. (These are the floor framing members along the outside of the wall.) See Fig. 13-19. This section has little protection against heat loss.

BASEMENTLESS STRUCTURES

Floors over unheated space directly above the ground require the same degree of insulation as walls in the same climate zone. This space, called

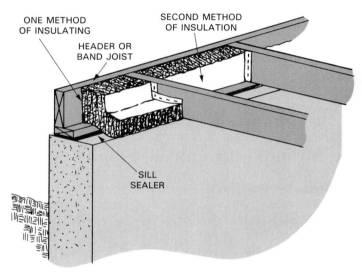

Fig. 13-19. Insulate behind headers atop basement walls. As you can see, there is little protection from loss of heat through the header.

a crawl space, if enclosed by foundation walls, is ventilated and will, therefore, approach the outside temperature.

Fig. 13-20 shows crawl space insulation. A vapor barrier should be positioned either on top of the insulation, as shown, or between the rough and finish floor.

Moisture coming up through the ground can be controlled by covering it with 6 mil (.006) polyethylene plastic film or roll roofing weighing at least 55 lb. per square. The material should be laid over the surface of the soil with edges lapping at least 4 in. If the ground is rough, it is advisable to put down a layer of sand or fine gravel before laying the vapor barrier. Covers of this kind greatly restrict the evaporation of water and less ventilation is needed than when no covers are used.

The soil surface beneath the building should be above the outside grade if there is a chance that

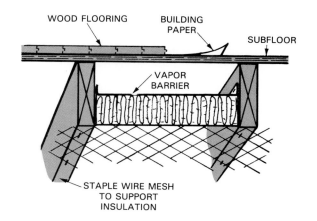

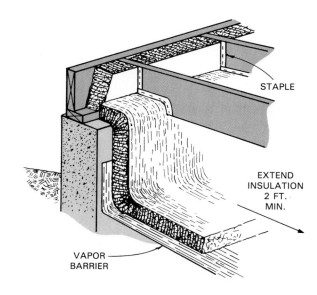

Fig. 13-21. Proper insulation of crawl space walls and ground. Staple fiber blankets to wood members. Beads of mastic adhesive can be used to secure blankets to masonry wall.

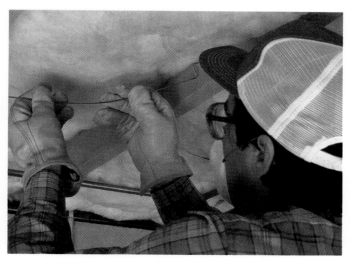

Fig. 13-20. Top. Basic construction and insulation requirements over a crawl space. R-19 is required by the FHA Minimum Property Standards. Insulation can be held in place with bowed wire, chicken wire, or even fishing line. Bottom. Wear gloves and protective clothing when working with insulation materials. (Manville Building Products, Inc.)

water might get inside the foundation wall. The soil cover is especially valuable where the water table is continually near the surface, or the soil has high capillarity (absorbs water easily). Be sure the covering is carried well up along the foundation wall.

Crawl space may be unvented and heated along with other parts of the structure. Insulation is installed along the inside of the foundation wall and floor frame. Fig. 13-21 shows the general installation of fiber blankets. Extruded polystyrene can provide high R-values for this type of installation. Sections should be carefully cut and fitted. Use a compatible mastic adhesive to hold insulation in place.

Closed crawl spaces are often used as a plenum (large duct) to distribute warm (or cold) air throughout the structure. See Unit 7.

Today many homes as well as other structures, are built on concrete slab floors. Such floors should contain insulation and a vapor barrier to prevent

heat loss along the perimeter. Very little heat is lost into the ground under the central part of the floor. The vapor barrier must be continuous under the entire floor. Only the perimeter needs to be insulated. The insulation can be installed horizontally (about 2 ft.) under the floor or vertically along the foundation walls as shown in Fig. 13-22.

INSULATING EXISTING FOUNDATIONS

Insulating walls and ceilings of existing structures is fairly simple. However, insulating floors or foundations of basementless structures can be a difficult task. Crawl spaces may not be easy to enter or may not provide room enough to work. Attaching a rigid insulation board to the outside surface of the foundation is usually the best solution.

Fig. 13-23 shows extruded polystyrene applied to the exterior wall of a crawl space. Dig a trench along the wall so that the insulation can be installed to a depth of at least 2 ft. below the finished grade. The wall surface must be cleaned. Cracks should be repaired.

The insulation board and protective cover will extend outward beyond the wall siding. To waterproof this joint, install a metal flashing as shown in the drawing. Extend the flashing at least 1 in. upward behind the siding.

Cut the polystyrene panels to size and then attach them to the wall with beads (approx. 3/8 in. dia.) of mastic adhesive. Follow the manufacturer's directions. Only a few minutes are required for the mastic to make an initial set.

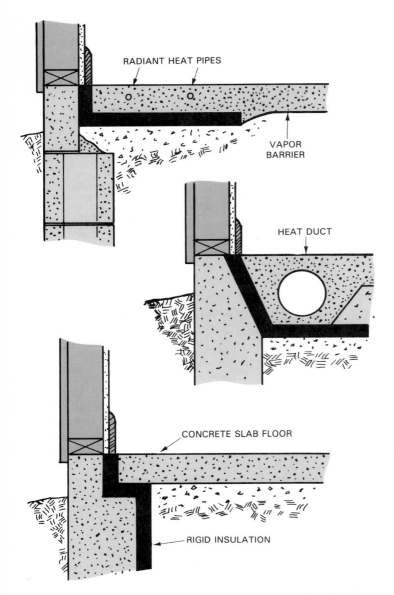

Fig. 13-22. Slab floors need insulation along their perimeters. Vapor barriers should be continuous under the entire floor.

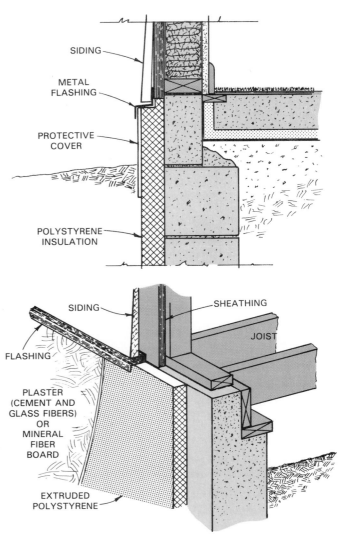

Fig. 13-23. Extruded polystyrene can be installed to the outside surface to provide insulation for all types of existing foundations. Top. Slab foundation. Bottom. Crawl space or basement foundation.

Polystyrene panels must be protected against ultraviolet light, wear, and impact forces. An approved method is to cover the surface with special plaster made from cement, lime, glass fibers, and a water resistant agent. This mixture is troweled onto the surface 1/8 in. to 1/4 in. thick. It should be extended downward to a level of about 4 in. below finished grade. The buried portion of the polystyrene panel does not require protection.

Mineral fiber panels, exterior plywood, or other weatherproof panels can also be used to protect the insulation. These panels can usually be attached with an approved type of mastic adhesive. Their use is practical when only a small area of the insulation board projects above the finished grade.

Polystyrene insulation is available in panels that have a weatherproof coating of fiber-reinforced cement on one side. They are especially designed to insulate the outside surface of foundations. These panels must be cut with a circular saw equipped with a masonry blade.

Existing concrete slab floors that require additional insulation can be handled in about the same manner. Fig. 13-23 shows a general detail of the basic design. Insulation panels should extend downward at least 2 ft. or to the top of the footing.

CONDENSATION

Water vapor is always present in the air. It acts like a gas and penetrates wood, stone, concrete, and most other building materials. Warm moisture-laden air within a heated building forms a vapor

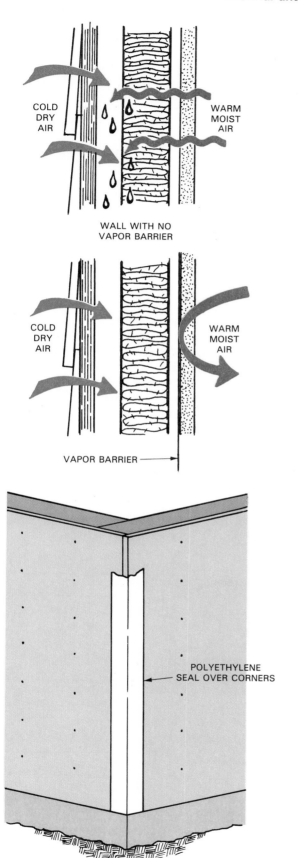

WALL WITH NO
VAPOR BARRIER

COLD
DRY
AIR

WARM
MOIST
AIR

COLD
DRY
AIR

WARM
MOIST
AIR

VAPOR BARRIER

POLYETHYLENE
SEAL OVER CORNERS

Fig. 13-24. Top. Vapor barrier is installed on warm side of a wall to protect the wall from condensation. Bottom. Codes may require seals over sheathing at outside corners.

pressure which constantly seeks to escape and mix with the colder, drier outside air.

Water vapor comes from many sources within a living space. It is generated by cooking, bathing, clothes washing, and drying, or by humidifiers which are often used to maintain a comfortable level of humidity.

When warm air is cooled, some of its moisture will be released as condensation. The temperature at which this occurs is called the DEW POINT. If you live in a cold climate—any region where the January temperature is 25 °F or colder—the dew point can occur within the wall structure or even within the insulation itself. The resulting condensation will reduce the efficiency of the insulation and may eventually damage structural members.

Moisture that collects within a wall during the winter months usually finds its way to the exterior finish in the spring and summer. It causes deterioration of siding and/or paint peeling. The siding is usually a porous material and will allow a considerable amount of moisture to pass. The paint is non-porous and the moisture gathers under the paint film, causing blisters and separation from the wood surface. Recent developments in paint manufacturing have provided products that are somewhat porous. They permit moisture to pass through.

During warm weather, condensation may occur in basement areas or on concrete slab floors in contact with the ground. When warm, humid air comes in contact with cool masonry walls and floors, some of the moisture will condense causing wet surfaces. Covering these surfaces with insulation will reduce or stop condensation. Operating a dehumidifier in areas surrounded by cool surfaces will also help.

The "dew point" is the temperature at which the air is completely saturated with moisture. Any lowering of the air temperature will cause condensation to occur.

VAPOR BARRIERS

A vapor barrier is a membrane through which water vapor cannot readily pass. When properly installed, it will protect ceilings, walls, and floors from moisture originating within a heated space. See Fig. 13-24. If you could check the temperature inside an insulated wall you would find that it is warm on the room side and cool on the outside. The vapor barrier must be located on the WARM SIDE to prevent moisture from moving through the insulation to the cool side where it could condense.

Fig. 13-25. Aluminum foil is a good vapor barrier when joints are carefully lapped.

Fig. 13-26. Install vapor barrier over insulation if the insulation does not already have it. Staple it securely at top and bottom plates and around door and window openings. (Owens-Corning Fiberglas Corp.)

Many insulation materials have a vapor barrier already applied to the inside surface. Also, many interior wall surface materials are backed with vapor barriers. When these materials are properly applied they usually provide satisfactory resistance to moisture penetration.

If the insulating materials do not include a satisfactory vapor barrier, then one should be installed. Vapor barriers in wide continuous rolls include:

1. Asphalt-coated paper.
2. Aluminum foil, Fig. 13-25.
3. Polyethylene films and various combinations of these materials.

To prevent accidental puncturing, vapor barriers should be installed after heat ducts, plumbing, and electrical wiring are in place. Cut and carefully fit the barrier around openings such as outlet boxes.

Some architects specify the use of 4 or 6 mil polyethylene film, Fig. 13-26. It is applied just before the plaster base or drywall and forms a continuous cover over walls, ceilings, and windows. The covering protects the window unit during plastering operations and permits light to enter. It can easily be trimmed out of the opening just before the finished wood trim is installed.

VENTILATION

Proper placement of vapor barriers alone will not protect the structure against moisture. Steps must be taken to provide good attic ventilation. The cold side (outside) of walls should be weathertight but still permit the wall to "breathe." Building paper can be used over wood sheathing to reduce infiltration if it is not waterproof and will permit moisture in the wall to escape.

It is especially important to ventilate an unheated attic or space directly under a low pitched or flat roof. Fig. 13-27 shows the most common systems for ventilation and the recommended size of the vent area.

Insufficient insulation or ventilation directly under low pitched roofs used in modern construction may cause a special problem, Fig. 13-28. In winter, heat escaping from rooms below may cause snow on the

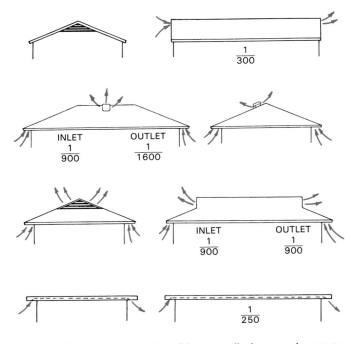

Fig. 13-27. Each type of roof has ventilation requirements. Numbers indicate ratio of vent area to total ceiling area.

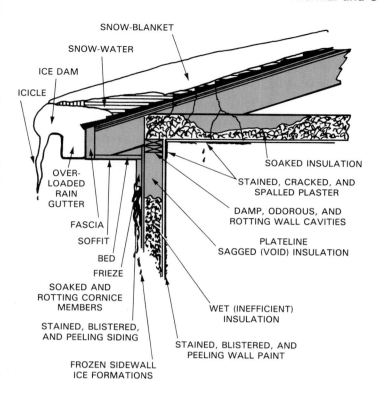

SNOW-BLANKET

SNOW-WATER

ICE DAM

ICICLE

OVER-LOADED RAIN GUTTER

FASCIA

SOFFIT

BED

FRIEZE

SOAKED AND ROTTING CORNICE MEMBERS

STAINED, BLISTERED, AND PEELING SIDING

FROZEN SIDEWALL ICE FORMATIONS

SOAKED INSULATION

STAINED, CRACKED, AND SPALLED PLASTER

DAMP, ODOROUS, AND ROTTING WALL CAVITIES

PLATELINE SAGGED (VOID) INSULATION

WET (INEFFICIENT) INSULATION

STAINED, BLISTERED, AND PEELING WALL PAINT

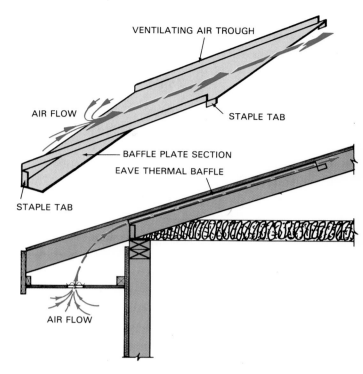

VENTILATING AIR TROUGH

AIR FLOW

STAPLE TAB

BAFFLE PLATE SECTION

EAVE THERMAL BAFFLE

STAPLE TAB

AIR FLOW

Fig. 13-29. Maintaining airway under eaves with special baffle which is attached to rafters. It prevents loose or blanket insulation from shutting off air flow. (Pease Co., Builders Div.)

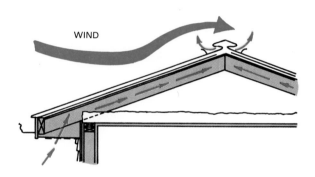

WIND

Fig. 13-28. How ice dams form on roof with too little insulation and not enough attic ventilation. Top. Escaping heat melts snow, causing runoff. Water freezes in overloaded gutter damming up water. Bottom. Insulation and good ventilation should solve the problem.
(Agricultural Extension Service, University of Minnesota)

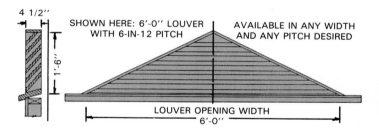

4 1/2''

1'-6''

SHOWN HERE: 6'-0'' LOUVER WITH 6-IN-12 PITCH

AVAILABLE IN ANY WIDTH AND ANY PITCH DESIRED

LOUVER OPENING WIDTH
6'-0''

Fig. 13-30. Prefabricated gable-end vents of wood or metal can be purchased assembled and ready for installation. (Ideal Co.)

roof to melt and water to run down the roof. At the overhang, the water may freeze again causing a ledge or dam of ice to build up. Water may back under the shingles and leak into the building.

In using thicker insulation near the cornice use care not to block the airway from the soffit vent into the attic. Fig. 13-29 shows a method of assuring adequate ventilation.

Gable roofs usually have ventilators in the gable ends. Fig. 13-30 describes one model that is manufactured in many sizes.

Ventilators located on the roof may leak if not properly installed. Whenever possible, these ventilators should be installed on a section of the roof

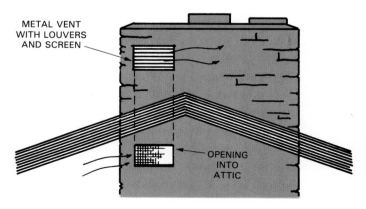

METAL VENT WITH LOUVERS AND SCREEN

OPENING INTO ATTIC

Fig. 13-31. A section of a chimney can sometimes be used for attic ventilation.

that slopes to the rear. Sometimes it may be possible to utilize a false flue or a section of a chimney for attic ventilation, Fig. 13-31.

Walls of existing frame buildings that have inadequate "cold side" ventilation can be corrected by installing patented ventilators as shown in Fig. 13-32. Most units are simply a metal tube with a cover that has tiny louvers. To install simply press them into a hole bored through the siding and sheathing. For maximum ventilation, install one at the bottom and top of every stud space.

SAFETY WITH INSULATION

Installing insulation is not particularly hazardous. However, there are some health safeguards to be observed when working with fiberglass. Tiny fibers of resin, if breathed in, could cause lung problems. Further, the fibers will cause itching and discomfort where they contact the skin.

Precautions are simple. Wear a mask over the nose and mouth if you will be working with insulation for any extended period. Wear a tight-fitting cap, gloves, long sleeves, and long trousers to avoid skin contact with the fibers. If you will be installing fiberglass insulation overhead, wear goggles to keep the fibers out of your eyes. Failure to observe this precaution could lead to vision problems.

INSTALLING BATTS AND BLANKETS

Insulation materials must be properly installed to perform efficiently. Even the best insulation will not provide its rated resistance to heat flow if the manufacturer's instructions are not followed or if materials are damaged.

Blankets or batts can be cut with a shears or a large knife. Measure the space and then cut the insulation 2 to 3 in. longer. On kraft faced batts

remove a portion of the insulation from each end so that you will have a flange of the backing or vapor barrier to staple to the framing.

When working with blanket insulation, it is usually best to mark the required length on the floor, and then unroll the blanket, align it with the marks, and cut the pieces. For wall installation, first staple the top end to the plate and then staple down along the studs, aligning the blanket carefully. Finally secure the bottom edge to the sole plate, Fig. 13-33.

To install batts in a wall section, place the unit at the bottom of the stud space and press it into place. Start the second batt at the top with it tight against the plate. Sections can be joined at the midpoint by butting them together. The vapor barrier should be overlapped at least 1 in. unless a separate one is installed. Some batts are designed without covers or flanges and are held in place by friction.

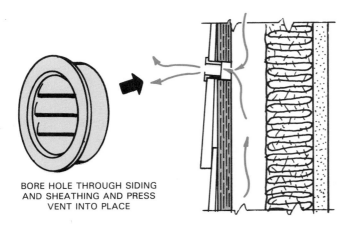

Fig. 13-32. Small vents can be installed in siding of existing structures. Sizes range from 1 to 4 in. A 1 in. size is usually big enough.

BORE HOLE THROUGH SIDING AND SHEATHING AND PRESS VENT INTO PLACE

Fig. 13-33. Installing blanket insulation. Top. To cut for length, follow marks laid out on the floor. Bottom. Staple to studs working from top to bottom. (Manville Building Materials Corp.)

Flanges, common to most blankets or batts, are stapled to the face or side of the framing members. Pull the flange smooth and space the staples no more than 12 in. apart. The interior finish, when applied, will serve to further seal the flange in place when it is fastened to the stud face. Some blankets have special folded flanges which enable them to be fastened to the face of the framing and also form an air space as shown in C, Fig. 13-34.

In drywall construction, specifications may require that the faces of the studs be left uncovered.

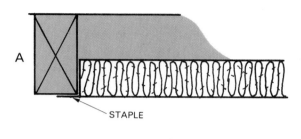

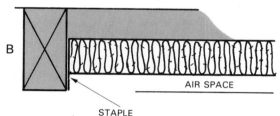

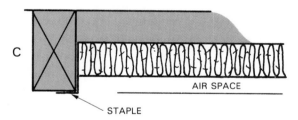

Fig. 13-34. Three methods of installing blankets and batts. A—Flush with inside surface of stud. B—With flange stapled along side of stud to form air space. C—With special flange providing air space.

The flanges of the insulation should fit smoothly along the sides of the framing and the staples should be spaced 6 in. or closer. Be sure there are no gaps or "fish mouths" (wrinkles). To secure maximum vapor protection, a separate vapor barrier should be applied over the entire wall or ceiling area. Avoid any perforations. Lap joints fully.

New construction will usually require that plumbing be installed in inside walls. On older dwellings you must thread the insulation carefully behind pipes that might be located in the wall, Fig. 13-35. If water service pipes (in cold climates they should never be located in outside walls) are present, it is well to add a separate vapor barrier between the pipe and interior surface to prevent moisture condensation on the cold pipe.

Some modern superinsulated homes use 2 x 6 wall studs. Fig. 13-36 shows a method of installing electrical service and insulating around it.

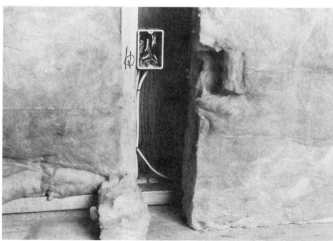

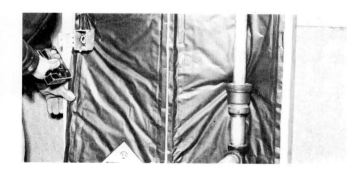

Fig. 13-35. On older construction, fit insulation carefully around plumbing and electrical service. (Acoustical and Board Products Assoc.)

Fig. 13-36. Mockup of new construction. Top. Method of notching 2 x 6 studs to carry electrical wiring where it will not interfere with insulation. Bottom. How insulation should be fitted around electrical boxes. (Owens-Corning Fiberglas Corp.)

Ceiling insulation can be installed from below or from above if attic space is accessible. When batts are used they are usually installed from below as shown in Fig. 13-37 following the same general procedure recommended for walls. Butt pieces snugly together at the ends and carry insulation right over the outside wall as shown in Fig. 13-38. In cold climates extra thicknesses are required. It may be necessary to raise the rafters of low pitched roofs by installing them on a secondary plate, Fig. 13-39. A raised roof truss design also provides extra space for insulation. This design is shown on the first page of this unit and in the Technical Information Section.

In multi-story construction, the floor frame should be insulated as shown in Fig. 13-40. Insulation should also be installed in the perimeter of the main floor even though the basement will be heated. Cut

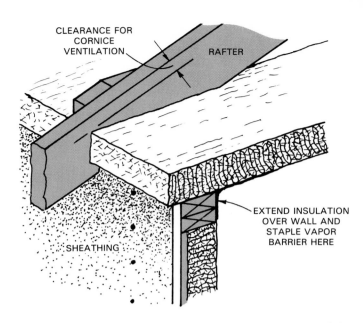

Fig. 13-38. Ceiling insulation should extend over top of the wall plate. Be sure to leave airway between cornice and attic.

Fig. 13-37. Top. Hammer tacker is used to install R-38 batt insulation. Vapor barrier has been applied to wall. Bottom. Vapor barrier is being added to ceiling to prevent condensation in insulation. (Owens-Corning Fiberglas Corp.)

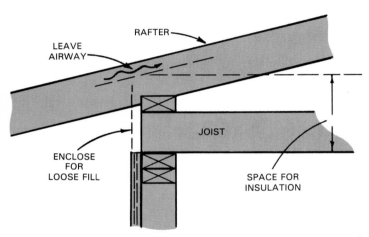

Fig. 13-39. One method of framing low-pitched roof to get extra space needed for thicker ceiling insulation. Rafter rests on 2 x 4 added over top of ceiling joist.

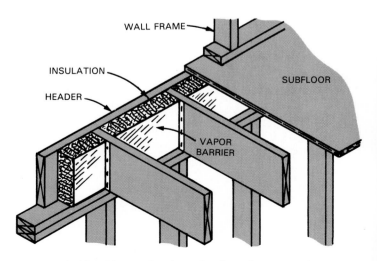

Fig. 13-40. Always insulate the floor frame as shown.

and fit pieces so they will fit snugly between the joists and against the header.

Complete large wall and ceiling areas first. Then insulate the odd sized spaces and areas above and below windows. Small cuttings remaining from the main areas can be used. Take the time necessary to carefully apply the insulation and vapor barrier around electrical outlets and other wall openings. Be careful that you do not cover outlet boxes or they may be missed when the wall surface is applied. (Refer, again, to Fig. 13-36.)

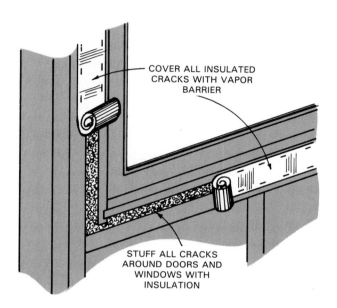

Fig. 13-41. Insulating around window and door frames. Carefully fill the cavities to stop heat loss. Use a stick to help place the insulation but use care not to compress the material so that it loses some of its insulating qualities.

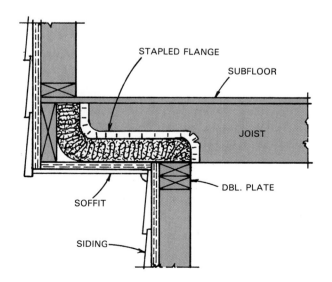

Fig. 13-42. Insulate to R-19 floors which project over an outside wall.

A thorough insulation job will require that all spaces be filled. For example: the space between window and door frames, and the rough framing as shown in Fig. 13-41. Use cuttings left over from larger spaces. Push the insulation into the small space with a stick or screwdriver. Cover the area with a vapor barrier.

Insulate floor projections which carry a chimney chase, bay window unit, or which extend the size of the room over an outside wall. See Fig. 13-42. Since there is no inside wall surface to form a seal against infiltration, the sheathing must be tight and the insulation flange and/or vapor barrier must be carefully stapled to the sides of the joist as shown. If weather conditions permit, this segment of insulation could be installed before the subfloor is laid. This would simplify installation since the work could be done from above.

INSTALLING FILL INSULATION

Fill insulation is placed by pouring or blowing. It is especially adaptable to existing structures because it avoids removing or cutting into wall surfaces.

Ceilings are easily insulated with loose insulation. It can be poured directly from bags into the joist spaces. Leveling is made simple with a straightedge, as shown in Fig. 13-43. A vapor barrier should be installed on the underside of the joists before the ceiling finish is applied. It will control the flow of moisture and also prevent fine particles (present in some forms of fill insulation) from sifting through cracks that might develop in the ceiling.

Ceiling insulation can be blown into place in either new or remodeled structures. Mineral fibers made from rock, slag, or glass are widely used, Fig. 13-44. Blown-in fill insulation is also made from cellulose fibers.

To contain the fill around the perimeter of a ceiling area, install 2 ft. lengths of thick batts next to outside walls. Be sure to provide for air circulation from cornice area. The National Electric Code requires a 3 in. space around heat producing devices, including recessed light fixtures and exhaust fans.

The installed R-value for fill insulation will vary depending on the method used (pouring or blowing). Manufacturers include these figures on bag or bale labels. R-values for a 20 lb. bag of mineral wood (pouring) are listed in Fig. 13-45.

Fill insulation is often poured into the core of block walls or in the cavity of masonry cavity walls. See Fig. 13-46. Thermal resistance is greatly increased. For example, the R-value of a standard concrete block (1.9) is increased to 2.8 when the cores are filled with insulation. A lightweight 8 in. block will be increased from R-3 to R-5.9.

Fig. 13-43. It is easy to level off poured insulation to a uniform depth with this setup. Strikeoff boards are installed permanently. (Vermiculite Institute)

Fig. 13-44. Blowing loose fill insulation into an attic area. Wear face mask, goggles, and gloves. A 14 in. depth will provide an R-30 factor. (Owens-Corning Fiberglas Corp.)

INSTALLING RIGID INSULATION

The installation of slab or block insulation varies with the type of product. Always study the manufacturer's specifications.

R-VALUE	MIN. THICKNESS	MAX. NET COVERAGE AREA (sq. ft.)	MIN. WT. /SQ. FT. (lbs.)
R-38	11 1/4"	8.5	2.344
R-33	9 3/4"	9.8	2.031
R-30	8 7/8"	10.8	1.849
R-26	7 3/4"	12.4	1.615
R-22	6 1/2"	14.8	1.354
R-19	5 5/8"	17.1	1.172
R-11	3 1/4"	29.5	0.677

Fig. 13-45. R-values for mineral wool pouring insulation.

Insulating board is widely used for exterior walls. Its application is covered in Unit 8. Sometimes it is used:
1. As the sheathing material for roofs.
2. As an insulating material installed over the roof deck.

A number of products are especially designed to insulate concrete slab floors. See Fig. 13-47. Waterproof materials, such as glass fibers or foamed plastic, provide desirable characteristics.

Fig. 13-48 shows a plastic foam insulation board applied to a masonry wall. It is bonded to the wall surface with a special mastic. It provides a permanent insulation and vapor barrier.

After the boards are installed, conventional

344

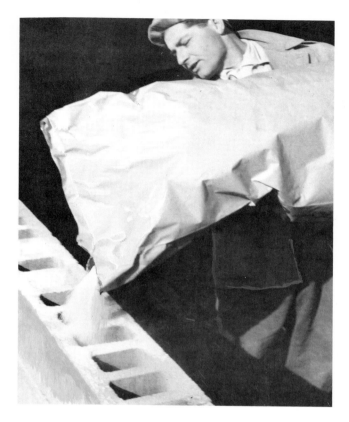

Fig. 13-46. Filling cores of concrete block will raise the R-value of the block wall. (Perlite Institute, Inc.)

Fig. 13-47. Perimeter of slab construction should be insulated with a rigid type of insulation. (Owens-Corning Fiberglas Corp.)

plaster coats can be applied to the surface. Plastic foam (Styrofoam) insulation is widely accepted as a rigid insulating material and has been successfully applied to a wide variety of constructions.

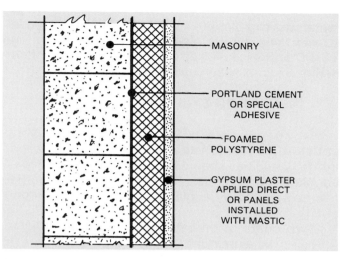

Fig. 13-48. Cross section of masonry wall insulated with rigid foamed polystyrene.

Fig. 13-49. In cold climates, basement walls may be given a framework of studs which provides a cavity for insulating batts. (Owens-Corning Fiberglas Corp.)

INSULATING BASEMENT WALLS

When basements will be used as living space, exterior walls should be insulated. The outside surface of masonry walls should be waterproofed below grade and should include a footing drain. (See Unit 6.)

The inside surface could be finished as shown in Fig. 13-48. Studs or furring strips could be used to form a cavity for the insulation and provide a nailing base for surface materials.

In cold climates a framework of 2 x 4 studs, spaced 24 in. O.C. is best. See Fig. 13-49. Use concrete nails or mastic to secure the sole plate and fasten the top plate to the joists or sill. Unfaced insulation will require a separate vapor barrier. Fig. 13-50 shows one method of insulating the floor perimeter in a basement.

INSULATING EXISTING STRUCTURES

Use special care when insulating an existing structure where no vapor barrier can be installed. Inside humidities should be controlled. ''Cold side'' ventilation is essential. A satisfactory vapor barrier can be secured by applying to room surfaces one of the following:
1. Two coats of oil base paint, rubber emulsion paint, or aluminum paint.
2. A vapor barrier wall paper.

The application should be carefully made on all surfaces of outside walls and ceiling.

AIR INFILTRATION

Infiltration refers to the air that leaks into buildings through cracks. It occurs around windows and doors, and through other small openings in the structure. During construction the problem of infiltration can be reduced by assembling materials properly and by sealing joints.

Caulking and sealing is usually required at the following locations:
1. Joints between sill and foundation.
2. Joints around door and window frames.
3. Intersections of sheathing with chimney and other masonry work.
4. Cracks between drip caps and siding.
5. Openings between masonry work and siding.

Be sure to caulk around the electrical service entrance and around hose bibs. The preferred types of caulking compound include polysulfide, polyurethane, or silicone materials.

Inside the structure, give special attention to recessed light fixtures and any built-in units located in outside walls. Also seal electrical conduit and plumbing that runs from the attic into walls and partitions located in living space. Seal conduit where it enters electrical boxes and seal the boxes to the inside wall surface.

Modern windows are built in a factory. Appropriate weather stripping is applied during fabrication. Outside doors, however, are often fitted to the door frame on the job and the weather stripping is installed by the carpenter. Each type will require a different method of installation. Follow the manufacturer's instructions. Fig. 13-51 shows a standard type of metal weather stripping.

Sometimes you may think that a window is leaking air when you are near it and feel a slight draft. This is usually due to the air in contact with the cold glass, becoming colder and heavier than the rest of

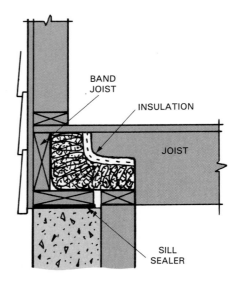

Fig. 13-50. Proper method for insulating space between basement wall and first floor.

the room air. It moves downward to the floor and across the room. Heating registers and convectors are usually positioned to offset or minimize these ''down-drafts.''

ESTIMATING MATERIALS

The amounts of insulating materials are figured on the basis of area (square feet). The thickness is then specified as separate data. The size of packages varies considerably depending on the type and thickness. For example: one manufacturer packages 1 1/2 in. blankets in rolls of 140 sq. ft. A 3 in. blanket in the same width comes in rolls of 70 sq. ft.

To determine the amount of insulation for exterior walls, first add up the total perimeter of the structure. Then multiply by the ceiling height. Deduct from the total the area of doors and windows. Many carpenters will deduct only large windows or window-walls and disregard doors and smaller openings. This extra allowance will:
1. Make up for loss in cutting and fitting.
2. Provide for additional material needed around plumbing pipes, recessed lighting boxes, and other items.

Here is an example:

Perimeter $= 30 + 40 + 36 + 20 + 6 + 20$
$= 152$
Area $= 152 \times 8 = 1216$
Window Wall $= 12 \times 8 = 96$
Net Area $= 1216 - 96 = 1120$ sq. ft.

Batts are furnished in packages (sometimes called

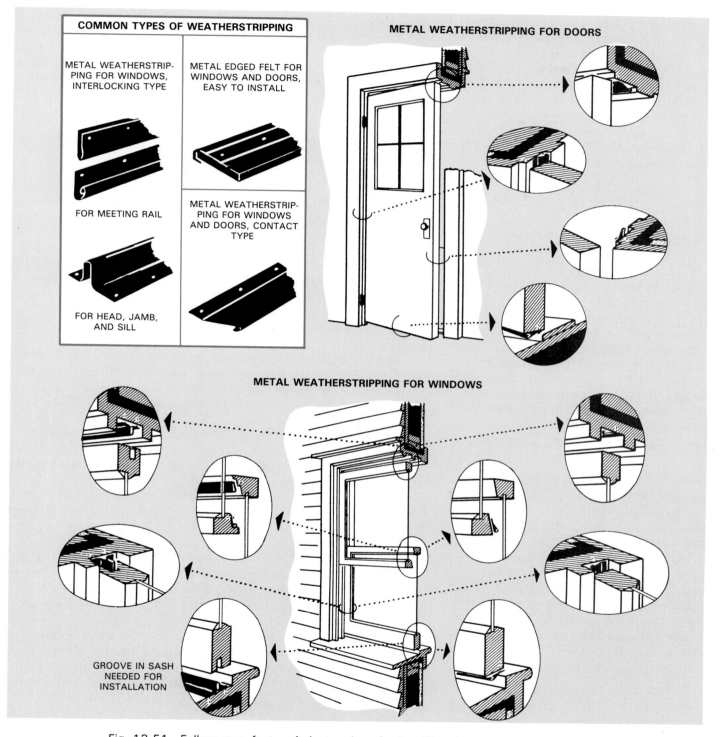

Fig. 13-51. Follow manufacturer's instructions for installing door and window stripping.

tubes) that contain as much as 100 sq. ft. A 6 in. batt usually contains 50 sq. ft.

When estimating the amount for floors and ceilings, use the same figures that were listed in calculating the subfloor area. Stairwells and openings for large fireplaces could be deducted. However, here again, these amounts may provide the extra needed for waste and special packing around fixtures.

The same procedure can be used to estimate reflective insulation. Make a greater allowance for cutting and waste especially when using accordion type. (This type cannot be spliced effectively.) Rolls of reflective insulation hold from 250 to 500 sq. ft.

Rigid insulation is also estimated on the basis of area and can be calculated from dimensions shown on the working drawings.

For a perimeter insulation strip in a concrete slab floor, multiply the perimeter of the building by the width of the strip. There is little waste on such an installation since even small pieces can be utilized.

Fill insulation comes in bags that usually contain 3 or 4 cu. ft. The cubic feet required can be calculated as follows:

Area	= 1200 sq. ft.
Thickness	= 4 in. = 1/3 ft.
Cu. ft. required	= 1200 × 1/3 = 400
*Less 10%	= 400 − 40
Net Amount	= 360
Number of bags (4 cu. ft.)	= 90
*Allowance for joists 16'' O.C.	

Manufacturers' directions and specifications will usually list tables that provide a direct reading of the number of bags required for a certain thickness of application. These are especially helpful in estimating amounts needed for such items as filling the core of concrete blocks. See Fig. 13-52.

ACOUSTICS AND SOUND CONTROL

Noise is unwanted sound. A by-product of our modern world, it has reached a magnitude that demands sound control in every home and building. Noise is unpleasant, reduces human efficiency, and can cause undue fatigue.

Houses, especially those built in northern climates, will have wall and roof structures heavy enough to repel average outside noises. Therefore, noise or sound control will apply mainly to interior partitions, floors, and surface finishes.

Sounds in an average home are generated by conversation, television, radios, record players, typewriters, and musical instruments. Vacuum cleaners, washing machines, food mixers, and garbage disposers are examples of mechanical equipment that creates considerable noise. Plumbing, heating, and air conditioning systems may be a source of excessive noise if poorly designed and installed. The activities in play rooms and workshops may create sounds that will be undesirable if transmitted to relaxing and sleeping areas.

The solution to problems of sound and noise control can be divided into three parts:
1. Reducing the source.
2. Controlling sound within a given area or room.
3. Controlling sound transmission to other rooms.

The carpenter should have some understanding of how the latter two can be accomplished and be familiar with sound conditioning material and constructions. As in the case of thermal insulations, the carpenter must appreciate the importance of careful work and proper installation methods.

ACOUSTICAL TERMS

SOUND. A vibration or wave motion that can be heard. It usually reaches the ear through air. The air itself does not move but vibrates back and forth in tiny molecular motions of high and low pressure.

DECIBEL. The unit of measurement used to indicate the loudness or intensity of sound. Comparable to the ''degree'' as a measurement of heat or cold. See Fig. 13-53.

REVERBERATION SOUNDS. These are airborne sounds which continue after the actual source has ceased. They are caused by reflections from floors, walls, and ceiling.

FREQUENCY. Rate at which sound-energized air molecules vibrate. The higher the rate, the more cycles per second (cps). Examples are the low frequency of a bass drum and the high frequency of a piccolo.

IMPACT SOUNDS. These are the sounds that are carried through a building by the vibrations of the structural materials themselves. Footsteps heard through the floors of a structure are an example of impact sounds.

MASKING SOUNDS. These are the normal sounds within habitable rooms which tend to ''mask'' some of the external sounds entering the room.

DECIBELS REDUCTION. An expression used to indicate the sound insulating properties of a wall or floor panel.

SOUND TRANSMISSION LOSS (TL). Sound insulating efficiency of wall or floor construction, measured in decibels. Transmission loss is the number of decibels which sound loses when transmitted through a wall or floor. Transmission loss of any wall or floor depends on the materials and techniques used in construction.

WALL AREA (SQ. FT.)	CORE FILL BLOCK SIZE			CAVITY FILL CAVITY WIDTH			
	6 IN.	8 IN.	12 IN.	1 IN.	2 IN.	2 1/2 IN.	3 IN.
100	5	7	12	2	4	5	6
500	23	33	58	10	21	26	31
1000	46	65	118	21	42	52	62
2000	91	130	236	42	84	104	125
3000	137	195	354	62	124	155	187

APPROXIMATE COVERAGE
NUMBER OF BAGS (4 CU. FT.) REQUIRED

Fig. 13-52. Table provides estimates of fill insulation needed for masonry walls. (Perlite Institute, Inc.)

SOUND ABSORPTION. The capacity of a material or object to reduce sound waves by absorbing them. Acoustical materials, such as acoustical ceiling tile, are designed to absorb sounds within a given area. These are sounds that otherwise would be reflected and cause excessive reverberation and build-up of intensity within that area.

NOISE REDUCTION COEFFICIENT (NRC). The sound absorption of acoustical materials is expressed as the average percentage absorption at the four frequencies which are representative of most household noises. These noises are 250, 500, 1000, and 2000 cycles per second.

SOUND TRANSMISSION CLASS (STC). The STC is a single number which represents the minimum performance of a wall or floor at all frequencies. The higher the STC number, the more efficient the wall or floor will be in reducing sound transmission.

SOUND INTENSITY

The number of decibels indicates the loudness or intensity of the sound. Roughly speaking, the decibel unit is about the smallest change in sound that is audible to the human ear. Actually, the decibel has the same relationship to a scale of loudness as the degree has to a thermometer. Reference to Fig. 13-53 will show that the rustle of leaves or a low whisper is on the threshold of audibility. That is, the sound is barely heard by the human ear. At the top of the scale are painfully loud sounds of over 130 decibels, often referred to as the "threshold of pain."

COMMON DESCRIP.	DECI-BELS	THRESHOLD OF FEELING
Threshold of Pain	— 130 — — 120 —	Space Shuttle (180+) (Lift-off) Concorde Boeing 747
Deafening	— 110 — — 100 —	Thunder, Artillery Nearby Riveter Elevated Train Boiler Factory
Very Loud	— 90 — — 80 —	Loud Street Noise Noisy Factory Truck Unmuffled Police Whistle
Loud	— 70 — — 60 —	Noisy Office Average Street Noise Average Radio Average Factory
Moderate	— 50 — — 40 —	Noisy Home Average Office Average Conversation Quiet Radio
Faint	— 30 — — 20 —	Quiet Home or Private Office Average Auditorium Quiet Conversation
Very Faint	— 10 — — 0	Rustle of Leaves Whisper Sound Proof Room Threshold of Audibility

Fig. 13-53. Levels of sounds and noises. Those above 90 decibels tend to be disagreeable.

One sound level is 10 decibels greater than another if its intensity is 10 times greater than the other. There is a logarithmic relation, on the decibel scale, to the amount of sound energy involved. If the sounds differ by 20 decibels, the ratio of their intensities is 10^2 or 100 times greater. If by 30 decibels, the ratio is 10^3 or 1,000 times greater, and so on up the scale.

TRANSMISSION

When sound is generated within a room, the waves strike the walls, floor, and ceiling. Much of this sound energy is reflected back into the room. The rest is "absorbed" by the surfaces. If there are cracks or holes through the wall (no matter how minute), a part of these sound waves will travel through as airborne sounds. The sound waves striking the wall will cause it to vibrate as a diaphragm, reproducing these waves on the other side of the wall.

Sound transmission through theoretically airtight partitions is the result of such diaphragm action. The sound insulation values of such substances is therefore almost entirely a matter of their relative weight, thickness, and area. In partitions of normal dimensions, this value depends mostly upon weight.

As the sound moves through any type of wall or barrier its intensity will be reduced. This reduction is called Sound Transmission Loss (TL) and is expressed in decibels. A wall with a TL of 30 dB. will reduce the loudness level of sound passing through it from 70 dB. to 40 dB. See Fig. 13-54. The transmission loss of any floor or wall will be determined by the materials, design, and quality of constructional techniques.

Although transmission loss rated in decibels is still used, another system of rating sound-blocking efficiency is widely accepted. It is called the Sound Transmission Class (STC) system. Standards have been established through extensive research by such associations as the Insulation Board Institute, and the National Bureau of Standards.

STC numbers have been adopted by acoustical engineers as a measure of the resistance to sound transmission of a building element. Like the

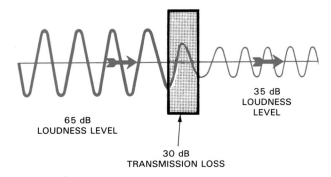

Fig. 13-54. Sound transmission loss through a wall. Values will vary somewhat depending upon the frequency of the sound waves.

65 dB
LOUDNESS LEVEL

30 dB
TRANSMISSION LOSS

35 dB
LOUDNESS
LEVEL

30 STC
LOUD SPEECH CAN BE UNDERSTOOD
FAIRLY WELL

resistance in thermal insulation (R), the higher the number, the better the sound barrier. Fig. 13-55 shows how a composite STC rating is applied to a given wall construction which has various TL values in decibels through a specified range of frequencies. Fig. 13-56 further describes these STC ratings by making a simple application to a wall separating two apartments.

Actually there is another factor present which should be considered when designing any sound insulating panel. These are usually referred to as MASKING SOUNDS. In theory, an inaudible sound of zero (0) on the decibel scale is for a perfectly quiet room. Since there are noises in every habitable room which tend to mask the sound entering, it is only

35 STC
LOUD SPEECH AUDIBLE BUT NOT
INTELLIGIBLE

42 STC
LOUD SPEECH AUDIBLE
AS A MURMUR

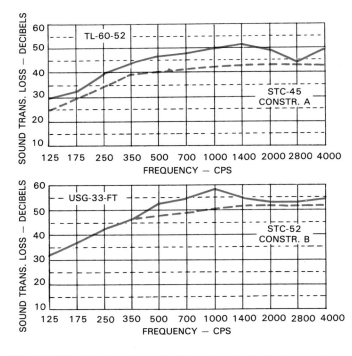

Fig. 13-55. Graphs show the TL values in decibels at various frequencies for two STC rated constructions.

50 STC
LOUD SPEECH NOT AUDIBLE

Fig. 13-56. How different STC ratings apply to a partition between two apartments.

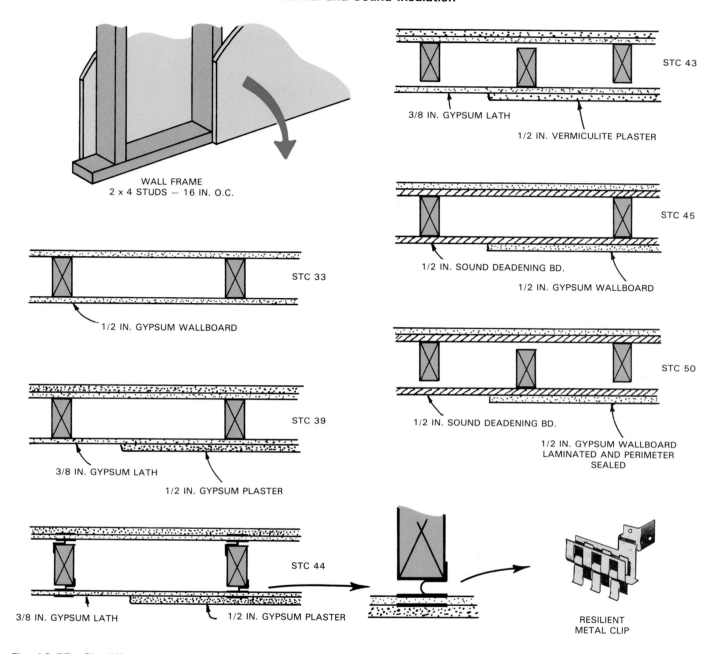

Fig. 13-57. Six different methods of constructing interior walls. Sound transmission class (STC) is given for each. An insulation blanket will usually increase the STC rating.

necessary to reduce sound to the level of the more or less maintained sound level within the space to be insulated. Assume that there is a radio playing soft music in the listening room (about 30 dB.). Thus, sound of less than 30 dB. entering the room would be completely masked.

WALL CONSTRUCTION

How high must an STC rating be for a given wall? This will depend largely on the type of areas it separates.

For example, partitions between bedrooms in an average home usually will not require special sound-proofing while those between bedrooms and activity or living rooms should have a high STC rating. Partitions surrounding a bathroom should also have a high STC number. Extra attention should be given to the placement of insulation around pipes.

Fig. 13-57 shows a number of practical constructions for partitions and their STC ratings. Some of them include sound deadening board. This is a structural insulation board product designed especially for use in sound control systems. It is made principally from wood and cane fibers in a nominal 1/2 in. thickness. Standard sizes of sound deadening board are the same as regular insulation board materials. It is usually identified with the words: IBI - RATED SOUND DEADENING BOARD on each sheet or package to distinguish it from other insulation board products.

In order to secure the high STC ratings in a wall structure when using a sound deadening board, application details supplied by the manufacturer should be carefully followed. Where nails are used, the size, type, and application patterns are critical. In drywall construction, the joints should be taped and finished and the entire perimeter sealed. Openings in the wall for convenience outlets and medicine cabinets require special consideration. For example, electrical outlets on opposite faces of the partition should not be located in the same stud space.

DOUBLE WALLS

Partitions between apartments are often constructed to form two separate walls as shown in Fig. 13-58. Standard blanket insulation is installed in about the same manner as for thermal insulation purposes. It should be stapled to only one row of the framing members.

For economical and space-saving construction, strips of special resilient channel are nailed to standard stud frames as shown in Fig. 13-59. The base layer of gypsum board is attached with screws. The surface layer is bonded with an adhesive. Since laminated systems like this minimize the use of

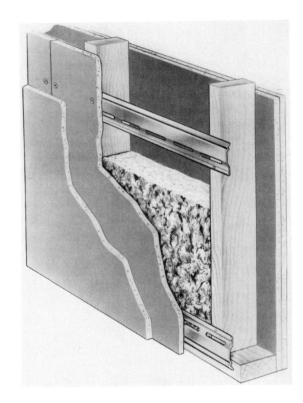

Fig. 13-59. Resilient channel can be added to wood studs on one side. Stud spacing is 24 in. O.C. Double drywall layer with 3 in. insulation batt gives an STC rating of 50. (United States Gypsum Co.)

metal fasteners, they result in a finer appearance along with better sound and fire resistance.

FLOORS AND CEILINGS

In general, the considerations for soundproofing that were applied to walls can also be applied to floors and ceilings. Floors are subjected to impact sounds. These are the noises you get from activities such as walking, moving furniture, operating vacuum cleaners and other vibrating equipment. Sound control is somewhat more difficult. Often an impact sound may cause more annoyance in the room below than it does in the room where it is generated. The addition of carpeting or similar material to a regular hardwood floor will be effective in minimizing impact sounds.

Sound deadening board, properly installed will increase the STC rating. See Fig. 13-60. The use of patented metal clips to attach the ceiling material is a practical solution. Various suspended ceiling systems also provide high levels of sound control.

Fig. 13-61 shows a cutaway view of a floor-ceiling system with an STC rating over 52. It consists of 2 x 10 joists placed 16 in. O.C. with standard wood subfloor and finished floor. The floor is

Fig. 13-58. A double wall provides a high STC rating for a partition between apartments. Insulation increases the STC. (Owens-Corning Fiberglas Corp.)

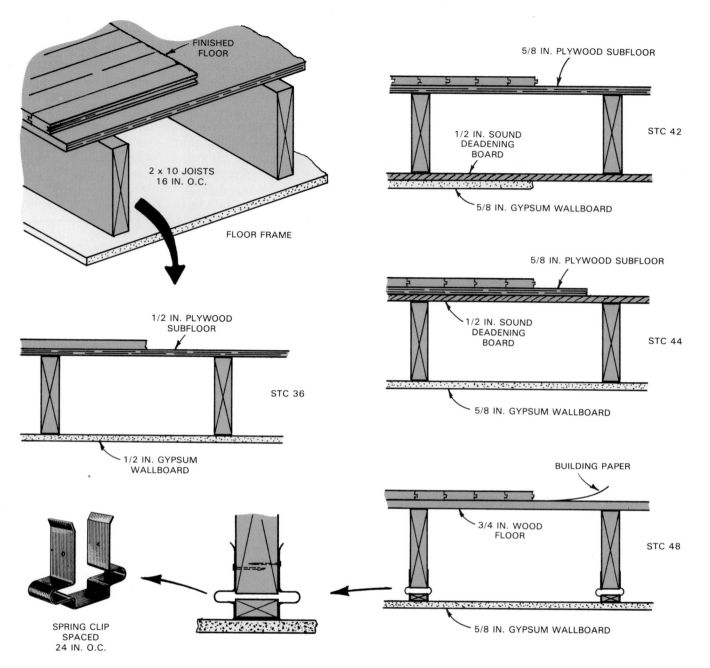

FINISHED FLOOR

2 x 10 JOISTS
16 IN. O.C.

FLOOR FRAME

1/2 IN. PLYWOOD SUBFLOOR

STC 36

1/2 IN. GYPSUM WALLBOARD

SPRING CLIP SPACED 24 IN. O.C.

5/8 IN. PLYWOOD SUBFLOOR

1/2 IN. SOUND DEADENING BOARD

STC 42

5/8 IN. GYPSUM WALLBOARD

5/8 IN. PLYWOOD SUBFLOOR

1/2 IN. SOUND DEADENING BOARD

STC 44

5/8 IN. GYPSUM WALLBOARD

BUILDING PAPER

3/4 IN. WOOD FLOOR

STC 48

5/8 IN. GYPSUM WALLBOARD

Fig. 13-60. Floor and ceiling constructions. Different methods are used to get certain STC ratings.

Fig. 13-61. Wood frame floor system includes resilient channels and 3 inches of blanket insulation. This system provides a high STC rating. (United States Gypsum Co.)

covered with carpet and pad. Resilient metal channels are attached to joists with 1 1/4 in. screws. *Nails must not be used.* Gypsum panels are attached to the channels with screws. The system includes a 3 in. insulation blanket.

An existing floor can be soundproofed through a change shown in Fig. 13-62. Sleepers of 2 x 3 in. wood are laid over a glass wool blanket but are not nailed to the old floor. When the new floor is laid, be sure the nails do not go all the way through the sleepers. The only contact between the new and old floor is the glass wool blanket. It will compress to about 1/4 in. under the sleepers. The system makes the floor resilient in addition to reducing sound transmission.

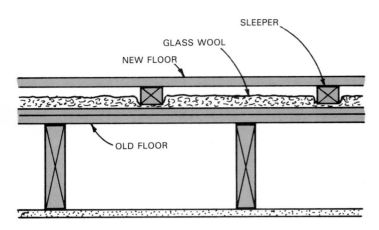

Fig. 13-62. Sleepers laid over insulation will soundproof an existing floor.

DOORS AND WINDOWS

Sound has a tendency to spread out after passing through an opening. Thus, cracks and holes should be avoided in every type of construction where sound insulation is important. Doors between rooms are probably the greatest transmitters of sound. A 1/4 in. crack around a wood door (1 3/4 in. thick) would admit four times as much sound of medium intensity as the door itself. Felt, rubber, or metal strips around the jambs and head are desirable. Conditions can be further improved by some form of ''draft excluder'' at the sill, such as a threshold or felt weatherstop.

Hollow-core interior doors that are well fitted will have a sound reduction value from 20 to 25 dB. Similar double doors, hung with at least 6 in. air space between, will have a sound reduction factor as high as 40 dB. This factor can be increased by using felt or rubber strips around the stops.

For special installations, ''soundproof'' doors are available and can be built to suit almost any condition. Special hardware is used on this type of door to prevent sound transmission through the doorknobs.

Similar precautions should be observed with glazed openings. Cracks around these openings may cancel out other efforts to cut down sound transmissions. Windows or glazed openings should be as airtight as is practical. Double or triple glazing will greatly reduce the amount of sound passing through the opening. Glass block have a sound reduction factor of about 40 dB. They are effective where transparent glazing is not needed.

NOISE REDUCTION WITHIN A SPACE

While it is important to design walls and floors that will reduce sound transmission between

spaces, it is also advisable to treat the enclosure so that sound will be trapped or reduced at its source. Reducing the noise level within the room will not only cut sound transmission to other rooms, but will improve living conditions within the room. Areas in homes where noise reduction is most important include kitchens, utility rooms, family rooms, and hallways.

There are a number of different types of acoustical material available to the builder. These come in a wide range of sizes from a 12 by 12 in. tile to boards 4 by 16 ft. Those with the best acoustical properties will absorb up to 70 percent of the sound that strikes them. Some are distinctly an applicator or ''acoustical engineer'' product. The latter require either special equipment or trained persons to install. The most common types and probably the most used methods are perforated or porous fiberboard units, perforated metal pan units, cork acoustical material, and acoustical plaster. All of these materials provide high absorption qualities and, except for the acoustical plaster, have a factory applied finish. Since the sound absorbing properties of any of these materials depends on its sponge-like quality, they are relatively light and do not require any building reinforcement or structural changes.

HOW ACOUSTICAL MATERIALS WORK

The sound absorbing value of most materials depends to a greater or lesser degree on its porous surface. Sound waves entering these pores, or holes, get ''lost'' and are said to be dissipated (scattered) as heat energy. See Fig. 13-63.

Other materials depend on a similar absorption action to reduce sound. The material used has a vibration point which approaches zero. Heavy draperies or hangings, hair felt, and other soft flexible materials function in this manner.

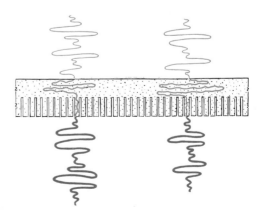

Fig. 13-63. Porous materials absorb sound waves. The sound absorption coefficient of a material is the fractional part of the energy of a sound wave that is absorbed at each reflection.

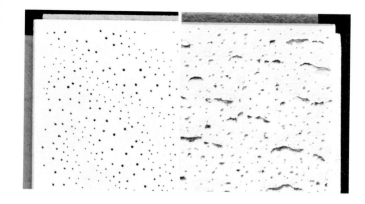

Fig. 13-64. Ceiling tile are designed to absorb sound. Left. Drilled holes. Right. Fractured surface. (Wood Conversion Co.)

The efficiency of an acoustical unit or product is measured by its ability to absorb sound waves. Since noise is a mixture of confused sounds of many frequencies, this efficiency is measured by the noise-reduction coefficient (NRC) of a material for the average middle range of sounds. For most installations, this figure can be used to compare the values of one material over another. However, some materials are designed to do a better job for high or low frequency sounds. For special cases such as music studios, auditoriums, and theaters, an acoustical engineer should be consulted. Most manufacturers of acoustical materials furnish this service.

Perforated fiberboard acoustical materials are made in tile shapes of a low density fibrous composition. Holes of various sizes are drilled almost through the tile. The sounds which strike these units are trapped in the holes. The walls of the holes, being relatively soft and fibrous in nature, form tiny pockets which absorb the sound. See Fig. 13-64.

There are also some fibrous products which are not drilled, but depend on the porosity of the surface and its low vibration point to absorb the sound. Most have a relatively smooth surface, providing a high degree of light reflection without glare.

They are usually installed on the ceiling or upper wall surface by gluing or stapling. Where the surface is in poor condition, furring strips can be used and the material nailed or glued to them. Each manufacturer has specifications on installation.

The perforated metal-pan type performs in somewhat the same manner. The sound enters through the holes in the surface of the pans and is trapped by the backing material which is of a soft resilient nature. It is commonly used in institutional and commercial buildings where ease of maintenance and fire resistance are important factors.

SUSPENDED CEILINGS

These factory finished units are installed on suspended metal runners which form a grid. They allow the large panels to simply be dropped into place, Fig. 13-65. These systems are used where suspended ceilings are advisable to conceal pipes, electrical wiring, and structural beams. The entire ceiling, or any part of it, may be removed and relocated without damage to the material.

Panels are made from various materials. Some are made from ground cork or glass fibers. The material is pressed into acoustical panels of various sizes and thicknesses. Because of the resistance of both these materials to moisture, they are ideal for use in indoor swimming pools, commercial kitchens, or any place where humidity is a problem. Additional information on suspended ceilings is included in Unit 14.

ACOUSTICAL PLASTER

A number of lightweight or fibrous materials are used in the preparation of acoustical plaster. For best appearance and highest sound-absorption qualities, the plaster should be sprayed on the surface,

Fig. 13-65. Two ft. by four ft. panels install rapidly in a suspended ceiling. (Armstrong World Industries Inc.)

Fig. 13-66. Acoustical plaster is being sprayed on a ceiling. (Vermiculite Institute)

Fig. 13-66. Acoustical plasters are generally used where an unlined or plain surface wall or ceiling is desired or where curved or intricate planes make the use of sheet materials impractical.

To install acoustical plaster omit the finished plaster coat and instead apply two coats of the acoustical plaster.

INSTALLATION OF MATERIALS

Manufacturers' recommendations on applying all acoustical materials should be carefully followed. If they are not, the sound-deadening materials may not do the job. In many cases, the amount of air space back of the material is a factor in its sound absorption qualities.

Because most acoustical materials are soft, it is the usual practice to install them on the ceiling or upper portion of sidewalls. For sounds originating in the average room, the ceiling offers a sufficient area for sound absorption materials. Directions concerning the methods and procedures for installing ceiling tile are included in Unit 14.

MAINTENANCE

Painting of perforated boards usually does not lower the efficiency, if properly done. The dirt clogging the pores of the material should be removed first. This may often be done with a vacuum cleaner or by brushing with a soft-hair brush. Some acoustical material can be cleaned by washing.

Spray painting is usually preferable to brush painting. A thinner mixture has less tendency to clog the pores of the material. Improperly applied paint will soon fill the pores of the material and destroy its efficiency.

Manufacturers of acoustical tiles and other products have prepared detailed instructions for installation and maintenance. Be sure to follow them carefully.

TEST YOUR KNOWLEDGE — UNIT 13

1. When heat moves from one molecule to another within a given material, the method of heat transmission is called _____.
2. Heat can be transmitted by wave motion. This method is referred to as _____.
3. The selection of an insulation material should be based on its cost, effectiveness,_____, and the adaptation to the construction.
4. The resistance of a material to transmission is represented by the letter _____.
5. To get an R-value of 23.3, a 2 x 6 insulated stud wall would need a combination of materials. Can you list the materials and give their thicknesses?
6. A standard insulation batt is usually _____ or _____ inches long.
7. The temperature at which condensation occurs for a given sample of air is called the _____ _____.
8. The vapor barrier in a wall structure should be located on the _____ (cold, warm) side of the insulation.
9. When stapling blanket insulation between studs start at the _____ (top, bottom).
10. Fill insulation can be poured or _____ into place.
11. Ceilings of residential structures, heated by electricity, should contain at least _____ inches of insulation.
12. Heat loss through a double-glazed window may be as much as _____ times the loss through a well insulated wall.
13. Air leaking in around windows and doors is called _____.
14. The unit of measure used to indicate the loudness or intensity of a sound is called the _____.
15. A wall with a TL of 40 dB. will reduce a loudness level passing through it from 70 dB. to _____ dB.
16. A method of rating the sound-blocking efficiency of a wall, floor, or ceiling structure is called the _____ _____ _____ system.
17. Sound deadening board is made largely from wood and _____ fibers.
18. To insure a high STC rating, electrical outlets on opposite faces of a partition should not be located in the same _____ _____.
19. In slit-stud construction, the cut should extend to within about _____ inches of the ends of the studs.
20. The efficiency of an acoustical ceiling unit is expressed in a NRC rating which is an abbreviation for _____ _____ _____.

OUTSIDE ASSIGNMENTS

1. Secure samples of various thermal insulating materials such as glass, mineral, organic fibers, and foamed plastic materials. Include loose fill insulations made from such material as vermiculite. Enclose the fibrous and granular materials in small envelopes made of polyethylene plastic film. Mount the samples on a display board with descriptive titles that include k and R factors.

2. Prepare a brief study of the most common types of heating and cooling systems used in your region. Make several drawings showing how they operate and how they are controlled. Study local building codes and outline the basic requirements that must be followed in the installation. Visit with a heating contractor and secure basic information concerning the heating and cooling load calculations for residential structures. Get information about manufacturers, approximate costs, and installation procedures. Give a report in your class.

3. From a study of reference books and trade magazines, prepare a report on radiant heating systems. Include information about panels that are heated with electricity as well as those heated with hot water. Place special emphasis on the methods of installation since structural design may need to be modified when these systems are employed. Also, secure information concerning types and amounts of insulation materials. Discuss special application procedures that may be required.

4. Using a tape recorder, carefully record the various sounds produced by equipment and devices found in a modern home. Include laundry equipment, dishwashers, food mixers, garbage disposers, plumbing fixtures, and vacuum cleaners. Also include such noises as walking on hard surfaced floors and closing of passage or cabinet doors. Play the recording for your class. Then lead a discussion on how to control each sound through proper design and the use of special materials and construction.

STC—47

Unbalanced wall, 2 1/2" metal studs 24" o.c.; double layer 1/2" Type X gypsum board one side, single layer 1/2" Type X gypsum board other side; one thickness R-8 Fiberglas insulation.

STC—53

Unbalanced wall, staggered wood studs 24" o.c.; double layer 1/2" gypsum board one side, single layer other side; one thickness R-11 Fiberglas insulation.

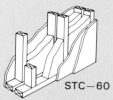

STC—60

Unbalanced wall, double wood studs 16" o.c.; double layer 1/2" gypsum board one side, single layer other side; two thicknesses R-11 Fiberglas insulation.

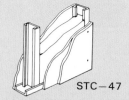

STC—47

3 5/8" metal studs 24" o.c.; double layer 1/2" Type X gypsum board one side, single layer other side; one thickness R-11 Fiberglas insulation.

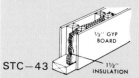

STC—43

Slit-stud wall with 1 1/2" blanket insulation hung from top plate offers greater resistance to noise transmission.

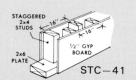

STC—41

Staggered studs provide complete separation between wall faces. Staggering 2 × 3s on 2 × 4 plate is almost as effective.

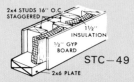

STC—49

Blanket insulation between staggered studs performs as well as weaving continuous insulation behind studs.

STC—52

Sound deadening board absorbs noise, dampens wall vibration, adds surprisingly to effectiveness of staggered-stud wall.

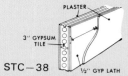

STC—38

Fireproof wall of 3" gypsum tile has been considered a good acoustical wall, yet its rating is surprisingly low.

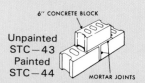

Unpainted **STC—43**
Painted **STC—44**

Concrete block wall becomes more effective when pores are sealed with two coats of paint.

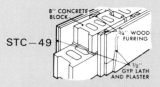

STC—49

Concrete block with furred gypsum lath and plaster both sides has excellent properties in mid and high frequencies.

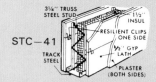

STC—41

Steel truss studs, 3 1/4" wide with gypsum lath and plaster both sides, require insulation to be effective.

STC—52

Steel channel studs with insulation plus sound deadening board on one side rate high. Caulking seals gaps.

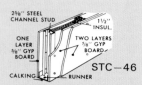

STC—46

Extra layer of 5/8" gypsum board on one side of 2 5/8" channel studs gives fair performance with insulation.

STC—55

Laminated 5/8" gypsum board on both sides of 2 5/8" steel channel studs plus insulation gets very high rating.

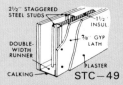

STC—49

Staggered steel studs with 3/8" gypsum lath and plaster compares favorably with staggered wood studs and insulation.

Structural designs of walls and their sound transmission factors (STC). Other designs using wood or steel framing are shown in the Technical Information Section.

Left. Cedar shakes produced this unusual wall treatment. Ceiling is a prefinished composition tile. (Shakertown Corp.) Right. Prefinished plywood 1/2 in. paneling is grooved to resemble planking. It is produced in standard 4 x 8 sheets. (Georgia-Pacific Corp.)

Standard grid suspended ceiling. Panels are made from materials that will not burn. Edges of the panels are rabbeted to create a shadow line. (USG Acoustical Products Co.)

Drywall ceiling surface is finished with a textured paint. Drywalled walls are covered with wallpaper. (U.S.Gypsum)

Unit 14

INTERIOR WALL AND CEILING FINISH

Interior finishing is the installation of cover materials to walls and ceilings. This stage of construction can start after utilities, heating, and insulation are in. Exterior doors must be hung and windows installed. They will protect the finishing materials from the weather.

Interior walls can be covered with any one of a number of materials:

1. GYPSUM WALLBOARD. Commonly called "drywall," gypsum wallboard is a laminated material with a gypsum core and paper covering on either side. It usually comes in 4 x 8 ft. sheets. However, it is also available in 7, 9, 10, 12, and 14 ft. lengths. It is sold in the following thicknesses: 1/4, 5/16, 3/8, 1/2, and 5/8 in. Gypsum wallboard is used on both walls and ceilings.

2. GYPSUM WALLBOARD for plaster veneering. This is a base of gypsum board, usually 1/2 in. thick. It is applied as a backing for a thin coat of plaster.

3. PREDECORATED GYPSUM PANELING. This is the same as gypsum wallboard. However, decorative vinyl finishes have been applied and edges have received special treatment so that no other finishing work need be done after the panels have been installed. The finishes are tough and easily cleaned.

4. PLYWOOD AND PARTICLE BOARD, Fig. 4-1. Plywood is fabricated in 4 ft. widths. Lengths

A

B

C

Fig. 14-1. Plywood and other wood materials are used as wall coverings. A—Plywood fabricated to look like diagonal planking. (American Plywood Assoc.) B—Plywood panel cut to look like planking. Surface is prefinished. C—Waferwood paneling used as wall covering in a den. (Louisiana-Pacific)

include 7, 8, 9, and 10 ft. Usually, the sheets are prefinished in a variety of colors and patterns. Surface material may be either a hardwood or a softwood. Panels are manufactured in thicknesses of 1/4, 3/8, 7/16, 1/2, 5/8, and 3/4 in. Surfaces can be embossed, antiqued, stained, or color toned. Some have veneers of paper with a wood grain printed on them. Flakeboard or waferwood is sometimes used for rustic finishes.

5. HARDBOARD and FIBERBOARD. These are produced from wood fibers in sizes and thicknesses similar to plywood. The face finish is simulated to look like wood. Other decorative patterns are also applied. Sheets may be embossed and grooved to simulate random planking, leather, or wallpaper. Surfaces may also be coated with plastic. Variations of fiberboard are used as ceiling coverings.

6. SOLID WOOD PANELING. These are boards or pieces of solid wood. Widths of boards will vary from 2 to 12 in. and thicknesses are either 1 or 2 in. Faces may be rough-sawed, plain, or molded in a variety of patterns. Lengths vary from 4 to 10 ft. Shingles, usually considered a siding or roofing material, are occasionally used on interior walls, Fig. 14-2.

7. PLASTER. For many years the most popular wall covering, plaster is made of powdered gypsum to which other materials are added to improve drying time. A plastered wall system includes a base support, such as metal or gypsum lath, over which is applied coats of wet plaster.

8. CLAY FINISHES. Clay wall finishing products include: brick, brick veneers, glazed tile, ceramic tile, and ceramic mosaic tile. They are generally used either as an accent material or to protect wall surfaces from damage. They are often found in kitchens and bathrooms.

9. PLASTIC LAMINATES. Sheets of hard synthetic material are bonded on the building site to a base of hardboard or other backing material. The laminates are durable and easily cleaned.

Ceilings can be covered with gypsum wallboard, wood, plaster, or a variety of composition materials. Tiles are especially suitable for remodeling work because they are easy to install. The prefinished tiles are simply stapled or cemented to the old ceiling, or hung below the old ceiling in a lightweight grid.

DRYWALL CONSTRUCTION

Drywall materials, such as gypsum wallboard, are the most common coverings in use in modern construction. Most builders prefer to use drywall because of the time-saving factor. Regular plaster requires considerable drying time. It may be difficult to find other work for carpenters and other tradespeople during the drying period.

Either type of finish presents some advantages and disadvantages. Drywall construction, for example, requires that studs and ceiling joists be perfectly straight and true, otherwise the wall surface will be uneven. The wood framing material must also have

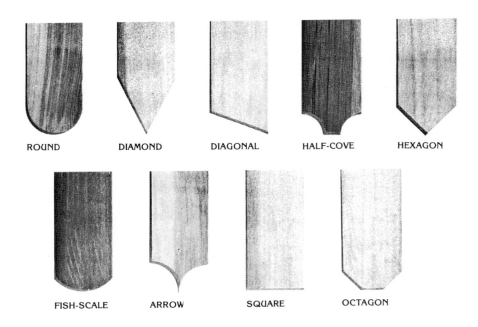

| ROUND | DIAMOND | DIAGONAL | HALF-COVE | HEXAGON |

| FISH-SCALE | ARROW | SQUARE | OCTAGON |

Fig. 14-2. Wood shingles may be used to produce durable interior wall surfacing. These patterns, known as fancy cuts, are sold in 5 x 8 in. panels. (Shakertown Corp.)

a moisture content very near to that which it will eventually attain in service to prevent "nail pops" and joint cracks.

Gypsum wallboard has a fireproof core. Fig. 14-3 lists a variety of thicknesses, edge joint designs, and types.

One type, BACKING BOARD, has a gray liner paper on both sides. It is used as the base sheet on multilayer applications. It is not suited for finishing and decorating.

Fig. 14-4 shows several standard edge designs for gypsum wallboard. Its tapered edges form a shallow channel between adjacent sheets. This depression is brought level with the tape and filler. The result is a smooth, uninterrupted surface.

SINGLE LAYER CONSTRUCTION

In single layer construction, use 1/2 or 5/8 in. gypsum wallboard. The 5/8 in. sheet is preferred for high quality construction. Apply covering to ceilings first, then to the walls.

There are two methods of arranging the drywall sheets: long edges parallel to studs and joists or long edges at right angles to studs and joists.

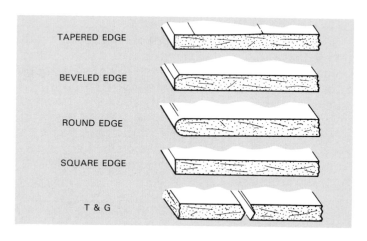

Fig. 14-4. Gypsum wallboard is manufactured with several different edge styles.

The second method is generally preferred because it is stronger. In either method, vertical wall joints must fall over a stud. Fig. 14-5 shows both methods of application.

MEASURING AND CUTTING

All measurements should be carefully taken from the spot where the wallboard will be installed. Usually it is best to make two readings, one for each side of the panel. Following this procedure will eliminate errors. It also allows for openings and framing that are not plumb or square. Use a 12 ft.

TYPE	THICKNESS (IN.)	EDGES
Regular (ASTM C36, FS SSL30d)	1/4 5/16 3/8 1/2 5/8	Tapered Square Square Tapered Bevel
Fire Resistant Type "X" Wallboard	1/2 5/8	Square Tapered Bevel
Insulating Wallboard (Aluminum Foil on Back Surface)	3/8 1/2 5/8	Square T & G Tapered Round
Regular Backing Board (ASTM C442, FS SSL30d)	1/4 3/8 1/2	Square Square T & G
Foil-Backed Backing Board	3/8	Square
Fire Resistant Type "X" Backing Board	1/2 5/8	T & G
Coreboard (Homogeneous or Laminated)	3/4 1	Square T & G Ship Lap
Predecorated	3/8	Bevel Round Square
	1/2 5/8	Bevel Square

Fig. 14-3. Main types of gypsum wallboard. Water resistant types are also made for shower areas in bathrooms.

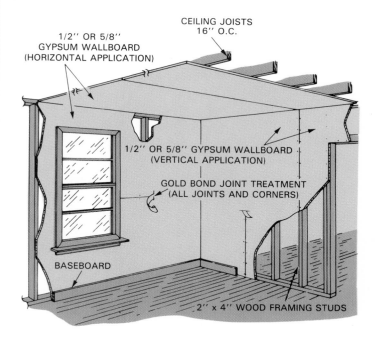

Fig. 14-5. Single layer drywall application. Lefthand wall shows horizontal application; righthand is a vertical application.

steel tape to take measurements.

Straight cuts across the width or length of a board are made by first scoring the face with a knife pulled along a straightedge. The scoring cut should penetrate the paper and go into the gypsum core.

With the main section of the board supported close to the scored line, snap the core by pressing downward on the overhang. Cut the back paper as shown in Fig. 14-6. When necessary the cut can be smoothed with a file or coarse sandpaper mounted on a block of wood.

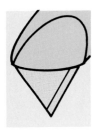 *When scoring wallboard, always use a sharp knife that will make a "clean" cut through the paper face and penetrate slightly into the core.*

Irregular shapes and curves can be cut with a coping saw, compass saw, or electric sabre saw. Fig. 14-7 shows an opening for a convenience outlet being made with a portable power saw.

NAILS AND SCREWS

Nail spacing will vary depending on the materials being used. For single layer construction they are spaced no farther apart than 7 in. on ceilings and 8 in. on walls. Annular ring nails with a 1/4 in. head and 1 1/4 in. long are generally recommended. See Fig. 14-8 for drywall fasteners.

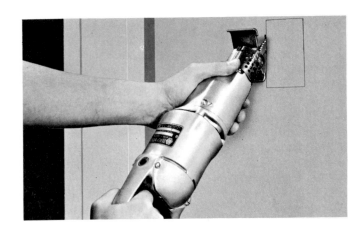

Fig. 14-7. Inside cuts, curves, and irregular shapes in drywall can be cut with a portable saber or reciprocating saw.

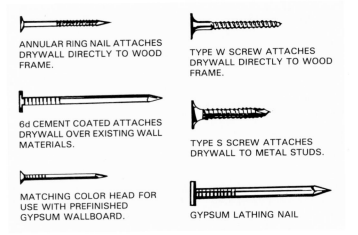

ANNULAR RING NAIL ATTACHES DRYWALL DIRECTLY TO WOOD FRAME.

TYPE W SCREW ATTACHES DRYWALL DIRECTLY TO WOOD FRAME.

6d CEMENT COATED ATTACHES DRYWALL OVER EXISTING WALL MATERIALS.

TYPE S SCREW ATTACHES DRYWALL TO METAL STUDS.

MATCHING COLOR HEAD FOR USE WITH PREFINISHED GYPSUM WALLBOARD.

GYPSUM LATHING NAIL

Fig. 14-8. Gypsum wallboard and gypsum lath fasteners. Others are available.

Fig. 14-6. Cut the paper on the back side of the drywall after breaking the gypsum core. Large steel T-square in the background is used as a guide while scoring the face.

All wallboard must be drawn tightly against the framing so there can be no movement of the board on the nail shank. During nailing, hold the board tightly against the stud or joist to avoid breaking through the wallboard face. Fig. 14-9 shows proper countersinking of nails. Fig. 14-10 shows how large panels can be held with braces while nailing them to a ceiling.

Drive nails straight and true. Use extra care during the final strokes so the nailhead rests in a slight dimple formed by the crowned head of the hammer or stapler. Be careful not to break the paper face of the board.

When applying drywall to ceilings it will be necessary to stand on a platform or use stilts to reach your work. See Fig. 14-11.

Fig. 14-9. Drywall nailer forms proper dimple and drives nail below surrounding surface without breaking paper facing. Recommended air pressure is 80 psi. (Paslode Co., Div. of Signode Corp.)

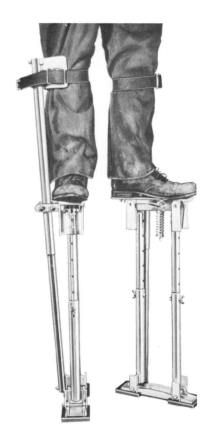

Fig. 14-11. Adjustable stilts are useful when applying drywall. They allow worker to reach upper walls and ceilings. (Goldblatt Tool Co.)

Fig. 14-10. Braces, also called shores, will hold ceiling panel in place while it is being nailed. (The Flintkote Co.)

Fig. 14-12. An electric drill with a special depth-adjusting clutch is used to drive wallboard screws. Clutch disengages when nose strikes panel surface.

The double nailing method of attachment insures firm contact with framing. Panels are applied as required for conventional nailing, except that nails in the field of the board should be spaced 12 in. on center. After the panel is secured, another nail is driven approximately 2 in. from the first. If necessary, the first nail should receive another blow to assure snug contact.

Gypsum wallboard screws were shown in Fig. 14-8. They provide firm, tight attachment to wood or metal framing, Fig. 14-12. Special self-tapping screws are used for metal-framed wallboard systems. Fasteners of this type must be driven so the screwhead rests in a slight dimple formed by the driving tool. The paper face of the wallboard should not be cut. Neither should the gypsum core be fractured.

Since screws hold the wallboard more securely than nails, ceiling spacing can be extended to 12 in. and side walls to 16 in.

ADHESIVE FASTENING

When wallboard is fastened with a special adhesive there are no depressions to be filled later. Adhesive also produces a sturdier wall that is more resistant to impact sounds.

Apply a continuous bead of adhesive to the center of all studs, joists, or furring as shown in Fig. 14-13. Where two pieces of wallboard join on a framing member, use a zig-zag bead pattern. The bead should be approximately 1/4 to 3/8 in. wide. Then, when the board is in place it will be held by a band at least 1 in. wide and 1/16 in. thick.

Use temporary nailing or bracing to insure full contact of the wallboard. This allows the adhesive to develop proper bonding strength.

JOINT AND FASTENER CONCEALMENT

To conceal joints, first apply a bedding coat of joint compound into the depression formed by the tapered edges of the board over all butt joints. Use a 5 or 6 in. joint knife. Center the reinforcing tape over the joint and smooth it out to avoid wrinkling or buckling. Press tape into compound by drawing the knife along the joint with enough pressure to remove excess compound. Apply a skim coat over

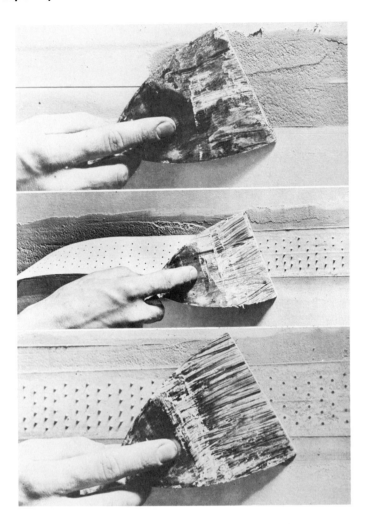

Fig. 14-14. Taping wallboard joint with broad taping knives. Top. First apply compound to channel at joint. Middle. Embed tape. Be sure it is centered over the joint. Bottom. Immediately apply a skim coat over tape and smooth edges. Use broader knife. (Gold Bond Bldg. Products)

the tape. See Fig. 14-14.

After the embedding coat is completely dry, apply a second coat over the tape. Feather the edges approximately 1/2 to 3/4 in. beyond the edges of the first coating. When this coat is completely dry, a third coat is applied with the edges feathered out about 2 in. beyond the second coat. After the last coat is dry, sand lightly if necessary. Fasteners are also concealed with compound - each coat being applied at the same time the joints are covered.

A mechanical taping tool like the one shown in Fig. 14-15 will speed up taping operations. Special compound applicators are also used to apply the various coats of joint compound, Fig. 14-16.

Pressure-sensitive glass-fiber tape reduces the time required to conceal and reinforce joints and in-

Fig. 14-13. Apply a bead of adhesive about 1/4 in. wide to framing members.

terior angles. It has an open weave (100 meshes per sq. in.) which provides excellent reinforcing and keying of plaster or compound coats. Simply use hand pressure to attach it to the wall. Bond it with a finishing knife or trowel as shown in Fig. 14-17. The tape is also easily applied to inside corners.

Next, apply two coats of fast-setting joint compound over the tape and sand smooth. When the surface is to be covered with a texture paint, you can finish the joints with a single coat of compound. Fig. 14-18 shows a texture paint being applied with a spray gun.

Fig. 14-15. Using a mechanical taping tool speeds up drywall construction. (U.S. Gypsum)

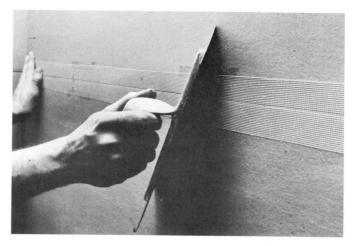

Fig. 14-17. Pressure sensitive fiberglass tape is permanently bonded with a trowel or knife. (U.S. Gypsum)

Fig. 14-16. Compound applicator can be used to put on successive coats of compound. Sometimes it is called a "finisher."

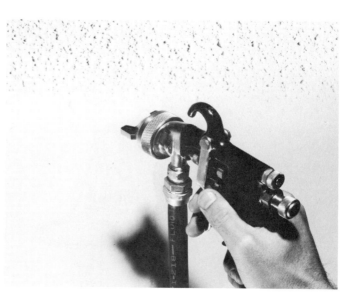

Fig. 14-18. Textured paints can be applied to a wall or ceiling with a spray gun. (Gold Bond Bldg. Products)

Rollers and brushes can also be used to make the application. A wide variety of finish coatings are available for drywall construction. Some are in a powder form ready to be mixed with water. Others are ready to use from the containers. Always read and follow the manufacturer's recommendations.

AUTOMATIC TAPE APPLICATOR

An automatic taping machine is available which dispenses tape and compound continuously. The compound is fed into the dispenser under pressure from a hopper. Tape, mounted on the dispenser, is threaded into the nose and feeds automatically from a control in the handle. See Fig. 14-19.

CORNERS

Outside corners are reinforced with a metal corner bead. Fasten the bead to the framing through the wallboard. One type of bead is made of metal with paper flanges, Fig. 14-20. After installation they are concealed with joint compound in about the same manner as regular joints.

At internal corners, both horizontal and vertical reinforcing tape is used. First apply a bedding coat of joint compound to both sides of the corner. Then fold the tape along the centerline and smooth it into

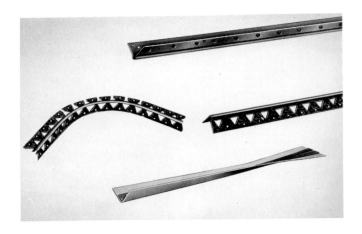

Fig. 14-20. Metal corner beads are made in many different styles. Bead at bottom of photograph has paper flanges.

place. Remove excess compound and finish surfaces along with the other joints.

Metal channel trim is available to finish and reinforce edges around doors, windows, and other openings. Fig. 14-21 shows several shapes and sizes.

DOUBLE LAYER CONSTRUCTION

Double layer (also called two-ply) wallboard applications over wood framing insure a strong wall surface. Fire protection and sound insulation qualities are also improved. This method is adaptable to either the use of predecorated panels or stan-

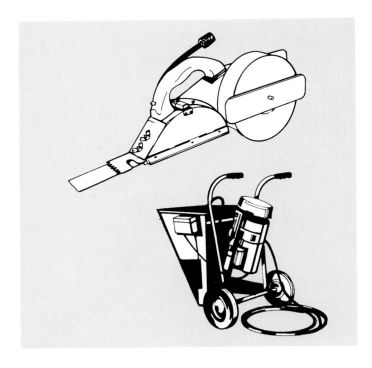

Fig. 14-19. Automatic tape "banjo" feeds tape and compound for drywall joint finishing. Unit holds 500 ft. of tape. Compound is fed from hopper of pump unit at bottom. (Goldblatt Tool Co.)

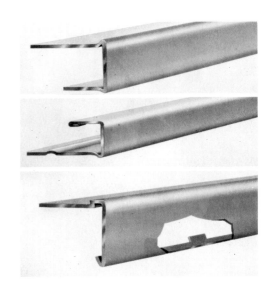

Fig. 14-21. Metal channel trim is installed to protect drywall edges around openings such as doors and windows.

dard beveled wallboard with treated joints.

The base layer, Fig. 14-22, may be a regular gypsum wallboard or a specially designed base called BACKING BOARD or backer board. It is made up about the same as regular gypsum board except that the coverings are a gray liner paper. It is not suitable for decorating and should never be used as a top surface.

For areas where there is likely to be moisture coming into direct contact with the wall, there is a highly water resistant backing board. Its use is recommended in shower areas as a base for tile or other protective coverings.

Sound deadening backing board is sometimes used for the base of double layered walls. It is specified where high STC (sound transmission control) ratings are needed.

ATTACHING THE LAYERS

Base layers are applied to framing with power-driven screws, staples, or nails. The finish layer is laminated to the base layer with an adhesive. Joints of the finish layer should be offset at least 10 in. from the joints of the base layer. Finish layers can be applied parallel to the base layer or at right angles.

Adhesive is usually applied to the entire surface. However, strip lamination is used in some applications. (This method has ribbons of adhesive spaced at regular intervals.)

Many methods of applying adhesive are acceptable. Trowels and powered devices are available. Whatever method is used, the spacing and size of the bead of adhesive must provide the required spread when panels are pressed into position. Fig. 14-23 shows a notched spreader being used to apply adhesive to the entire back surface of a finish layer panel. Strip lamination is often used for sidewall panels, Fig. 14-24. The application can be made either on the base surface or on the face panel. Temporary bracing may be used to hold panels in position until bonding has taken place.

When nails are used, they should provide a minimum penetration of 3/4 in. into the wood framing members. Consult a nail chart.

FINISHING DOUBLE LAYER WALLBOARD

If the wallboard is to receive other covering material, joints should be taped, nails concealed, and corners finished in the same way as single layer construction. If a veneer plaster is to be applied, use reinforcing tape and a single bedding coat of compound over joints. Inside corners should be taped

FINISH LAYER
3/8 OR 1/2 IN.
TAPERED EDGE
GYPSUM WALLBOARD

CEILING JOISTS
16 IN. O.C.

BASE LAYER 3/8 OR 1/2 IN.
GYPSUM BACKING BOARD
OR GYPSUM WALLBOARD

LAMINATING ADHESIVE
(APPLY WITH NOTCHED
TROWEL OR MECHANICAL
SPREADER)

2 x 4 STUDS
16 IN. O.C.

BASEBOARD

Fig. 14-22. Cutaway of a double layer gypsum wallboard construction. Finish layer may be applied at right angles to the base (as shown) or running parallel.

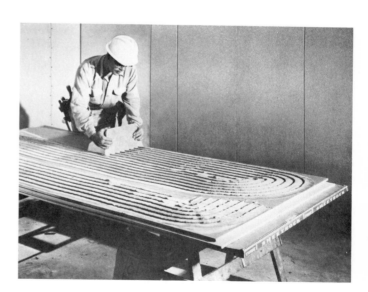

Fig. 14-23. Drywaller is using notched spreader to apply adhesive for double layer construction. Spreader forms 1/4 in. beads spaced 2 in. apart. (U.S. Gypsum)

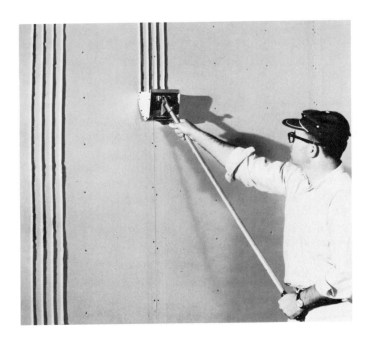

Fig. 14-24. Strip laminating with a mechanical spreader. Adhesive may be applied to the base, as shown, or to the back of the finish layer.

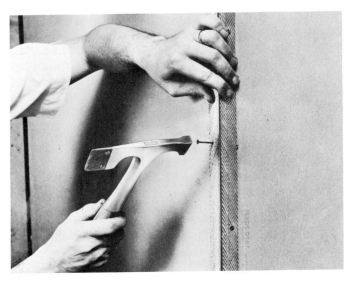

Fig. 14-25. Installing special corner bead for veneer plaster coating. Bead provides 1/16 in. grounds for one-coat system. For two-coat veneer, use bead with 3/32 in. grounds. (U.S. Gypsum)

and a special bead applied to outside corners as shown in Fig. 14-25. Always use a single length that extends from floor to ceiling.

MOISTURE RESISTANT (MR) WALLBOARD

This type of gypsum wallboard is processed to withstand the effects of moisture and high humidity. Its facing paper is light green so it can be easily identified. Standard thicknesses include 1/2 in. and 5/8 in. Standard width is 4 ft. and lengths range from 8 ft. through 12 ft.

Moisture resistant wallboard should be used as a base under ceramic tile or other nonabsorbent finishing materials in showers and tub alcoves. Fig. 14-26 shows a detail of the construction at the edge of a tub. Stud spacing should not be greater than 16 in. O.C. Note the furring strip. It insures alignment between the tub lip and the wallboard. Also note the 1/4 in. space. This must be maintained.

Panels are installed in about the same way as regular wallboard, Fig. 14-27. Before setting the panels in place, apply a coating of water resistant adhesive to exposed edges, holes, and joints. In areas to be tiled, apply a coat of tile adhesive over nail or screw heads. *Do not apply regular drywall joint compound in any area to be tiled.* Fig. 14-28 shows moisture resistant wallboard in place with openings and edges caulked. One section of tile has been installed.

VENEER PLASTER

Veneer plaster is a high-strength material for application as a very thin coat less than 1/8 in. thick. Because of the composition and thinness of the coat, it dries very rapidly. Trim and decoration work may proceed after a minimum drying time of about 24 hours.

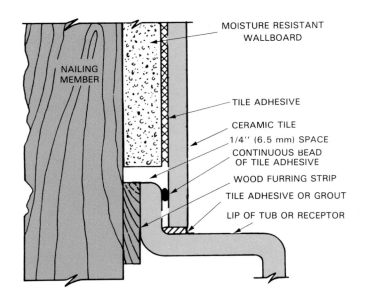

Fig. 14-26. Moisture resistant (MR) gypsum wallboard is designed for tub and shower areas. Note detail around tub edge. (Gold Bond Bldg. Products)

Fig. 14-27. Drywall worker is installing MR gypsum wallboard around tub area. Spacer block is used to assure proper clearance with tub flange.

of veneer plaster being applied over gypsum board that has been bonded to the wall with adhesive. The horizontal joint has been taped and covered with compound.

Veneer plaster can be applied as a one or two-coat system. Either system can be given a smooth or textured surface. Corner bead, trim, and grounds must be carefully set for a 1/16 in. thickness in one-coat applications and 3/32 in. for two-coat applications.

For best performance and workability of veneer plaster, the manufacturer's directions should be carefully followed. Proper mixing is especially important and is usually accomplished with a cage-type paddle mounted in an electric drill.

In the application of any drywall system, be sure to study and follow the recommendations of the manufacturer of each of the products used.

The base for veneer plaster is a gypsum board with a special surface. Other than this, the materials and methods are nearly the same as those for regular drywall construction. Fig. 14-29 shows a coat

Fig. 14-28. Setting ceramic tile in tub alcove. Special elastomeric caulking applied around pipes and tub lip. (Gold Bond Bldg. Products)

Fig. 14-29. Applying veneer coats of plaster takes considerable skill. A special gypsum board base is used. (U.S. Gypsum)

PREDECORATED WALLBOARD

A variety of predecorated gypsum wallboard is available. This is usually applied vertically because of the difficulty involved in successfully matching and finishing butt joints. The use of an adhesive to bond the panels to a base layer is common practice. However, matching colored nails are available. Be sure to drive colored nails with a plastic-headed hammer, rawhide mallet, or a special cover placed over the face of a regular hammer. Nails should never be spaced closer than 3/8 in. from the ends or edges of the wallboard.

To trim edges and joints of predecorated panels, you can use aluminum moldings made to match the finished surface of the wallboard, Fig. 14-30. Cut them with a hacksaw and attach with flat-head wire nails spaced 8 to 10 in. apart.

When attaching divider strips, first place the molding on one panel that is carefully aligned, nail the exposed flange in place, then insert the next panel.

WALLBOARD ON MASONRY WALLS

Gypsum wallboard can be installed over metal or wood furring strips attached to a masonry wall. Where the structure consists of an interior wall that is straight and true, the panels can be laminated directly to the masonry surface with a special adhesive.

Exterior walls must be thoroughly waterproofed and insulation should be included if the structure is located in a cold climate. Fig. 14-31 shows a masonry wall application made with furring strips. The insulation may be rigid plastic foam or batts. When wood furring strips are used, they should be

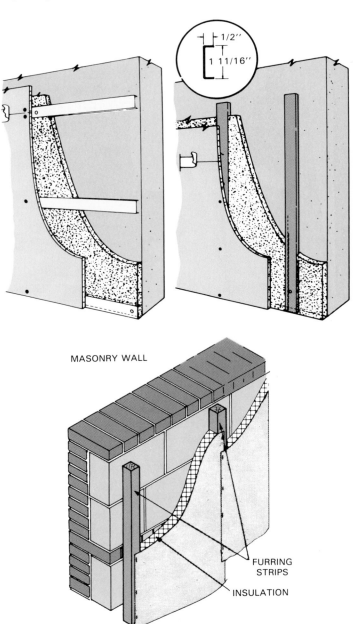

MASONRY WALL

FURRING STRIPS

INSULATION

Fig. 14-31. Two methods of preparing masonry walls for interior finish. Top. Wallboard can be attached to metal furring channels. Rigid insulation is used. Bottom. Wood furring strips and blanket insulation.

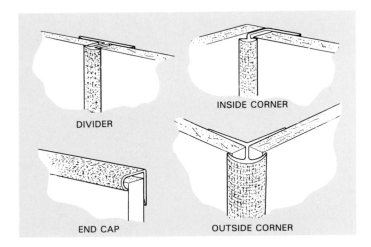

DIVIDER

INSIDE CORNER

END CAP

OUTSIDE CORNER

Fig. 14-30. Moldings can be used to cover raw edges of predecorated gypsum wallboard.

a nominal 2 in. wide and 1/32 in. thicker than the insulation. The plastic foam slabs are usually bonded to the masonry surface with adhesive. Wallboard joints and nail holes are concealed, following the finishing steps previously described.

Installations of wallboard on exterior masonry walls, especially those below grade, must be carefully done. Always follow recommendations furnished by manufacturers.

INSTALLING PLYWOOD

Most of the plywood used for interior walls has a factory-applied finish that is tough and durable. Manufacturers can furnish matching trim and molding that is also prefinished and easy to apply. Color-coordinated putty sticks are used to conceal nail holes.

Joints between plywood sheets can be treated in a number of ways. Some panels are fabricated with machine shaped edges that permit almost perfect joint concealment. Usually it is easier to accentuate the joints with grooves or use battens and strips, Fig. 14-32.

Before installation, the panels should become adjusted (conditioned) to the temperature and humidity of the room. Prefinished plywood should be removed from cartons and carefully stacked horizontally. Place 1 in. spacer strips between each pair of face-to-face panels. Do this at least 48 hours before application.

Plan the layout carefully to reduce the amount of cutting and number of joints. Fig. 14-33 shows two application designs. It is important to align panels with openings whenever possible. If finished panels will have a grain, stand the panels around the walls and shift them until you have the most pleasing effect in color and grain patterns. To avoid mixups, number the panels in sequence after their position has been established.

When cutting plywood panels with a portable saw mark the layout on the back side. Support the panel carefully and check for clearance below. Make the cut as shown in Fig. 14-34. The cutting action of the saw blade will be upward against the panel face. Splintering will be minimal (less). This is even more important when working with prefinished panels.

Plywood can be attached directly to the wall studs with nails or special adhesives. Use 3/8 in.

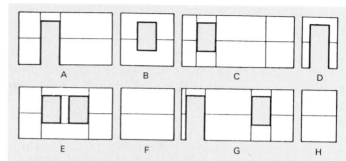

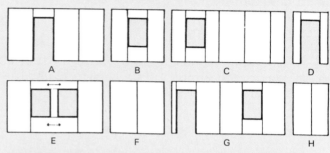

DESIGN 1 is a simple two-panel horizontal arrangement. The single continuous horizontal joint is placed midway between door and ceiling. Vertical joints at openings (elevations A, C, E, and G), again, key panel design. However, they may be omitted where panel length exceeds wall element width as in Elevations B and D.

DESIGN 2, a vertical panel arrangement is another illustration of the basic principle of initiating panel design by lining up vertical joints with wall openings. The plain wall space is then divided vertically in widths proportionate to that of openings. In a vertical panel arrangement where width of a door or window opening exceeds panel width, panels may be placed horizontally as shown by arrows in Elevation E. Such combinations of vertical and horizontal arrangements may be used in the same room with pleasing effect.

Fig. 14-33. Two plywood wall paneling arrangements commonly used. The basic rule is to "work from the openings." First, line up vertical joints above doors and above and below windows. Divide the remaining plain wall space into an orderly pattern as stud location allows.

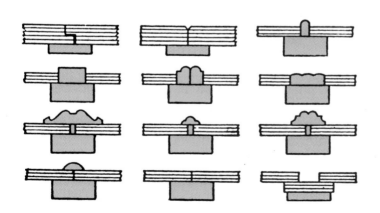

Fig. 14-32. Different styles of battens can be used to conceal joints in plywood paneling.

Fig. 14-34. A portable power saw can be used to cut prefinished plywood panels. Work should be performed with the back of the sheet facing up.

plywood for this type of installation. When studs are poorly aligned or when the installation is made over an existing surface in poor condition, it is usually advisable to use furring. Nail 1 x 3 or 4 in. furring strips horizontally across the studs. Start at the floor line and continue up the wall. Spacing depends on the panel thickness. Thin panels need more support. Install vertical strips every 4 ft. to support panel edges. Level uneven areas by shimming behind the furring strips. Fig. 14-35 shows furring strips being attached to a concrete block wall. Prefinished plywood panels are being installed. A special panel adhesive can be used, Fig. 14-36. The panels are simply pressed into place with no sustained pressure being required.

Begin installing panels at a corner. Scribe and trim the edge of the first panel so it is plumb. Fasten it in place before fitting the next panel. Allow about 1/4 in. clearance at the top and bottom. After all panels are in place, molding is used to cover the

Fig. 14-36. Applying adhesive to furring strips. Use a 1/4 in. bead. Follow adhesive manufacturer's directions. (Borden Inc.)

space along the ceiling. Baseboards will conceal the space at the floor line.

On some jobs 1/4 in. plywood is installed over a base of 1/2 in. gypsum wallboard. This backing is recommended for several reasons:
1. It tends to bring the studs into alignment.
2. It provides a rigid finished surface.
3. It improves the fire resistant qualities of the wall. The plywood is bonded to the gypsum board with adhesive. In general, you should follow the same procedure as was described in double-layer drywall construction. See Fig. 14-37.

HARDBOARD

Through special processing, hardboard, also called fiberboard, can be fabricated with a very low moisture absorption rate. This type is often scored to form a tile pattern and is used in bathrooms and kitchens. Panels for wall application are usually 1/4 in. thick.

Since hardboard is made from wood fibers, the panels will expand and contract slightly with changes in humidity. They should be installed when they are at their maximum size. There will be a tendency for them to buckle between the studs or attachment points if installed when moisture con-

Fig. 14-35. Paneling can be attached directly to masonry walls but furring strips produce a better looking wall. Strips where panels join should be spray painted a color close to the tones in the paneling. This avoids unsightly off-color cracks at joints.

Fig. 14-37. Plywood paneling produces an attractive wall covering and is available in many different patterns. This style simulates board paneling. (American Plywood Assoc.)

Fig. 14-38. Prefinished interior hardboard wall paneling is made to look like pine. Note the wainscoting in the dining area. (Masonite Corp.)

tent is low. Manufacturers of prefinished hardboard panels recommend that they be unwrapped and then placed separately around the room for at least 48 hours before application.

Procedures and attachment methods are similar to those previously described for plywood. Special adhesives are available as well as metal or plastic molding in matching colors. Drill nail holes for the harder types. Fig. 14-38 shows an attractive application of hardboard.

When applying any of the various factory finished wallboard, plywood, or hardboard materials, always follow the recommendations furnished by the manufacturer.

PLASTIC LAMINATES

Plastic laminates are sheets of synthetic material that is hard, smooth, and highly resistant to scratching and wear. Although basically designed for table and countertops, they are also used for wainscoting and wall paneling in homes and commercial buildings.

Since the material is thin (1/32 to 1/16 in.), it must be bonded to other supporting panels. Contact bond cement is commonly used. Recently, manufacturers have developed prefabricated panels with the plastic laminate already bonded to a base

or backer material. Fig. 14-39 shows the installation of a prefabricated panel consisting of a 1/32 in. plastic laminate mounted on 3/8 in. particle board. Edges are tongue and grooved so that units

Fig. 14-39. These wall panels are manufactured from plastic laminate with a particle board base. (Formica Corp.)

can be blindnailed into place. Various matching corner and trim moldings are available. Fig. 14-40 shows a completed installation.

SOLID LUMBER PANELING

Solid wood paneling makes a durable and attractive interior wall surface and may be appropriately used in nearly any type of room. A number of different species of hardwood and softwood are available. Sometimes, grades that contain numerous knots are used to secure a special appearance. Defects, such as the deep fissures in pecky cypress, can provide a dramatic effect.

Softwood species most commonly used include pine, spruce, hemlock, and western red cedar. Boards range in widths from 4 in. to 12 in. (nominal size) and are dressed to 3/4 in. Board and batten or shiplap joints are used, but tongue and groove joints combined with shaped edges and surfaces are more popular.

When solid wood paneling is applied horizontally, furring strips are not required and the boards are nailed directly to the studs. Inside corners are formed by butting the paneling units flush with the other walls.

If random widths are used, boards on adjacent walls must match and be accurately aligned. Vertical installations require furring strips at the top and bottom of the wall and at various intermediate spaces, Fig. 14-41. Sometimes 2 x 4 in. pieces are installed between the studs to serve as a nailing

Fig. 14-41. Installing solid wood paneling. Boards are edge and end matched.   (Forest Products Laboratory)

base. Even when heavy tongue and groove boards are used, these nailing members should not be spaced more than 48 in. apart.

Narrow widths (4 to 6 in.) of tongue and groove (T&G) paneling are blind nailed (nailed so heads of nails do not appear on finished surface). This eliminates the need for countersinking and filling nail holes and provides a smooth blemish-free surface—especially important when clear finishes are used. Use a 6d finish nail and drive it at a 45 deg. angle into the base of the tongue and on into the bearing point.

Exterior wall constructions, where the interior surface consists of solid wood paneling, should include a tight application of building paper located close to the back side of the boards. This will prevent the infiltration of wind and dust through the joints. In cold climates, insulation and vapor barriers are important.

Interior solid wood paneling may cause problems resulting from expansion and shrinkage. Be sure the material used has a moisture content about equal to that which it will attain in service. This should be about 8 to 10 percent for most parts of the United States.

PLASTER

Through the years, gypsum plaster, Fig. 14-42, has provided qualities that are desired in a wall and ceiling finish: beauty, durability, economy, fire protection, structural rigidity, and resistance to sound

Fig. 14-40. An office wall. Plastic laminated paneling is used for its durability.

Fig. 14-42. Plaster is always applied over a supporting base. Old style wood lath has been replaced by several new products. The one used here is called perforated gypsum lath.

transmission. It is also highly adaptable since it can be readily applied to curved or irregular surfaces.

Plaster is made from gypsum, one of the common minerals found in the earth. Fire protection engineers, recognizing the fire resistance qualities of plaster, have developed accurate ratings which are listed by the National Bureau of Standards and the National Bureau of Fire Underwriters. These ratings are used as a basis for establishing requirements in various building codes.

When plaster is used for the interior wall and ceiling surface, the carpenter usually applies the plaster base and installs the grounds that serve as guides for the plasterer. Carpenters, in addition to a thorough knowledge of the requirements and methods of making this installation, should also have a general knowledge of how the plaster coats are applied.

PLASTER BASE

A plaster finish requires some type of base. For many years, wood lath was used for this purpose. However, in modern construction, sheet materials and metal lath have eliminated its use. Plaster is sometimes applied directly to masonry surfaces or special gypsum block units.

Today, commonly used plaster bases include gypsum lath and expanded metal, Fig. 14-43. A standard size gypsum panel measures 16 by 48 in. and is applied horizontally to the framing members of the structure. It consists of a rigid gypsum filler with a special paper cover.

For a stud or ceiling joist spacing of 16 in. O.C., 3/8 in. lath is used. For a 24 in. spacing 1/2 in. is required.

Gypsum lath is also made with an aluminum foil vapor barrier. Lath with perforations improves the plaster bond and extends the time the wall surface will remain intact when exposed to fire. Some building codes specify this type.

Insulating fiberboard lath is also used as a plaster base. It comes in a 3/8 in. and 1/2 in. thickness with a shiplap edge. Fiberboard lath has considerable insulation value and is often used on ceilings or walls adjoining exterior or unheated areas. Fig. 14-44 shows sizes available.

Expanded metal lath consists of a copper alloy sheet steel, slit and expanded to form openings for keying the plaster. The two most common types are diamond mesh and flat rib. Standard size pieces are 27 in. wide by 96 in. long. Metal lath is usually dipped in black asphaltum paint, or is galvanized, to help it resist rust.

INSTALLING LATH

Before lath is applied, the framing should be inspected for proper spacing and alignment. Check to see that corners and openings have nailing

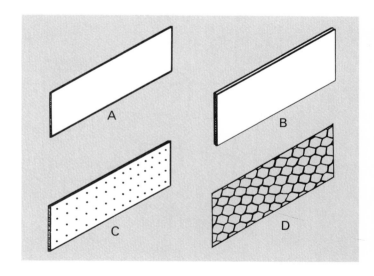

Fig. 14-43. Plaster base materials. A—Gypsum lath. B—Perforated gypsum lath. C—Insulating fiberboard lath. D—Expanded metal lath.

TYPE	THICKNESS (Inches)	WIDTH (Inches)	LENGTH (Inches)
Plain	3/8 1/2	16 16	48 or 96 48
Perforated	3/8 1/2	16 16	48 or 96 48
Insulating	3/8 3/8 or 1/2	16 24	48 or 96 as requested to 12 ft.
Long Length	1/2	24	as requested to 12 ft.

Fig. 14-44. Gypsum and insulating lath sizes.

TYPE FRAMING	BASE THICKNESS		FASTENER	MAX. FRAME SPACING		MAX. FASTENER SPACING	
	in	mm		in	mm	in	mm
Wood	3/8	9.5	Nails—13 ga., 1 1/8'' long, 19/64'' flat head, blued	16	406	5	127
			Staples—16-ga. galv. flattened wire flat crown 7/16'' wide, 7/8'' divergent legs				
	1/2	12.7	Nails—13 ga., 1 1/4'' long, 19/64'' flat head, blued	24	610	4	102
			Staples—16-ga. galv. flattened wire, flat crown 7/16'' wide, 1'' divergent legs				
USG Steel Stud	3/8	9.5	1'' Type S Screws	16 24	406 610	12	305
TRUSSTEEL Stud	3/8	9.5	Clips	16	406	16	406

Fig. 14-46. Nailing requirements for gypsum lath.

members to support the ends of the lath.

Units should be applied with the long dimension at right angles to the framing. Stagger the end joints between adjacent courses, Fig. 14-45. Turn the folded or lapped paper edges of gypsum lath toward the framing. Edges and ends of lath should be in moderate contact.

Gypsum lath 3/8 in. thick is usually applied with 13 ga. gypsum lathing nails 1 1/8 in. long. Nail sizes for other installations are given in Fig. 14-46. The lath must be nailed at each stud or joist crossing.

Insulating lath is installed in the same manner as gypsum lath, except that 13 ga. 1 1/4 in. blued nails should be used. Nails can be driven with pneumatic staplers as shown in Fig. 14-47. In modern construction, staples are often used to attach the lath.

Gypsum lath is easily cut to size by scoring one or both sides with a pointed or edge tool. Break the lath along the line. Be sure to make neatly fitted cut-outs for plumbing pipes and electrical outlets.

Fig. 14-45. Application of gypsum lath to wood studs. Joints must be staggered. Avoid joints, also, at corners of window and door openings.

Fig. 14-47. Pneumatic powered nailer used in attaching gypsum lath.

When applying gypsum lath, drive the nails "home" so the head will be just below the paper surface. Avoid additional hammer blows that will crush the gypsum core.

METAL LATH

In commercial construction, metal lath is common. In residential work it may be used only in shower stalls or tub alcoves. The metal lath provides a rigid wall surface when the construction is properly designed.

Metal lath must be of the proper type and weight for the support spacing. Sides and ends are lapped and corners are returned (overlapped). Studs are usually covered first with a 15 lb. asphalt-saturated felt, Fig. 14-48. Portland cement plaster is often used as the first coat when the surface will be finished with ceramic tile. Gypsum plaster is used for top coats.

REINFORCING

Since some drying will nearly always occur in wood frame structures, shrinkage can be expected. This will likely cause plaster cracks to develop around openings, in corners, or wherever there is a concentration of cross-grain wood.

To minimize cracking, expanded metal lath is often used in key positions over the plaster base. Strips about 8 in. wide are applied at an angle over

the corners of doors and windows as shown in Fig. 14-49. Tack or staple these into place lightly so they become a part of the plaster base only. If nailed securely to the framing, warping, shrinking, and twisting of the frame will be transmitted into the plaster coats and cause cracks.

Metal lath should be used under and around wood beams that will be covered with plaster. Be sure to extend the edges of the reinforcing well beyond the structural element being covered.

Inside corners may be reinforced with a specially formed metal lath or wire fabric called Cornerite, Fig. 14-50. Minimum widths should be 5 in., 2 1/2 in. on each surface of the internal angle. In some

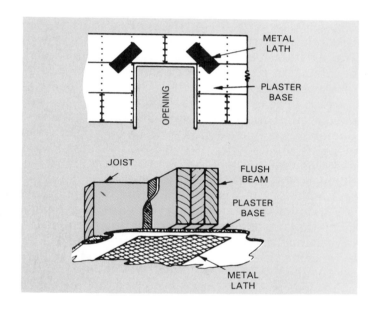

Fig. 14-49. Reinforcing the plaster base. Top. Place metal lath around the jamb area of large openings especially where large headers are used. Bottom. Providing reinforcing under flush beams.

Fig. 14-48. Metal lath is installed over layer of waterproof felt paper in areas subject to moisture.

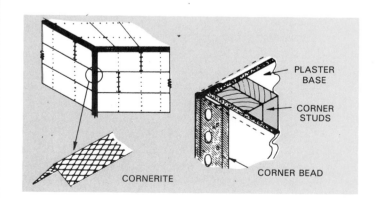

Fig. 14-50. Prefabricated reinforcing is made for both inside and outside corners.

plaster base systems that employ a special attachment clip, Cornerite is not recommended.

Outside corners, where no wood trim will be applied, are reinforced with metal corner beads. They must be carefully applied and plumbed or leveled since they serve not only as a reinforcement for the plaster but as a ground (guide) for its application.

When applying corner bead, use a straightedge and level or plumb line. Also use a spacer block to check the distance between the corner of the bead and the surface of the plaster base. This distance must be equal to the thickness of the plaster coats.

PLASTER GROUNDS

Plaster grounds are usually wood strips as thick as the plaster base and plaster. For average residential construction this would be 3/8 in. plus 1/2 in. Grounds are installed before the plaster is applied. The plasterer uses them as a gauge for thickness of the plaster. They also help keep the plaster surface level and even. Later the grounds may become a nailing base for attaching trim members. They are used around doors, windows, and other openings. Sometimes they are included at the bottom of walls along the floor line, Fig. 14-51. In some wall systems, especially large commercial and institutional buildings, metal edges and strips serve as grounds.

Windows are usually equipped with jamb extensions that are adjusted for the various thicknesses of materials used in the wall structure. These jamb extensions serve as plaster grounds and the carpenter seldom needs to make any changes.

Grounds around some openings are removed after the plastering is complete. Those used at door openings must be carefully set (plumbed) and conform to the width of the door jamb to insure a good fit

of the casing. The width of standard interior door jambs is usually 5 1/4 in.

Carpenters often construct a jig or frame that is temporarily attached to the door opening. Grounds can then be quickly nailed in place along the straightedges of the jig. Instead of using two strips, some carpenters prefer to use a single piece of 3/8 in. exterior plywood ripped to the same width as the door jamb. Such grounds, if carefully removed, can be reused on future jobs, Fig. 14-52.

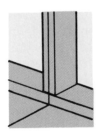

When the plaster base is complete, mark lines on the subfloor at the center line of each stud. After the plaster has been applied and is dry, these marks are transferred to the wall and are helpful when installing baseboards, cabinets, and fixtures.

PLASTER BASE ON MASONRY WALLS

Gypsum plaster can be applied directly to most masonry surfaces. However, to prevent excessive heat loss through outside walls, furring strips are usually installed. These carry the plaster base materials and provide an air space. Strips are attached with:

1. Case hardened nails driven into the mortar joints.
2. Metal or wood nailing plugs placed in the wall during construction.
3. Newly developed adhesive materials and patented fasteners.

Fig. 14-53 shows strips being attached with a pneumatic nailer.

When heat loss needs to be further reduced,

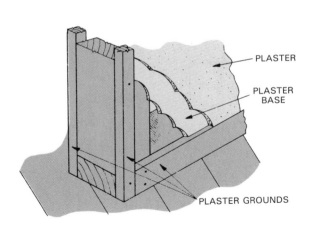

Fig. 14-51. Plaster grounds are installed around openings and sometimes along the floor.

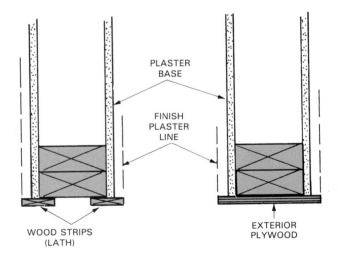

Fig. 14-52. Method of installing removable grounds at door openings.

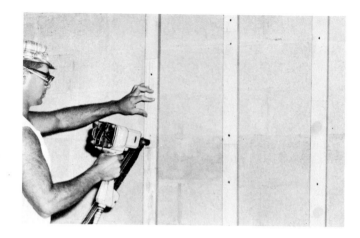

Fig. 14-53. Usually, furring strips are attached to concrete or masonry walls to form a base for lath. Pneumatic nailer with 6d round headed nails is being used.
(Duo-Fast Fastener Corp.)

Fig. 14-54. Base coat of plaster being applied with a trowel. Care must be used around electrical outlets.

blanket insulation can be installed. Some carpenters use plastic foam panels that are attached directly to the wall with adhesives. If furring strips are not used, the plastic foam can serve as a plaster base. Follow manufacturer's recommendations when making an installation of this kind.

PLASTERING MATERIALS AND METHODS

Plaster is applied in two or three coats. The base coats are prepared by mixing gypsum with an aggregate, either at the gypsum plant or on the job. These aggregates may consist of wood fibers, sand, perlite, or vermiculite. Sand is the most commonly used material.

Discard plaster that has started to set. Wash out the mixer with clean water after each batch has been prepared. Keep tools and equipment clean.

In three-coat work, the first application, called the SCRATCH COAT, is applied directly to the plaster base, Fig. 14-54. It is cross raked or scratched after having ''taken up'' (stiffened). It is then allowed to set and partially dry.

The second or BROWN COAT is then applied and leveled with the grounds and screeds. A long flat tool called a darby and a rod (straightedge) are used. After the brown coat has set and is somewhat dry, the third or finish coat is applied.

In two-coat work the scratch coat and brown coat are applied almost at the same time. The cross raking of the scratch coat is omitted. The brown coat of plaster is usually applied (doubled-back) within a few minutes. This application method is the one most frequently used over gypsum or insulating lath plaster bases commonly used in residential construction.

Minimum plaster thickness for all coats should not be less than 1/2 in. when applied to regular gypsum or insulating lath bases. A 5/8 in. thickness is usually required over brick, tile, or masonry. When plaster is applied to metal lath it should measure 3/4 in. in thickness from the back side of the lath.

A plaster job should be inspected constantly for basecoat thickness. Unless grounds and screeds are used on ceilings or large wall areas, it is extremely difficult to keep the thickness uniform. Should the thickness be reduced, the possibility of checks and cracks is much greater. A 1/2 in. thickness of plaster possesses almost twice the resistance to bending and breaking as a 3/8 in. thickness.

Recent developments in the plastering trade center around the plastering machine, Fig. 14-55. It not only saves a great deal of labor but also improves the quality of the plaster application. The lapsed time between mixing and application is shortened. The machine also makes possible the control of the plaster coat's density.

The final or finish coat (about 1/16 in. thick) consists of two general types: the sand-float or textured surface and the putty or smooth finish. In the sand-float finish, special sand is mixed with gypsum or lime and cement. After the plaster is applied to the surface it is smoothed with a float to produce various effects depending on the floating method and the coarseness of the sand.

A smooth finish is produced by applying a putty-

Fig. 14-55. Plastering machine is being used to apply a base coat. (Gypsum Assoc.)

Fig. 14-56. A puttylike mixture of lime and gypsum or cement is used for the final or finish coat of plaster.

When selecting a product, consider its appearance, light reflection, fire resistance, sound absorption, maintenance, cost, and ease of installation.

A standard size tile is 12 x 12 in. However, the tiles are available in larger sizes; for example, 24 x 24 in. and 16 x 32 in. A wide range of surface patterns and textures are manufactured. Fig. 14-57 shows a typical fiberboard tile design that would be appropriate for a residential installation.

Fig. 14-58 illustrates an overlapping, tongue and groove edge that provides a wide flange to receive staples. This type of joint also permits efficient installation when an adhesive is used to hold the tile in place.

Fig. 14-57. Typical fiberboard ceiling tile is a foot square with a face that has acoustical properties.

like material consisting of lime and gypsum or cement, Fig. 14-56. It is troweled perfectly smooth, somewhat like concrete.

CEILING TILE

Ceiling tiles are used for both old and new construction. They can be installed over furring strips, solid plaster, drywall, or any smooth, continuous surface.

There is a considerable range in the types of material used to make ceiling tile. Generally available are: fiberboard tile, mineral tile, perforated metal tile, and glass-fiber tile.

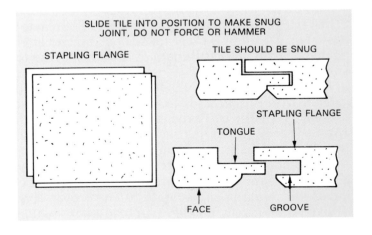

Fig. 14-58. Tongue and groove joint used on standard ceiling tile. Flanges will receive staples while groove holds one edge of the next tile.

LAYOUT PROCEDURE

First, measure the two short walls and locate the midpoint of each one. Snap a chalk line, Fig. 14-59, to establish the centerline. In the middle of this line establish a chalk line that is at right angles to the long centerline. All tiles are installed with their edges parallel to these lines.

For an even appearance, the border courses along opposite walls should be the same width. For example, in a room 10 ft. 8 in. wide use nine full tile and a border tile trimmed to 10 in. on each side.

After determining the width of the border tile, snap chalk lines parallel to the centerlines that will provide a guide for the installation of these border tile or the furring strips that will support them.

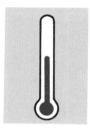

Most manufacturers recommend that fiberboard tile be unpacked in the area where they will be installed, at least 24 hours before application. This will allow the tile to adjust to room temperature and humidity.

INSTALLING FURRING

Furring should be nailed to the ceiling joists. If applied over an old ceiling surface, be sure to locate and mark the joists before attaching the strips. Place the first furring strip flush against the wall at right angles to the joists and nail it with two 8d nails at each joist, Fig. 14-60. Nail the second strip in place so that it will be centered over the edge of the

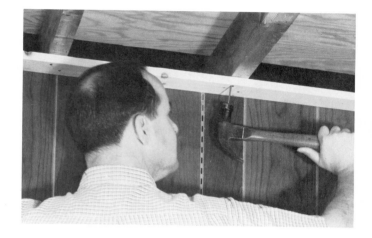

Fig. 14-60. Install the first furring strip. Use two 8d nails at each joist.

border tile. Use the line previously laid out. All other strips are then installed O.C. (For 12 x 12 in. tile, locate them every 12 in. apart.) Fig. 14-61 shows furring complete.

Some carpenters prefer to start from the center and work each side toward the walls. If the furring strips are uniform in width, a spacer jig, Fig. 14-62, may speed up the job. Double check the position of the furring strips from time to time. Make sure they will be centered over the tile joints.

It is essential that the faces of the furring strips be level with each other. Check alignment with a carpenter's level or straightedge. To align strips, drive tapered shims between the strips and the joist as shown in Fig. 14-63. If only one or two joists

Fig. 14-59. Use a chalk line to mark the center of the ceiling tile layout.

Fig. 14-61. Completed installation. Note that two strips along wall are closer together. This is to take care of the narrower border tile.

Fig. 14-62. A spacer jig can be used to locate furring strips.

Fig. 14-64. Doubling the furring provides space for conduit located below ceiling joists.

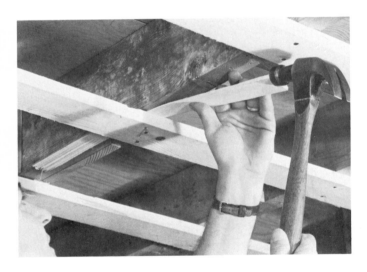

Fig. 14-63. Furring strips can be leveled by driving tapered shims between the joist and the strip.

extend below the plane of the others, it may be best to notch the joists before installing the furring strips.

If pipes or electrical conduit are located below the ceiling joists, it may be necessary to double the furring strips, Fig. 14-64. The first course is spaced 24 to 32 in. on center and attached directly to the joists. The second course is then applied at right angles to the first. Space according to the width of the tiles as previously described. Large pipes or ducts that project below the ceiling joists should be boxed in with furring strips before the tile is installed. Wood or metal trim can be used to finish corners and edges.

There is an alternate furring method, which could be called "sheet furring." It consists of installing gypsum wallboard to the joists or an existing ceiling surface. The tile is then attached to the wallboard either with adhesive or special staples.

INSTALLING TILE

Check over the furring carefully and then snap chalk lines on the strips to provide a guide for setting the border tile. Do this along each wall and be certain the lines are correct and in accord with the initial layout previously described. Make a double check to see that the chalk lines form 90 deg. angles at the corners.

To cut the border tile, first score the face deeply with a knife drawn along a straightedge. Break it along this line by placing it over a sharp edge. See Fig. 14-65. Irregular cuts around light fixtures or other projections may be made with a coping or compass saw. Power tools can also be used. Some tiles are made of mineral fiber which rapidly dulls regular cutting edges.

Start the installation with a corner tile and then set border tile out in each direction. Fill in full size tile. When you reach the opposite wall the border tile will have the stapling flanges removed, and will need to be face-nailed. Locate these nails close to the wall so they will be covered by the trim molding.

Fig. 14-66 shows tile being attached with a stapler. Be sure the tile is correctly aligned. Hold it firmly while setting the staples.

For 12 x 12 tile use three staples along each flanged edge. Use four staples for 16 x 16 tile. Staples should be at least 9/16 in. long.

There are several types of adhesive designed especially for installing ceiling tile. The thick putty type is applied in daubs about the size of a walnut, Fig. 14-67. Apply adhesive to each corner (12 x 12 tile) and about 1 1/2 in. away from each edge. Now position the tile. Slide it back against the other tile

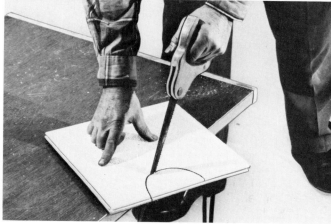

Fig. 14-65. Above. Snapping a border tile after it has been scored. Remove the tongue edge. Leave the wide flange for stapling. Below. Using a compass saw for cutting curves. (Flintkote Co.)

as shown in Fig. 14-68. This motion, along with firm pressure, will spread the adhesive so the layer will be about 1/8 in. thick.

Some adhesives are thinner and are applied with a brush. In remodeling work be sure the old ceiling is clean and that any paint or wallpaper is adhering well. Always follow the recommendations and directions provided by the manufacturer of the products being used.

Be sure to keep your hands clean while handling ceiling tile.

Fig. 14-67. Installing tile directly to a finished ceiling with putty type adhesive.

Fig. 14-66. Staples can be used to attach tile to furring strips. (Duo-Fast Corp.)

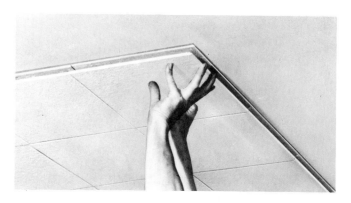

Fig. 14-68. Carefully align new tile with previously set tile.

383

METAL TRACK SYSTEM

Another system uses 4 ft. metal tracks to replace the wood furring strip. The track is nailed to the old ceiling or to joists at 12 in. O.C. intervals, Fig. 14-69. Tongue and grooved panels are slipped into place. A clip, snapped into the track slides over the tile lip. No other fasteners are used, Fig. 14-70.

Fig. 14-69. Attach predrilled track to ceiling at 12 in. intervals.

Fig. 14-70. Track system uses clips to hold ceiling panels. Use one clip for each panel. Border panels must have two clips. (Armstrong World Industries, Inc.)

SUSPENDED CEILINGS

When heating ducts and plumbing lines interfere with the application of a finished surface, a suspended ceiling is practical. Modern installations consist of a metal framework especially designed to support tile or panels, Fig. 14-71.

The height of the ceiling must first be determined. Then a molding must be attached to the perimeter of the room, Fig. 14-72. Use a level chalk line as a guide.

Carefully calculate room dimensions, lay out positions of main runners, and then install screw eyes (4 ft. O.C.) in existing ceiling structure. Panels next

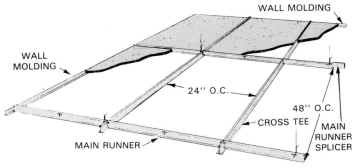

WALL MOLDING

WALL MOLDING

24" O.C.

48" O.C.

CROSS TEE

MAIN RUNNER SPLICER

MAIN RUNNER

GRID SYSTEM DESIGNED FOR 24" x 48" PANELS

Fig. 14-71. This sketch is typical of the framework used for suspended ceilings.

Fig. 14-72. Attach metal molding to the wall. Make sure it is level and straight. (Armstrong World Industries, Inc.)

to the wall (border panels) may need to be reduced in width to provide a symmetrical (same on both sides of room) arrangement. Plan the layout the same way as previously described for regular ceiling tile.

Install the main runners by resting them on the wall molding and attaching to the wires tied to the screw eyes, Fig. 14-73. Use chalk lines or string, stretched between the wall molding to insure a level assembly. Sections of runners are easily spliced. Odd lengths can be cut with a fine-tooth hacksaw or aviation snips.

After all main runners are in place, check the level and adjust the wires. Install cross tees next. See Fig. 14-74. Check the required spacing for the tiles or panels. Simply insert the end tab of the cross tee into the runner slot.

When the suspension framework is complete, panels can be installed. Each panel is tilted upward and turned slightly on edge so it will ''thread'' through the opening as illustrated in Fig. 14-75. After the entire panel is above the framework, turn it flat and lower it onto the grid flanges.

A concealed suspension system is shown in Fig. 14-76. The tongue and groove joint is similar to the joint on regular ceiling tile. Joints are interlocked with the flange of the special runner when the installation is made, Fig. 14-77.

ESTIMATING MATERIALS

The amount of wall or ceiling covering materials required is estimated by first calculating the total number of square feet of wall and ceiling area. Regu-

Fig. 14-74. Cross tees are installed between main runners. (Armstrong World Industries, Inc.)

Fig. 14-75. Ceiling panels are tilted, slipped through the framework, and lowered into place.

lar size door and window openings are disregarded. These provide allowances for waste. However, larger window walls and picture windows may be subtracted.

DETERMINING AREA OF ROOMS

For materials that do not come in sheets you need to know the area you are covering. Ceiling area is

Fig. 14-73. Attach eye screws to joists and hang metal runner from joists. Use a chalk line to check level.

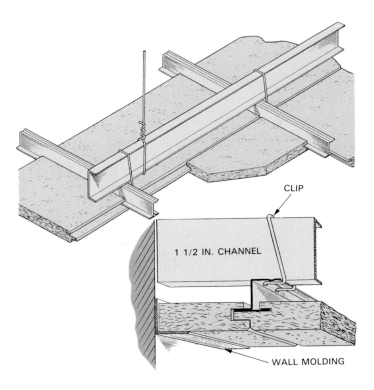

1 1/2 IN. CHANNEL

CLIP

WALL MOLDING

Fig. 14-76. Above. Channels of suspension system are concealed by the ceiling panels. Below. Tongue and groove joint detail.

Fig. 14-77. Ceiling panel joints interlock with supporting flange. (Wood Conversion Co.)

usually the same as the floor area. It is much easier to take floor dimensions and then multiply the length times the width.

To find area of walls, add all the lengths together and multiply by the wall height.

SHEET MATERIALS

When ordering sheet materials such as wallboard or paneling, be sure to specify the length. Always plan to use the longest practical sheet. This holds butt joints to a minimum or eliminates them. Divide the total length of the walls by the width of the sheets to find the number of sheets needed.

Estimate each room separately. Take dimensions directly from the walls or carefully scale the plan. Consult tables and charts of suppliers for estimates of joint compound, adhesives, and nails.

When working with expensive hardwood panels, it is usually advisable to make a scaled layout. Look at the diagrams in Fig. 14-33 for application procedure.

SOLID PANELING

Estimates for solid paneling are based on its nominal and unfinished size. Seasoning and planing bring it to its dressed size. Forming of joints

may further reduce the actual face widths.

For example, a 1 x 6 tongue and groove board has a face width of 5 1/8 in. First calculate the square footage of the wall to be covered. Then multiply by various factors taken from lumber tables:
1. For 1 x 6 tongue and groove boards, use 1.17.
2. For 1 x 8 tongue and groove boards, use 1.16.

On standard vertical applications add about 5 percent for waste when required lengths can be selected or when the lumber is end matched.

ESTIMATING GYPSUM LATH

Since gypsum lath is produced in smaller sections than full sheets you will need to use different methods of figuring quantity needed. Figure the area of the ceiling and add to this the area of the walls (length of walls × height).

For example:

Ceiling area (same as floor area)	= 1250 sq. ft.
Total length of walls	= 14 + 12 + 14 + 12 + 10
	= 62
Wall area	= 62 × 8
	= 496 sq. ft.
Total area	= 496 + 1250
	= 1746 sq. ft.
Standard lath bundle	= 64 sq. ft.
Bundle estimate	= 28

Plasterers usually base their prices and estimates on the number of square yards. Convert square footage to square yards by dividing by 9 (1 sq. yd. = 9 sq. ft.).

Quantities of ceiling tile are estimated by figuring the area (square footage) to be covered. Round out any fractional parts of a foot in width and length to the next larger full unit when making this calculation.

Add extra units when it is necessary to balance the installation pattern with border tile along each wall. When using 12 x 12 tile the number required will equal the square footage plus the extra allowance described. Standard 12 x 12 ceiling tile are packaged 64 sq. ft. to a carton.

TEST YOUR KNOWLEDGE — UNIT 14

1. List nine different materials that are used as interior wall and ceiling coverings.
2. In single layer dry wall construction, the gypsum wallboard thickness should be _____ or _____ in.
3. When nailing wallboard in place, the nails are spaced closer together on the _____ (walls, ceilings).
4. In addition to the skim coat applied over the tape, wallboard joints usually require _____ (1, 2, 3) coats of joint compound.
5. When making a double layer wallboard application, joints running parallel to each other should be offset at least _____ in.
6. Solid wood wall paneling should have a moisture content of about _____ to _____ percent for most areas of the United States.
7. Prefinished plywood wall paneling should be "room conditioned" for at least _____ hours before application.
8. Using a base layer of gypsum wallboard under plywood paneling improves alignment and rigidity, and also makes the construction more _____ _____.
9. A standard size panel of gypsum lath measures 3/8 x 16 x _____.
10. To reinforce the plaster base at the corners of window and door openings, a strip of _____ _____ is used.
11. Strips of wood or metal installed at the edge of openings to provide a guide for the plasterer are called _____.
12. The most common aggregate used in plaster mixes is _____.
13. In standard three-coat plaster applications, the brown coat is applied _____ (first, second, last).
14. When applied to regular gypsum lath, the total thickness of the plaster coats should be not less than _____ in.
15. Standard ceiling tile have a wide stapling or nailing flange on _____ (1, 2, 3, 4) edges.
16. For a symmetrical ceiling tile pattern in a room 11 ft.-6 in. wide, use 10 full (12 x 12) tile and a _____ in. border tile.
17. The metal framework of a modern suspended ceiling is supported mainly by _____ tied to the building structure directly above.
18. It is general practice for plasterers to base their cost estimates on the number of _____ _____ of ceiling and wall areas.

OUTSIDE ASSIGNMENTS

1. Secure and study literature from companies that manufacture gypsum products. Learn about the history and development of plastering and modern processes used in making plaster, gypsum lath, and gypsum wallboard. Get samples of the products commonly used in your locality, along with approximate prices and costs of installation. Prepare a written report or outline the information carefully and report to your class.
2. Visit a local drywall contractor and learn about some of the special problems and remedies typical to this type of wall finish. Secure such information as: care and handling materials; repairing and adjusting warped studs and framework; repairing damaged boards and surfaces; cause and remedies of tape blisters; and definition and prevention of nail pops. Organize the information you secure under appropriate headings and make a presentation to your class.
3. From various reference books, including manuals on architectural standards, learn about methods and construction details for ceiling systems that include radiant heating. Cover both hot water and electrical systems. Prepare scaled drawings (larger than actual size) of sections through various ceiling constructions, showing the heating pipes and elements. Include notes concerning material specifications and critical temperatures for the various constructions.
4. Suspended ceilings in institutional and commercial buildings often include a ventilation system that provides both heating and cooling. The space above the ceiling serves as a plenum. Air enters the room through small slots or holes either in the tile or at special joints between the tile. Study this method of air distribution and prepare a written report with drawings and other illustrations.

Flooring materials of wood and composition are very popular in residential construction.
(Memphis Hardwood Flooring Co. and Armstrong World Industries, Inc.)

Unit 15

FINISH FLOORING

"Finish flooring" is any material used as the final surface of a floor. A wide selection of materials are made for this purpose.

Hardwoods and softwoods are available as strip flooring in a variety of widths and thicknesses, and as random width planks or unit blocks. Many new and improved materials are being used in the production of composition (resilient) flooring. Notable among these are the vinyl plastics which have largely replaced asphalt tile, rubber tile, and linoleum.

Flagstone, slate, brick, and ceramic tile are frequently selected for special areas such as entrances, bathrooms, or multipurpose rooms. Floor structures usually need to be designed to carry this type of finish surface.

When the finish flooring is wood, it is usually laid after wall and ceiling surfaces are completed, and before interior door frames and other trim are added. The surface should be covered to protect it during other inside finish work. Sanding and finishing of the floor surface becomes the last major operation as the interior is completed.

Where prefinished wood flooring is used, it must be covered and protected. As with wood, the installation of resilient flooring and prefinished wood flooring must be among the last steps in interior finishing. Select and install finish flooring carefully.

WOOD FLOORING

Wood is popular as flooring in residential structures. Wood flooring, especially hardwood, has the

Fig. 15-1. Natural oak strip flooring is durable and has a natural beauty which is brought out by clear finishes and wax.

strength and durability to withstand wear while providing an attractive appearance, Fig. 15-1.

Oak is a widely used species. However, maple, birch, beech, and other hardwoods also have desirable qualities. Softwood flooring includes such species as pine and fir. These are farily durable when produced with an edge-grain surface.

TYPES OF WOOD FLOORING

Three general types of wood flooring are used in residential structures: strip, plank, and block. As the name implies, strip flooring consists of pieces cut into narrow strips. It is laid in a random pattern of end joints.

Most strip flooring is tongue and groove both on the sides and ends. This design is also referred to as side-and-end matched. Another feature of modern strip flooring is the undercut. See Fig. 15-2. This is a wide groove on the bottom of each piece which enables it to lay flat and stable even when the subfloor surface is slightly uneven.

Plank flooring, Fig. 15-3, provides an informal atmosphere. It is particularly appropriate for colonial and ranch style homes.

Plank floors are usually laid in random widths. The pieces are bored and plugged to simulate the wooden pegs originally used to fasten them in place.

Today this type of floor has tongue and groove edges. It is laid in about the same manner as regular strip floors.

Block flooring looks like conventional parquetry. This is an elaborate design formed by small wood blocks. Modern unit blocks consist of short lengths of flooring, held together with glue, metal splines, or other fasteners. Square and rectangular units are

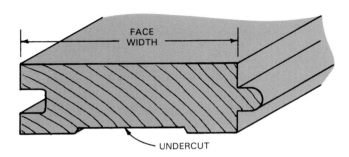

Fig. 15-2. Typical section of strip flooring. Ends are also tongue and grooved.

Fig. 15-3. Two types of wood flooring. Left. Random planks have walnut plugs. Pieces are 3/4 in. thick and are 2 1/4 and 3 1/4 in. wide. Lengths are random. Right. Unit block parquet flooring. Prefinished units are 12 in. square and 5/16 in. thick. (Robbins Inc.)

produced. Generally, each block is laid with its grain at right angles to the surrounding units, Fig. 15-3. Blocks, called laminated units, are produced by gluing together several layers of wood.

SIZES AND GRADES

Today, hardwood strip flooring is generally available in widths ranging from 1 1/2 to 3 1/4 in. Standard thicknesses include 3/8, 1/2, and 3/4 in.

Solid planks are usually 3/4 in. thick. However, greater thicknesses are available. Widths range from 3 to 9 in. in multiples of 1 in. Unit blocks are also commonly produced in a 3/4 in. thickness. Dimensions (width and length) are in multiples of the widths of the strips from which they are made. For example, squares assembled from 2 1/4 in. strips will be 6 3/4 x 6 3/4 in., 9 x 9 in., or 11 1/4 x 11 1/4 in.

Uniform grading rules are established by manufacturers working with the U.S. Bureau of Standards and such organizations as the National Oak Flooring Manufacturers' Association or the Maple Flooring Manufacturers Association. Grading is based largely on appearance. Consideration is given to knots, streaks, color, pinworm holes, and sapwood. The percentage of long and short pieces is also a consideration.

Oak, for example, is separated into two grades of quarter-sawed stock and five grades of plain-sawed. In descending order, plain-sawed grades are: clear, select and better, select, No. 1 common, and No. 2 common.

White and red oak species are ordinarily separated in the highest grades. A chart on grades for hardwood flooring is included in the reference section.

Manufacturers recommend that wood flooring be delivered four or five days before installation. It should be piled loosely throughout the structure. An inside temperature of at least 70 °F should be maintained. This period of "conditioning" permits the wood to match its MC (moisture content) with that in the building.

SUBFLOORS

In conventional joist constructions, most building codes specify a sound subfloor. It adds considerable strength to the structure and serves as a base for attaching the finish flooring.

For regular strip or plank flooring, the subflooring should consist of good quality boards about 1 in. thick and not more than 6 in. wide. Space the boards about 1/4 in. apart and face nail them solidly at every bearing point. Inadequate or improper nailing of subfloors usually results in squeaky floors.

In modern construction, plywood is commonly used for the subfloor. It must be installed according to recognized standards. Refer to Unit 7 for sizes and installation methods.

INSTALLING WOOD STRIP FLOORING

Check over the subfloor to make certain it is clean and that nailing patterns are complete. Put down a good quality building paper. It should extend from wall to wall, with a 4 in. lap as shown in Fig. 15-4. The location of the joists should be chalk lined on the paper. This is even more important when plywood is used as the subfloor.

Over the heating plant or hot air ducts, it is advisable to use a double-weight building paper. Insulation may be attached directly to the underside of the subfloor. This extra precaution will prevent excessive heat from reaching the finish flooring causing cracks and open joints.

Lay strip flooring at a right angle to the floor joist. The flooring will look best when this direction aligns with the longest dimension of a rectangular room. Since floor joists will normally span the shortest dimension of the living room, this will establish the direction the flooring runs in other rooms.

% *MC usually recommended for flooring at the time of installation is 6 percent for the dry southwestern states; 10 percent for the more humid southern states; and about 7 or 8 percent for the remainder of the country.*

NAILING

Floor squeaks or creaks are caused by the movement of one board against the other. The problem may be in either the subfloor or finished floor. Using adequate number and proper size of nails will reduce these undesirable noises. When possible, the nails used for the finish floor should go through the subfloor and into the joist. When plywood is used for the subfloor, place a nail at each joist and one in between. Nail sizes are specified for different types of floors in Fig. 15-5.

Start the installation in one of the rooms. Lay the first strip along either sidewall. Select long pieces. If the wall is not perfectly straight and true, set the first course with a chalk line. Place the groove edge next to the wall and leave at least 1/2 in. for expansion. This space will be covered later by the baseboard and baseshoe.

Make sure the first strip is perfectly aligned. Then face nail it as illustrated in Fig. 15-6. In face nailing, the nail head must be set (sunk below face) and the hole filled.

Succeeding strips are blind nailed with the nail

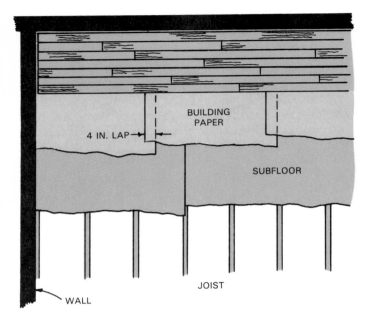

Fig. 15-4. Cutaway view showing various layers of material under strip flooring.

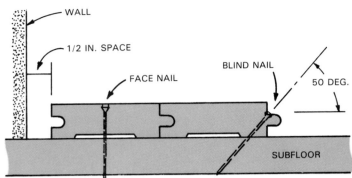

Fig. 15-6. Cutaway drawing shows how to nail the starter strip. Note, also, how to nail succeeding strips.

penetrating the flooring where the tongue joins the shoulder. The nail is driven at an angle of about 50 deg. Use a nail set to finish the driving so that the edge of the strips will not be damaged.

Each strip should fit tightly against the preceding strip. When it is necessary to drive strips into position, use a piece of scrap flooring as a driving block.

Cut and fit a number of pieces and lay them ahead of the installed strips as shown in Fig. 15-7. Use different lengths. Match them so they will extend from wall to wall with about 1/2 in. clearance at each end.

Joints in successive courses should be 6 in. or more from each other. Try to arrange the pieces so the joints are well distributed. Blend color and grain for a pleasing pattern. Pieces cut from the end of a course should be carried back to the opposite wall to start the next course. This is the only place it can be used since its leading end has no tongue or groove.

LAYING AROUND PROJECTIONS

Flooring strips should run uninterrupted through doorways and into adjoining rooms. When there is

Flooring Nominal Size, Inches	Size of Fasteners	Spacing of Fasteners
3/4 × 1 1/2 3/4 × 2 1/4 3/4 × 3 1/4 3/4 × 3'' to 8'' plank	2'' machine driven fasteners; 7d or 8d screw or cut nail.	10''-12''* apart 8'' apart into and between joists.
Following flooring must be laid on a subfloor.		
1/2 × 1 1/2 1/2 × 2	1 1/2'' machine driven fastener; 5d screw, cut steel or wire casing nail.	10'' apart
3/8 × 1 1/2 3/8 × 2	1 1/4'' machine driven fastener, or 4d bright wire casing nail.	8'' apart
Square-edge flooring as follows, face-nailed— through top face		
5/16 × 1 1/2 5/16 × 2	1'', 15-gauge fully barbed flooring brad. 2 nails every 7 inches.	
5/16 × 1 1/3	1'', 15-gauge fully barbed flooring brad. 1 nail every 5 inches on alternate sides of strip.	

*If subfloor is 1/2 inch plywood, fasten into each joist, with additional fastening between joists.

Fig. 15-5. Nail chart for application of strip flooring.

Fig. 15-7. A portable nailer is useful for installing strip flooring.

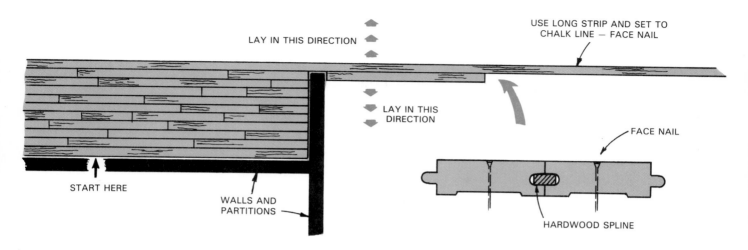

Fig. 15-8. Follow this procedure when laying strip floors around a wall or partition.

a projection into the room, such as a wall or partition, follow the procedure described in Fig. 15-8.

First, lay the main area to a point even with the projection and then extend the next course all the way across the room. Set the extended strip to a chalk line. Then face nail it in place. Form a tongue on the grooved edge by inserting a hardwood spline. Next, install the flooring in both directions from this strip.

If a large area is to be laid with strip or plank flooring, it sometimes helps to set up a starter strip at or near the center. Use a spline in the groove of the starter strip as previously described. Be sure to measure and accurately align the starter strip with the walls on each side of the room.

When the floor has been brought to within 2 or 3 feet of the far wall, the room should be checked again to find out if the strips are parallel to the wall. If not, dress off the grooved edges slightly at one end until the strips have been adjusted to run parallel. This is necessary in order that the last piece may be aligned with the baseboard.

MULTIROOM LAYOUT

When flooring installation is carried throughout the major part of a building, study the floor plan to determine the most efficient procedure to follow.

Fig. 15-9 shows a typical residential floor plan with a method for the installation of strip flooring.

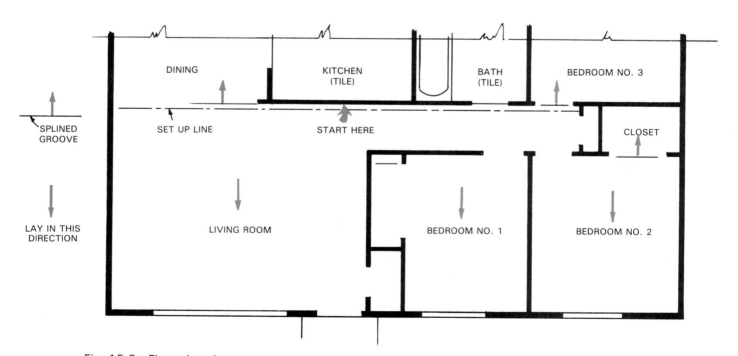

Fig. 15-9. Floor plan shows correct procedure for laying strip flooring throughout a number of rooms.

A setup line (chalk line) is first laid out two flooring widths (plus 1/2 in.) from the partition as shown. The starter courses are aligned with the setup line before face nailing. Installation is continued across the living room, across the hall, into and through bedrooms No. 1 and No. 2.

A spline is set in the groove of the starter strip and the floor is laid from this point into the dining room and bedroom No. 3. In bedroom No. 2, a splined groove is also used to lay the floor back into the closet.

In small closet areas, such as shown in bedroom No. 1, it is usually impractical to reverse the direction of laying with a splined groove and the pieces are simply face nailed in place.

The last strip laid along the wall of a room will usually need to be ripped so it will have the required clearance (1/2 in. minimum). The last several strips must be face nailed. Do not use ripped strips where they might detract from the appearance. It is recommended that full length strips always be used around entrances and across doorways.

In general, plank flooring is installed by following the same procedures used for strip flooring. In addition to the regular blind nailing, screws are set and concealed in the face of wide boards.

ESTIMATING STRIP FLOORING

To determine the number of board feet of strip flooring needed to cover a given area, first calculate the area in square feet. Then add the percentage listed for the particular size being used. See Fig. 15-10. The figures listed are based on laying flooring straight across the room. They provide an allowance for side-matching, plus about 5 percent for end matching. Where there are many breaks and projections in the wall line, add additional amounts.

For example:

Total area = 900 sq. ft.
Flooring size = 3/4 × 2 1/4
Bd. ft. of flooring = 900 + 38 1/3% of 900
　　　　　　　　　= 900 + 345
　　　　　　　　　= 1245
Number of bundles = 52 (24 bd. ft. to bundle)

WOOD FLOORING OVER CONCRETE

Finished wood flooring systems can be successfully installed over a concrete slab. When the concrete floor is suspended, with an air space below, a moisture barrier is usually not required. For slab-on-grade installation, put down sleepers, also called screeds. These are wood strips attached to or embedded in the concrete surface. The sleepers serve as a nailing base for the flooring material.

Watch out for damp floors! Moisture from a con-

55% for	3/4'' × 1 1/2''
42 1/2% for	3/4'' × 2''
38 1/3% for	3/4'' × 2 1/4''
29% for	3/4'' × 3 1/4''
38 1/3% for	3/8'' × 1 1/2''
30% for	3/8'' × 2''
38 1/3% for	1/2'' × 1 1/2''
30% for	1/2'' × 2''

Fig. 15-10. Add above percentages to total area to be covered with strip flooring.

crete floor not completely dry or cured will damage wood flooring.

To test for moisture presence, lay down a flat, noncorrugated rubber mat on the slab. Weight it down so the moisture cannot escape. Allow the mat to remain overnight. Moisture in the concrete will show as water marks when the mat is removed.

If the concrete is placed directly on grade either at or below ground level, an approved membrane moisture barrier must be installed between the flooring and the concrete. Fig. 15-11 illustrates, in a general way, a system of waterproofing and sleeper installation that will provide a base for 3/4 strip flooring. A coat of asphalt is first evenly applied over the concrete surface. This is covered with polyethylene film. Another coat of a special asphalt mastic is then applied as shown. Then wood sleepers are set in place. The sleepers are 2 × 4 lumber about 30 in. long, laid flat and running at a right angle to the flooring. Lap them at least 3 in.

Fig. 15-11. Above. Moisture proofing a concrete slab-on-grade with asphalt mastic and polyethylene film. Below. 2 × 4 sleepers set in asphalt mastic provide nailing base for wood flooring. (E.L. Bruce Co.)

for 2 1/4 in. and 4 in. for 3 1/4 in. flooring. Spacing between centers is usually 12 to 16 in.

A newer system of laying strip floors over a concrete slab-on-grade offers savings over older methods. It consists of a double layer of 1 × 2 wood sleepers nailed together, with a moisture barrier of 4-mil polyethylene film placed between them, Fig. 15-12. This is accepted by FHA.

The sequence of photographs in Fig. 15-13 illustrates the basic steps for installation. First clean and prime the floor. Then snap chalk lines 16 in. apart. Put down bands (rivers) of a special adhesive along the layout lines.

Treated wood sleepers are embedded in the adhesive and secured with 1 1/2 in. concrete nails, about 24 in. apart. Strips could be attached with explosion-actuated equipment.

Place a layer of polyethylene film over the strips. Join sheets by forming a lap over a sleeper.

After the vapor barrier, a second layer of untreated strips are nailed to the bottom courses with 4d nails about 16 in. apart. Install strip or plank flooring as previously described. No two adjoining flooring strips should break joints between the same two sleepers.

Installation of wood flooring over concrete must be carefully designed and installed. Secure detailed specifications from manufacturers of the products to be used, or such organizations as the National Oak Flooring Manufacturers' Association.

WOOD BLOCK (PARQUET) FLOORING

Block or parquet flooring requires different application methods than strip flooring. Blocks are usually made up on squares of 6, 8, or 12 in.

Block flooring is produced in two different ways:
1. Unit blocks are glued up from several short lengths of flooring.
2. Laminated blocks are made by bonding three plies of hardwood with moisture resistant glue.
3. Parquet flooring uses more intricate patterns of small pieces in producing the squares.

Each block is tongue and grooved. See Fig. 15-14.

Installation of all block type flooring is basically the same. Allow some clearance on unit blocks for expansion. Rubber strips are sometimes used for this purpose. Generally, it is recommended that unit blocks be installed with a 1 in. space at the wall.

Any type of block flooring can be fastened with nails or laid in mastic. Nail through the tongue edges as with strip or plank floors. When laid in mastic a thin coat is first applied over a smooth, level, dry

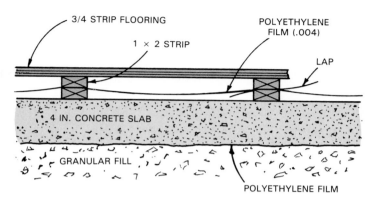

Fig. 15-12. Strip flooring system over concrete slab. Sleepers are doubled 1 × 2 strips.

base. See Fig. 15-15. Put down a layer of 30 lb. asphalt-saturated felt. Spread another coat of mastic, about 3/32 in. thick, over the felt. Lay the blocks in this top coat as illustrated in Fig. 15-16.

Installation can be made either on a square or a diagonal pattern. Chalk lines are first snapped equal distances from the sidewalls if laying is done from a center point. Layout procedures are similar to those described for resilient flooring materials included later in this Unit. See Fig. 15-17.

Since block flooring and adhesive materials vary from one manufacturer to another, detailed information concerning installation procedures should always be secured from the manufacturer of the product.

PREFINISHED WOOD FLOORING

Most of the flooring materials described are available with a factory applied finish, Fig. 15-18. They are generally considered to be superior to that which can normally be applied on-the-job. Another advantage of prefinished flooring is that it speeds construction. Floors are ready for service immediately.

Disadvantages arise from the fact that the installation must be made with care. Although face nailing can be covered with special filler materials, it must be held to a minimum when working with prefinished materials. Hammer marks caused by careless blind nailing are extremely difficult to repair.

A prefinished floor must be the last step in the interior finish sequence. Other trim must be set beforehand with proper allowances. For example, interior door jambs and casing require spacer blocks equal to the floor thickness. Place these under the units while they are being installed.

Unless prefinished baseboards are installed after the floor is laid, they must also be placed to allow necessary clearance. Other special provisions will need to be made around built-in cabinets, and at stairways and entrances.

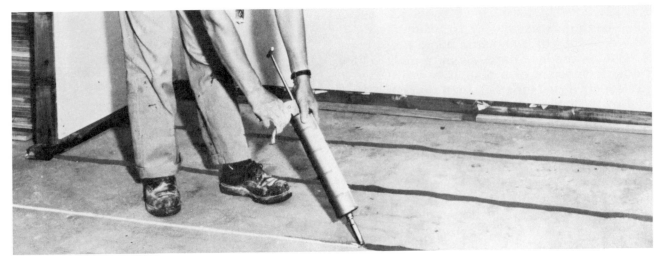

APPLY RIVERS OF MASTIC ALONG CHALKED LAYOUT LINES.

NAIL TREATED SLEEPERS OVER MASTIC.

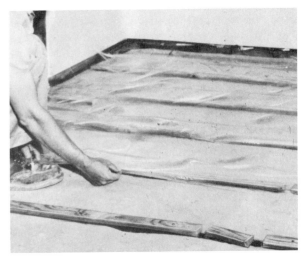

SPREAD POLYETHYLENE FILM OVER BOTTOM SLEEPERS. LAP FILM
ONLY OVER A SLEEPER.

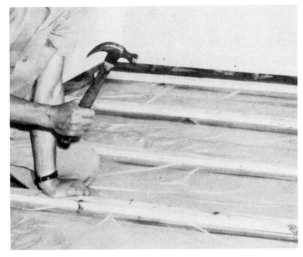

NAIL SECOND LAYER OF SLEEPERS ON TOP OF FIRST SLEEPERS.

NAIL STRIP FLOORING TO SLEEPERS.

Fig. 15-13. Steps for installing strip hardwood flooring over concrete slab. (National Oak Flooring Manufacturers Assoc.)

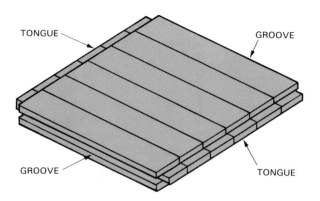

Fig. 15-14. Typical block or parquet flooring. Edges are always tongue and grooved.

Fig. 15-16. Attaching wood blocks with mastic. Same method can be used to lay 3/8 in. planks of hardwood veneer. (E.L. Bruce Co.)

Fig. 15-15. Spreading mastic over plywood subfloor. Floor should be smooth, clean, and dry.

UNDERLAYMENT

Flooring materials such as asphalt, vinyl, linoleum, and rubber will usually reveal rough or irregular surfaces in the flooring structure upon which they are laid. Conventional subflooring does not provide a satisfactory surface. An underlayment of plywood or hardboard is required. On concrete floors a special mastic material is sometimes used when the surface condition does not meet the requirements of the finish flooring.

An underlayment also prevents the finish flooring materials from checking or cracking when slight movements take place in a wood subfloor. When used for carpeting and resilient materials, the underlayment is usually installed as soon as wall and ceiling surfaces are complete.

HARDBOARD AND PARTICLEBOARD

Both of these products meet the requirements of an underlayment board. The standard thickness for hardboard is 1/4 in. Particleboard thicknesses range from 1/4 in. to 3/4 in.

This type of underlayment material will bridge small cups, gaps, and cracks. Larger irregularities should be repaired before application. High spots should be sanded down and low areas filled. Panels should be unwrapped and placed separately around the room for at least 24 hours before they are installed.

To apply, start in one corner and fasten each panel securely before laying the next. Some manufacturers print a nailing pattern on the face of the panel. Allow at least a 1/8 to 3/8 in. space along an edge next to a wall or any other vertical surface.

Stagger the joints of the underlayment panel. See Fig. 15-19. The direction of the continuous joints should be at right angles to those in the subfloor. Be especially careful to avoid alignment of any joints in the underlayment with those in the subfloor. Leave a 1/32 in. space at the joints between hardboard panels. Particleboard panels are butted lightly.

Underlayment panels should be attached to the subfloor with approved fasteners like those in Fig. 15-20. For hardboard, space the fastener in 3/8 in. from the edge.

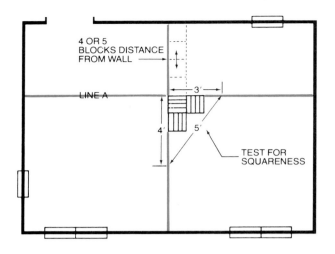

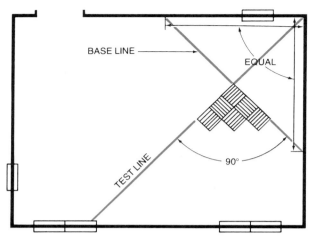

Fig. 15-17. Laying out chalk lines for parquet flooring patterns. Left. Square pattern. Right. Diagonal pattern. (Oak Flooring Institute)

Fig. 15-18. Prefinished flooring. Three piece units are 3/4 x 4 1/2 x 9 in. Edges have a slight bevel. Units can be installed in mastic over concrete or wood subfloors. (Memphis Hardwood Flooring Co.)

Fig. 15-19. Installing particleboard underlayment. Keep nails between 1/2 and 3/4 in. from edge. End joints must be staggered. (Weyerhaeuser Co.)

7/8 in. long. Space staples not over 4 in. apart along panel edges.

Special adhesives will also bond underlayment to subfloors. They eliminate the possibility of nail-popping under resilient floors.

PLYWOOD

Plywood is preferred by many carpenters for underlayment. It is dimensionally stable and spacing between joints is not critical. Since a range of thicknesses are available, alignment of the surfaces of various finish flooring materials is easy, Fig. 15-21.

Follow the same general application procedures described for hardboard. Turn the grain of the face ply to run at a right angle to the framing supports. Stagger end joints. Nails may be spaced farther apart for plywood but should not exceed a field

Spacing for particleboard varies for different thicknesses. Be sure to drive nail heads flush with the surface. When fastening underlayment with staples, use a type that is etched or galvanized and at least

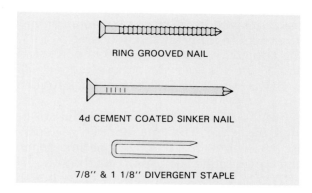

Fig. 15-20. Approved fasteners for underlayment.

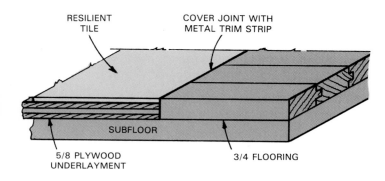

Fig. 15-21. Alignment of resilient floor tile with strip flooring is easier when using a plywood underlayment.

spacing of 10 in. (8 in. for 1/4 and 3/8 in. thicknesses) and an edge spacing of 6 in. on center. Use ring groove or cement coated nails.

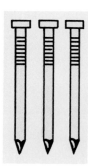

When preparing a base for resilient flooring materials by applying an underlayment over a solid board subfloor—do not drive long nails through the underlayment, subfloor, and on into the joists. Subsequent shrinkage of the subfloor and frame (especially in new construction) will likely cause these nails to rise above the surface of the underlayment and form "blisters" in the floor surface.

RESILIENT FLOOR TILE

After the underlayment is securely fastened, sweep and vacuum the surface carefully. Check to see that surfaces are smooth and joints level. Rough edges should be removed with sandpaper or a block plane.

The smoothness of the surface is extremely important, especially under the more pliable materials (vinyl, rubber, linoleum). Over a period of time these materials will "telegraph" (show on the surface)

even the slightest irregularities and/or rough surfaces. Linoleum is especially susceptible and for this reason a base layer of felt is often applied over the underlayment when this material, either in tile or sheet form, is to be installed.

Because of the many resilient flooring materials on the market, it is essential that each application be made according to the recommendations and instructions furnished by the manufacturer of the product.

INSTALLING RESILIENT TILE

Start a floor tile layout by locating the center of the end walls of the room. Disregard any breaks or irregularities in the contour. Establish a main centerline by snapping a chalk line between the two points, Fig. 15-22. When snapping long lines, remember to hold the line at various intervals and snap only short sections.

Fig. 15-22. Snapping a chalk line to establish the centerline of a room. This is needed before installing resilient tile. (Armstrong World Industries, Inc.)

Fig. 15-23. Use a carpenters' square to lay out a centerline at right angles to the first centerline.

Next, lay out another centerline at right angles to the main one. Use a carpenters' square as illustrated in Fig. 15-23 or set up a right triangle (base 4 ft., altitude 3 ft., hypotenuse 5 ft.). A chalk line can be used or you can draw the line along a straightedge.

With the centerlines established, make a trial layout of tile along the centerlines as shown in Fig. 15-24. Measure the distance between the wall and last tile. If the distance is less than 2 in. or more than 8 in., move the centerline 4 1/2 in. (for 9 × 9 tile) closer to the wall. This adjustment will eliminate the need to install border tiles that are too narrow. Check the layout along the other centerline in the same way. Since the original centerline is moved exactly half the tile size, the border tile will remain uniform on opposite sides of the room.

SPREADING ADHESIVE

Remove the loose tile. Reclean the floor surface and spread the adhesive over one-quarter of the total area, Fig. 15-25. Make the spread even with the chalk line but do not cover it. Be sure to use the type of spreader (trowel or brush) recommended by the manufacturer of the adhesive.

The spread of adhesive is very important. If it is too thin, the tile will not adhere properly. If too heavy, the adhesive will creep up between the joints.

Allow the adhesive to take an initial set before a single tile is laid. The time required will vary from about 15 minutes to a much longer time depending on the type of adhesive used. Test the surface with your thumb. It should feel slightly tacky but should not stick to your thumb.

LAYING TILE

Start laying the tile at the center of the room. Make sure the edges of the tile align with the chalk line. If the chalk line is partially covered with the adhesive, snap a new one or tack down a thin, straight strip of wood to act as a guide in placing the tile.

Fig. 15-25. Carefully spread adhesive up to the centerlines on one quadrant (1/4) of the floor area.

Fig. 15-24. Making a trial layout. Top. Lay down tiles along both centerlines to walls. Bottom. Measure width of border tile and adjust centerline, if necessary.

Fig. 15-26. Laying tile. Align joining edges first and then lower the rest of the tile.

Butt each tile squarely to the adjoining tile, with the corners in line, Fig. 15-26. Carefully lay each tile in place; do not slide, it may cause the adhesive to work up between the joints and prevent a tight fit. Take sufficient time so each tile will be positioned correctly. There is usually no hurry since most adhesives can be "worked" over a period of several hours.

Asphalt and vinyl-asbestos tile do not need to be rolled. Rubber, vinyl, and linoleum are usually rolled after a section of the floor is laid. Be sure to follow the manufacturer's recommendations.

After the main area is complete, set the border tile as a separate operation. To lay out a border tile, place a loose tile (the one that will be cut and used) over the last tile in the outside row. Now take another tile and place it in position against the wall and mark a sharp pencil line on the first tile, Fig. 15-27.

Cut the tile along the marked line, using heavy-duty household shears or tin snips. Some types of tile require a special cutter or they may be scribed and broken. Asphalt tile, if heated, can be readily cut with snips.

Various trim and feature strips are available to customize a tile installation. They are laid by following the general procedures previously described for the regular tile, Fig. 15-28.

After all sections of the floor have been completed, cove base can be installed along the wall and around fixtures as shown in Fig. 15-29. A special adhesive is available for this operation. Cut the proper lengths and make a trial fit. Apply the adhesive to the cove base and press it into place.

Check over the completed installation carefully. Remove any spots of adhesive. Work carefully using cleaners and procedures approved by the manufacturer.

SELF-ADHERING TILES

Self-adhering tiles are easy to install. Remove the paper from the back of the tile, place the tile in position on the floor, and press it down.

It is very important that floors be dry, smooth, and completely free of wax, grease, and dirt. Generally, tiles can be laid over smooth-faced resilient floors. Embossed floors, urethane floors, or cushioned floors should be removed. Fig. 15-30 shows basic steps in the installation of a self-adhering tile floor over an existing sheet linoleum floor.

Fig. 15-28. Setting a feature strip creates a custom effect.

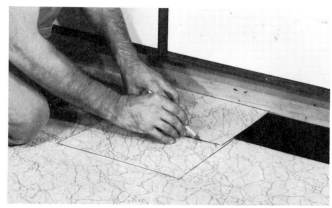

Fig. 15-27. Finishing off the borders. Top. Mark the border tile. Bottom. Cut tile with shears.

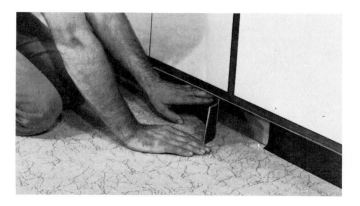

Fig. 15-29. Cove base is installed last. Apply special adhesive to the cove.

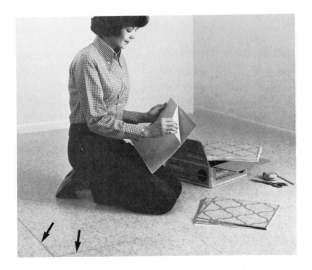

REMOVE RELEASE PAPER FROM TILE BACK. ROOM CENTERLINES (ARROW) HAVE BEEN LAID OUT.

Tiles should be kept in a warm room, at least 65 °F, for 24 hours before and during installation. The room should be kept warm for one week after installation. This will insure a firm bond to the sub-floor surface. Always study and follow the manufacturer's directions.

SHEET VINYL FLOORING

Recent developments in vinyl flooring have produced a material that is extremely flexible. This property makes installation much easier. Since sheets are available in 12 ft. widths, many installa-

LOCATE POSITION AND LAY DOWN THE TILE. PRESS FIRMLY. IF REVERSE SIDE CARRIES ARROWS, ALL SHOULD POINT IN SAME DIRECTION.

Fig. 15-31. Trim flexible vinyl sheet linoleum with a utility knife drawn along a straightedge. Be sure straightedge is parallel to wall.   (Armstrong World Industries, Inc.)

BORDER TILES CAN BE CUT WITH HOUSEHOLD SHEARS. MARK CUTTING LINE AS SHOWN IN FIG. 15-27.

Fig. 15-30. How to install self-adhering vinyl tile. (Armstrong World Industries, Inc.)

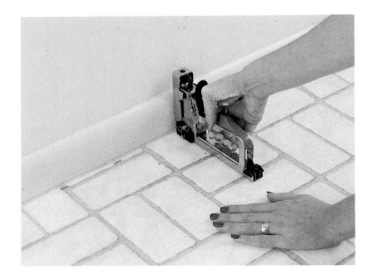

Fig. 15-32. Attach trimmed edge to floor with a staple gun. Base shoe will be used to cover the staples.

Fig. 15-33. Completed installation of a sheet vinyl floor in a kitchen-dining area. (Armstrong World Industries, Inc.)

tions can be made free of seams.

Flexible vinyl flooring is fastened down only around the edges and at seams. It can be installed over concrete, plywood, or old linoleum.

To install, spread the sheet smoothly over the floor. Let excess material turn up around the edges of the room. When there are seams, carefully match the pattern. Fasten the two sections to the floor with adhesive. Trim edges to size as shown in Fig. 15-31.

After all edges are trimmed and fitted, secure them with a staple gun as shown in Fig. 15-32, or use a band of double-faced adhesive.

Always study the manufacturer's directions carefully before starting the work. Fig. 15-33 shows a completed installation.

TEST YOUR KNOWLEDGE — UNIT 15

1. The species of wood most commonly used for finish flooring in residential structures is
_____.
2. Standard thicknesses of hardwood flooring include 3/8, _____, and 3/4 in.
3. When laying strip floors that start against a wall, lay the first course with the _____ (tongue edge, groove edge) turned toward the wall.
4. When the starter strip is located in the center of an area, install a hardwood _____ in the grooved side to permit laying in both directions.
5. How many board feet are there in a standard bundle of hardwood flooring?
6. The two types of block flooring most common-

ly used in residential installations are unit-blocks and _____ blocks.
7. Hardboard panels used for underlayment are laid with a space of _____ in. at each joint.
8. Approved metal fasteners for underlayment include ring-groove nails, divergent staples, and _____ nails.
9. A resilient flooring material that is most likely to show small irregularities in the base is
_____.
10. Before cutting asphalt tile, they should be
_____.

OUTSIDE ASSIGNMENTS

1. Study reference books located in the library or pamphlets secured from a local building supply firm. Prepare a written report describing the procedures and processes used in the manufacturing of hardwood flooring. Include kiln drying requirements and moisture content standards. Also include grading rules that are applied to the species and qualities commonly used in your geographical area.
2. Prepare an oral report on resilient flooring materials. Secure samples from a floor covering contractor or home furnishing store. Include information about thicknesses, colors, tile sizes, and approximate costs. Include a list of the advantages and disadvantages of the various types. If time permits, include information about the composition (basic materials used in manufacture) of each type and general requirements for installation.

403

Fig. 16-1. A main stairs can provide an attractive architectural feature to a residence. Its design and construction has long been considered one of the highest forms of joinery.

Unit 16
STAIR CONSTRUCTION

A stair is a series of steps leading from one level of a structure to another. When the series is a continuous section without breaks formed by landings or other constructions, the term "run of stairs" or "flight of stairs" is sometimes used. Other terms that can be properly substituted for stairs include "stairway" and "staircase."

In residential buildings, the popularity of the one-story structure has, for many years, minimized requirements for stair construction. Regular carpenters could usually handle the relatively simple task of constructing the service stairs leading from the first floor to the basement level. However, revival of traditional styling along with split-level and multi-level designs have made fine stair construction an important skill.

Main stairs are often made the chief architectural feature in an entrance hallway or other area. Its construction requires a high degree of skill. See Fig. 16-1. The quality of the work should compare with that found in fine cabinetwork.

Today, the parts for main stairways are usually made in millwork plants and then assembled on the job. Even so, the assembly work must be performed by a skillful carpenter who understands the basic principles of stair design and who knows layout and construction procedures.

Main stairways are usually not made or installed until after interior wall surfaces are complete and finish flooring or underlayment has been laid. Basement stairs should not be installed until the concrete floor has been placed.

Carpenters build temporary stairs from framing lumber to provide access until the permanent stairs are installed. These are usually designed as a detachable unit so they can be moved from one project to another.

Sometimes carriages are installed during the rough framing and temporary treads are attached. Later, as the interior is finished, these treads are replaced with finished parts.

TYPES OF STAIRS

Basically stair types are divided into service stairs and main stairs. Either of these may be closed, open, or a combination of open and closed. See Fig. 16-2. The types usually listed are straight run, plat-

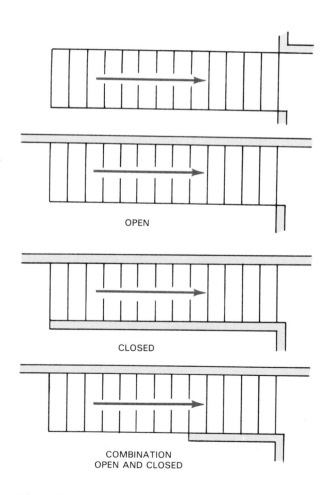

OPEN

CLOSED

COMBINATION
OPEN AND CLOSED

Fig. 16-2. Simplified drawings of open and closed stairways. Heavy colored lines represent walls. A stairs is called "open" even though one side is enclosed by a wall.

form, and winding. The platform type includes landings where the direction of the stair runs is usually changed. Such descriptive terms as L type (long L and wide L), double L type and U type are commonly used. See Fig. 16-3.

The straight run stairway is continuous from one floor level to another without landings or turns. It is the easiest to build. Standard multi-story designs require a long stairwell. This often presents a problem in smaller structures. A long run of 12 to 16 steps also has the disadvantage of being tiring. It offers no chance for a rest during ascent.

In modern split-level designs, the runs are short and generally straight. They connect directly to the next floor level. Usually stair runs of this type are located so that headroom is automatically provided by the stair run directly above, Fig. 16-4.

Winding stairs, also called "geometrical," are circular or elliptical. They gradually change directions as they ascend from one level to another. These often require curved wall surfaces that are difficult to build. Because of the expense, such stairs are seldom used except in prestigious homes.

PARTS AND TERMS

Stairs are basically sets of risers and treads supported by stringers. The height of the riser is called the unit rise and the width of tread, nosing excluded, is called the unit run. The sum of all the risers is the total rise and the sum of all the tread is the total run, Fig. 16-5.

Vertical space above the stair is called headroom. It is measured from a line along the front edges of

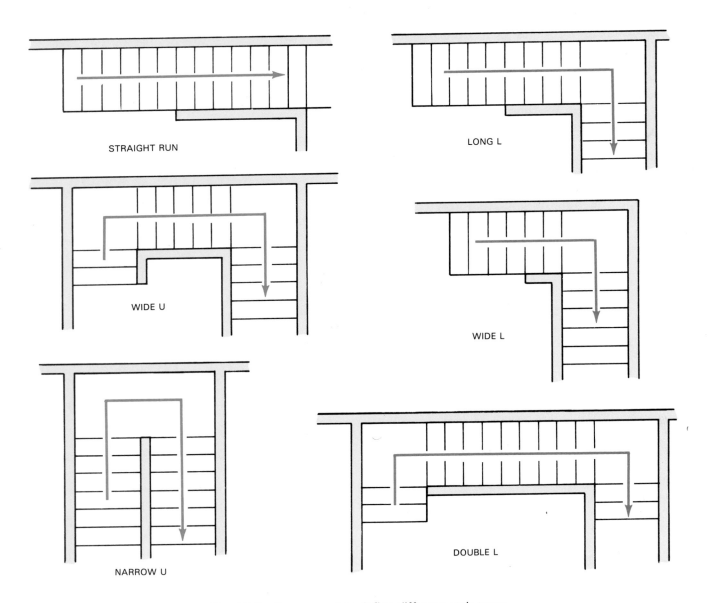

Fig. 16-3. Terms used to define different stair types.

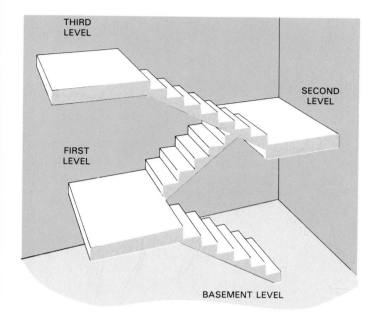

Fig. 16-4. Stair runs are often made one above the other to get headroom. This one is designed for a split-level home.

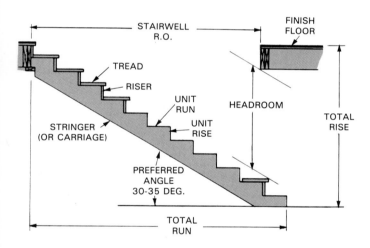

Fig. 16-5. Basic stair parts and terms. Total number of risers is always one greater than the total number of treads.

the tread to the enclosed surface or header above. This distance is usually specified in local building codes. FHA requires a minimum headroom of 6 ft.-8 in. for main stairs and 6 ft.-4 in. for basement or service stairs.

STAIRWELL FRAMING

Methods of stair building differ from one locality to another. One carpenter may cut and install a carriage (stringers) during the wall and floor framing, Fig. 16-6. Another may put off all stairwork until

the interior finishing stages.

Regardless of procedures followed, the rough openings for the stairwell must be carefully laid out and constructed. If the architectural drawings do not include dimensions and details of the stair installation then the carpenter will need to calculate the sizes. She or he must follow recognized standards and local code restrictions.

Trimmers and headers in the rough framing should be doubled, especially when the span is greater than 4 ft. Headers more than 6 ft. long should be installed with framing anchors unless supported by a beam, post, or partition. Tail joists over 12 ft. long should also be supported by framing anchors or a ledger strip. See Unit 7 for additional information on framing rough openings.

Providing adequate headroom is often a problem, especially in smaller structures. Installing an auxiliary header close to the main header, Fig. 16-7, will permit a slight extension in the floor area above a stairway. When a closet is located directly above, the closet floor is sometimes raised for additional headroom.

STAIR DESIGN

Most important in stair design is the mathematical relationship between the riser and tread. There are three generally accepted rules for calculating the rise-run or riser-tread ratio as follows. It is wise to observe them:

1. The sum of two risers and one tread should be 24 to 25 in.
2. The sum of one riser and one tread should equal 17 to 18 in.
3. The height of the riser times the width of the tread should equal between 70 and 75 in.

A riser 7 1/2 in. high would, according to Rule 1, require a tread of 10 in. A 6 1/2 in. riser would require a 12 in. tread.

In residential structures, treads (excluding nosing) are seldom less than 9 in. or more than 12 in. wide. In a given run of stairs it is extremely important that all of the treads and all risers be the same size. A person tends to measure (subconciously) the first few risers and will probably trip on subsequent risers that are not the same.

When the rise-run combination is wrong, the stair will be tiring and cause extra strain on the leg muscles. Further, the toe may kick the riser if the tread is too narrow.

A unit rise of 7 to 7 5/8 in. high with an appropriate tread width will combine both comfort and safety. Main or principal stairs are usually planned to have a rise in this range. Service stairs are often steeper but risers should be no higher than 8 in. As stair rise is increased the run must be decreased. See Fig. 16-8.

STAIR CARRIAGE HAS THREE STRINGERS CUT AND INSTALLED DURING ROUGH FRAMING. THE 2 × 4 SPACER (ARROW) PLACED ON EACH SIDE PROVIDES CLEARANCE FOR THE APPLICATION OF WALL FINISH.

NEWEL POST HAS BEEN INSTALLED BY CUTTING THROUGH THE SUBFLOOR AND ANCHORING THE BOTTOM END SECURELY TO THE FLOOR JOISTS. CLOSED STRINGER (ARROW) IS IN PLACE. OPEN STRINGER AT LEFT IS ATTACHED TO SUPPORTING STRINGER WITH BLOCKING. PLYWOOD UNDERLAYMENT HAS BEEN INSTALLED ON THE FLOOR.

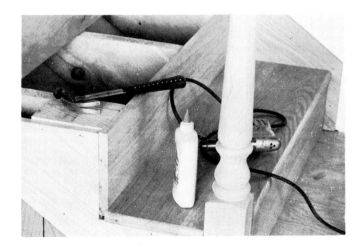

FIRST TREAD INSTALLED. SINCE THE STAIR PARTS, PURCHASED FROM A MILLWORK PLANT, ARE OAK, IT IS NECESSARY TO DRILL NAIL HOLES. GLUE IS ALSO APPLIED TO EACH JOINT. IN ADDITION TO FACE NAILING OF RISERS AND TREADS, SEVERAL NAILS ARE DRIVEN THROUGH THE BACK LOWER EDGE OF THE RISER INTO THE TREAD.

GYPSUM WALLBOARD AND VENEER PLASTER HAVE BEEN APPLIED TO WALL. THE BOTTOM SIDE OF THE STAIR HAS BEEN SEALED WITH A HEAVY SHEET OF GYPSUM BOARD ACCORDING TO LOCAL CODE REQUIREMENTS.

INSTALLING THE FIRST TWO RISERS. TOP EDGE OF RISER MUST BE PERFECTLY ALIGNED WITH SUPPORTING MEMBERS SO NO CRACK WILL SHOW WHEN TREADS ARE INSTALLED.

BALUSTRADE ASSEMBLY. HANDRAIL IS TEMPORARILY INSTALLED TO LAY OUT POSITION OF BALUSTERS BETWEEN THE TREAD AND RAIL. HOLES FOR THE BALUSTERS ARE BORED IN THE HANDRAIL AND TREAD. AFTER FINAL ASSEMBLY OF THE BALUSTRADE, TRIM PIECES, CALLED BRACKETS, (SEE FIG. 16-25) ARE INSTALLED ALONG THE OPEN STRINGER TO COVER END GRAIN OF RISERS.

Fig. 16-6. This is typical of the sequence to follow in constructing a main stair.

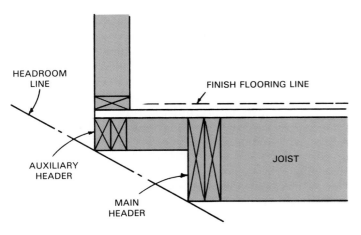

HEADROOM
LINE

FINISH FLOORING LINE

AUXILIARY
HEADER

JOIST

MAIN
HEADER

Fig. 16-7. Extending upper floor area with a shallow auxiliary header. Partition over the auxiliary must be nonsupporting.

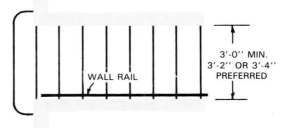

WALL RAIL

3'-0'' MIN.
3'-2'' OR 3'-4''
PREFERRED

Fig. 16-9. A main stair should be at least 3 ft. wide for easy movement of people and furniture.

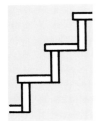

In a given run of stairs, be sure to make all of the risers the same height and all of the treads the same width. An unequal riser, especially one that is too high, may cause a fall.

A main stair should be wide enough to allow two people to pass without contact. Further, it should provide space so furniture can be moved up or down. A minimum width of 3 ft. is generally recommended, Fig. 16-9. FHA permits a minimum of width, measured clear of the handrail, of 2 ft.-8 in. On service stairs, the requirement is reduced to 2 ft.-6 in. Furniture moving is an important consideration and extra clearance should be provided in closed stairs of the L and U type; especially those that include winders, Fig. 16-10.

Stairs should have a continuous rail along the side for safety and convenience. A handrail (also called a stair rail) is used on open stairways that are constructed with a low partition or banister.

In closed stairs the support rail is called a wall rail. It is attached to the wall with special metal brackets. Except for very wide stairs, a rail on only one side is sufficient. Fig. 16-11 illustrates the correct height for a rail.

A complete set of architectural plan should include detail drawings of main stairs, especially when the design includes any unusual features. For example, the stair layout in Fig. 16-12 shows a split-level entrance with open-riser stairs leading to upper and lower floors. An exact description of tread mountings, overlap or nosing requirements, and

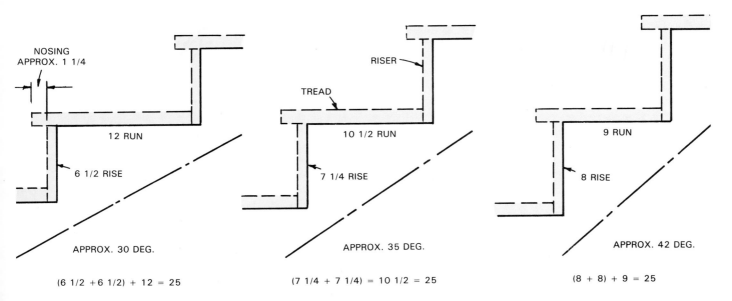

NOSING
APPROX. 1 1/4

12 RUN

6 1/2 RISE

APPROX. 30 DEG.

(6 1/2 + 6 1/2) + 12 = 25

RISER

TREAD

10 1/2 RUN

7 1/4 RISE

APPROX. 35 DEG.

(7 1/4 + 7 1/4) = 10 1/2 = 25

9 RUN

8 RISE

APPROX. 42 DEG.

(8 + 8) + 9 = 25

Fig. 16-8. Be careful about rise-run relationships in stair design.

Fig. 16-10. This L-shaped stairs is spacious enough for moving furniture up and down. (C.E. Morgan)

STAIR CALCULATIONS

To calculate the number and size of risers and treads (less nosing) for a given stair run, first divide the total rise by 7. (Some divide by 8. Either number is accurate enough.) For example: if the total rise for a basement stairway is 7'-10'' or 94'', the answer will be 13.43. Since there must be a whole number of risers, select the one closest to 13.43 and divide it into the total rise:

$$94'' \div 13 = 7.23 \text{ or } 7\ 1/4''$$
$$\text{Number of risers} = 13$$
$$\text{Riser height} = 7\ 1/4''$$

In any stair run the number of treads will be one less than the number of risers. A 10 1/2 in. tread will be correct for the example and the total run would be calculated as follows:

$$\text{Number of treads} = 12$$
$$\text{Total run} = 10\ 1/2'' \times 12$$
$$= 126''$$
$$= 10'\text{-}6''$$

The stairs in the example will have 13 risers 7 1/4 in. high, 12 treads 10 1/2 in. wide and a total run of 10 ft.-6 in.

Since the example was assumed to be a basement stairs, the total run could be shortened by using a steeper angle. Decrease the number or risers and shorten the treads. The calculations are:

$$94'' \div 12 = 7.83 = 7\ 5/6''$$
$$\text{Number of risers} = 12$$
$$\text{Height of risers} = 7\ 5/6''$$
$$\text{Tread width selected} = 9''$$
$$\text{Number of treads} = 11$$
$$\text{Total run} = 9'' \times 11$$
$$= 99''$$
$$= 8'\text{-}3''$$

Some manufacturers supply tables for arriving at rise and run, riser and tread ratios. See Fig. 16-13.

STAIRWELL LENGTH

The length of the stairwell opening must be known during the rough framing operations. If not included in the architectural drawings it can be calculated from the size of the risers and treads.

It is necessary to know the headroom required. Add to this the thickness of the floor structure and divide this total vertical distance by the riser height. This will give the number of risers in the opening.

When counting down from the top to the tread from which the headroom is measured, there will be the same number of treads as risers. Therefore

height of the handrail is not included. These items of construction become the responsibility of the carpenter who must have a thorough understanding of basic stair design and how to lay out and make the installation.

All stairs, whether main or service, will be shown on floor plans. When details of the stair design are not included in the complete set of plans, the architect will usually specify on the plan view the number and width of the tread for each stair run. Sometimes the number of risers and the riser height is also included.

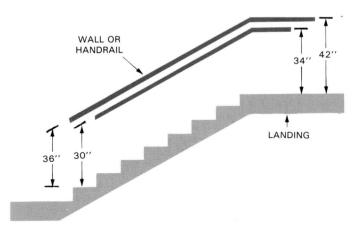

WALL OR HANDRAIL

34'' 42''

LANDING

36'' 30''

Fig. 16-11. A handrail height of 30 in. at rake (slope) and 34 in. at landings have been an accepted standard. Recently building codes in some places have been adopting heights of 36 in. at the rake and 42 in. at landings. (C.E. Morgan)

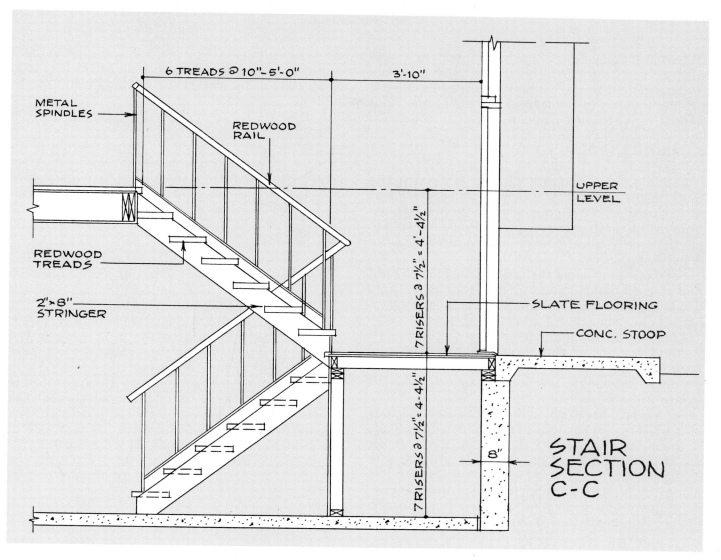

Fig. 16-12. Architectural drawings will show stair layouts like this. Note information given for riser-tread ratios.

Total Rise Floor to Floor H	Number of Risers	Height of Riser R	Number of Treads	Width of Run T	Total Run L	Well Opening U	Length of Carriage	Use Stock Tread Width	Dimension of Nosing Projection
WELL OPENINGS BASED ON MIN. HEAD HGT. OF 6'-8" **DIMENSIONS BASED ON 2" x 10" FLOOR JOIST**									
8'-0"	12	8"	11	9-1/2"	8'-8 1/2"	9'-1"	11'-4 5/8"	10-1/2"	1"
	14	6-7/8"	13	10-5/8"	11'-6-1/8"	10'-10"	13'-8-1/2"	11-1/2"	0-7/8"
8'-4"	13	7-11/16"	12	9-13/16"	9'-9-3/4"	10'-0"	12'-5-1/2"	10-1/2"	11/16"
	14	7-1/8"	13	10-3/8"	11'-2-7/8"	11'-0"	13'-7-5/8"	11-1/2"	1-1/8"
8'-6"	13	7-7/8"	12	9-5/8"	9'-7-1/2"	9'-2"	12'-5-1/4"	10-1/2"	0-7/8"
	14	7-5/16"	13	10-3/16"	11'-0-1/2"	10'-8"	13'-7"	11-1/2"	1-5/16"
8'-9"	14	7-1/2"	13	9-1/4"	10'-0-1/4"	9'-5"	12'-10-3/4"	10-1/2"	1-1/4"
	14	7-1/2"	13	10"	10'-10"	10'-1"	13'-6-1/2"	11-1/2"	1-1/2"
8'-11"	14	7-5/8"	13	9-3/8"	10'-1-7/8"	9'-5"	13'-1-1/4"	10-1/2"	1-1/8"
	14	7-5/8"	13	9-1/16"	9'-9-7/8"	9'-0"	12'-10"	10-1/2"	1-7/16"
	14	7-5/8"	13	10-1/4"	11'-1-1/4"	10'-2"	13'-10-1/4"	11-1/2"	1-1/4"
9'-1"	14	7-13/16"	13	9-11/16"	10'-6"	9'-5"	13'-5-3/4"	10-1/2"	13/16"
	15	7-1/4"	14	10-1/4"	11'-11-1/2"	10'-8"	14'-7-3/4"	11-1/2"	1-1/4"

Fig. 16-13. A chart such as this can be used to determine number of risers and treads and their dimensions. (C.E. Morgan)

to find the total length of the rough opening, multiply the tread width by the number of risers previously determined. Some carpenters prefer to make a scaled drawing (elevation) of the stairs and floor section to check the calculations.

STRINGER LAYOUT

To lay out the stair stringer, first determine the riser height. Place a story pole (straight strip of 1 × 4 lumber) in a plumb position from the finished floor below through the rough stair opening above. On the pole, mark the height of the top of the finished floor above.

Set a pair of dividers to the calculated riser height and step off the distances on the story pole. There will likely be a slight error in the first layout so adjust the setting and try again.

Continue adjusting the dividers and stepping off the distance on the story pole until the last space is equal to all the others. Measure the setting of the dividers. This length will be the exact riser height to use in laying out the stringers.

For a simple basement stair, select a straight piece of 2 x 10 or 2 x 12 stock of sufficient length. Place it on sawhorses to make the layout. Begin at the end that will be the top and hold the framing square in the position shown in Fig. 16-14. Draw a line along the outside edge of the blade and tongue. Now move the square to the next position and repeat. The procedure is similar to that described

for rafter layout in Unit 9. Accuracy can be assured in this layout by using framing square clips (Fig. 9-9, page 201) or by clamping a strip of wood to the blade and tongue.

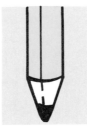

Extreme accuracy is required in laying out the stringer. Be sure to use a sharp pencil or knife and make the lines meet on the edge of the stock.

Continue stepping off with the square until the required number of risers and unit treads have been drawn, Fig. 16-15.

The stair begins with a riser at the bottom, so extend the last tread line to the back edge of the stringer as shown. At the top, extend the last tread and riser line to the back edge.

One other adjustment must be made before the stringer is cut. Earlier calculations which gave the height of the riser did not take into account the thickness of the tread. Therefore the total rise of the stringer must be shortened by one tread thickness. Otherwise the top tread will be too high. The bottom of the stringer must be trimmed as shown in Fig. 16-16.

TREADS AND RISERS

The thickness of a main stair tread is generally 1 1/6 or 1 1/8 in. Hardwood or softwood may be used. FHA requires that stair treads be hardwood, vertical grain softwood, or flat-grain softwood

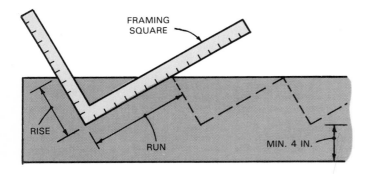

Fig. 16-14. Use a framing square to lay out a stringer. Method is almost identical to laying out rafters.

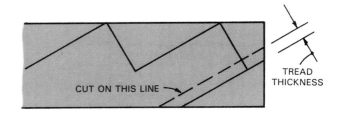

Fig. 16-16. Trim the bottom end of the stringer to adjust for tread thickness.

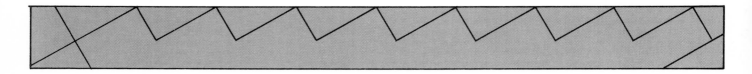

Fig. 16-15. Completed stringer layout will look something like this.

covered with a suitable finish flooring material.

Lumber for risers is usually 3/4 in. thick and should match the tread material. This is especially important when the stairs are not covered. In most construction, the riser drops behind the tread, making it possible to reinforce the joint with nails or screws driven from the back side of the stairs.

Where the top edge of the riser meets the tread, glue blocks are sometimes used. A rabbeted edge of the riser may fit into a groove in the tread. A rabbet and groove joint may also be used where the back edge of the tread meets the riser.

Stair treads must have a nosing. This is the part of the tread that overhangs the riser. Nosings serve the same purpose as toe space along the floor line of kitchen cabinets. They provide toe room.

The width of the tread nosing may vary from about 1 1/8 to 1 1/2 in. It should seldom be greater than 1 3/4 in.

In general, as the tread width is increased, the nosing can be decreased. Fig. 16-17 illustrates a number of nosing forms. Cove molding may be used to cover the joint between riser and tread as well as conceal nails used to attach the riser to the stringer or carriage.

Basement stairs are often constructed with an open riser (no riser board installed). Sometimes an open riser design is built into a main stair to provide a special effect. Various methods of support or suspension may be used. Often custom-made metal brackets or other devices are needed. Fig. 16-18 shown basic types of riser designs. A slanted riser is sometimes used in concrete steps since it provides an easy way to form a nosing.

TYPES OF STRINGERS

Treads and risers are supported by stringers or carriages that are solidly fixed to the wall or framework of the building. For wide stairs, a third stringer is installed in the middle to add support.

The simplest type of stringer is formed by attaching cleats on which the tread can rest. Another method consists of cutting dados into which the tread will fit, Fig. 16-19. This type is often used for basement stairs where no riser enclosure is called for.

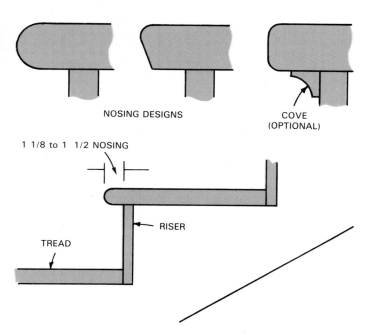

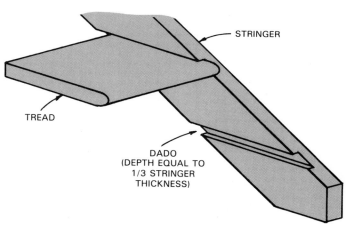

Fig. 16-17. Tread nosings commonly take these shapes.

Fig. 16-19. Open riser stairs. Treads are set into dados cut in the stringer.

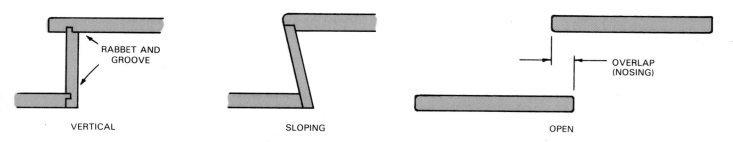

Fig. 16-18. Basic stair riser shapes. For the open riser, the tread should overlap the riser at least 2 in.

Standard cutout stringers (type used in layout description) are commonly used for either main or service stairs. Prefabricated treads and risers are often used for this type of support. An adaptation of the cutout stringer, called semihoused construction, is illustrated in Fig. 16-20. The cutout stringer and backing stringer may be assembled and then installed as a unit or each part may be installed separately.

A popular type of stair construction has a stringer with tapered grooves into which the treads and risers fit. It is commonly called housed construction. Wedges, with an application of glue, are driven into the grooves under the tread and behind the riser, Fig. 16-21. The treads and risers are joined with rabbeted edges and grooves or glue blocks.

This type of construction produces a stair that is strong and dust tight. It will seldom develop squeaks.

Housed stringers can be purchased completely cut and ready to install. They can be cut on the job, using an electric router and template.

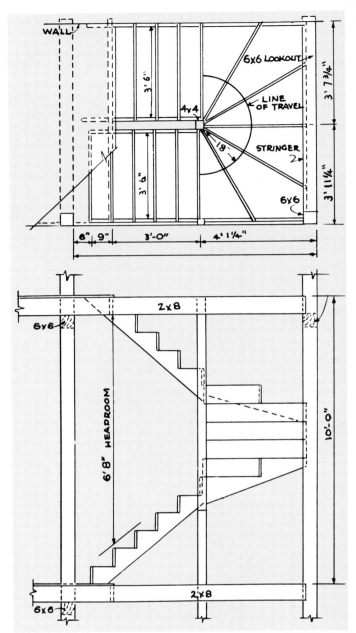

Fig. 16-22. Typical drawing of a winder stairs. Tread width on winding section should be the same at line of travel (near middle of stairs) as tread in straight run.

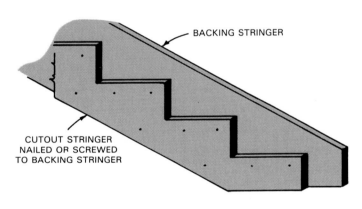

Fig. 16-20. This is a semihoused stringer.

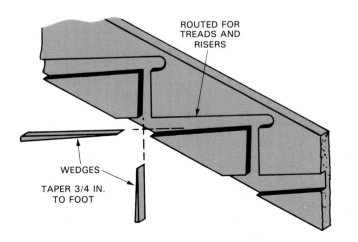

Fig. 16-21. In a housed stringer, risers and treads are let into the stringer. This type of housing is difficult to make.

To assemble the stairs, the housed stringer is spiked to the wall surface and into the wall frame. The treads and risers are then set into place. Work from the top downward, using wedges and glue. This type of stringer shows above the profiles of the treads and risers and provides a finish strip along the wall. The design should permit a smooth joint where it meets the baseboard of the upper and lower level.

WINDER STAIRS

Winder stairs, Fig. 16-22, present stair conditions that are frequently regarded as undesirable. Their use, however, may sometimes be necessary where space is limited.

It is important to maintain a winder-tread width along the line of travel that is equal to the tread width in the straight run.

An adaptation of the standard winder layout is illustrated in Fig. 16-23. Here, if you extend the lines of the riser, they meet outside the stairs. This provides some tread width at the inside corner. Before starting the construction of this type of stairs, the carpenter should make a full-size or carefully scaled layout (plan view). The best radius for the line of travel can then be determined.

OPEN STAIRS

Main stairs that are open on one or both sides require some type of decorative enclosure and support for a handrail. Typical designs consist of an assembly of parts called a balustrade, Fig. 16-24. The principal members of a balustrade are newels, balusters, and rails. They are usually manufactured and assembled on the job by the carpenter.

The starting newel must be securely anchored either to the starter step or carried down through the floor and attached to the building frame. Balusters are joined to the stair treads using either a round or square mortise. Two or three may be mounted on each tread.

USING STOCK STAIR PARTS

While many parts of a main staircase could be cut and shaped on the job, the usual practice is to use factory-made parts. These are available in a wide range of stock sizes and can be selected to fill requirements for most standard stair designs. See Figs. 16-25 and 16-26. Stair parts are ordered through lumber and millwork dealers and are shipped to the building site in heavy, protective cartons along with directions for fitting and assembly.

A completely prefabricated stairway is shown in Fig. 16-27. Stringers are made in two sections for easier shipping. The system is available in lengths up to 18 steps and widths of 36 and 48 in.

Fig. 16-28 shows some suggested assemblies of balustrades, using stock parts. Hardware, especially designed for stairwork, is illustrated in Fig. 16-29.

SPIRAL STAIRWAYS

Where space is limited, metal spiral stairways eliminate framing. Units are available in aluminum or steel in a variety of designs to fit requirements up to 30 steps and heights up to 22 1/2 ft. The Uniform Building Code permits use of the spiral stairway for exits in private dwellings or in some other

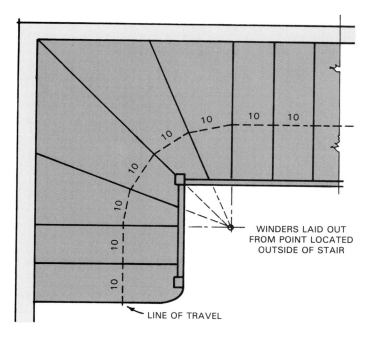

Fig. 16-23. Laying out a winder stairs with lines representing the tread nosings converging outside the construction. When winders must be used it is best to place them near the bottom of the run.

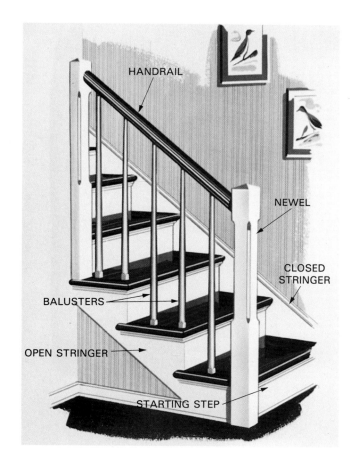

Fig. 16-24. Parts of an open stair. An assembly including a newel, balusters, and rail is called a balustrade. (C.E. Morgan)

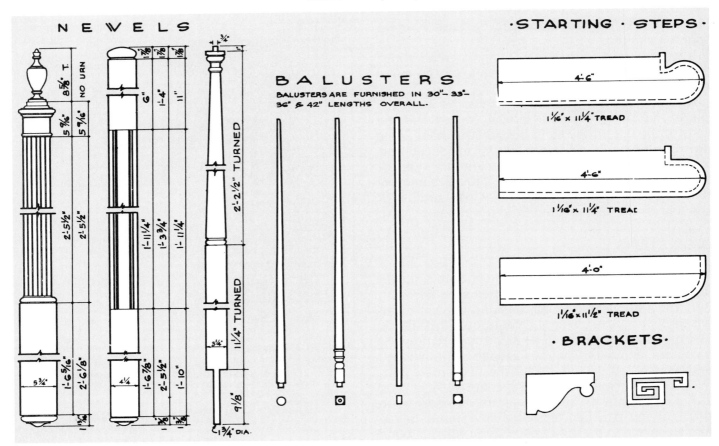

Fig. 16-25. Typical stock parts above are commonly available for stair construction.

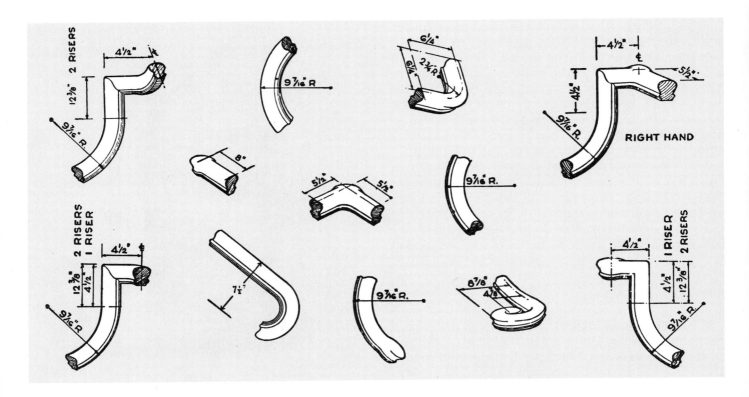

Fig. 16-26. Preformed handrails and stock parts for special shapes can be purchased.   (C.E. Morgan)

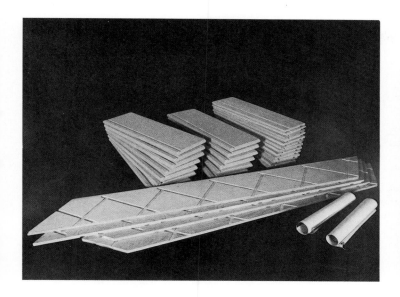

Fig. 16-27. Left. Parts for prefabricated stairway system. Sections of stringers lock together with a common tread. Treads and risers fit into dovetails in the stringers and lock together in grooves. Right. Installed mock-up shows assembled section mounted on substringers. (Visador Co.)

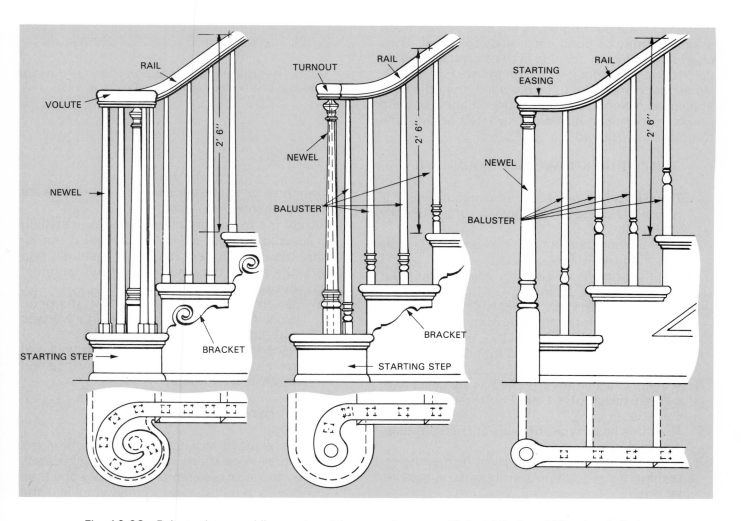

Fig. 16-28. Balustrade assemblies produced from stock parts. (Colonial Stair and Woodwork Co.)

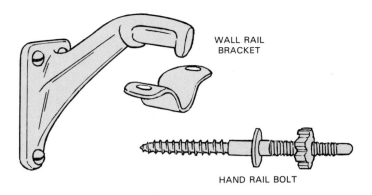

WALL RAIL
BRACKET

HAND RAIL BOLT

Fig. 16-29. Hardware for rails. Rail bolt is concealed in the center of a joint. Nut is engaged and tightened through a hole bored in the underside.

Fig. 16-30. Prefabricated metal stairway can be installed quickly in a finished opening. (Columns, Inc.)

situations when the area served is not more than 400 sq. ft. See Figs. 16-30 and 16-31.

DISAPPEARING STAIR UNITS

Where attics are used primarily for storage and where space for a fixed stairway is not available, hinged or disappearing stairs are often used. Such stairways may be purchased ready to install. They operate through an opening in the ceiling and swing up into the attic space when not in use, Fig. 16-32. Where such stairs are to be provided, the attic floor should be designed for regular floor loading and the rough opening should be constructed at the time the ceiling is framed.

TEST YOUR KNOWLEDGE – UNIT 16

1. The platform type of stairway includes _____ where the direction of the stair runs is usually changed.
2. The minimum headroom for a main stairway as specified by FHA is _____.
3. The riser height of a service stairs should not exceed _____ in.
4. One of the rules used to calculate riser-tread relationship states that the sum of two risers and one tread should be _____.
5. The front edge of the tread that overhangs the riser is called the _____.
6. A stairs in a split-level home has six risers with a tread width of 11 in. The total run of the stairs is _____.
7. The three basic types of risers are vertical, sloping, and _____.
8. A semi-housed stair stringer is formed by attaching a _____ stringer to a backing stringer.
9. Wedges used to assemble risers and treads in housed stringers should have a taper of _____ in. to the foot.
10. The three principal members of a balustrade are called newels, rails, and _____.
11. Study the following instructions for installing a housed stringer stairs. Determine if the instructions are correct for proper assembly and attachment of the stair parts. If not, suggest correct procedure.
    a. Set treads and risers into the stringers. Work from the bottom up using wedges and glue.
    b. Place the assembly into the stair well and spike the stringers to the wall surface and into the wall frame.

OUTSIDE ASSIGNMENTS

1. Secure a set of architectural plans where the main or service stairway is not drawn in detail. Study the stair requirements carefully and then prepare a detail drawing somewhat like the diagram in Fig. 16-5 or the drawing in Fig. 16-12. Use a scale of 1/2 in. equals 1 ft. Select

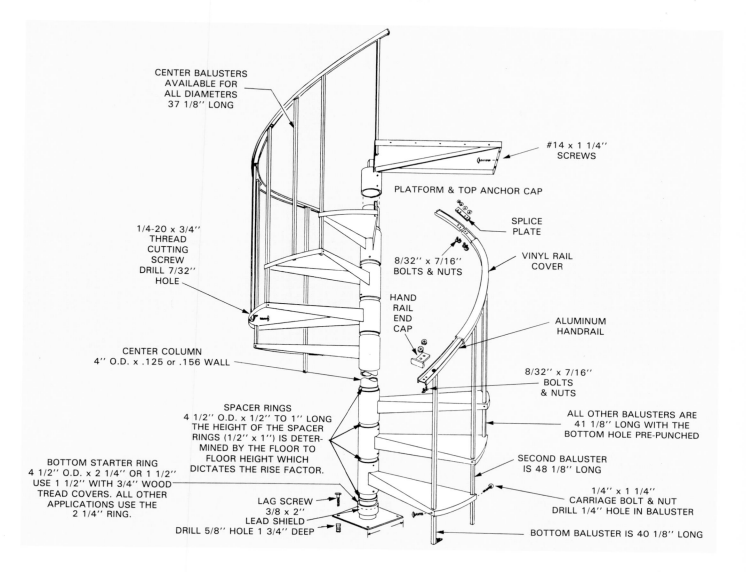

CENTER BALUSTERS
AVAILABLE FOR
ALL DIAMETERS
37 1/8'' LONG

#14 x 1 1/4''
SCREWS

PLATFORM & TOP ANCHOR CAP

SPLICE
PLATE

1/4-20 x 3/4''
THREAD
CUTTING
SCREW
DRILL 7/32''
HOLE

8/32'' x 7/16''
BOLTS & NUTS

VINYL RAIL
COVER

HAND
RAIL
END
CAP

ALUMINUM
HANDRAIL

CENTER COLUMN
4'' O.D. x .125 or .156 WALL

8/32'' x 7/16''
BOLTS
& NUTS

SPACER RINGS
4 1/2'' O.D. x 1/2'' TO 1'' LONG
THE HEIGHT OF THE SPACER
RINGS (1/2'' x 1'') IS DETER-
MINED BY THE FLOOR TO
FLOOR HEIGHT WHICH
DICTATES THE RISE FACTOR.

ALL OTHER BALUSTERS ARE
41 1/8'' LONG WITH THE
BOTTOM HOLE PRE-PUNCHED

SECOND BALUSTER
IS 48 1/8'' LONG

BOTTOM STARTER RING
4 1/2'' O.D. x 2 1/4'' OR 1 1/2''
USE 1 1/2'' WITH 3/4'' WOOD
TREAD COVERS. ALL OTHER
APPLICATIONS USE THE
2 1/4'' RING.

LAG SCREW
3/8 x 2''
LEAD SHIELD
DRILL 5/8'' HOLE 1 3/4'' DEEP

1/4'' x 1 1/4''
CARRIAGE BOLT & NUT
DRILL 1/4'' HOLE IN BALUSTER

BOTTOM BALUSTER IS 40 1/8'' LONG

Fig. 16-31. Assembly drawing of spiral stairway.

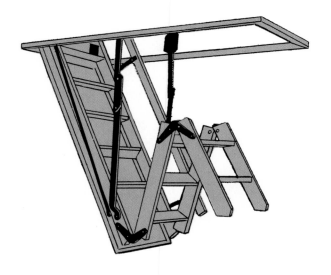

Fig. 16-32. Disappearing stair unit is designed to fold into the ceiling. (Rock Island Millwork)

and calculate the riser-tread ratio carefully and be sure the number and size of risers is correct for the distance between the two levels. Check the headroom requirements against your local building codes and determine the stairwell sizes. Submit the completed drawing and size specifications to your instructor.

2. Study a millwork catalog and become familiar with the stock parts shown for a main stairway. Working from a set of architectural plans or a stair detail that you may have drawn, prepare a list of all the stair parts you would need to construct the stairway. Include the number of each part needed and also its size, quality, kind of wood, and catalog number. Take your list to a building supply dealer and secure a cost estimate for the materials. Be prepared to discuss the materials and costs with your instructor and the class.

Today's housing economics requires better use of space in both new and remodeled homes. The folding doors shown above enclose wardrobes without the loss of space required by regular swinging doors.   (Rolscreen Co.)

Standard interior panel type passage door. In good residential planning, passage doors are located so they will swing open (90 deg.) against another wall as shown. (C.E. Morgan)

Replacing an exterior door with a modern prehung unit. Replacement unit has a steel door and frame. It is sized so it will fit into the existing door frame. See Fig. 23-17. (General Products Co., Inc.)

Unit 17

DOORS AND INTERIOR TRIM

This unit will deal with the methods and materials of an important part of interior finish. It will include:
1. Installing door frames.
2. Hanging doors.
3. Fitting trim around openings.
4. Fitting trim at intersections of walls, floors, and ceilings.

This aspect of carpentry requires great skill and accuracy. Well-fitted trim greatly enhances the appearance and desirability of the home.

MOLDINGS

Moldings are decorative wood or plastic strips. They are designed to provide essential functions as well as provide decoration. For example, window and door casings cover the space between the jamb and the wall covering. They also make the installation more rigid.

A wide range of types, patterns, and sizes of moldings are used in modern homes. See Fig. 17-1. Common shapes and where to use them are shown in Fig. 17-2 and Fig. 17-3. In addition to those shown, a complete list would also include: cove molding, brick molding, battens, glass beads, drip caps, picture molding, and screen mold. Information on molding patterns along with a numbering system and grading rules is included in a manual which is available from Western Wood Products Association, Portland, Oregon.

Fig. 17-1. A few of the many moldings used in a modern dwelling. The various shapes produce shadows and highlights. (E.L. Bruce Co.)

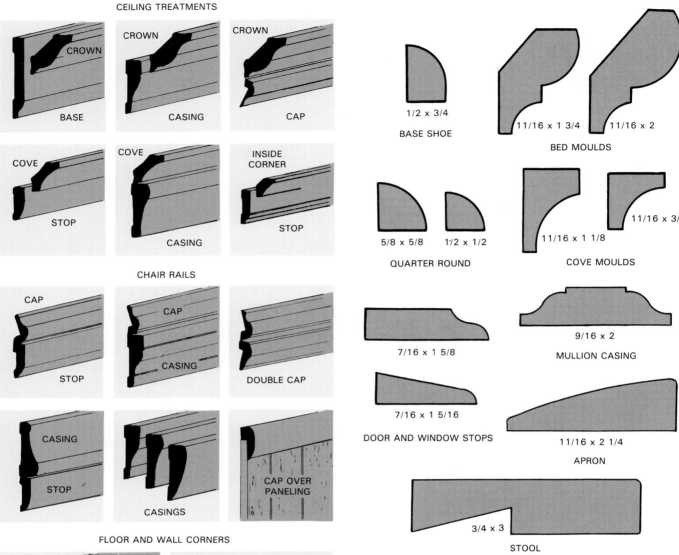

CEILING TREATMENTS

CROWN / BASE

CROWN / CASING

CROWN / CAP

COVE / STOP

COVE / CASING

INSIDE CORNER / STOP

CHAIR RAILS

CAP / STOP

CAP / CASING

DOUBLE CAP

CASING / STOP

CASINGS

CAP OVER PANELING

FLOOR AND WALL CORNERS

BASES

OUTSIDE CORNERS / INSIDE CORNER

Fig. 17-2. Moldings perform various functions depending upon where they are used. (Abitibi-Price Corp.)

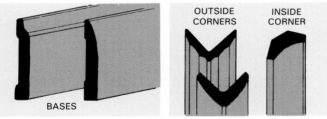

1/2 x 3/4
BASE SHOE

11/16 x 1 3/4   11/16 x 2
BED MOULDS

5/8 x 5/8   1/2 x 1/2
QUARTER ROUND

11/16 x 1 1/8   11/16 x 3/4
COVE MOULDS

7/16 x 1 5/8

9/16 x 2
MULLION CASING

7/16 x 1 5/16
DOOR AND WINDOW STOPS

11/16 x 2 1/4
APRON

3/4 x 3
STOOL

Fig. 17-3. Other typical molding patterns.

INTERIOR DOORFRAMES

The doorframe forms the lining of the door opening. It also covers the edges of the partition.

The frame consists of two side jambs and a head jamb, Fig. 17-4. Interior frames are simpler than exterior frames. The jambs are not rabbeted and no sill is included. See Unit 11.

Standard jambs for regular 2 x 4 stud partitions are made from nominal 1 in. material. For walls of plaster, the jambs are 5 1/4 in. wide. For drywall, the jambs are 4 1/2 in. wide. The back side is usually kerfed to reduce the tendency toward cupping (warping). The edge of the jamb is beveled slightly so the casing will fit snugly against it with no visible crack.

Side jambs are dadoed to receive the head jamb. The side jambs for residential doorways are made 6'-9'' long (measured to the head jamb). This provides clearance at the bottom of the door for flooring materials.

Interior doorjambs are sometimes made to be adjustable. They are designed to fit walls of different thicknesses. One, the three-piece type, depends upon a rabbet joint and a concealing doorstop. A second type is made in two pieces. See Fig. 17-5.

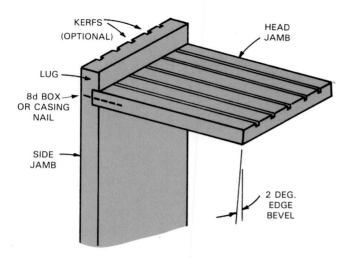

Fig. 17-4. Section of interior doorframe. Parts are listed. Note that edges of all jambs are beveled slightly so trim will fit snugly to it.

Modern doorframes are usually cut, sanded, and fitted in millwork plants. This allows quick assembly on the job. Doorframes should receive the same care in storage and handling as other finished woodwork.

INSTALLING DOORFRAMES

Before assembling the doorframes, check the length of the head and side jamb. Determine if they are correct for the opening. Nail the jambs together using 8d casing or box nails. If there is not enough vertical clearance in the rough opening, trim away part of the lug.

Place the frame in the opening. Let the side jambs rest on the finish flooring or on spacer blocks of the right thickness. (This spacer is needed only if the final flooring surface has not been laid.) Level the head jamb. Trim the bottom of the side jamb that is high.

Place a 1 x 6 spreader between the side jambs at the floor level, Fig. 17-6. Its length should equal

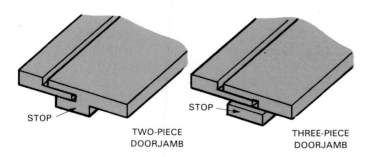

Fig. 17-5. Adjustable doorjambs will fit any thickness of wall. Stop will conceal joint.

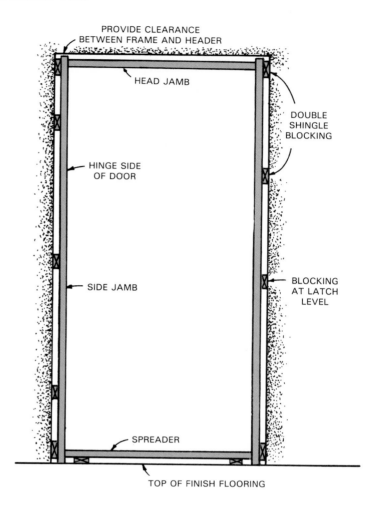

Fig. 17-6. Doorframes must be sent into rough opening using spreader and frequent shingle blocking.

the horizontal distance between the side jambs measured at the top.

On each side jamb, draw a light pencil line in from the edge of the door side, a distance equal to the door thickness plus 7/8 in. All nailing of the jambs is done along this line. Later it will be covered by the doorstop, Fig. 17-7.

Center the frame in the opening. Secure it with double-shingle wedges at the top and bottom on each side. Plumb the jambs with a straightedge and level, or a long carpenter's level. Make adjustments in the double-shingle blocking until each side is correct. Then fasten the top and bottom of each side jamb with an 8d casing nail.

Complete the blocking by placing more double-shingle wedges back of each jamb as illustrated in Fig. 17-6. On the hinge jamb, locate one block 11 in. up from the bottom and one 7 in. down from the top. Set a third block halfway between these two. Continue to check the jamb with a straightedge while adjusting the wedges; then nail through the

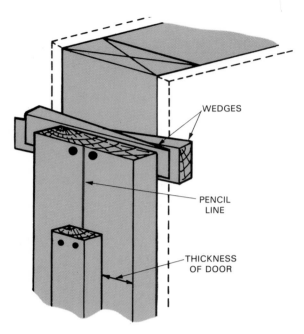

Fig. 17-7. Draw light pencil line on doorjamb measuring in the thickness of the door plus half the width of the doorstop. This provides a guide for concealed nailing of the jamb.

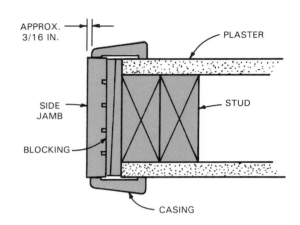

Fig. 17-8. Section shows position of casing which is attached to the doorjamb.

blocking into the studs. Generally, it is best to use two 8d nails, and stagger them about 1/2 in. on either side of the nailing line.

When setting a doorframe, do not drive any of the nails ''home'' until all blocking has been adjusted and the jambs are straight and plumb.

DOOR CASING

Door casing is applied to each side of the door-frame to cover the space between the jambs and the wall surface. This secures the frame to the wall structure and stiffens the jambs so they will carry the door. Fig. 17-8 shows a section view through a door jamb. Note that the casing covers the block-ing and is attached to the jamb and wall surface.

To apply the casing, first select the necessary pieces. Place them near the opening. Some carpen-ters prefer to draw a light pencil line on the edge of the jamb 1/4 in. back from the face. Check the bottom end of the side casing to see that it is square and will rest tightly on the finished floor.

With the side pieces held exactly in place, mark the position of the miter joint at the top. Use a miter box and/or a wood trimmer, Fig. 17-9, to make an

Fig. 17-9. Wood trimmer makes a smooth, accurate cut for mitered joints. This cut can also be made with the table saw miter or a small miter box.

accurate cut. Nail the side casing temporarily with casing or finish nails. Mark, cut, and fit the head casing. If the miters do not fit properly, trim them with a block plane.

Finally, drive nails home and complete the nailing pattern. Use 4d or 6d nails along the jamb edge and drive 8d nails through the outer edge into the studs. Each pair should be spaced about 16 in. O.C. Fig. 17-10 shows the completed casing at the miter joint. Nails have been set. When using hardwood casing, it is advisable to drill nail holes.

PANEL DOORS

There are two general types of doors: panel and flush. The panel door is also referred to as a stile-and-rail door. This type of construction is used in sash, louver, storm, screen, and combination doors. Sash doors are similar to panel doors in appearance

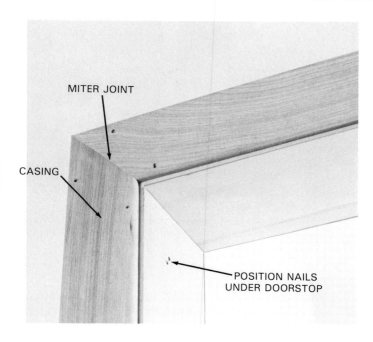

Fig. 17-10. Closeup of casing applied to a doorframe. Note how tightly miter joint fits.

and construction but have one or more glass lights which replace the wood panels.

A panel door, Fig. 17-11, consists of stiles and rails with panels of plywood, hardboard, or solid stock. The rails and stiles are usually made of solid material, however some are veneer over a lumber

core. A variety of designs are formed by changing the number, size, and shape of the panels.

Special effects are secured by installing raised panels which add line and texture, Fig. 17-12. This kind of a panel is formed of thick material which is reduced around the edges where it fits into the grooves in the stiles and rails.

FLUSH DOORS

A flush door consists of a wood frame with thin sheets of material applied to both faces. Improvements in the manufacture of plywood and adhesives have made it possible to produce a flush door that is strong and durable. Today, flush doors account for a high percentage of wood doors.

Face panels, also called skins, are commonly made of 1/8 in. plywood. However, hardboard, plastic laminates, fiberglass, and metal are also used. Fig. 17-13 shows a flush door constructed with metal face panels.

Flush doors are made with solid or hollow cores. Cores of wood or various composition materials are used in solid (or slab) construction. See Fig. 17-14. The frame is usually made of softwood that matches the color of the face veneers. The most common type of solid core construction uses wood blocks bonded together with the end joints staggered. See Fig. 17-15.

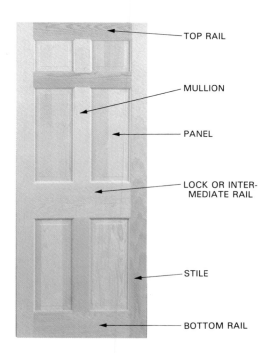

Fig. 17-11. Panel doors are constructed of several intricate parts which are carefully fitted and glued. (C.E. Morgan)

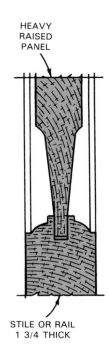

Fig. 17-12. Left. Panel door designed for use as an entrance. Raised panels are combined with carving to create interesting lines. Right. Section through stile and panel shows joint detail. (C.E. Morgan)

Fig. 17-13. Flush type entrance door with wood rails and stiles and a skin of 24 gauge steel. Core is foamed-in-place polyurethane. (Stanley Door Systems)

Fig. 17-15. Forming solid core for flush door with wood blocks bonded together in an electronic clamping and curing machine. (Andersen Corp.)

Flush doors with hollow cores are widely used for interior doors, and may also be used for exterior doors if the construction is bonded with waterproof adhesives. Hollow flush doors do not ordinarily provide as much thermal (heat) and sound insulation as solid core doors. Usually their fire resistance rating is lower. Some modern exterior doors (flush type) have compression-molded fiberglass face panels. They are attached to a wooden frame and can be formed to show various traditional and contemporary designs. The core is high-density polyurethane foam having a high R-value. Unlike steel-faced doors—the fiberglass door can be trimmed for a precision fit. The surface, textured like wood, can be stained or painted, Fig. 17-16.

SIZES AND GRADES

Standard thickness for exterior doors is 1 3/4 in.; interior passage doors are 1 3/8 in. Widths of 2'-8''

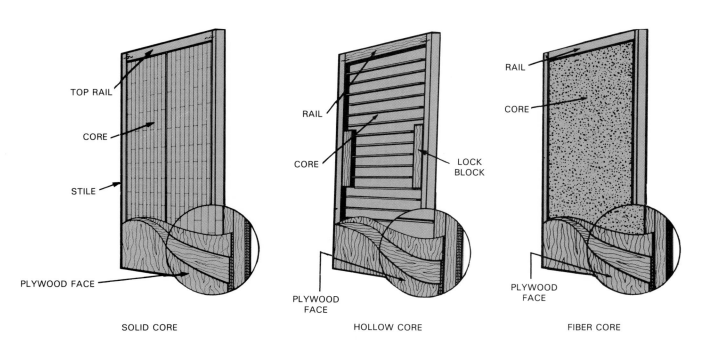

TOP RAIL

CORE

STILE

PLYWOOD FACE

SOLID CORE

RAIL

CORE

LOCK BLOCK

PLYWOOD FACE

HOLLOW CORE

RAIL

CORE

PLYWOOD FACE

FIBER CORE

Fig. 17-14. Basic types of cores used in modern wood flush door construction. Frames are usually made of softwood. (Mohawk Flush Doors, Inc.)

Fig. 17-16. Top. Plane is being used to trim edge of a fiberglass door. Bottom. Oil stain is used to highlight simulated woodgrain pattern and texture. (Therma-Tru)

stallations. A 7 ft. height is usually considered standard for commercial buildings.

Interior door widths vary with the installation. FHA specifies a minimum size of 2'-6'' for bedrooms and a 2 ft. width for bathrooms. Closet doors may also be 2 ft. Interior door sizes and patterns are illustrated in Fig. 17-18.

Grades and manufacturing requirements for doors are listed in Industry Standards developed by the National Woodwork Manufacturing Association (NWMA) and the Fir and Hemlock Door Association (FHDA). The purpose of these standards is to establish nationally recognized dimensions, designs, and quality specifications for materials and work.

DOOR INSTALLATION

Check the architectural drawings and door schedule to determine:

1. The correct type of door for the opening.
2. The direction it will swing.

Mark the door jamb that will receive the hinges. Also mark the edge on which they will be mounted.

Doors may be trimmed to fit but should not be cut to fit smaller openings. If a large amount of the perimeter is removed, the structural balance of the door may be disturbed. Warping may result. Cutouts

and 3'-0'' are most commonly used for exterior doors, Fig. 17-17.

Residential doors, both interior and exterior, have a standard height of 6'-8'' but 7 ft. doors are sometimes used for entrances or special interior in-

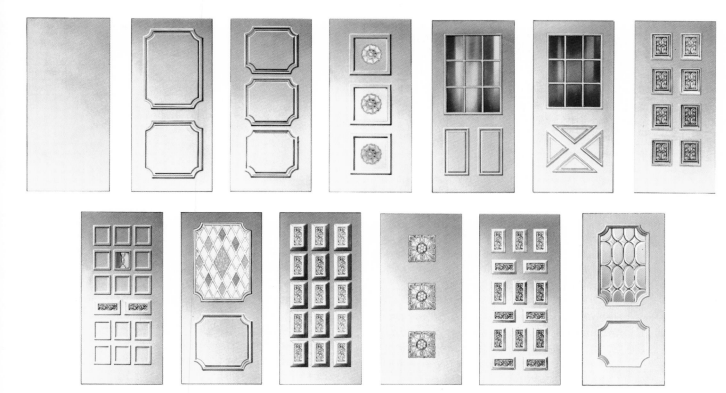

Fig. 17-17. Exterior door designs made with face panels of 24 gauge steel. Decorative moldings are plastic bonded to the steel surface. Standard widths include: 2'-6'', 2'-8'', 3'-0'', and 3'-6''. Standard heights are 6'-8'' and 7'-0''. (Stanley Door Systems)

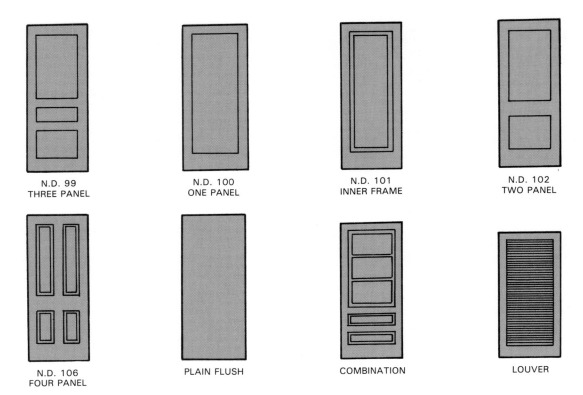

| N.D. 99 THREE PANEL | N.D. 100 ONE PANEL | N.D. 101 INNER FRAME | N.D. 102 TWO PANEL |

| N.D. 106 FOUR PANEL | PLAIN FLUSH | COMBINATION | LOUVER |

CONSTRUCTION DETAILS

Design No.	Stiles	Top Rail	Cross Rail	Lock Rail	Intermediate Rails	Mullions or Muntins	Bottom Rail	Panels
N.D. 99	4 3/4''	4 3/4''	4 5/8''	4 5/8''	. . . . . .	. . . . . .	9 5/8''	Flat
N.D. 100	4 3/4''	4 3/4''	. . . . . .	. . . . . .	. . . . . .	. . . . . .	9 5/8''	Flat
N.D. 101	4 1/4'' Face	4 1/4'' Face	. . . . . .	. . . . . .	. . . . . .	. . . . . .	9 1/4'' or 9 1/2'' Face	Flat
N.D. 102	4 3/4''	4 3/4''	. . . . . .	8''	. . . . . .	. . . . . .	9 5/8''	Flat
N.D. 106	4 3/4''	4 3/4''	. . . . . .	8''	. . . . . .	4 5/8''	9 5/8''	Raised
N.D. 107	4 3/4''	4 3/4''	. . . . . .	. . . . . .	4 5/8''	. . . . . .	9 5/8''	Raised
N.D. 108 (1)(2)(3)	4 3/4''	4 3/4''	. . . . . .	8''	3 7/8'' or 4 5/8''	3 7/8'' or 4 5/8''	9 5/8''	Raised
N.D. 111 (1)(2)(3)	4 3/4''	4 3/4''	. . . . . .	8''	3 7/8'' or 4 5/8''	3 7/8'' or 4 5/8''	9 5/8''	Raised

NOTES: (1) Height of Top Panels Overall 7 1/8''.
         (2) Doors 1'-6'' are made 1 Panel Wide.
         (3) Bottom and lock rails may be reversed when specified.
       *Also for exterior use.

Fig. 17-18. Sizes and patterns commonly used for interior doors. Table lists construction details of various parts. (National Woodwork Manufacturers Assoc.)

for glass inserts in flush doors should never be more than 40 percent of the face area and the opening should not be closer than 5 in. to the edge.

Handle doors carefully. Do not soil unfinished doors. If doors must be stored for more than a few days, stack them horizontally on a clean flat surface and keep them covered.
Doors should be conditioned for several days so they will reach the average prevailing humidity before hanging or finishing.

First, trim the door to fit the opening. Most doors are carefully sized at the millwork plant, leaving only a slight amount of on-the-job fitting and adjustment. The amount of planing necessary can be laid out with a rule.

Clearances should be 3/32 in. on the lock side and 1/16 in. on the hinge side, Fig. 17-19. A clearance of 1/16 at the top and 5/8 at the bottom is generally satisfactory. If the door is to swing across heavy carpeting, increase the bottom clearance. Thresholds are used under exterior doors. Where weatherstripping is used around exterior door openings to reduce infiltration, additional clearance is needed. About 1/8 in. on each side and on the top edge is enough. The threshold for exterior doors may be installed before or after the door is hung.

Door trimming can be done with a hand plane but the modern carpenter generally uses a powered plane. While planing, the door may be held securely on edge either by clamping to sawhorses or by using a special door holder, Fig. 17-20.

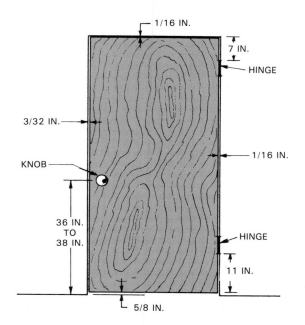

Fig. 17-19. Recommended clearances around interior doors. Some carpenters use a quarter to check clearance at top and lock side. In quality construction, a third hinge is located midway between top and bottom hinges.

After the door is brought to the correct size, plane a bevel on the lock side to provide clearance for the edge when it swings open. This bevel should be about 1/8 in 2 in. or approximately 3 1/2 deg. Narrow doors require a greater bevel than wide doors since the arc of swing is smaller. The type of hinge and the position of the pins should be considered in determining the exact bevel required.

After the bevel is cut and the fit of the door is checked, use a block plane to soften (round) corners on all edges of the door. Smooth with sandpaper.

Millwork plants can furnish prefitted doors that are machined to the size specified with the lock edge beveled and corners slightly rounded. Doors can also be furnished with gains (inset) cut for hinges, and/or holes bored for lock installation.

INSTALLING HINGES

Gains for hinges are usually cut with an electric router. A door-and-jamb template saves time and insures accuracy.

The template is positioned on the door as shown in Fig. 17-21. After the gains are routed, it is attached to the door jamb where matching cuts are made.

Adjustments can be made for various door thicknesses and heights, as well as for different sizes of butt hinges. See Fig. 17-22. The design of most templates makes it nearly impossible to mount them on the wrong side of the door or jamb.

Fig. 17-23 shows the cut being made with template and router. For this type of equipment, hinges with rounded corners may be used. It will save the time required to square the corner with a wood chisel.

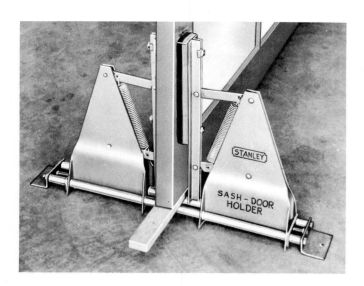

Fig. 17-20. Door holder with clamps. Cushioning is used to protect surface of door.

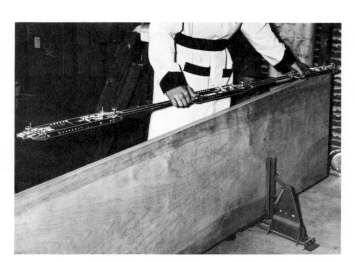

Fig. 17-21. Setting routing template over hinge edge of door in preparation for cutting hinge gain.

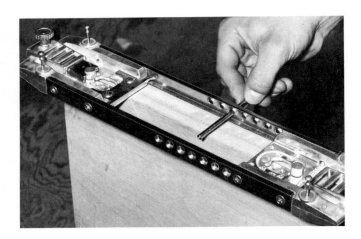

Fig. 17-22. Hinge templates can be adjusted for different door sizes as well as for different hinge sizes.

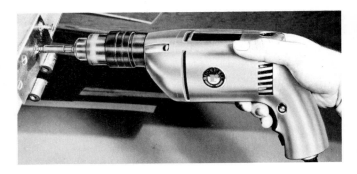

Fig. 17-24. Electric drill with special chuck for driving screws. Screws should be started slightly toward back of the hole to draw the leaf tightly into the gain.

Fig. 17-23. Above. Router and template are being used to cut a gain for a round cornered hinge. Below. Checking the completed gain. Depth should equal thickness of hinge leaf.

Place the hinge in the gain so the head of the removable pin will be up when the door is hung. Drive the first screw in slightly toward the back edge to draw the leaf of the hinge tightly into the gain. To speed up the work, use an electric drill with a special screwdriving attachment, Fig. 17-24.

Follow the same procedure to attach the free leaf of the hinge to the jamb. Some carpenters prefer to set only one or two screws in each hinge leaf. Then they will check the fit before installing the remaining screws.

After all hinges are installed, hang the door and check clearance on all edges. Required corrections should be made by planing the door edges or adjusting the depth of the hinge gains.

Minor adjustments may be made by applying cardboard shims behind the hinge leaf, Fig. 17-25. As the illustration shows, the center of the hinge can be shifted so more clearance is provided along the lock jamb, View A. The space can be closed with the shim applied on the outside edge, View B. When adjusting a door that has been in service it may be necessary to use longer screws.

DOORSTOPS

The doorstops are usually the last trim members to be installed. Many carpenters cut and tack the

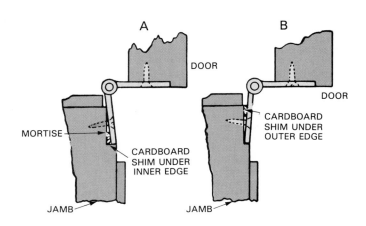

Fig. 17-25. Cardboard shims can sometimes be used to make minor adjustments in door clearance. Shims should be placed between leaf and jamb.

Fig. 17-26. Installing door stop. Side stops are attached first, then the head stop as shown.

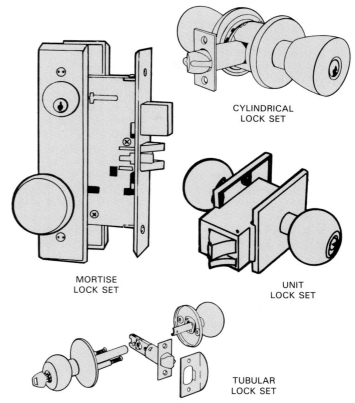

Fig. 17-27. Basic types of door lock sets. Mortise lock sets provide high security and are often found in apartment buildings. Cylindrical and tubular locks are most often chosen for residential with the cylindrical being more secure for entrance doors. Unit lock sets are best for apartments where locks are frequently changed.

stops in place before installing the lock. Permanent nailing comes after the lock installation has been completed.

With the door closed, set the stop on the hinge jamb with a clearance of 1/16 in. The stop on the lock side is set against the door except in the area around the lock. Here allow a slight clearance for humidity changes and decorating. Set the stop on the head jamb so it aligns with the stops on the side jambs. Use miter joints and attach with 4d nails spaced 16 in. O.C. Using power operated nailers, Fig. 17-26, will save time.

DOOR LOCKS

Four types of passage door locks are illustrated in Fig. 17-27. Cylindrical and tubular locks are used most often in modern residential work because they can be installed easily and quickly. Unit locks are installed in an open cutout in the edge of the door and need not be disassembled when installed. Such locks are commonly used on entrance doors for apartments and some commercial buildings where locks must be changed from time to time.

Cylindrical locks have a sturdy, heavy-duty mechanism that provides security for exterior doors, Fig. 17-28. They require boring a large hole in the door face, a smaller hole in the edge, and a shallow mortise for the front plate. The tubular lock is similar but requires a smaller hole in the door face.

When ordering door locks it is often necessary to describe the way in which the door swings. This is referred to as the ''Hand of the Door.'' Hand is determined by facing the outside of the door. The outside is the street side of an entrance door and

the corridor side of an interior door. Fig. 17-29 illustrates standardized procedure in determining this specification.

DEADBOLTS

Deadbolts or deadlocks are extra locks which provide additional security against unauthorized entry. Units are made with single cylinder ahd double cylinder action. (Double cylinder deadbolts require key use on both sides of the door. They offer more security than single cylinders since thieves cannot open the door even from the inside.) Fig. 17-30 shows two types of deadbolts.

LOCK INSTALLATION

Open the package that contains the lock set and check the content. Instructions furnished by the manufacturer should be carefully followed.

Open the door to a convenient position and block it with wedges placed underneath. Measure up from the floor a distance of 38 in. (36 in. is sometimes

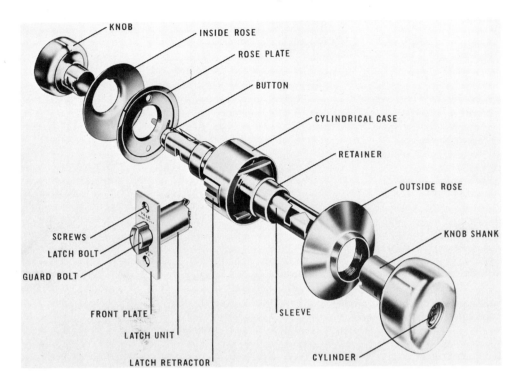

Fig. 17-28. Parts of a cylinder lock set. (Eaton Yale & Towne Inc.)

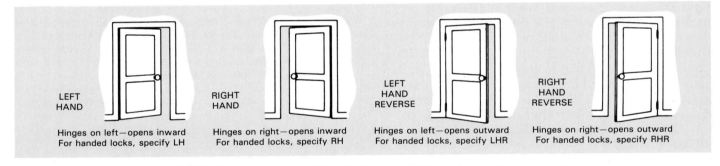

LEFT HAND	RIGHT HAND	LEFT HAND REVERSE	RIGHT HAND REVERSE
Hinges on left—opens inward For handed locks, specify LH	Hinges on right—opens inward For handed locks, specify RH	Hinges on left—opens outward For handed locks, specify LHR	Hinges on right—opens outward For handed locks, specify RHR

Fig. 17-29. Determining the "hand of a door." Use abbreviations suggested.

used). Mark a light horizontal line. This will be the center of the lock. Position the template furnished with the lock set on the face and edge of the door. Lay out the centers of the holes, Fig. 17-31. Continue to follow the instructions included with the lock set. Procedures for all cylindrical and tubular lock sets will be similar. The installation details will vary slightly, however.

Use of boring jigs, Fig. 17-32, assures accurate work. A template layout is not required since the jig is designed to make holes in the correct locations. Either hand operated or power driven bits can be used to bore the holes.

The shallow mortise on the edge of the door can be laid out and cut with standard wood chisels. A faceplate mortise marker also called a marking

chisel, Fig. 17-33, is faster and more accurate. After the perimeter is cut with this device the wood inside can be quickly removed with a standard wood chisel of appropriate width.

THRESHOLDS AND DOOR BOTTOMS

Exterior doors require a trim unit called a threshold to seal the space between the bottom of the door and the door sill. For many years oak or other hardwoods have been machined to a special shape for this purpose. Modern wooden thresholds are usually equipped with rubber or vinyl sealing strips.

A wide range of threshold designs are made from aluminum extrusions. These are available in a clear anodized finish or gold color. Special vinyl strips are

Fig. 17-30. Two styles of deadbolts. Both have bolts with full 1 in. throw. Left. Rim cylinder safety lock attaches to the inside of door. Right. Tubular deadbolt.
(National Lock Hardware)

inserted into the threshold, Fig. 17-34, or to matching units attached to the door. Thresholds of this type are effective in providing an under-the-door seal. Manufacturers furnish detailed instructions for making the installation.

Interior doors do not require a threshold but may be equipped with various sealing strips such as the automatic door bottom shown in Fig. 17-35. This prevents air movement, and reduces sound

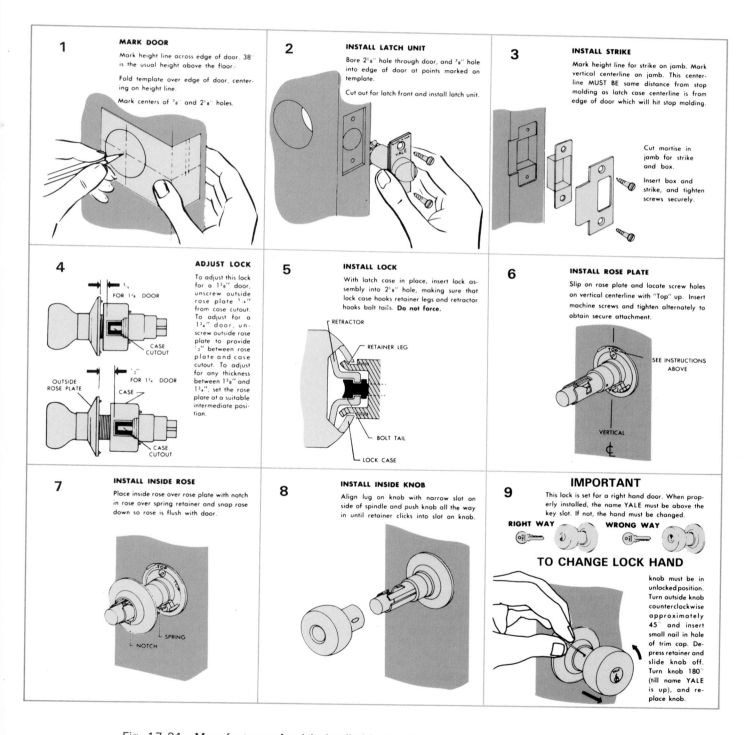

1 MARK DOOR
Mark height line across edge of door. 38" is the usual height above the floor.
Fold template over edge of door, centering on height line.
Mark centers of 7⁄8" and 2 1⁄8" holes.

2 INSTALL LATCH UNIT
Bore 2 1⁄8" hole through door, and 7⁄8" hole into edge of door at points marked on template.
Cut out for latch front and install latch unit.

3 INSTALL STRIKE
Mark height line for strike on jamb. Mark vertical centerline on jamb. This centerline MUST BE same distance from stop molding as latch case centerline is from edge of door which will hit stop molding.
Cut mortise in jamb for strike and box.
Insert box and strike, and tighten screws securely.

4 ADJUST LOCK
FOR 1 3⁄8" DOOR
CASE CUTOUT
OUTSIDE ROSE PLATE
CASE
FOR 1 3⁄4" DOOR
CASE CUTOUT
To adjust this lock for a 1 3⁄8" door, unscrew outside rose plate 5⁄8" from case cutout. To adjust for a 1 3⁄4" door, unscrew outside rose plate to provide 1⁄2" between rose plate and case cutout. To adjust for any thickness between 1 3⁄8" and 1 3⁄4", set the rose plate at a suitable intermediate position.

5 INSTALL LOCK
With latch case in place, insert lock assembly into 2 1⁄8" hole, making sure that lock case hooks retainer legs and retractor hooks bolt tails. **Do not force.**
RETRACTOR
RETAINER LEG
BOLT TAIL
LOCK CASE

6 INSTALL ROSE PLATE
Slip on rose plate and locate screw holes on vertical centerline with "Top" up. Insert machine screws and tighten alternately to obtain secure attachment.
SEE INSTRUCTIONS ABOVE
VERTICAL

7 INSTALL INSIDE ROSE
Place inside rose over rose plate with notch in rose over spring retainer and snap rose down so rose is flush with door.
SPRING
NOTCH

8 INSTALL INSIDE KNOB
Align lug on knob with narrow slot on side of spindle and push knob all the way in until retainer clicks into slot on knob.

9 IMPORTANT
This lock is set for a right hand door. When properly installed, the name YALE must be above the key slot. If not, the hand must be changed.
RIGHT WAY   WRONG WAY
TO CHANGE LOCK HAND
knob must be in unlocked position. Turn outside knob counterclockwise approximately 45° and insert small nail in hole of trim cap. Depress retainer and slide knob off. Turn knob 180° (till name YALE is up), and replace knob.

Fig. 17-31. Manufacturers furnish detailed instructions for installation of their lock sets.

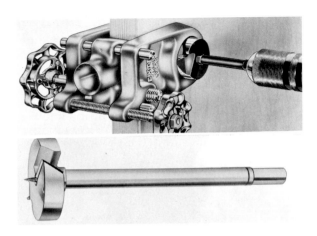

Fig. 17-32. Top. Boring jig for door hardware saves time and insures accuracy. Bottom. Boring bit. (Dexter Industries Inc.)

Fig. 17-33. Using a faceplate mortise marker is more accurate and faster than use of a template.

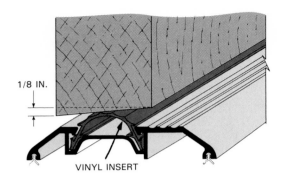

Fig. 17-34. Aluminum threshold with vinyl seal strip. These are frequently found in residential construction. Bottom edge of door should be beveled. (Pemko Mfg. Co.)

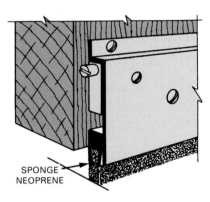

Fig. 17-35. Automatic door bottom. Sealing strip moves upward when the door is opened.

transmission. The unit has a movable strip which drops to the floor when the door is closed and lifts when the door is opened.

PREHUNG DOOR UNIT

A prehung door unit consists of a door frame with a door already installed. The frame includes both sides of casing. Lock hardware may or may not be installed although machining for its installation has usually been completed. Quite often the door is prefinished. The unit is carefully packaged and shipped to the job.

Several frame designs for prehung doors are available. One with a split jamb is shown in Fig. 17-36. This is adjustable for varying wall thicknesses and can be installed in a short time.

Inside doors normally swing into the room which they serve. Therefore, the prehung door package

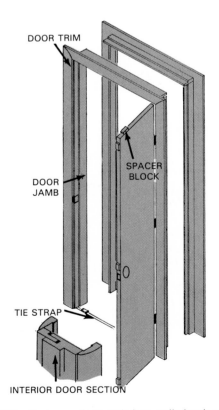

Fig. 17-36. Prehung door unit has split jamb which adjusts to different wall thicknesses.

should be opened on the room-side of the rough opening. Fig. 17-37 shows a general procedure to follow in making the installation. Always study and follow the manufacturer's directions which are included in the package.

SLIDING DOORS (POCKET-TYPE)

This type of door offers a space-saving feature since it is opened by simply sliding it into an opening in the partition. The doorframe consists of a split side jamb attached to a framework built into the wall. The rough opening in the structural frame must be large enough to include the finished door opening and the pocket, Fig. 17-38. The pocket framework and track is installed during the rough framing stages. Pocket-frame units are available from millwork plants in a number of standard sizes.

Manufacturers that specialize in builders hardware have developed steel pocket door frames, Fig. 17-39. They are easy to install and provide a firm base for wall surface materials.

Fig. 17-40 shows a typical track and roller assembly for a pocket door. The hanger (wheel

assembly) snaps into the plate attached to the top of the door. It can be easily adjusted up or down to plumb the door in the opening.

SLIDING DOORS (BYPASS TYPE)

Standard interior door frames can be used for a bypass sliding door installation. When the track is mounted below the head jamb the height of a standard door must be reduced and a trim strip installed to conceal the hardware, Fig. 17-41. Head jamb units are available with a recessed track that permits the doors to ride flush with the underside of the jamb.

Cutaway views of standard bypass track and hangers are shown in Fig. 17-42. Each type can be used with either 3/4 or 1 3/8 in. doors. Note the hanger adjustment that raises or lowers the door for alignment after installation.

Sliding door hardware for bypass doors is packaged complete with track, hangers (rollers), floor guide, screws, and instructions for making the installation.

A disadvantage of bypass sliding doors is that

1. Remove door unit from carton and check for damage. Separate the two sections. Place tongue side (not attached to door) outside of the room.

2. Slide frame section that includes the door into the opening. Side jambs should rest on finished floor or spacer blocks.

3. Carefully plumb door frame and nail casing to wall structure. Be sure all spacer blocks are in place between the jambs and door.

4. Move to other side of wall and install shims between side jambs and rough opening. Shims should be located where spacer blocks make contact with door edges. Nail through jambs and shims.

5. Install remaining half of door frame. Insert the tongue edge into the grooved section already in place. Nail casing to wall structure.

6. Nail through stops into jambs. Remove spacer blocks and check door operation. Make any adjustments required. Drive extra nails where shims are located. Install lock set.

Fig. 17-37. General procedure for installing prehung interior door units. (Frank Paxton Lumber Co.)

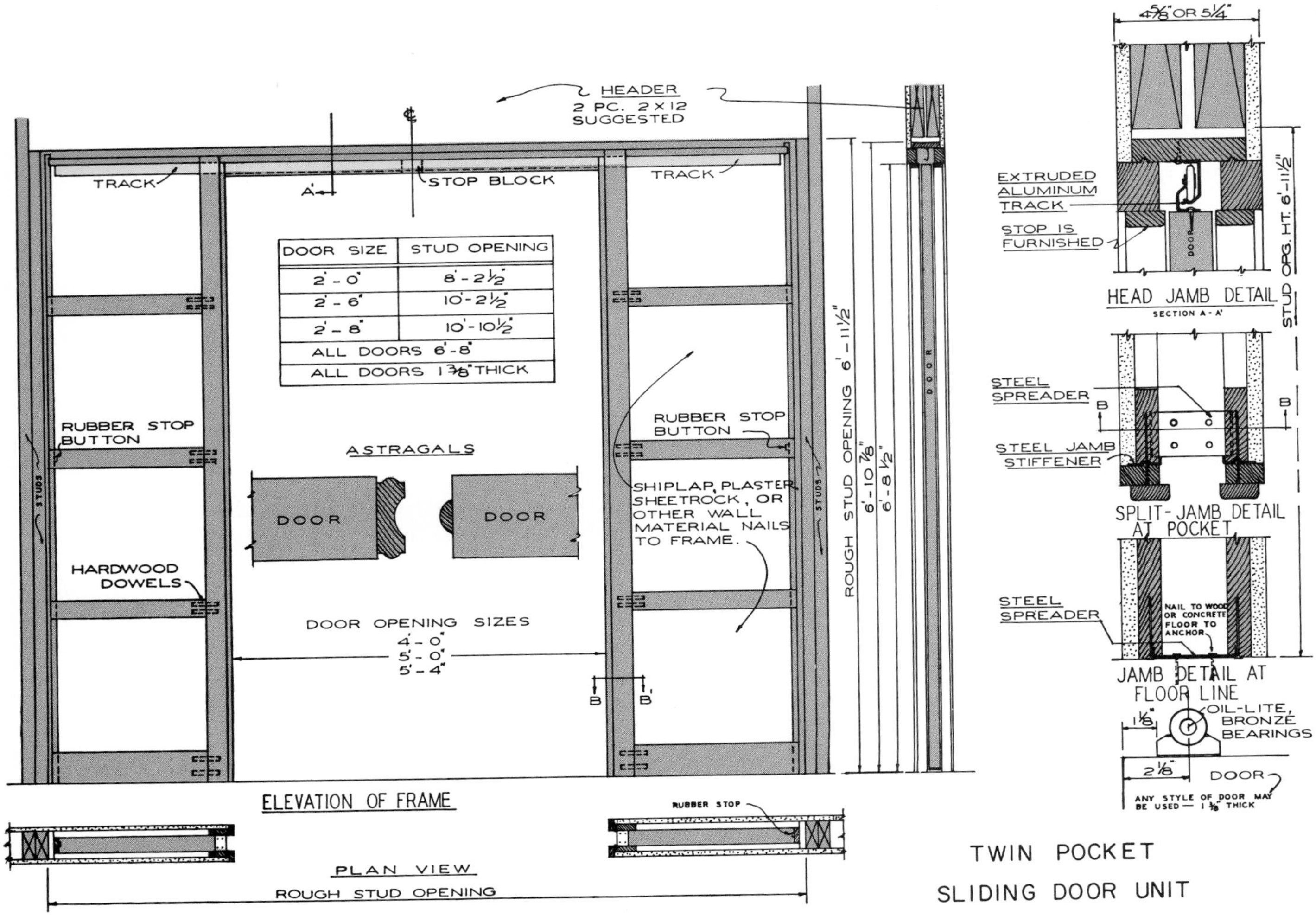
HEADER
2 PC. 2 X 12
SUGGESTED
TRACK
STOP BLOCK
TRACK
DOOR SIZE | STUD OPENING
2'-0" | 8'-2½"
2'-6" | 10'-2½"
2'-8" | 10'-10½"
ALL DOORS 6'-8"
ALL DOORS 1⅜" THICK
RUBBER STOP BUTTON
ASTRAGALS
DOOR
DOOR
RUBBER STOP BUTTON
SHIPLAP, PLASTER, SHEETROCK, OR OTHER WALL MATERIAL NAILS TO FRAME.
HARDWOOD DOWELS
DOOR OPENING SIZES
4'-0"
5'-0"
5'-4"
STUDS
ROUGH STUD OPENING 6'-11½"
6'-10⅞"
6'-8½"
ELEVATION OF FRAME
RUBBER STOP
PLAN VIEW
ROUGH STUD OPENING
4⅝" OR 5¼"
EXTRUDED ALUMINUM TRACK
STOP IS FURNISHED
HEAD JAMB DETAIL
SECTION A-A'
STUD ORG. HT. 6'-11½"
STEEL SPREADER
STEEL JAMB STIFFENER
SPLIT-JAMB DETAIL AT POCKET
STEEL SPREADER
NAIL TO WOOD OR CONCRETE FLOOR TO ANCHOR
JAMB DETAIL AT FLOOR LINE
OIL-LITE, BRONZE BEARINGS
1⅛"
2⅛"
DOOR
ANY STYLE OF DOOR MAY BE USED — 1⅜" THICK
TWIN POCKET
SLIDING DOOR UNIT

Fig. 17-38. Structural details of a pocket door. (Ideal Co.)

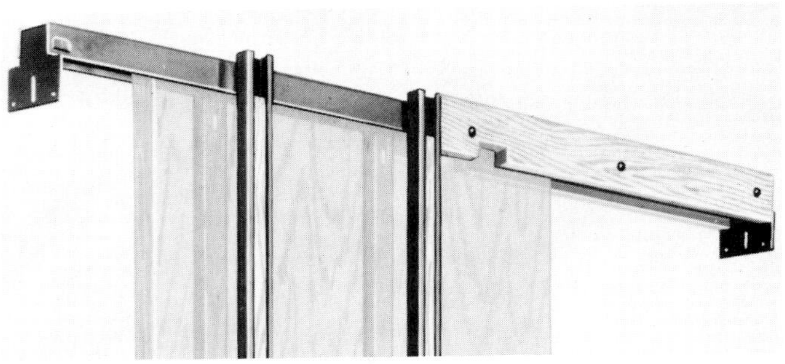

Fig. 17-39. Top section of steel framework unit for pocket door. Kit includes track, hangers, and guides. (Ekco Building Products Co.)

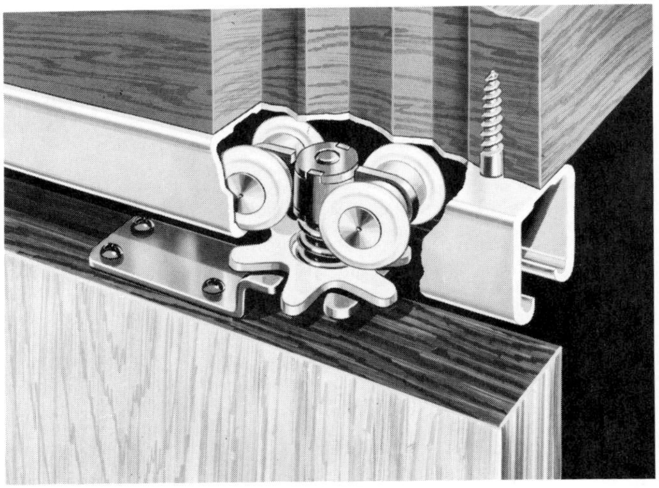

Fig. 17-40. Cutaway shows track and roller assembly for pocket door.

access to the total opening at one time is not possible. They are, however, easy to install and are practical for wardrobes and many other interior wall openings. Fig. 17-43 shows a completed installation of bypass sliding doors.

FOLDING DOORS (BIFOLD)

This type of folding door consists of pairs of doors hinged together. Folding action is guided by an overhead tract. A complete unit may consist of a single pair of doors or two or more pairs of doors. See Fig. 17-44. Folding door units are well suited

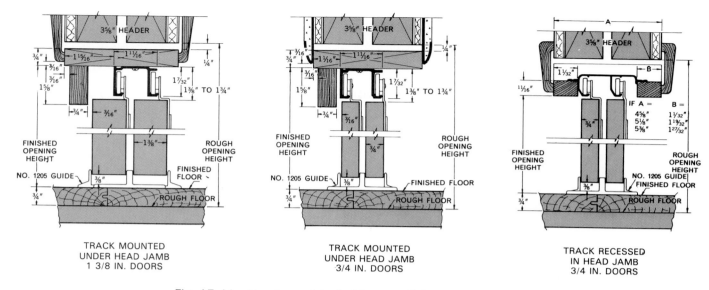

Fig. 17-41. Drawings of typical bypass sliding door installations.

Fig. 17-42. Cutaway views of bypass sliding door installations. Top. Track mounted below head jamb. Note trim member needed to conceal track. Bottom. Track is recessed into head jamb. Hangers include vertical adjustment. (Ekco Building Products Co.)

Fig. 17-43. Bypass doors used for closet space. The doors open from either side. (C.E. Morgan)

Fig. 17-44. Bifold doors are often used to close off wardrobe space. (C.E. Morgan)

to wardrobes, closets, and certain openings between rooms.

The opening for bifold units is trimmed with standard jambs and casing. Fig. 17-45 illustrates an installation of hardware. Pivot brackets and center guides have self-lubricating nylon bushings. The weight of the doors is supported by the pivot brackets and hinges between the doors, not by the overhead track and guide. Two-door units generally range from 2 to 3 ft. wide while four-door units are available in widths of from 3 to 6 ft.

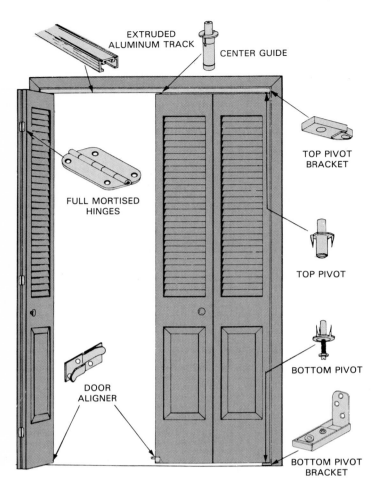

Fig. 17-45. Typical hardware used for folding door installation. The hinged units are supported by brackets. No weight is carried by center guide. (Ideal Co.)

to match the door. Fig. 17-46 shows how the track is installed in either a wood framed or plastered opening.

Door panel surfaces are available in a variety of materials and finishes. The best grade of panels are made from genuine wood veneers, bonded to wood cores. Panels are also made of stabilized particleboard wrapped with wood-grain enbossed vinyl film.

An important advantage of this type of folding door is its space-saving feature. As the door is opened, the panels fold together forming a "stack" that does not require clear room space. Bifold doors and regular passage doors must have clearance in the room as they are opened and closed. Fig. 17-47 shows general details of construction with the door in both opened and closed positions.

When open the door requires only a small amount of space. For example, the stack dimension for an 8 ft. opening is only 11 1/2 in. When it is desirable to clear the entire opening, the stack can be housed in a special wall cavity.

Folding doors are available in a variety of sizes. Fig. 17-48 shows an installation for a wardrobe that includes hanging space, dresser, and drawers. Folding doors can be used to separate room areas, laundry alcoves, and general storage space. Manufacturers furnish door units in a complete package that includes track, hardware, latches, and instructions for making the installation.

WINDOW TRIM

Interior window trim consists of casing, stool, apron, and stops, Fig. 17-49. Millwork companies select and package the proper length trim members to finish a given unit or combination of units.

When the total opening for a four-door unit is greater than 6 ft. (two-door unit over 3 ft.) it is usually necessary to install heavier hardware. Also, a supporting roller-hanger is used instead of a regular center guide. This type of heavy-duty hardware is also used to carry multi-folding door units that serve as room dividers.

Folding door hardware is supplied in a package that includes hinges, pivots, guides, bumpers, aligners, nails, and screws. Instructions for the installation are also included. Millwork plants can supply matching doors with prefitted hardware.

MULTIPANEL FOLDING DOORS

Multipanel folding doors are built from narrow panels with some type of hinge along the edges. One manufacturer produces a patented design where the hinge action is provided by steel springs threaded through the panels. The entire door assembly is supported by nylon rollers located in an overhead metal track. The track is wood-trimmed

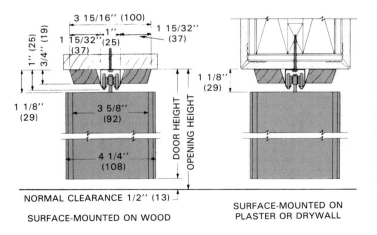

NORMAL CLEARANCE 1/2" (13)

SURFACE-MOUNTED ON WOOD

SURFACE-MOUNTED ON PLASTER OR DRYWALL

Fig. 17-46. Typical head sections show installation of overhead track for folding doors. Metric sizes (millimetres) are shown in parentheses. (Rolscreen Co.)

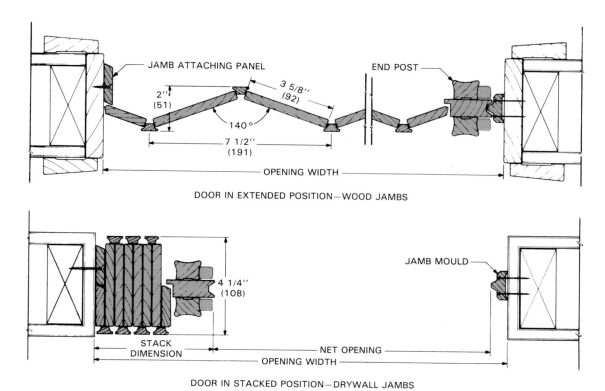

Fig. 17-47. General details show operation of folding door. Metric sizes are shown in parentheses.   (Rolscreen Co.)

Fig. 17-48. Application for use of folding door includes closing off a wardrobe. Doors stack on either side of the wardrobe opening.   (Rolscreen Co.)

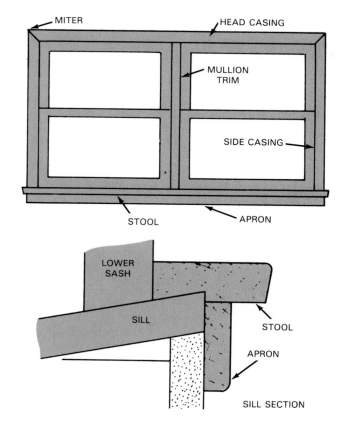

Fig. 17-49. Trim members used for the standard double-hung window.

To mark the stool for cutting, hold it level with the sill and mark the inside edges of the side jambs. Also mark a line on the face where it will fit against the wall surface. For a standard double-hung window this line will usually be directly above the square edge of the stool rabbet (notch where it fits over the top of the sill).

Carefully cut out the ends and check the fit. You will have to open the lower sash slightly to slide the stool into position. Bring down the window carefully on top of the stool and draw the cutoff line so the sash will clear the stool. Allow about 1/16 in. between the front edge of the stool and the window sash. Position a piece of side casing and measure beyond it about 3/4 in. Mark the cutoff lines for the ends of the stool.

Cut the ends, sand the surface, and nail the stool into place. Some installations require that the stool be bedded in caulking compound or white lead.

Install casing next. Set a length in position on the stool, Fig. 17-50, and mark the position of the miter on the inside edge. Some carpenters prefer to set the casing back from the jamb edge about 1/8 in. following about the same procedure as previously described for door casing. Cut the miter. Drill nail holes if the casing is made of hardwood. After both side casings are nailed, cut and install head casing.

The apron is the last member to be applied. Its length should be equal to the distance between the outside edges of the side casing. Sand all saw cuts that will show before nailing any trim in place, especially ends.

When the profile (edge) of the apron is curved, the ends should be returned or coped so that the shape of the ends will be the same as the sides. The returned end is commonly used and formed with miter cuts as illustrated in Fig. 17-51. Completed trim at the lower right-hand corner of a double-hung

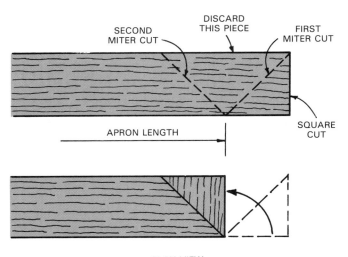

Fig. 17-51. How to mark and cut a returned end. Use glue to attach the end piece.

window is shown in Fig. 17-52. Note the spacing of the various members and the end contour of the apron which has been returned.

In some construction the stool and apron are sometimes eliminated. Instead, a piece of beveled sill liner is installed along with a standard piece of casing. This is commonly known as PICTURE FRAME trimming.

After all trim members are in place, complete the nailing pattern, Fig. 17-53. Nail holes are filled later as a step in the application of finishes.

Fig. 17-50. Measuring and fitting side casing. Bottom must be squared and should rest firmly on the stool before marking miter cut at top.

Fig. 17-52. View of finished right-hand corner of a double-hung window. Note nailing pattern especially for stool. Nail holes will be filled during finishing operations. Some carpenters nail the stool to the wall by toenailing it from underneath. This eliminates the nail hole.

Fig. 17-53. Finish nailer is being used to attach door casing. It drives nails up to 2 1/2 in. long. (Paslode Corp.)

BASEBOARD AND BASE SHOE

The baseboard covers the joint between the wall surface and the finish flooring. It is among the last of the interior trim members to be installed since it must be fitted to the door casings and cabinetwork. Base shoe is used to seal the joint between the baseboard and the finished floor, Fig. 17-54. It is usually fitted at the time the baseboard is installed but is not nailed in place until after surface finishes (lacquer, varnish, or paint) have been ap-

plied. Base shoe is often used to cover the edge of resilient tile or carpet.

Baseboards run continuously around the room between door openings, cabinets, and built-ins. The joints at internal corners should be coped. Those at outside corners are mitered. See Fig. 17-55.

Select and place the baseboard material around the sides of the room. Sort the pieces so there will be the least amount of cutting and waste. Where a straight run of baseboard must be joined, use a mitered-lap joint (also called a scarf joint). Be sure to locate the joint so it can be nailed over a stud. See Fig. 17-56.

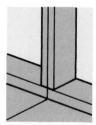

Baseboard installation is easy if stud locations have been recorded, first on the rough floor ahead of plaster and then on the wall surface before the finish flooring or underlayment is applied.

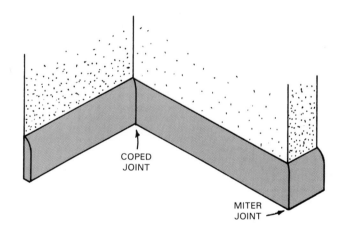

Fig. 17-55. Coped joints are suitable for inside corners; however, use mitered joint for outside corners.

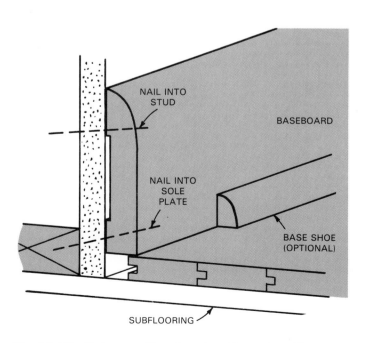

Fig. 17-54. Cutaway of baseboard and base shoe. They conceal the gap between flooring and finished wall.

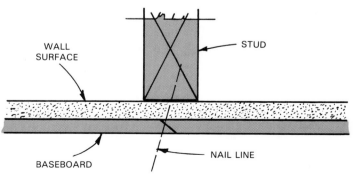

Fig. 17-56. Cut a scarf joint when joining a straight run of baseboard. Be sure to drill nail holes when using hardwood material.

If the studs have not been marked, tap along the wall with a hammer until a solid sound is heard or use a stud finder. Drive nails into the wall to locate the exact position of the stud, then mark the location of others by measuring the stud spacing (usually 16 in. O.C.).

Cut and fit the first piece of baseboard so it makes a tight butt joint with the intersecting wall surface. The next piece, running from an internal corner, is joined to the first with a coped joint, Fig. 17-57.

To form this type of joint, first cut an inside miter on the end of the baseboard. Then, using a coping saw, cut along the line where the sawed surface of the miter joins the curved surface of the baseboard, Fig. 17-58. Make the cut perpendicular

Fig. 17-59. Using a miter box to cut miter joint for outside corners. Hold piece securely against fence for an accurate cut.

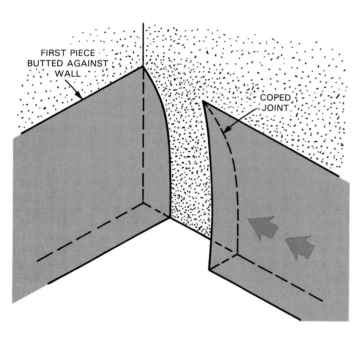

Fig. 17-57. Coped joint follows contour (shape) of the piece it will butt against.

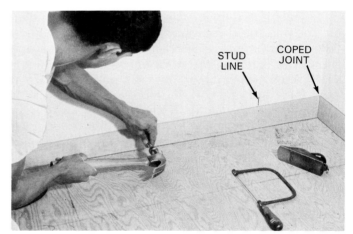

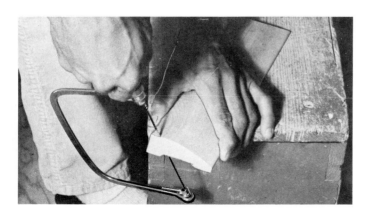

Fig. 17-58. Form a coped joint by making a perpendicular cut along the front edge of an inside miter joint.

Fig. 17-60. Top. Set nails after installing a section of baseboard. Right-handed workers usually prefer to install pieces in a counterclockwise direction. Bottom. Using a pneumatic nailer to attach shoe base. (Senco Products, Inc.)

442

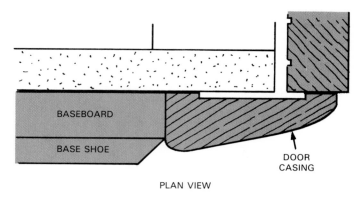

PLAN VIEW

Fig. 17-61. Baseboard is butted against door casing.

to the back face. This forms an end profile that will match the face of the baseboard. Some carpenters undercut the coped end slightly to insure a tight fit at the front edge.

The coped joint takes a little longer to make than a plain miter but it makes a better joint at an inside corner. It will not open when the baseboard is nailed into place. Neither will a noticeable crack appear if the wood shrinks after installation.

All outside corners are joined with a miter joint. Hold the baseboard in position and mark at the back edge. Make the cut using a miter box, Fig. 17-59.

Before installing a section of baseboard, check both ends to make sure the cut and fit is correct. To install, hold the board tightly against the floor and nail it in place with finishing nails long enough to penetrate well into the studs. The lower nail is angled slightly down so it is easier to drive and will enter the sole plate. Set the nail heads as shown in Fig. 17-60.

Baseboards are normally butted against the door casing as illustrated in Fig. 17-61. The edge of the casing is designed with sufficient surface to accommodate the slightly thinner dimension of the baseboard. Base shoe, if used, is ended at the casing with a miter cut as shown.

TEST YOUR KNOWLEDGE - UNIT 17

1. Interior door jambs for a standard framed wall with a plastered finish should be _____ in. wide.
2. If a door frame is slightly high for the opening, part of the _____ should be trimmed.
3. When nailing the door jambs, locate the nails so they will be covered by the _____.
4. To cover the space between the wall surface and the door jambs, door _____ is applied to each side of the frame.
5. A panel door consists of rails, panels, and _____.
6. The face panels of flush doors are usually made of 1/8 in. plywood and are commonly called door _____.

7. Three types of hollow core door construction are: lattice, _____, and implanted blanks.
8. A standard size bedroom door is _____ wide and _____ high.
9. When setting the door stop on the hinge jamb, provide a clearance of _____ with the door.
10. If you are facing the outside of an entrance door and it swings toward you with the hinges on your left, it is called a _____ (LH, RH, LHR, RHR).
11. The two most commonly used types of residential door locks are cylindrical and _____.
12. The type of sliding door that operates by moving it into an opening in the wall or partition is called a _____.
13. After the stool has been set, the next step in trimming a single window unit is to install the _____.
14. The last trim member applied to a window unit is the _____.
15. When installing baseboards, a _____ joint should be used at internal corners.

OUTSIDE ASSIGNMENTS

1. Visit a residential building site in your community where the inside finish work is in progress. Note the type of trim being applied and the procedures being followed. Possibly one of the carpenters would discuss with you the advantages and disadvantages of prehung door units. Prepare carefully organized notes of your observations and make an oral report to the class. Obtain permission from the supervisor or head carpenter before making the visit.

2. Develop a door schedule for a preliminary or presentation drawing of a residence. Sizes of doors will usually not be shown and you will need to make the selection. A door schedule should include the location, width, height, thickness, type and design, kind of material, and quality requirements.

   To extend the assignment, secure estimated costs. Refer to a builders supply catalogue or visit a local lumber dealer.

3. Build a full-size sectional mock-up of a partition with a doorway. Include all parts—studs, wall finish, door jamb, casing, baseboard, and door. Carry the section up about 16 in. above the floor and extend the partition out from the door only about the same amount. A 3/4 in. thickness of particleboard could serve as the base and represent the finish flooring.

   It will be best to glue most of the parts together since the structure may not withstand much nailing.

Traditional styled cabinets feature solid oak-framed doors with raised panels. Working surfaces are covered with matching plastic laminate. Porcelain knobs are used for door and drawer pulls. (Brammer Mfg. Co.)

Contemporary styling features melamine plastic laminate surface, in almond color, on doors and drawers. Solid oak accent strips serve as finger pulls. Doors have fully concealed hinges. (Kitchen Kompact, Inc.)

Unit 18

CABINETMAKING

Cabinetwork, as used in the interior finish of a residence, refers to kitchen and bathroom cabinets, and, in a general way, to such work as closet shelving, wardrobe fittings, desks, bookcases, and dressing tables. The term "built-in" emphasizes that the cabinet or unit is located within or attached to the structure. The use of built-in cabinets and storage units is an important development in modern architecture and design.

Modern kitchen cabinets, Fig. 18-1, present an attractive appearance and help to increase kitchen efficiency. Here, and in other areas of the home, storage units should be designed for the items that will be stored.

Space must be carefully allocated. Drawers, shelves, and other elements should be proportioned to satisfy specific needs.

Three types of cabinetwork used in homes are:
1. That which is built on the job by the carpenter.
2. Custom-built units constructed in local cabinet shops or millwork plants.
3. Mass-produced cabinets from factories that specialize in this area of manufacturing.

Except in giant housing projects, combinations of these types of cabinetwork are usually found on most jobs. Even when most of the cabinets are factory produced, the carpenter is responsible for the installation. This task requires skill and careful attention to detail.

DRAWINGS FOR CABINETWORK

Architectural plans usually include details of built-in cabinetwork. The floor plan shows cabinetwork location. Elevations, usually drawn to a larger scale, provide detailed dimensions. A typical drawing of a base cabinet, desk, and room divider is illustrated in Fig. 18-2.

Drawings of this type are scaled and, thus, the need for extensive dimensioning is eliminated. The drawings serve as a construction guide to the carpenter. They are also followed in the selection of factory-built components. When built on the job, the detail of joints and structure becomes the responsibility of the carpenter, who should be skilled in cabinetmaking.

Architectural drawings may include more specific details concerning the cabinetwork. This is justified on large commercial contracts. Here a number of individuals will be associated with the project. It also becomes important when the cabinetwork contains special materials and constructions.

Fig. 18-3 shows kitchen cabinets designed for an apartment complex. Written specifications will define the type of joinery and quality of materials.

STANDARD SIZES

Overall heights and other dimensions of built-in units are usually included in the architectural

Fig. 18-1. Modern kitchen cabinets are attractive and provide efficient storage space in shelving and drawers. They also provide important working counter surface around sinks, cooking tops, and ovens. (Riviera Kitchens, an Evans Products Co.)

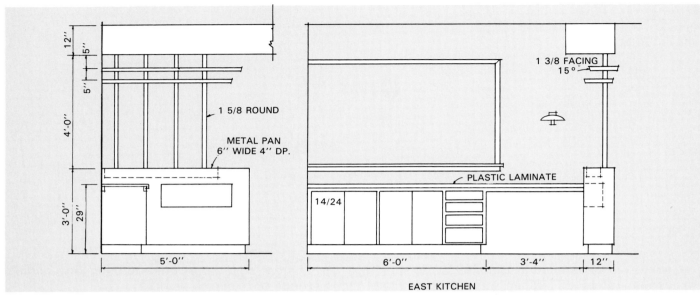

12''
5''
5''
4'-0''
1 5/8 ROUND
METAL PAN
6'' WIDE 4'' DP.
3'-0''
29''
5'-0''

1 3/8 FACING
15°
PLASTIC LAMINATE
14/24
6'-0''
3'-4''
12''

EAST KITCHEN

Fig. 18-2. Above. Typical built-in cabinetwork detail. Since the drawings are carefully scaled, many dimensions have been eliminated. Below. Completed cabinetwork. Drawers include a recess under the front edge that serves as a finger pull.

plans. However, the carpenter should be familiar with basic design requirements.

Base cabinets for kitchens are usually 36 in. high and 24 in. deep. The countertop extends about 1 inch, Fig. 18-4. The vertical distance between the top of the base unit and the bottom of the wall unit may vary from 15 to 18 in. FHA specifies a minimum height of 24 in. when the wall cabinet is located over a cooking unit or sink.

Cabinets with built-in lavatories in bathrooms or dressing rooms are normally 31 in. high. Depth varies depending on the type of fixture. Knee room is especially important in this type of cabinet.

Standards for closets and wardrobes vary widely. The determining factors are the items to be stored. The minimum clear depth for clothing on hangers is 24 in. When hooks are mounted on doors or the rear wall, this distance must be ex-

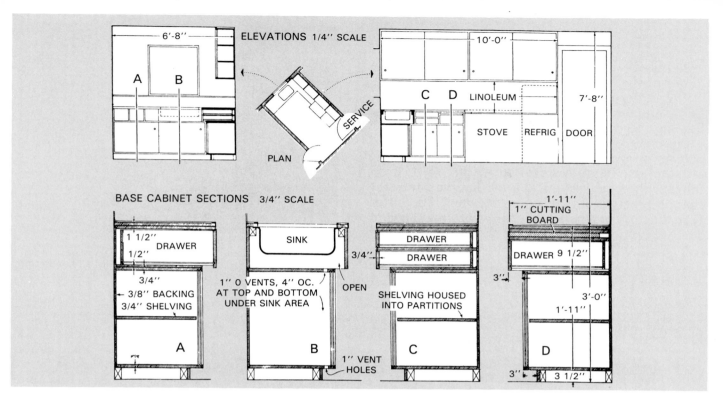

Fig. 18-3. These drawings are for cabinetwork in a large apartment building. (Architectural Woodwork Institute)

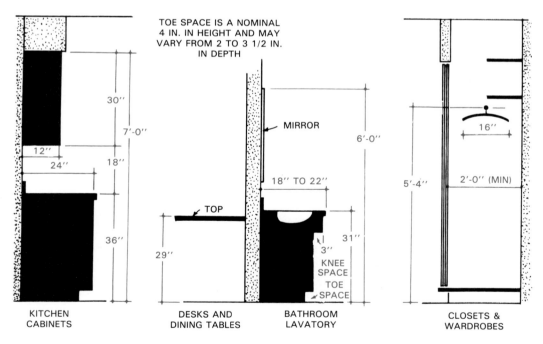

Fig. 18-4. Dimensions are standardized for cabinetwork.

tended. Refer to architectural standards manuals for additional information.

BUILDING CABINETS

Two different procedures are commonly used when building cabinets on the job. The first con-

sists of cutting the parts and assembling them "in place" a piece at a time. The carpenter or cabinetmaker attaches each piece to the floor, wall, or to other members. After the basic structure is assembled, facing strips, doors, and other fittings are marked, cut to size, and attached.

In the second procedure, the entire unit is first

assembled. Then the finished piece is fixed to the floor or attached to the wall.

The structure is formed with end panels, partitions, and backs. Horizontal frames join the parts. See Fig. 18-5.

Basically, this is the method used in custom cabinet shops or for mass-production of cabinets in factories. When units of this type are built on the job, they can be much larger (longer sections) than is practical in shops or factories. The latter type must be of a size easily moved to the building site.

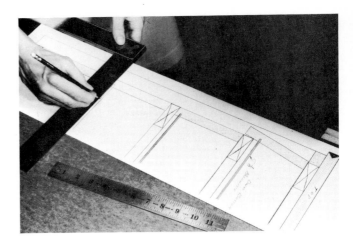

Fig. 18-6. Making a full-size master layout of a base cabinet.

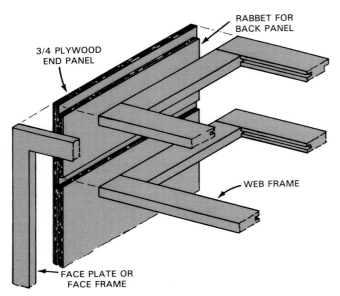

Fig. 18-5. Typical frame construction of cabinets built on site as a separate unit. Terminology is recommended by Architectural Woodworking Institute.

MASTER LAYOUTS

Before cutting out the parts for a cabinet, it will be helpful to prepare a master layout on plywood or cardboard, Fig. 18-6. This is especially true when following the second procedure described.

Several layouts may be required. These should be drawn as section views, where the structure is complicated with drawers, pull-out boards, and special shelving.

Follow the overall dimensions provided in the architectural plans and details. Draw each member full size. Show all clearances that may be required.

This master layout will be valuable when cutting side or end panels to size and locating joints. It can also be referred to for:

1. Exact sizes and locations of drawer parts.
2. Other detailed dimensions not included in the regular drawings.

BASIC FRAMING

When following the assembled-in-place procedure, some of the basic layout can be made directly on the floor and/or wall surface. Each end panel and partition is represented with two lines. Be sure these lines are plumb since they can be used to line up the panels when they are installed.

Construct the base first, as illustrated in Fig. 18-7. Use straight 2 x 4s and nail them to the floor and to a strip attached to the wall. If the floor is not level, place shims under the various members of the base. Exposed parts can be faced with a finished material or the front edge can be made of a finished piece such as base molding.

After the base is completed, cut and install the end panels, Fig. 18-8. Attach a strip along the wall between the end panels and level with the top edge. Be sure the strip is level throughout its length. Nail it securely to the wall studs.

Next, cut the bottom panels and nail them in place on the base. Follow this with the installation of the partitions which are notched at the back

Fig. 18-7. Constructing the base. Note the layout lines marked on the wall for an end panel and partition.

Fig. 18-8. Marking an end panel while it is held in place. This is typical of the method used when constructing cabinets by the assembled-in-place procedure.

Fig. 18-9. Base cabinet partially completed. "Lazy Susan" (also called carousel shelving) unit must be installed at this stage.

The carpenter may prefer to install the wall cabinets before the base units. This makes it easier to work on the wall cabinets since he or she can work directly below them.

corner of the top edge so they will fit over the wall strip.

Plumb the front edge of the partitions and end panels. Secure them with temporary strips nailed along the top as shown on the left-hand cabinet in Fig. 18-9.

Wall units are constructed about the same way as the base units. Make layout lines directly on the wall and ceiling. Attach mounting strips by nailing through the wall surface into the studs. At inside corners, end panels can be attached directly to the

wall, Fig. 18-10.

During the basic framing operations, use care in constructing the openings for built-in appliances. They must be the correct size. Secure rough-in drawings and specifications, Fig. 18-11. These are furnished by the manufacturer of the appliances. Follow them carefully. Fig. 18-12 shows a completed cabinet with a built-in range.

FACING

Finished facing strips are applied to the front of the cabinet frame. In factories and cabinet shops, these strips are often assembled into a framework (called a face plate or face frame) before they are attached to the basic cabinet structure. The vertical members are called stiles and the horizontal members are called rails.

For assembled-in-place cabinets, each piece is cut and installed separately. The size is laid out by

Fig. 18-10. Left. End panels of a wall cabinet are in place. Right. Completed framing with facing partially applied.

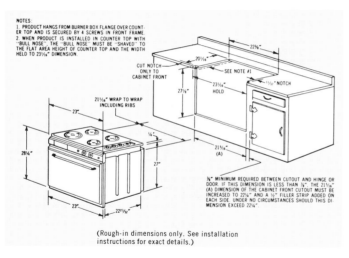

Fig. 18-11. Typical drawing shows rough-in dimensions for a built-in range unit. (Whirlpool Corp.)

Fig. 18-12. Installation of a built-in range has been completed. (I-XL Furniture Co.)

Fig. 18-13. Lengths of facing strip are marked while held in place against the cabinet frame.

positioning the facing stock on the cabinet and marking them as shown in Fig. 18-13. Then the finished cuts are made. A marked part can be used to lay out duplicate pieces.

Generally, stiles are installed first and then the rails, Fig. 18-14. Sometimes an end stile is attached plumb and rails are installed to determine the position of the next stile.

The parts are glued and nailed in place with finishing nails. When nailing hardwoods, drill nail holes where splitting is likely to occur.

Many kinds of joints can be used to join the stiles and rails. A practical design is illustrated in Fig. 18-15. The depth of the gain or dado is maintained at about 3/8 in. so it will be covered by the lip of the door or drawer. Lap joints and dowel joints are also commonly used to make this attachment.

DRAWER GUIDES

Three common types of drawer guides are used:
1. Corner guides.
2. Center guides.
3. Side guides.

Fig. 18-14. Facing pieces are being applied to a base cabinet. Members are glued and nailed in place.

Fig. 18-16 shows all three. The corner guide may be formed in the cabinet by the side panel and frame. It may be necessary to add a spacer strip to hold the drawer in alignment with the front facing.

An adaptation of the corner guide is used in the cabinet frame shown in Fig. 18-17. Note that the guide unit located in the center is designed to carry a drawer on each side.

Center guides are often used in cabinetwork and consist of a strip or runner fastened between the front and back rails. A guide which is attached to

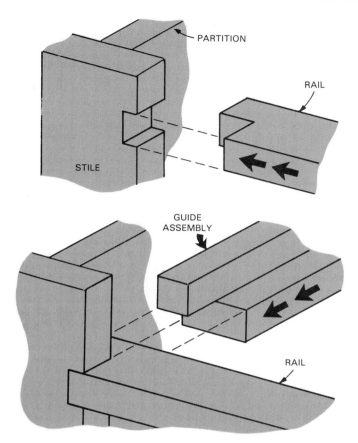

Fig. 18-15. Practical joinery for assembled-in-place cabinet facing. Above. Stile and rail joint. Below. Drawer guide assembly fits behind facing and rests on rail.

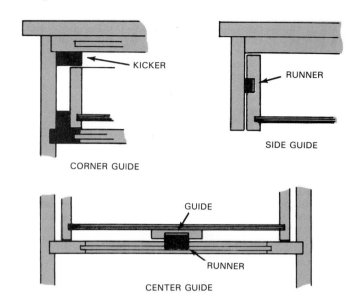

Fig. 18-16. There are three basic types of drawer guides.

the underside and back of the drawer, rides on this runner. The runner is attached to the frame or rails with screws. It can be adjusted so the clearance on each side is equal and the face of the drawer aligns with the front of the structure.

In drawer openings where there is no lower frame, a side guide may be used. Grooves are cut in the drawer side before it is assembled and matching strips are fastened to the side panels. This type of guide can be used for shallow drawers or trays in wardrobe units where framing between each drawer opening is unnecessary and would waste space.

The drawer carrier arrangement may require a "kicker." One is shown in Fig. 18-17. It keeps the drawer from tilting downward when it is opened. The kicker may be located over the drawer side or centered to hold down the drawer back.

Fig. 18-17. Lavatory cabinet in a bathroom area. Left. General view of framing with facing applied. Right. Close-up view showing drawer guides. The back of the drawer will slide along the kicker. It prevents the drawer from tilting downward when opened.

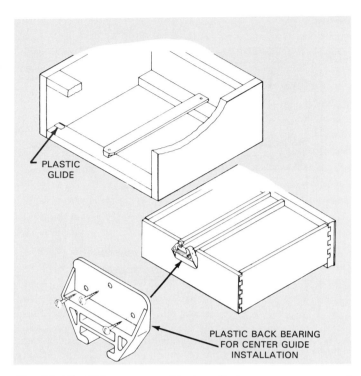

PLASTIC
GLIDE

PLASTIC BACK BEARING
FOR CENTER GUIDE
INSTALLATION

Fig. 18-18. This center guide installation uses a patented back bearing.   (Ronthor Plastics Div., U.S. Mfg. Corp.)

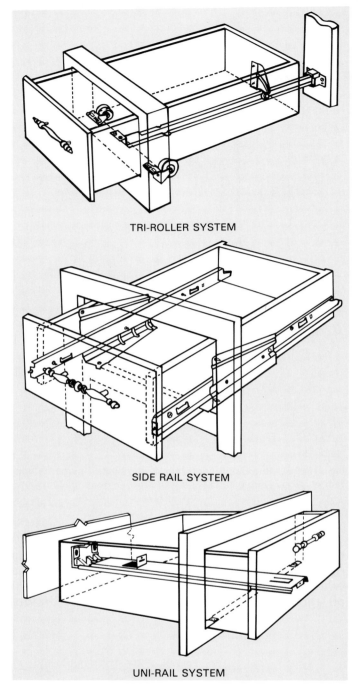

TRI-ROLLER SYSTEM

SIDE RAIL SYSTEM

UNI-RAIL SYSTEM

Fig. 18-19. Manufactured drawer slide assemblies are available in several designs.   (Amerock Corp.)

Wooden drawer guides should be carefully fitted. The parts should be given a coat of sealer. When the sealer on the moving parts is sanded lightly and waxed, the drawer will work smoothly.

Fig. 18-18 shows how to attach a commercially made back bearing for a center slide. For large drawers, or those that will carry considerable weight, special drawer slides, Fig. 18-19, may be used. Some of these support weights up to 50 lb. Units are available in several sizes.

DRAWERS

There are two general types of drawers:
1. Lip.
2. Flush.
Flush drawers, which must be carefully fitted, are commonly used in furniture. Lip drawers have a rabbet along the top and sides of the front. This style overlaps the opening and is much easier to construct. The rabbet is used mainly in built-in cabinetwork.

Sizes and designs in drawer construction vary widely. Fig. 18-20 shows standard construction with several types of joints commonly used.

The joint between the drawer front and the drawer side receives the greatest strain. It should be carefully designed and fitted. Often, in high quality work, the corners are dovetailed and the bottom grooved into the back, front, and sides. The top edges of the drawer sides are rounded.

Flush drawer fronts may be constructed with the sides set about 1/16 to 1/8 in. deeper into the

fronts so that a slight lug (projection) is formed. After the drawer is assembled, this lug is trimmed to form a good fit.

DRAWER CONSTRUCTION

Drawers are usually constructed after the cabinetwork has been assembled. The following procedure is generally recommended:
1. Select the material for the fronts. Grain patterns should match or blend with each other.

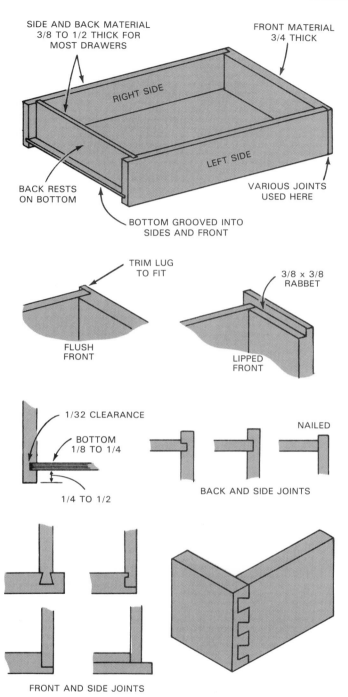

SIDE AND BACK MATERIAL 3/8 TO 1/2 THICK FOR MOST DRAWERS

FRONT MATERIAL 3/4 THICK

RIGHT SIDE

LEFT SIDE

BACK RESTS ON BOTTOM

VARIOUS JOINTS USED HERE

BOTTOM GROOVED INTO SIDES AND FRONT

TRIM LUG TO FIT

FLUSH FRONT

3/8 x 3/8 RABBET

LIPPED FRONT

1/32 CLEARANCE

BOTTOM 1/8 TO 1/4

1/4 TO 1/2

NAILED

BACK AND SIDE JOINTS

FRONT AND SIDE JOINTS

Fig. 18-20. Standard drawer construction. Variety of joints may be used.

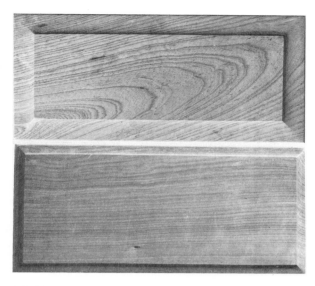

Fig. 18-21. Prefabricated drawer fronts are made in many designs. Above. Raised panel effect. Below. Ogee edge. (Frank Paxton Lumber Co.)

clearance. For lip drawers, add the depth of the rabbets to the dimensions.
3. Select and prepare the stock for the sides and back. A less expensive hardwood or a softwood can be used. Plywood is usually not satisfactory for these parts. The surfaces should be sanded either before or after cutting to length.
4. Select material for the bottom. Hardboard or plywood should be used. Trim to final size after the joints for the other drawer members are cut.
5. Cut groove for the bottom in the front and sides.
6. Cut joints in the drawer fronts that will hold the drawer sides. Fig. 18-22 shows a sequence for cutting a locked joint. For drawer fronts with a lip, cut the rabbet first, then the joint.
7. Cut the matching joint in the drawer sides. *Be sure to cut a left and right side for each drawer.*

Solid stock or plywood may be used. Today, a variety of prefabricated drawer fronts are offered in many standard sizes. See Fig. 18-21.
2. Cut the drawer fronts to the size of the opening (if flush type) allowing a 1/16 in. clearance on each side and on the top. This clearance will vary depending on the depth of the drawer and the kind of material. Deep drawers with solid fronts will require greater vertical

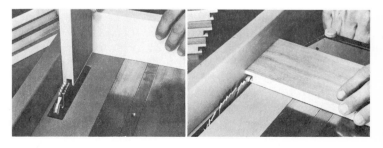

Fig. 18-22. Cutting lock joint in drawer front. Left. Cut groove first. Right. Trim tongue to length in second cut.

8. Cut the required joints for the drawer sides and backs.
9. Trim the bottom to the correct size. Make a trial assembly as shown in Fig. 18-23. Make any adjustments in the fit that may be needed. The parts should fit together smoothly. If they are too tight they will be difficult to assemble after glue is applied.

Fig. 18-24. Gluing drawer parts together. Left. Bottom in place and glue being applied to a front corner. Right. Placing the second side in position after drawer has been turned on its side.

Fig. 18-23. Above. Make trial assembly of the drawer parts to be certain they fit. Below. Parts are laid out and ready to be glued.

12. Carefully check the drawer for squareness. Then drive one or two nails through the bottom into the back. If the bottom was carefully squared, you should have little trouble with this operation. Wipe off the excess glue.

After the glue has cured, fit each drawer to a particular opening. Trim and adjust drawer guides. Place an identifying number or letter on the underside of the bottom. This ''label'' will make it easy to return the drawer to its proper opening after sanding and finishing operations. The inside surfaces of quality drawers should always be sealed and waxed.

10. Disassemble and sand all parts. Rounding of the top edge of the sides can be done at this point. Stop the rounding about 1 in. from each end.
11. Make the final assembly. Any of a number of procedures can be used.

One method is to first glue the bottom onto the front and then glue one of the front corners as shown in Fig. 18-24. Be sure the bottom is centered. The side grooves are usually not glued. Turn the drawer on its side and glue the back into the assembled side. Finally glue on the remaining side as shown. Clamps or a few nails can be used to hold the joints together until the glue hardens.

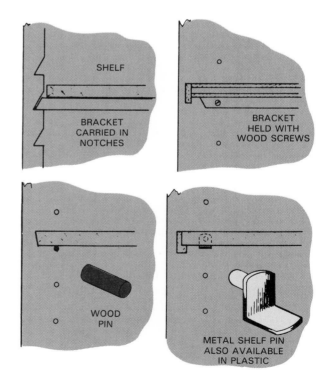

Fig. 18-25. Methods for supporting shelves. When plywood is used for shelving materials, it should be faced with solid stock.

SHELVES

Shelves are widely used in cabinetwork, especially in wall units. In some designs, it may be necessary to fit and glue the shelves into dados cut into the sides of the cabinet. (Refer to Fig. 18-10.) This adds strength to the joint.

If possible, make the shelves adjustable. Then the storage space can be used for various purposes.

Fig. 18-25 presents several methods of installing adjustable supports. Lay out the shelf support system carefully and accurately so the shelves will be level. Usually, it is best to do the cutting and drilling before the cabinet is assembled. If shelf standards are the type set in a groove, it is absolutely necessary to cut the groove before assembly. Some patented adjustable shelf supports are designed to mount on the surface.

Standard 3/4 in. shelving should be supported every 42 in. or closer. This applies especially to shelves that will carry heavy loads. The front edge of plywood shelving should be overlaid with a strip of wood material that matches the cabinet wood. This may be solid stock or thin strips which may be glued to the plywood edge.

DOORS

Either swinging or sliding doors may be used with built-in cabinets. Swinging doors are commonly used for kitchen cabinets.

Doors may be made of plywood, preferably with a lumber core, or they may consist of a frame with a panel insert. Today, fine cabinet doors often have a frame covered on each side with thin plywood. This construction is similar to a hollow core door. Because of its tendency to warp, solid stock is seldom used except on very small doors.

Many carpenters use prefabricated cabinet doors. These are available in various styles, Fig. 18-27, and in a wide range of stock sizes. These

Fig. 18-27. Door inserts of simulated lead glass and expanded metal allow display of items stored on cabinet shelves. (Brammer Mfg. Co.)

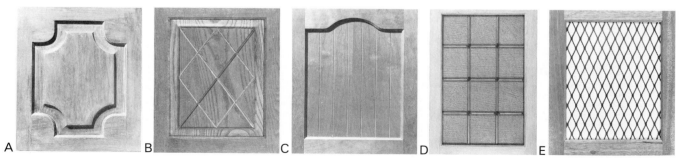

Fig. 18-26. Cabinet doors are constructed in various styles: A—Raised panel. B—Raised panel with straight rails. C—Ply paneled with curved and straight rails. D—Simulated lead glass panel. E—Straight rail frame with diamond design expanded metal insert. (Century Wood Products, Inc.)

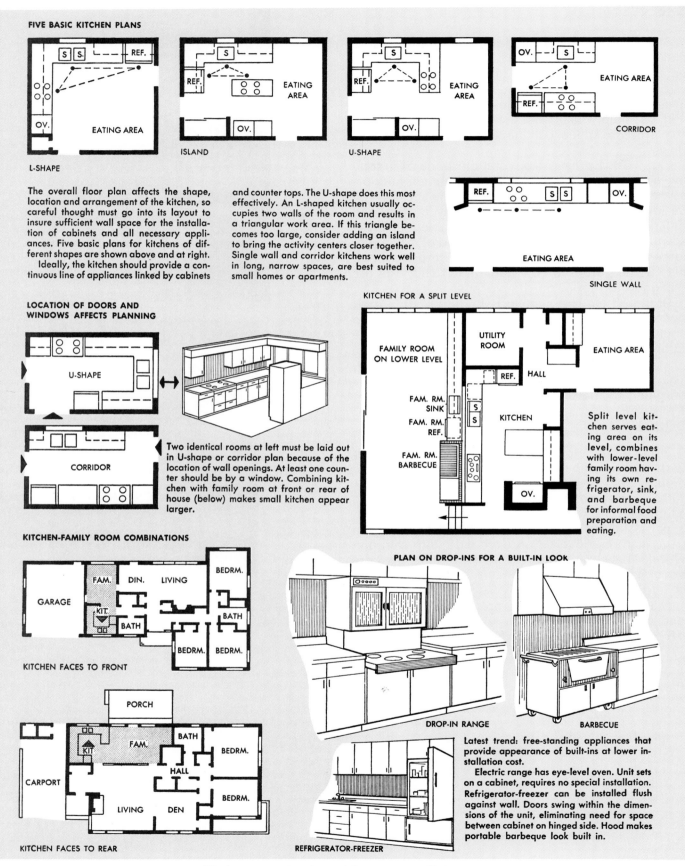

FIVE BASIC KITCHEN PLANS

L-SHAPE

ISLAND

U-SHAPE

CORRIDOR

EATING AREA

EATING AREA

EATING AREA

EATING AREA

SINGLE WALL

The overall floor plan affects the shape, location and arrangement of the kitchen, so careful thought must go into its layout to insure sufficient wall space for the installation of cabinets and all necessary appliances. Five basic plans for kitchens of different shapes are shown above and at right.

Ideally, the kitchen should provide a continuous line of appliances linked by cabinets and counter tops. The U-shape does this most effectively. An L-shaped kitchen usually occupies two walls of the room and results in a triangular work area. If this triangle becomes too large, consider adding an island to bring the activity centers closer together. Single wall and corridor kitchens work well in long, narrow spaces, are best suited to small homes or apartments.

LOCATION OF DOORS AND WINDOWS AFFECTS PLANNING

U-SHAPE

CORRIDOR

Two identical rooms at left must be laid out in U-shape or corridor plan because of the location of wall openings. At least one counter should be by a window. Combining kitchen with family room at front or rear of house (below) makes small kitchen appear larger.

KITCHEN FOR A SPLIT LEVEL

FAMILY ROOM ON LOWER LEVEL

UTILITY ROOM

EATING AREA

REF.

HALL

FAM. RM. SINK

FAM. RM. REF.

KITCHEN

FAM. RM. BARBECUE

OV.

Split level kitchen serves eating area on its level, combines with lower-level family room having its own refrigerator, sink, and barbeque for informal food preparation and eating.

KITCHEN-FAMILY ROOM COMBINATIONS

GARAGE

FAM.

DIN.

LIVING

BEDRM.

KIT.

BATH

BATH

BEDRM.

BEDRM.

KITCHEN FACES TO FRONT

PORCH

CARPORT

KIT.

FAM.

BATH

HALL

BEDRM.

BEDRM.

LIVING

DEN

KITCHEN FACES TO REAR

PLAN ON DROP-INS FOR A BUILT-IN LOOK

DROP-IN RANGE

BARBECUE

REFRIGERATOR-FREEZER

Latest trend: free-standing appliances that provide appearance of built-ins at lower installation cost.

Electric range has eye-level oven. Unit sets on a cabinet, requires no special installation. Refrigerator-freezer can be installed flush against wall. Doors swing within the dimensions of the unit, eliminating need for space between cabinet on hinged side. Hood makes portable barbeque look built in.

Kitchens must be planned for greatest efficiency.

When specifying the size of the opening or the size of a cabinet door, always list width first and height second. This is standard practice among carpenters and is commonly followed by lumber supply stores and millwork plants.

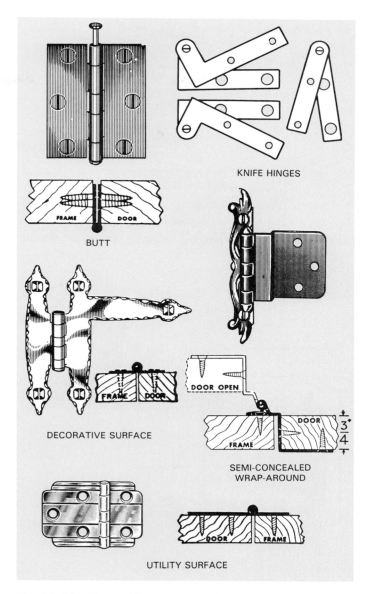

KNIFE HINGES

BUTT

DECORATIVE SURFACE

DOOR OPEN

SEMI-CONCEALED
WRAP-AROUND

UTILITY SURFACE

Fig. 18-29. These hinges are designed for flush cabinet doors. (Amerock Corp.)

are delivered to the job, cut to the exact dimensions with lipped edges (if specified) already machined.

Some manufacturers offer panel doors with special "see-through" inserts. These are useful for display of items on cabinet shelves. Simulated lead glass and expanded metal inserts shown in Figs. 18-26 and 18-27 are of this type.

Swinging doors for cabinets are of three basic types depending on how they fit into or over the door opening. Fig. 18-28 shows the basic shapes of the three types:
1. Flush door. The door fits into the opening and

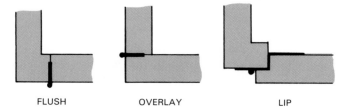

FLUSH OVERLAY LIP

Fig. 18-28. Cabinet doors are of three basic types.

does not project outward beyond the frame.
2. Overlay door. Though its edges are square like the flush door, it is mounted on the outside of the frame wholly or partly concealing it.
3. Lipped door. The door is rabbeted along all edges so that part of the door is inside the door frame. A lip extends over the frame on all sides concealing the opening.

INSTALLING FLUSH DOORS

The flush door is usually installed with butt hinges. However, surface hinges, wrap-around hinges, knife hinges, or various semiconcealed hinges can be used. Fig. 18-29 shows several styles that are good for cabinetwork.

The size of a hinge is determined by its length and width when open. Select the size to suit the size of the door. Large wardrobe or storage cabinet doors should have three hinges.

Flush doors must be carefully fitted to the opening with about 1/16 in. clearance on each edge. The total gain (space) required for the hinge can be cut entirely in the door. However, for fine work, it

is best to cut equally into door and stile. On large doors, it is practical to use the portable router for this operation.

Install the hinges on the door and then mount the door in the opening. Use only one screw in each hinge leaf. Adjust the fit and then set the remaining screws. It may be necessary to plane a slight bevel on the door edge opposite the hinges.

Stops are set on the door frame so the door will be held flush with the surface of the opening when closed. The stops may be placed all around the opening or only on the lock or catch side.

Overlay doors provide an attractive appearance in some contemporary styles of cabinetwork. Butt hinges can be used; however, it is usually best to make the installation with a type that is designed for the purpose, Fig. 18-30.

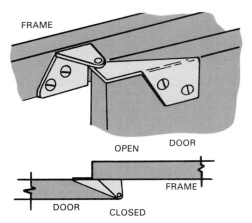

Fig. 18-30. This pivot hinge is designed for use on overlay doors.

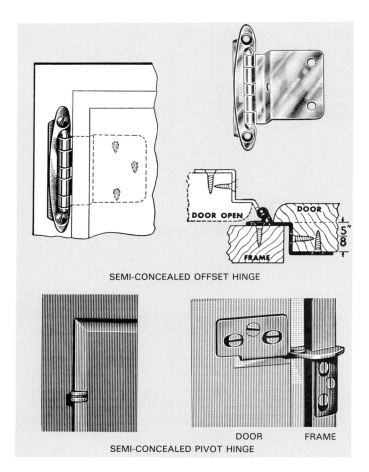

SEMI-CONCEALED OFFSET HINGE

SEMI-CONCEALED PIVOT HINGE

Fig. 18-31. Hinges for lip doors. Note how pivot hinges conceal all but the pin.

CUTTING AND FITTING LIP DOORS

Lip doors are easier to cut and fit than flush doors because the clearance is covered. They must be installed with a special off-set hinge. See two styles in Fig. 18-31.

Select the door blanks for a given series of openings. Try to match grain and color when a natural finish will be applied. Cut the doors to the size of the openings plus the amount required for the lip all around. Allow clearance of about 1/16 in. on each edge. Make an additional allowance for the hinges which are not gained into the door or frame.

The lip is formed by cutting a rabbet along the edge. This can be accomplished on the table saw or with a shaper as shown in Fig. 18-32.

Check the fit of each door in its proper opening and mark its position in small letters on the back. Remove all machine marks with abrasive paper.

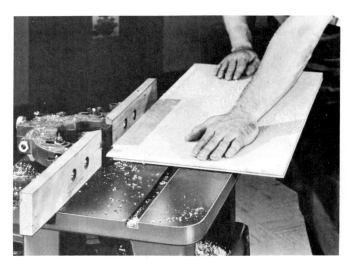

Fig. 18-32. Using a shaper to form the lip on a cabinet door. (Rockwell International)

Soften edges and corners. Hardware may be pre-fitted and then removed before the finishing operations. Fig. 18-33 shows lip doors in place after finish has been applied.

SLIDING DOORS

Sliding doors are often used where the swing of regular doors would be awkward or cause interference. They are adaptable to various styles and structural designs, Fig. 18-34.

Construction details of a sliding door arrangement are shown in Fig. 18-35. Grooves are cut in the top and bottom of the cabinet before assembly. The doors are sometimes rabbeted so the edge formed will match the groove with about 1/16 in. clearance. Cut the top rabbet and groove deep enough that the door can be inserted or removed by simply raising it into the extra space. The doors will slide smoothly if the grooves are carefully cut, sanded, sealed, and waxed. Avoid too much finish.

Fig. 18-33. Lip doors are installed on this base cabinet.

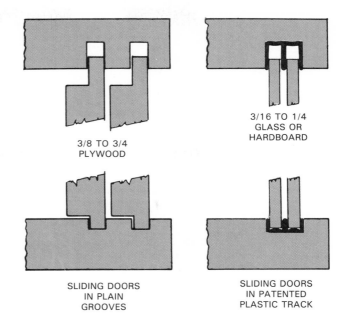

3/8 TO 3/4
PLYWOOD

3/16 TO 1/4
GLASS OR
HARDBOARD

SLIDING DOORS
IN PLAIN
GROOVES

SLIDING DOORS
IN PATENTED
PLASTIC TRACK

Fig. 18-35. Sliding door details. Thicker door stock must be rabbeted to fit the track.

Fig. 18-34. Sliding doors in a storage cabinet. They can be adapted to various styles.

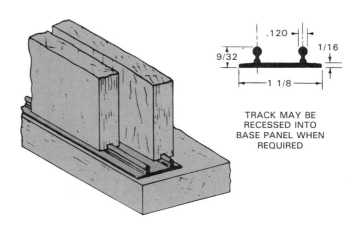

.120
9/32
1/16
1 1/8

TRACK MAY BE
RECESSED INTO
BASE PANEL WHEN
REQUIRED

Fig. 18-36. Plastic door track is self-lubricating.
(Kentron Div. North American Reiss)

Sliding glass doors are sometimes specified in cabinets. They are heavy and require a special plastic or roller track. Follow the manufacturer's recommendations for installation.

A wide range of sliding door track and rollers are available. Fig. 18-36 illustrates a self-lubricating plastic track. It is easy to install and provides smooth operation for furniture and cabinet doors. Overhead track and rollers are used for large wardrobe doors and passage doors.

Sliding glass doors are adaptable to wall units and add variety to an installation. Fig. 18-37 shows bypass (1/4 in. plate glass) sliding doors for a cabinet in a kitchen-dinette area.

COUNTERS AND TOPS

In modern cabinetwork, a high-pressure type of plastic laminate is commonly used as a surface for counters and tops. This material, usually 1/16 in.

thick, offers high resistance to wear and is unharmed by boiling water, alcohol, oil, grease, and ordinary household chemicals, Fig. 18-38.

Although the laminate is very hard, it does not possess great strength and is serviceable only when bonded to plywood, particle board, waferwood, or hardboard. This base or core material must be smooth and dimensionally stable. Hardwood plywood (usually 3/4 in. thick) makes a satisfactory base. However, some plywoods, especially fir, have a coarse grain texture which may telegraph (show through). Particle board, which is less expensive than plywood, provides a smooth surface and adequate strength.

When the core or base is free to move and is not supported by other parts of the structure, the laminated surface may warp. This can be counteracted by bonding a backing sheet of the laminate

Fig. 18-37. Sliding plate glass doors are featured in a wall cabinet.

to the second face. It will minimize moisture gain or loss and provides a balanced unit with identical materials on either side of the core. For a premium grade of cabinetwork, Architectural Woodwork Institute standards specify that a backing sheet be used on any unsupported area exceeding 4 sq. ft. Backing sheets are like the regular laminate without the decorative finish and are usually thinner. A

standard thickness for use opposite a .060 in. (1/16) face laminate is .020 in.

WORKING LAMINATES

Plastic laminates can be cut to rough size with a handsaw, tablesaw, portable saw, or portable router, Fig. 18-39. Use fine-tooth blades and support the material close to the cut. Laminates 1/32 in. thick, which are used on vertical surfaces, can be cut with tin snips.

It is best to make the roughing cuts 1/8 in. to 1/4 in. oversize. Trim the edges after the laminate has been mounted. Handle large sheets carefully because they can be easily cracked or broken. Be careful not to scratch the decorative side.

Contact bond cement is used to apply the plastic laminate because no sustained pressure is required. It is applied with a spreader, roller, or brush to both surfaces being joined. On large horizontal surfaces, it is best to use a spreader, Fig. 18-40.

For soft plywoods, particle board, or other porous surfaces, the spreader is held with the serrated edge perpendicular to the surface. On hard, nonporous surfaces, and plastic laminate, hold the edge at a 45 deg. angle, as shown. A single coat should be sufficient.

Fig. 18-38. Heat resistance factors make plastic laminates an ideal surface material for counters with built-in cooking units. (Formica Corp.)

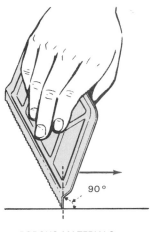
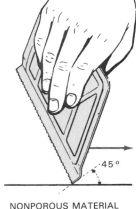

90°

POROUS MATERIALS

45°

NONPOROUS MATERIAL AND PLASTIC LAMINATE

Fig. 18-40. Metal spreader may be used to apply contact bond cement.

An animal hair or fiber brush may be used to apply the adhesive to small surfaces or those in a vertical position. Apply one coat, let it dry thoroughly and then apply a second.

All of the surface should be completely covered with a glossy film. Dull spots, after drying, indicate that the application was too thin and that another coat should be applied.

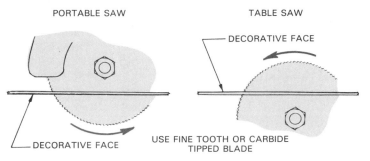

PORTABLE SAW   TABLE SAW

DECORATIVE FACE

DECORATIVE FACE   USE FINE TOOTH OR CARBIDE TIPPED BLADE

Fig. 18-39. Plastic laminate can be cut with circular saws.

Stir the adhesive thoroughly before using and follow the manufacturer's recommendations. Usually, brushes and applicators must be cleaned in a special solvent.

Some types of contact cement are extremely flammable. Nonflammable types may produce harmful vapors. Be sure the work area is well ventilated. Follow the manufacturer's directions and observe precautions.

After the adhesive has been applied to both surfaces, let them dry (usually at least 15 minutes). You can test the dryness by pressing a piece of paper lightly against the coated surface. If no adhesive sticks to the paper, it is ready to be bonded. This bond can usually be made any time within an hour. Time varies with different manufacturers. If the assembly cannot be made within this time, the adhesive can be reactivated by applying another thin coat.

ADHERING LAMINATES

Bring the two surfaces together in the EXACT POSITION required because they cannot be shifted once contact is made. When joining large surfaces, place a sheet of heavy wrapping paper, called a slip-sheet, over the base surface. Then slide the laminate into position. Withdraw the paper slightly so one edge can be bonded and then remove the entire sheet and apply pressure.

Total bond is secured by the application of momentary pressure. Hand rolling provides satisfactory results if the roller is small (3 in. or less in length), Fig. 18-41. Long rollers apply less pressure per square inch. Work from the center to the outside edges and be certain to roll every square inch of surface. In corners and areas that are hard to roll, hold a block of soft wood on the surface and tap it with a rubber mallet.

Fig. 18-42. Using an electric router to trim an overhanging edge. The guide insures an accurate cut.
(Black & Decker Mfg. Co.)

Trimming and smoothing the edges is an important step in the application of a plastic laminate. A plane or file may be used but many carpenters prefer an electric router, equipped with an adjustable guide, Fig. 18-42. In making cutouts for sinks or other openings, an electric saber saw, as illustrated in Fig. 18-43, is practical.

The corner and edges of a plastic laminate application should be beveled, Fig. 18-44. This makes it smooth to touch and more durable. The bevel angle can be formed with a smooth mill file, Fig. 18-45. Stroke the file downward and use care not to damage the surfaces of the laminate. Some routers can be equipped with an adjustable guide and a special bit that will make this cut, Fig. 18-46. Final smoothing and a slight rounding of the bevel should be done with a 400 wet-or-dry abrasive paper.

When working with plastic laminates, be especially careful that files, edge tools, or abrasive

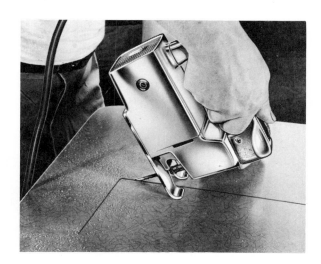

Fig. 18-43. Using an electric saber saw to cut an opening for a sink. View shows start of cut.

Fig. 18-41. Apply pressure to the laminate with a hand roller.

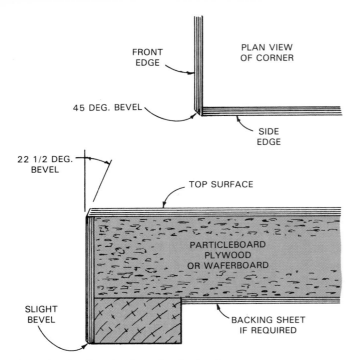

Fig. 18-44. Bevels for plastic laminate corners are very important in the production of quality work.

Fig. 18-46. Special router bit also makes a beveled corner. (Black & Decker Mfg. Co.)

papers do not damage the finished surfaces. Such damage cannot easily be repaired.

A completed countertop for the bathroom built-in unit pictured previously in construction views is shown in Fig. 18-47. A special metal cove and cap strip have been used to trim the "back splash." FHA requires a minimum back splash height of 4 in. where kitchen counters join the wall surface.

CABINET HARDWARE

After counters and tops have been covered and surface finishes applied, hardware is installed. This consists of knobs, pulls, and various other metal fittings. Care should be used in the selection of an appropriate size, style, material, and finish.

Fig. 18-48 shows several designs of quality hardware for drawers and doors.

Some hardware, mainly hinges, should be prefitted before the final surface finish is applied. Pulls and catches can be installed afterward. Use care in the layout. Drill holes with sharp bits that will not splinter the surface. Drilling jigs, like the one shown in Fig. 18-49, are fast and accurate.

Drawer pulls usually look best when they are located slightly above the centerline of the drawer front. Normally, they should be centered horizontally and level. Door pulls are convenient when located on the bottom third of wall cabinets and the top third of base cabinets.

Swinging doors require some type of catch to keep them closed. Several common types are illustrated in Fig. 18-50. An important consideration in their selection is the noise they produce. In general, the catch should be placed as near as practical to the door pull. The package usually contains instructions for installation. Follow these carefully.

OTHER BUILT-IN UNITS

Previous descriptions have been largely directed toward kitchen cabinets. The same general procedures can be applied to wardrobes, room dividers, and various built-in units for living rooms and family rooms. See Fig. 18-51.

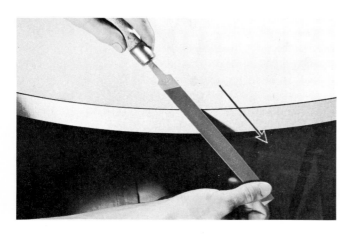

Fig. 18-45. A mill file can be used to produce a beveled corner.

Fig. 18-47. Plastic laminate counter surface is completed for a bathroom built-in unit.

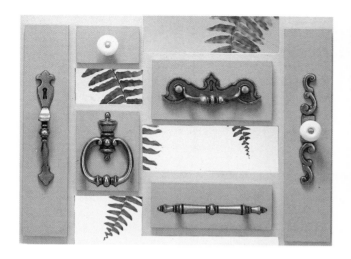

Fig. 18-48. Cabinet drawer and door hardware. (Amerock Corp.)

Fig. 18-49. A drilling jig can be used to locate holes for door pulls.

Built-in features are popular largely because of:
1. The demand for storage space.
2. The general requirement that all space be organized and used as efficiently as possible.
3. The attractive customized appearance that can result from well designed and carefully constructed units.

Study the details of the construction provided in the architectural plans or develop a carefully prepared working drawing using the actual dimensions secured from the wall, floor, and ceiling surfaces. When drawings are provided, check the space available to make sure that it agrees with the drawings.

Wall and floor surfaces are seldom perfectly level or plumb. Slight adjustments will be needed as the cabinetwork is constructed. Room corners or the corners of alcoves designed for built-in units will likely not be square. Here, the cabinetwork will need to be trimmed or strips and wedges used to bring the work into proper alignment.

Keep a square and level constantly at hand especially during the rough-in stages, Fig. 18-52. Do not carry inaccuracies from the wall or floor

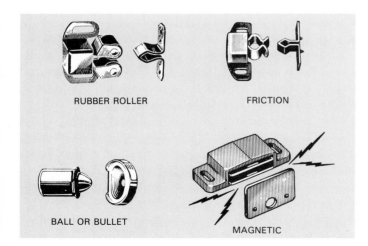

RUBBER ROLLER   FRICTION

BALL OR BULLET

MAGNETIC

Fig. 18-50. Cabinet doors catches use various devices to hold doors shut.

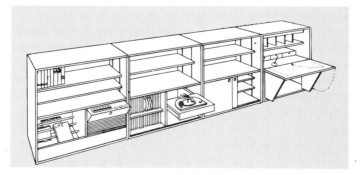

Fig. 18-51. Built-in units for living rooms add storage and work areas.

Fig. 18-52. This built-in unit will serve as a wardrobe and room divider in a bedroom area. (American Plywood Assoc.)

into the cabinet framework. Doing so will make it difficult to keep the cabinet facing plumb and level. Errors will also be a source of annoyance during the hanging of doors and fitting of drawers.

Built-in units vary in so many ways that no particular procedure can be recommended for their construction. For example: a bookcase that also serves as a room divider, Fig. 18-53, could be constructed during the regular interior finishing stages, as shown. It could also be built as a detachable unit and installed just before interior painting and decorating. Fig. 18-54 shows a partial view of built-in units located in a bedroom area. Here the various components (headboard, bookcase, desk) were constructed and surface finishes applied in a basement area. Then they were moved into the room and attached to the walls after walls were painted and the carpet was laid.

SEQUENCE OF INTERIOR FINISH

Interior finish, after wall, ceiling, and floor surfaces are complete, involves a wide variety of work as described in this Unit and Unit 17. The se-

quence in which the work is performed may vary considerably. Much depends upon the type of materials and construction used. The carpenter should organize the work to prevent bottlenecks (delays). Appropriate times should be established

Fig. 18-53. Attaching facing strips to a bookcase. This unit could serve as a room divider, as well.

Fig. 18-54. Built-in bookcase, desk, and storage in a bedroom area. Plastic laminate top matches grain pattern and color of wood. Door features a woven wood pattern.

Fig. 18-55. Sequence of interior finish. Top. Basic framing for a built-in desk. Center. Facing and top complete. Bottom. Drawers fitted, plastic laminate applied, surface finish applied, and floor covering laid.

for the delivery of materials. When delivered too far ahead, they will interfere with other work and may be damaged. Work schedules need to be carefully planned and followed as closely as possible. The application of trim and the construction of cabinetwork generally proceed at the same time.

Fig. 18-55 illustrates the sequence on a job where assembled-in-place cabinetwork was used. The top view shows the basic framework for a built-in desk in a kitchen area. It is installed on the underlayment and before the baseboard is set. In the center view, the facing and top have been added. Note that the baseboard is in position. Drawers are built and fitted and then surface finishes are applied to the desk, baseboard, and other trim. The plastic laminate top surface is installed. Then, the floor covering is laid. Finally, as shown in the bottom view, hardware is fitted and the base shoe is set in place.

FACTORY-BUILT CABINETS

Today, a major part of the cabinetwork for residential and commercial buildings is constructed in factories that specialize in this work. Modern production machines and tools can save time and produce high-quality work. Mass produced parts are assembled with the aid of jigs and fixtures. See Figs. 18-56 and 18-57.

Factory-built cabinets may be obtained in one of three forms:
1. Disassembled.
2. Assembled but not finished (in-the-white).
3. Assembled and finished.

Disassembled or "knocked-down" cabinets consist of parts cut to size and ready to be assembled on the job. The assembled but unfinished cabinet is ready to set in place. Hardware is included but not installed. All surfaces are sanded and ready for finish. After installation, finishing materials and procedures can be coordinated with doors and inside trim—thus insuring an exact match.

The assembled and finished cabinets are widely used because they save time during finishing stages of construction. Manufacturers offer a variety of shades and colors which are applied by experts, Fig. 18-58. Because of the controlled conditions and special equipment, finishing materials

Fig. 18-56. Mass production of cabinet work. View shows final assembly operations as cabinets are carried on a conveyor. (Kitchen Kompact, Inc.)

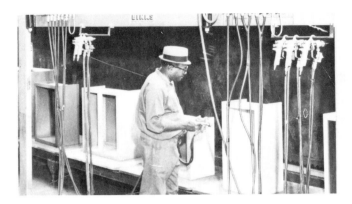

Fig. 18-59. Spray finishing kitchen cabinet units as they move through a modern spray booth. (Binks Mfg. Co.)

Fig. 18-57. Operator removes workpiece from automatic shaper. Revolving table carries workpieces past a cutterhead that forms profile and shapes edge in a single pass. (Conestoga Wood Specialties, Inc.)

with high resistance to moisture, acids, and abrasion can be applied, Fig. 18-59. After the finishing process is complete and hardware has been installed, the units are carefully packaged and shipped to the distributor or directly to the construction site.

Manufacturers of cabinetwork offer a variety of standard units, especially in kitchen cabinets. Most kitchen layouts can be made entirely from these units. However, when necessary, special custom built units can be ordered. Sometimes factory-built cabinets are combined with units constructed in custom cabinet shops, Fig. 18-60, or with various units built on the job. Fig. 18-61 shows sketches of standard base and wall units offered by one cabinet manufacturer. In addition to these, other standard units include oven and utility cabinets, peninsular base units, and bathroom vanity and sink cabinets.

Fig. 18-58. Highly skilled woodfinishers hand rub stain coating on premium quality cabinet doors. (Riviera Kitchens, an Evans Products Co.)

Fig. 18-60. Built-in storage cabinet with sliding doors constructed in a cabinet shop. (Weyerhaeuser Co.)

Cabinetmaking

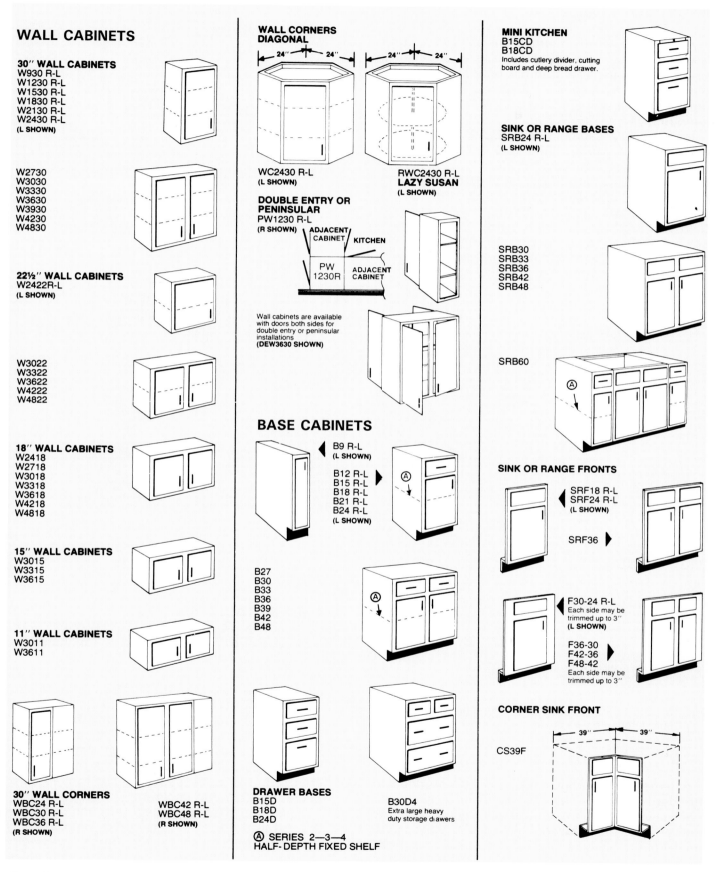

Fig. 18-61. Cabinet manufacturers provide information about their standard cabinet units. Dimensions are given in code numbers. (Brammer Mfg. Co.)

467

CABINET MATERIALS

A wide range of materials are used in factory-built cabinets. Low-priced cabinets are usually made from panels of particle board with a vinyl film applied to exposed surfaces. The vinyl is printed with a wood grain pattern and has the appearance of genuine wood.

High quality cabinets are made from veneers and solid hardwoods including such species as oak, birch, ash, and hickory. Hardboard, particle board, and waferboard may be used for certain interior panels, drawer bottoms, and as the base for plastic laminate countertops. Frames are assembled with accurately made joints. Dovetail joints are generally used in drawer assemblies. See Fig. 18-62. A completed installation along one side of a kitchen is shown in Fig. 18-63.

Fig. 18-63. Installation view of modern factory-built cabinets along one wall. Note the built-in cooking top and oven. (Riviera Kitchens, an Evans Products Co.)

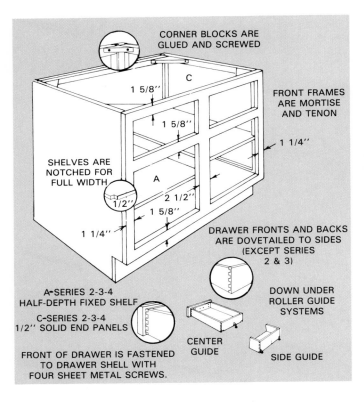

Fig. 18-62. Drawings show construction details of factory-built kitchen base cabinet. (Brammer Mfg. Co.)

A variety of storage features can be added to standard cabinet units. Examples include revolving shelves (Lazy Susan), special compartments and dividers in drawers, slide-out breadboards, and slide-out shelves, Fig. 18-64. A canned goods storage unit, Fig. 18-65, provides extra convenience with swing-out shelving. A wire rack provides special lid storage, Fig. 18-66.

Fig. 18-64. Slide-out shelves can be located in base cabinet. (Riviera Kitchens, an Evans Products Co.)

CABINET INSTALLATION

There are two basic procedures for installing factory-built cabinets:

1. Some manufacturers recommend that the wall cabinets be installed first. When layouts are made and wall studs located, the wall units are lifted into position. They are held with a padded T-brace that allows the worker to stand close to the wall while making the installation. After the wall cabinets are securely attached and checked, the base cabinets are moved into place, leveled, and secured.

2. Following a second procedure, base cabinets are installed first. The tops of the base cabinets can then be used to support braces that hold the wall units in place. This procedure is illustrated in Fig. 18-67.

Floors and walls are seldom exactly level and

Fig. 18-65. Can good storage unit has swing-out shelves. (Riviera Kitchens, an Evans Products Co.)

Fig. 18-66. Wire rack stores pan lids under cook surface. (Amerock Corp.)

Factory-built cabinet units are attractive in appearance and offer organized storage, shelving, and desk space for an office or family room. Shown are four standard base units and a desk insert. (Haas Cabinet Co., Inc.)

1.

Locate the position of all wall studs where cabinets are to hang by tapping with a hammer. Mark their position where the marks can easily be seen when the cabinets are in position.

2.

Find the highest point on the floor with a level. This is important for both base and wall cabinet installation later. Remove the baseboard from all walls where cabinets are to be installed. This will allow them to go flush against the walls.

3.

Start the installation with a corner or end unit. Slide it into place then continue to slide the other base cabinets into the proper position.

4.

When all base cabinets are in position, fasten the cabinets together. This is done by drilling a 1/4'' diameter hole through the face frames and using the 3'' screws and T-nuts provided. To get maximum holding power from the screw, one hole should be close to the top of the end stile and one should be close to the bottom.

5.

Check the position of each cabinet with a spirit level, going from the front of the cabinet to the back of the cabinet. Next shim between the cabinet and the wall for a perfect base cabinet installation.

6.

Starting at the high point in the floor, level the leading edges of the cabinets. Continue to shim between the cabinets and the floor until all the base cabinets have been brought to level.

7.

After the cabinets have been leveled, both front to back and across the front, fasten the cabinets to the wall at the stud locations. This is done by drilling a 3/32'' diameter hole 2 1/4'' deep through both the hanging strips for the 2 1/2'' x 8 screws that are provided.

8.

Fit the counter top into position and attach it to the base cabinets by predrilling and screwing through the front corner blocks into the top. Use caution not to drill through the top. Cover the counter top for protection while the wall cabinets are being installed.

9.

Position the bottom of the 30'' wall cabinets 19'' from the top of the base cabinet, unless the cabinets are to be installed against a soffit. A brace can be made to help hold the wall cabinets in place while they are being fastened. Start the wall cabinets installation with a corner or end cabinet. Use care in getting this cabinet installed plumb and level.

10.

Temporarily secure the adjoining wall cabinets so that leveling may be done without removing them. Drill through the end stiles of the cabinets and fasten them together as was done with the base cabinets.

11.

Use a spirit level to check the horizontal surfaces. Shim between the cabinet and the wall until the cabinet is level. This is necessary if doors are to fit properly.

12.

Check the perpendicular surface of each frame at the front. When the cabinets are level, both front to back and across the front, permanently attach the cabinets to the wall. This is done by predrilling a 3/32'' diameter hole 2 1/4'' deep through the hanging strip inside the top and below the bottom of the cabinets at the stud location. Enough Number 8 screws should be used to fasten the cabinets securely to the wall.

Fig. 18-67. Basic steps for installing factory-built cabinets. Some installers prefer to hang wall cabinets first. (Haas Cabinet Co., Inc.)

plumb. Therefore, shims and blocking must be used so the cabinets are not racked or twisted. Doors and drawers cannot be expected to operate properly if the basic cabinet is distorted by improper installation.

Screws should go through the hanging strips and into the stud framing. Never use nails. Toggle bolts are required when studs are inaccessible. Join units by first clamping them together and then, while aligned, install bolts and T-nuts.

CABINETS FOR OTHER ROOMS

Some manufacturers of kitchen cabinets build storage units for other areas of the home. Built-in units provide drawers and cabinets that save space through organized storage. They improve efficiency and also offer greater convenience. For example, the cabinets shown in Fig. 18-68 include storage for towels, soap, dental and pharmaceutical supplies, cosmetics, and many other items

Fig. 18-68. Ready-built storage unit for a bathroom includes a lavatory counter top, a vanity, base cabinets, and wall cabinets. Such units are designed for quick installation by the carpenter.

Fig. 18-69. Typical modular wall storage system. It was formed with four base units, two end panels, two center panels, three valance units, a shelf cabinet, and three back panels. (Haas Cabinet Co., Inc.)

needed in the bath area.

In other rooms, carefully planned built-in cabinets provide efficient storage and display for items used in work, recreation, and hobby activities. Their custom-built appearance usually adds to room decor. They may even eliminate the need for some movable pieces of furniture.

Fig. 18-69 shows a wall unit installed in a family room. The term, modular, refers to units assembled from parts having standard sizes. Thus, a variety of finished assemblies can be made from the same parts.

TEST YOUR KNOWLEDGE — UNIT 18

1. The base unit of a standard kitchen cabinet is _____ in. high.
2. When carpenters draw a full-size sectional view of a cabinet, they generally refer to it as a _____ _____.
3. When building assembled-in-place base cabinets, the partitions are installed _____ (before, after) the bottom is secured to the base.
4. The vertical members used to face a cabinet are called _____.
5. Generally, stiles are installed _____ (before, after) the rails on assembled-in-place cabinets.
6. Three common types of drawer guides include: side guides, corner guides, and _____ _____.
7. Stock for kitchen cabinet drawer sides is usually _____ to _____ in. thick.
8. Standard shelving that is 3/4 in. thick should be carried on supports spaced no greater than _____ in.
9. Three types of swinging doors that may be used in cabinetwork include: lip, flush, and _____.

10. When making a plastic laminate installation on a table or counter, it is recommended that a _____ be used on any unsupported area greater than four square feet.
11. When exerting pressure to bond a plastic laminate by hand, it is best to use a _____ (long, short) roller.
12. Drawer pulls are usually located slightly _____ (above, below) the centerline.

OUTSIDE ASSIGNMENTS

1. Secure a set of architectural plans that include detail drawings of the kitchen cabinets. After a study of the details, prepare a master layout of a typical base cabinet. Since a complete view would be rather large, you may prefer to show only the front 4 - 6 in. Use a sharp pencil. Show all the parts and clearances. Present your drawing to the class with explanations of the various structural parts.
2. If you have access to shop equipment, construct a typical cabinet drawer. Select appropriate joints that can be used for on-the-job built cabinets. Cut and fit the parts carefully. You may prefer not to glue the drawer together so that when you make a presentation to your class, it will be easier to show the joints. Another alternative is to glue the drawer, then cut it apart in several places so joint sections could be easily viewed.
3. Study the various methods and devices used to install sinks in counter tops. From manufacturer's literature, prepare a sectional drawing about four times actual size that you can use in making a presentation to your class. Check into the possibility of borrowing sample rims and brackets from a building supply store that can be passed around to the group during your explanation.

Island cabinetry is featured in this modern kitchen design. Carpenters are often called upon to lay out and install similar factory-built systems.

Built-in bathroom unit includes lavatory counter top, base cabinets, and wall cabinets for plenty of storage. (Haas Cabinet Co., Inc.)

Modern prefabricated fireplace. Room air is drawn from both sides into sealed heating chamber and flows back into room through top louvers. Air for fire is piped in from outside. Bi-fold glass doors limit the loss of heated room air through the chimney. (Majestic Co.)

The framework for a modern chimney is erected while the wall and roof framing are being built. No masonry work is needed for such a chimney. (American Plywood Assoc.)

Masonry fireplace with random stone facing and rustic mantel. Note the vent holes on the sides. Warmed air enters the room through these vents. (C.E. Morgan)

Unit 19

CHIMNEYS AND FIREPLACES

A chimney is a vertical shaft which exhausts the smoke and gases from heating units, fireplaces, and incinerators. When properly designed and built, a chimney may also improve the outside appearance of a home.

The fireplace once was the only source of heat for dwellings. Today the fireplace is popular even though its efficienty as a source of heat is low compared with modern heating systems. The desire for a fireplace results from the cheerful, homelike atmosphere it creates and the value it adds to the home.

The rising cost of energy has led to many improvements in fireplace design. Masonry fireplaces are often equipped with glass doors to save heat. They may also include ductwork to heat room air as it is circulated around the firebox. A wide variety of prefabricated fireplaces are available. They are easy to install and are usually less expensive than masonry units.

MASONRY CHIMNEYS

A masonry chimney usually has its own footing and is built in such a way that it provides no support to, nor receives support from, the building frame. Footings should extend below the frost line and project at least 6 in. beyond the sides. Walls of a chimney with clay-flue lining should be at least 4 in. thick. Foundations for a chimney or fireplace, especially when located on an outside wall, may be combined with those used for the building structure.

The size of a chimney will depend on the number, arrangement, and size of the flues (vertical openings in the chimney). The flue for a heating plant should have enough cross-sectional area and height to create a good draft. This permits the equipment to develop its rated output. Always follow the heating equipment manufacturer's recommendations when deciding on flue sizes.

Building codes require that chimneys be constructed high enough to avoid downdrafts caused by the turbulence of wind as it sweeps past nearby obstructions or over sloping roofs. Minimum heights generally required are illustrated in Fig. 19-1. The chimney should always extend at least 2 ft. above any roof ridge that is within a 10 ft. horizontal distance.

Combustible materials such as wood framing members should be located at least 2 in. away from the chimney wall, Fig. 19-2. The open spaces between the framework and the chimney can be filled with mineral wool or other noncombustible material.

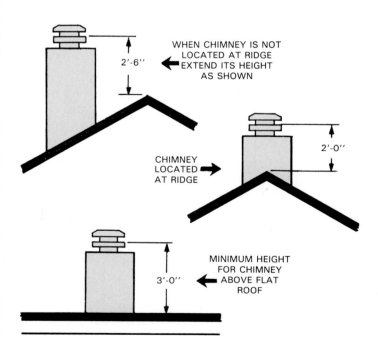

Fig. 19-1. Maintain these minimum chimney heights above roof. Check building codes in your area.

475

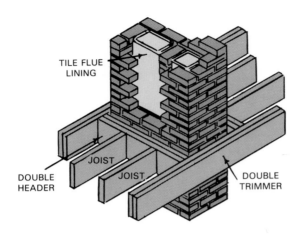

Fig. 19-2. Floor framing around chimney. Note tile flue lining. When a flue column is next to another, joints of flue lining should be staggered vertically a distance of at least 7 in.

FLUE LININGS

The National Board of Fire Underwriters recommends fireclay flue linings for masonry chimneys. Most local building codes also require them.

Linings are available in square, rectangular, and round shapes. Each shape is available in several sizes as listed in Fig. 19-3. Note that different methods of measurement are used for the three types:
1. Outside measurements for the old standard.
2. Nominal outside dimensions for the modular.
3. Inside dimensions for the round linings.
Small and medium flue linings are usually available in 2 ft. lengths. Flue rings are placed during the assembly of the chimney to provide a form for the masonry.

Although unlined fireplace chimneys with masonry walls (minimum thickness 8 in.) will operate well, glazed flue lining is usually recommended. The smooth inside surface of a glazed lining is less likely to attract pitch and tar which may eventually restrict the passage.

As a rule, a single flue should be used for only one heating unit. However, it is often permitted to connect the vent from a gas-fired hot water heater to a furnace flue. Do not combine larger flues.

CONSTRUCTION

Allow the chimney footing to cure for several days to give it proper strength. Use care in starting the masonry work. Make all lines level and plumb.

Each section of flue lining is set in place before each part of the chimney wall is built. The lining is a guide for the brick work. Joints in the flue lining are bedded in mortar or fireclay. Use care in placing the lining units so they are square and plumb. Brick work is carried up along the lining. Then another lining unit is placed.

Where offsets or bends are necessary in the chimney, miter the ends of the abutting sections of lining. This prevents reduction of the flue area. The angle of offset should be limited to 60 deg. The center of gravity of the upper section must not fall beyond the centerline of the lower wall.

Chimneys are often corbeled (extended outward) just before they project through the roof, Fig. 19-4. The larger exterior appearance is often more attractive. Also, breakage due to wind is less likely for thicker masonry. Corbeling should not exceed a 1 in. projection in each course. The final size should be reached at least 6 in. below the roof framing.

Openings in the roof frame should be formed

STANDARD LINERS		MODULAR LINERS		ROUND LINERS	
Outside Dimensions	Area of Passage	Nominal Outside Dimensions	Area of Passage	Inside Diameter	Area of Passage
8 1/2x8 1/2	52.56	8x12	57	8	50.26
		8x16	74		
8 1/2x13	80.50			10	78.54
		12x12	87		
8 1/2x18	109.69	12x16	120	12	113.00
13x13	126.56	16x16	162	15	176.70
13x18	182.84	16x20	208		
18x18	248.06	20x20	262	18	254.40
		20x24	320	20	314.10
		24x24	385	22	380.13
				24	452.30

Fig. 19-3. Flue liner dimensions and clear cross-sectional areas in square inches. The National Building Code requires that the wall of the flue liner be a minimum of 5/8 in. thick.

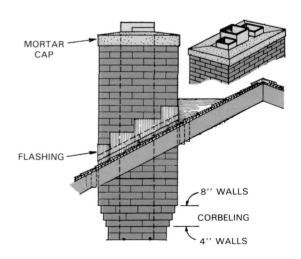

Fig. 19-4. Enlarging the top of a chimney helps it resist wind breakage and usually provides a more attractive outline.

before constructing the chimney. Refer to Unit 9 for information and procedure. Water leakage around the base can be prevented by proper flashing. Corrosion resistant metal such as sheet copper is often used for this purpose. The flashing is built into the roof surface and extends up along the masonry. Cap or counter flashing is bonded into the mortar joints and is then lapped down over the base flashing. Refer to Unit 10 for further details concerning the installation of flashing.

At the top of the chimney, the flue lining should project at least 4 in. above the top brick course or capping. Surround the lining with cement mortar at least 2 in. thick. Slope the cap so wind currents are directed upward and so water will drain away from the center. When several flues are located in the same construction, it is best to extend them to different heights. Space them no closer than 4 in. apart horizontally.

MASONRY FIREPLACES

Fig. 19-5 shows a cutaway view that reveals the parts of a masonry fireplace. The chimney and fire-

place are combined in a single unit which often includes a flue for the regular heating equipment.

The hearth consists of two parts: one is located in front (front hearth) and the other is below the fire area (back hearth). The latter, along with the sides and back wall, are lined with firebrick which will withstand direct contact with flame. The side and back walls are sloped to reflect heat into the room.

A damper is located above the fire to control it. The damper also prevents loss of heat from the room when the fireplace is not being used. The throat, smoke shelf, and smoke chamber are all important parts of the fireplace and must be carefully designed to insure good operation.

Architectural plans often include details of the fireplace construction. See Fig. 19-6. The drawings

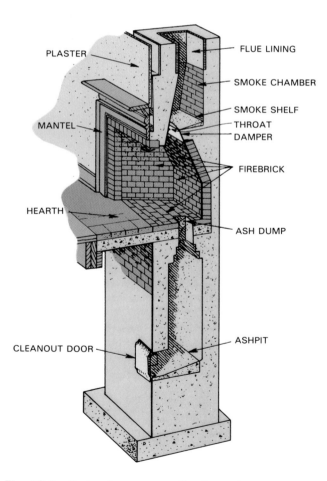

Fig. 19-5. Parts of a masonry fireplace. The carpenter who does remodeling should know about many kinds of structures.

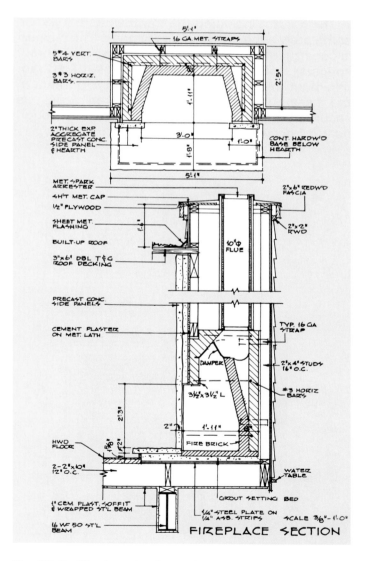

Fig. 19-6. Fireplace details are included in the architectural plans. Note the 2 x 4 framework on the outer edge in both views. A chimney frame of wood which juts out from a wall is called a chase.

include overall dimensions and can be scaled (measured) to find other lengths.

DESIGN DETAILS

The size of the fireplace opening should be based on the size of the room and be matched with the style of architecture. Some authorities suggest that it accommodate firewood 2 ft. long (standard cordwood cut in half). Fig. 19-7 illustrates standard procedure in listing fireplace dimensions.

Masonry structure and wood framing details are shown in the detail drawings of Fig. 19-8. This design includes a flue for the furnace. The table in Fig. 19-9 lists some recommended dimensions for openings and parts for many fireplace sizes.

HEARTH

The hearth, including the front section, must be completely supported by the chimney. The support is constructed by first building shoring (temporary support) and forms. Then concrete (3 1/2 in. minimum thickness) with reinforcing is poured. In the best construction, a cantilevered design is secured by recessing the back edge of the rough hearth into the rear wall of the chimney.

An ash dump may be provided in the rear hearth for clearing ashes, if there is space below for an ashpit. The dump consists of a metal frame with pivoted cover. The basement ashpit should be of tight masonry and include a clean-out door.

When the floor structure consists of a slab-on-grade, an ashpit may be formed by a raised hearth

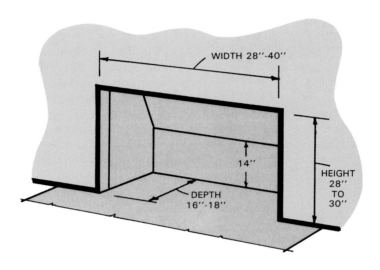

Fig. 19-7. Fireplace opening dimensions with range of sizes commonly used.

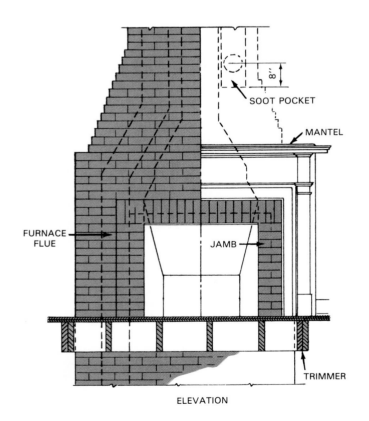

ELEVATION

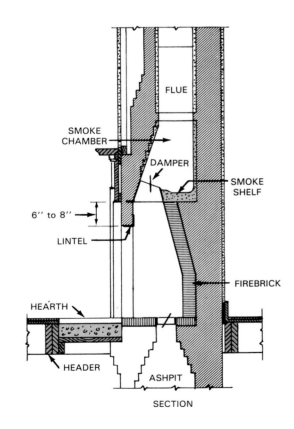

SECTION

Fig. 19-8. Elevation and section views show details of construction for typical masonry fireplace. The carpenter must understand how wood is joined to masonry before doing remodeling or new construction.

Fireplace Opening		Depth	Minimum Back (Horizontal)	Vertical Back Wall	Inclined Back Wall	Outside Dimensions of Standard Rectangular Flue Lining	Inside diameter of Standard Round Flue Lining
Width	Height						
Inches	Inches	Inches	Inches	Inches	Inches	Inches	Inches
24	24	16-18	14	14	16	8 1/2 by 8 1/2	10
28	24	16-18	14	14	16	8 1/2 by 8 1/2	10
24	28	16-18	14	14	20	8 1/2 by 8 1/2	10
30	28	16-18	16	14	20	8 1/2 by 13	10
36	28	16-18	22	14	20	8 1/2 by 13	12
42	28	16-18	28	14	20	8 1/2 by 18	12
36	32	18-20	20	14	24	8 1/2 by 18	12
42	32	18-20	26	14	24	13 by 13	12
48	32	18-20	32	14	24	13 by 13	15
42	36	18-20	26	14	28	13 by 13	15
48	36	18-20	32	14	28	13 by 18	15
54	36	18-20	38	14	28	13 by 18	15
60	36	18-20	44	14	28	13 by 18	15
42	40	20-22	24	17	29	13 by 13	15
48	40	20-22	30	17	29	13 by 18	15
54	40	20-22	36	17	29	13 by 18	15
60	40	20-22	42	17	29	18 by 18	18
66	40	20-22	48	17	29	18 by 18	18
72	40	22-28	51	17	29	18 by 18	18

Fig. 19-9. Recommended dimensions for a wide range of fireplace sizes.

as illustrated in Fig. 19-10. This design may be used when the fireplace is located on an outside wall. In some designs, especially when no ashpit is included, the rear hearth is lowered several inches so ashes are contained in this area.

SIDE AND BACK WALLS

The side and back walls of the combustion chamber continue upward to the level of the damper. These must be lined with firebrick at least 2 in. thick. The firebrick is set in a special clay mortar that will withstand the heat. The total thickness of the walls, including the firebrick, should not be less than 8 in.

Side walls are built at an angle to reflect heat into the room. This angle (also called ''splay'') is usually laid out at 5 in. per foot. The back wall is extended vertically from the base a distance slightly less than one-half of the opening height and then sloped forward. This slope directs the smoke into the throat of the fireplace. The slope also keeps an area clear for the smoke shelf which will be located above.

DAMPER AND THROAT

Part of the fireplace throat is formed by the damper, Fig. 19-11. A stationary front flange is angled a small amount away from the masonry of the front wall. The back flange is movable.

The damper affects a flow called the downdraft. In Fig. 19-11, arrows indicate the upward flow of hot air and smoke from the throat into the front side of the smoke chamber. The rapid upward passage of hot gases creates a downward current (downdraft) on the opposite side. One purpose of the damper is to change the direction of the downdraft so smoke is not forced into the room.

The damper is installed so that masonry work above will not interfere with its full operation. Also the ends have a slight clearance in the masonry to

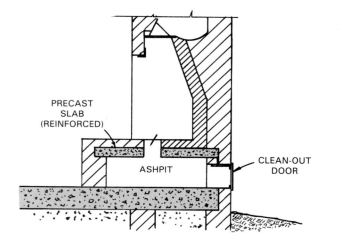

PRECAST SLAB (REINFORCED)

ASHPIT

CLEAN-OUT DOOR

Fig. 19-10. A raised hearth provides space for an ashpit in slab-on-grade construction. This type of ashpit is not used in regions where snow reaches door height.

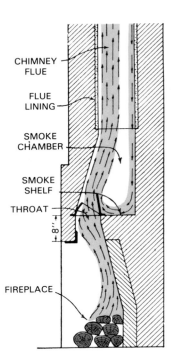

CHIMNEY
FLUE

FLUE
LINING

SMOKE
CHAMBER

SMOKE
SHELF

THROAT

8"

FIREPLACE

Fig. 19-11. Details of operation of fireplace. Arrows indicate air and smoke currents. Note how damper keeps downward current from entering room.

permit expansion. A standard damper design is shown in Fig. 19-12. Manufacturers provide data and recommendations concerning model and size for specific installations.

Correct throat size controls the efficiency of a fireplace. The highest efficiency occurs when the cross-sectional area of the throat is equal to that of the flue. The length of the throat along the face of the fireplace is the same as the opening width. Horizontal depth is much less to get the proper cross-section.

SMOKE SHELF AND CHAMBER

The smoke shelf helps the damper to change the flow direction of the downdraft. The deeper the shelf behind the damper, the better the fireplace works. The depth may vary from 4 in. to 12 in.,

Fig. 19-12. Top view of a standard fireplace damper.

depending on the depth of the fireplace. Some smoke shelves are curved to reduce turbulence in the air flow. The length along the face for all types is equal to the full width of the throat.

The smoke chamber is the space extending from the top of the throat up to the bottom of the flue. The area at the bottom of the chamber is quite large, since its width includes that of the throat plus the depth of the smoke shelf. This space holds accumulated smoke temporarily if a gust of wind across the top of the chimney momentarily cuts off the draft. Without this chamber, smoke would likely be forced out into the room. A smoke chamber also lessens the force of the downdraft by increasing the area through which it passes. Side walls are generally drawn inward one foot for each 18 in. of rise. The surfaces of most smoke chambers are plastered with about 1/2 in. thick cement mortar.

FLUE SIZE

The cross-sectional area of the flue is based on the area of the fireplace opening. A general rule states that the flue area should be at least 1/10th of the total opening when a lining is used. This applies to chimneys that are at least 20 ft. high. A somewhat larger flue may be required in the lower chimney heights normally used in modern single-story construction. The upward movement of smoke in a low chimney does not reach a high velocity—thus a greater cross-sectional area is required.

One recommended method of calculating the area for a flue is to allow thirteen square inches of area for the chimney flue to every square foot of fireplace opening. For example: if the fireplace opening equaled 8.25 sq. ft., then the flue area should equal at least 107 sq. in. If the flue is to be built of brick and unlined, it would probably be made 8 in. x 16 in., or 128 sq. in., because brickwork can be laid to better advantage if the dimensions of the flue are multiples of 4 in. If the flue is lined, and lining is strongly recommended, the lining should have an inside area of at least 107 sq. in.

CONSTRUCTION SEQUENCE

Masonry fireplaces are nearly always built in two stages. The first begins during the rough framing of the structure. Masonry work is carried up from the foundation and the main walls of the fireplace are formed. After a steel lintel is set above the opening, the damper is installed and the smoke chamber built. Then the chimney is carried on through the roof and the exterior masonry is completed. These steps usually occur before the roof surface is laid.

The second and finishing stage of the fireplace takes place after plastering or other type of wall surface is complete and during the application of interior trim. Decorative brick or stone may be set over the exposed front face. The surface of the front hearth can be finished at this time, since the reinforced concrete base was placed during the rough masonry construction. Install wood trim (mantel) when masonry work is complete.

SPECIAL DESIGNS

Contemporary fireplace designs often have openings on two or more sides. For these, follow the same principles in planning as previously described for conventional designs. When calculating the flue area, the SUM OF THE AREA OF ALL FACES must be used. Fig. 19-13 illustrates a corner fireplace design in which the flue area is based on the total of the front face opening, plus the end face opening. In this particular construction, the side walls are not splayed. However the rear wall is sloped in the usual manner.

Multi-face fireplaces must incorporate a throat and damper with requirements similar to standard designs. Special dampers with square ends and sides are available for two-way fireplaces that serve adjoining rooms.

BUILT-IN CIRCULATORS

The heating capacity of a fireplace can be increased by using a factory-built metal unit as shown in Fig. 19-14. The sides and back consist of a double wall within which air is heated. Cool air

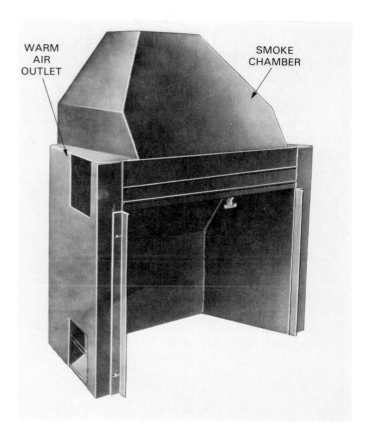

Fig. 19-14. Modern fireplace circulator made of heavy gauge metal. The sides and back consist of a double wall. (Majestic Co.)

enters this chamber near the floor level. When the air is heated, it rises and returns to the room through registers at a higher level.

Modern built-in circulators (also called modified fireplaces) include not only the firebox and heating chamber, but also the throat, damper, smoke shelf, and smoke chamber. Since all of these parts are carefully engineered, proper flue draw is assured when the installation is made according to the manufacturer's directions and flue size is adequate.

To install a circulator unit, first place it in position on the hearth and then build the brick and masonry work around the outside, Fig. 19-15. Steel angles (lintels) are required across the top of the opening as shown and may be required in other locations to provide support. The unit itself should not be used for support of any masonry work.

When installing, follow specifications furnished by the manufacturer. Some type of fireproof insulating material is usually placed around the metal form not only to prevent the movement of heat but also to provide some expansion space between the metal and masonry.

In any type of fireplace, correct operation will depend somewhat on an adequate flow of air into the

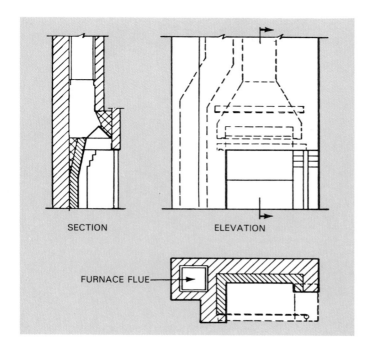

SECTION   ELEVATION

FURNACE FLUE

Fig. 19-13. General design for a projecting corner fireplace.

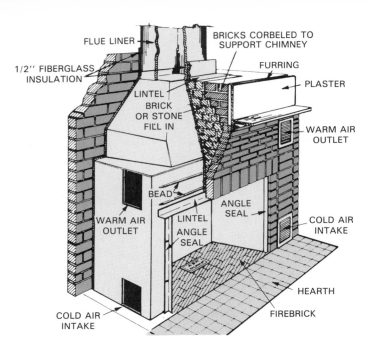

Fig. 19-15. Cutaway view shows how masonry is installed around circulator unit. (Majestic Co.)

building to replace the air exhausted through the flue. Usually infiltration around doors and windows is sufficient. However, where weather stripping is tight, some inlet should be provided.

PREFABRICATED CHIMNEYS

Lightweight chimney units are available that require no masonry work. They provide flues for heating equipment or fireplaces and can be installed in one-story or multi-story structures, Fig. 19-16.

Fig. 19-16. Prefabricated chimneys serve fireplaces and furnaces in a multiple-unit housing complex. The carpenter can often install the chimney sections in addition to building the support framework. (Preway Inc.)

Prefabricated chimneys usually consist of double- or triple-walled sections of pipe that are assembled to form the flue. Special flanges and fittings are used to fasten the flue to the building frame and provide proper clearance from wood members. Fig. 19-17 shows details of a typical installation.

The roof-top section of a prefabricated chimney unit must be sealed to prevent roof leaks. Fig. 19-18 shows the basic parts of a simple pipe projection. Bed the flashing unit in mastic with the roofing material overlapping the top and side edges. The storm collar diverts rain water from the pipe

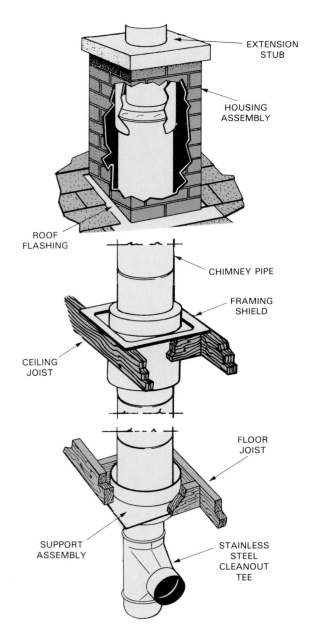

Fig. 19-17. Prefabricated chimney. Pipe sections consist of a triple metal wall. Inside diameters range from 6 to 14 in. Various types of raincaps can be mounted on extension stub. (Wallace Murray Corp.)

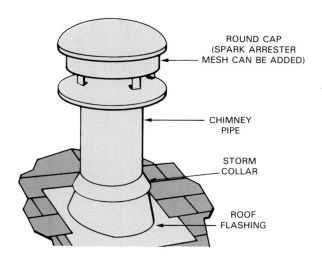

ROUND CAP
(SPARK ARRESTER
MESH CAN BE ADDED)

CHIMNEY
PIPE

STORM
COLLAR

ROOF
FLASHING

Fig. 19-18. Basic parts of a rooftop projection for a standard flue pipe.

to a conical section of the flashing. Some type of cap should be installed on the top of the flue to prevent the entry of rain or snow.

Many types of prefabricated termination tops (part of the chimney that extends above the roof) are available. For the best appearance they should blend with the architectural style of the building. Fig. 19-19 shows several typical designs. Be sure to follow the manufacturer's directions for assembly and installation.

When a prefabricated chimney is used for venting smoke and gases that may contain corrosive acids (incinerators or solid fuel boilers), the inside pipe

should be made of stainless steel or have a porcelain coated surface.

A prefabricated chimney should have a label showing that it has been tested and listed by the Underwriters' Laboratories, Inc. or other nationally recognized testing organizations. It must also conform to local building codes. Be sure to make the installation in strict accordance with the manufacturer's instructions.

PREFABRICATED FIREPLACES

As the cost of energy continues to rise, an increasing number of homes are being equipped with fireplaces. Many of the fireplaces are factory-built units which are generally less expensive than masonry units.

Manufacturers of prefabricated fireplaces provide a wide range of designs that are easy to install with ordinary tools, Fig. 19-20. In some cases, multiple steel wall construction and special firebox linings permit ''zero-clearance.'' This means that outside housings can rest directly on wood floors and touch wood framing members. Follow the manufacturer's recommendations when installing.

Other precautions should be taken if gas or electric starters are used or if gas logs are installed. Check local fire codes before starting work.

Prefabricated fireplaces operate like the metal circulator built into masonry fireplaces. Room air enters intakes at floor level and flows through chambers around the firebox. As the air is heated,

Fig. 19-19. Rooftop housings and termination caps (raincaps) for prefabricated chimneys. Left. Traditional (simulated brick). Center. Contemporary (simulated stucco). Right. Traditional (simulated wood). (Majestic Co.)

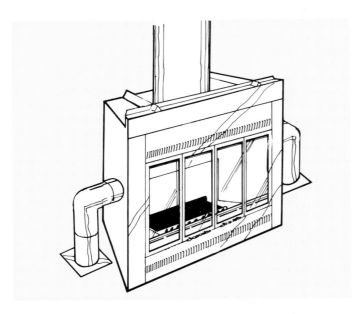

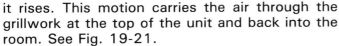

Fig. 19-20. A prefabricated fireplace consists of a metal shell, firebrick, ductwork, and a face panel. Two persons can lift most units. Installation may take as little as 1/2 hr. (Preway Inc.)

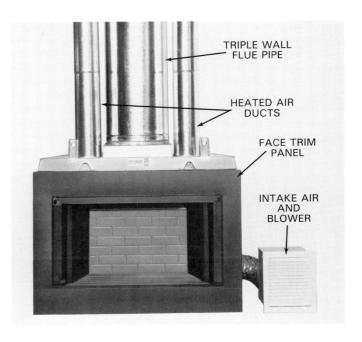

Fig. 19-22. Prefabricated fireplace with ducts to return heated air to one or more rooms. Blower is optional item. (Superior Fireplace Co.)

it rises. This motion carries the air through the grillwork at the top of the unit and back into the room. See Fig. 19-21.

Some units can be attached to vertical ducts that carry the heated air to various locations within the room or to an adjoining room. See Fig. 19-22. Some prefabricated fireplaces have blowers that increase the flow of air and thus improve heating efficiency.

Fig. 19-23 shows framing built around a prefabricated unit. In new construction, the basic frame is usually built at the same time as exterior walls and partitions. The facing side of the frame should remain open until the fireplace is installed. If support is required above this opening, it should be framed

Fig. 19-21. Cutaway view with glass removed shows basic operation of prefabricated fireplace. Room air (blue arrows) is drawn from floor level into heating chamber. There it is warmed and returned to the room (red arrows). Air for burning can be piped from outside and connected to inlet at lower left. Note triple-walled flue. (Preway Inc.)

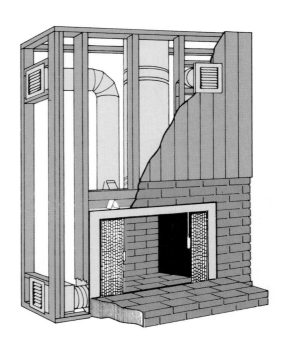

Fig. 19-23. Cutaway view shows basic wood framing for fireplace and ductwork. (Superior Fireplace Co.)

with a header like that for a door. After the fireplace unit is installed, any front framing below the header can be added.

Manufacturers provide detailed drawings and instructions for the installation of their fireplace units. These should be carefully studied and followed throughout installation.

Many materials including both wood and masonry can be used to finish the wall and trim the fireplace opening. According to FHA specifications, wooden parts should not be placed closer than 3 1/2 in. to the edge of the opening. Greater clearance is required when the parts project more than 1 1/2 in. For example: a wooden mantel shelf must be 12 in. above the opening.

Free-standing fireplaces are usually fabricated with double walls somewhat like built-in units, Fig. 19-24. Proper support exists in many rooms in the home and the units are easy to install. A single-walled pipe runs from the unit to the ceiling where it is fitted to a triple-walled section.

Fig. 19-25 shows a wall-mounted unit which is partially supported by the floor. Triple-wall construction is used around the firebox so the fireplace unit can be placed against any combustible material. Free-standing fireplaces are available with many of the same features of built-in models. They can be equipped with a blower to force the circulation of air around the firebox. Air for combustion can be drawn from the outside.

Fig. 19-25. Wall-mounted fireplace. Hearth corners are cantilevered. It includes a 300 CFM (cubic foot per minute) blower and ceramic glass door. (Malm Fireplaces Inc.)

Before selecting and installing any type of factory-built fireplace, be sure to check local building codes. Failure to comply with these requirements can result in the installation being disapproved by the local building inspector.

CHIMNEYS

Chimney (flue pipe) systems for prefabricated fireplaces are designed for specific units and usually are sold as a package along with the fireplace. Fig. 19-26 shows the general assembly of standard components which run through the ceiling and roof structure.

Special attention must be given to clearance and support. Openings through the ceiling and roof should be carefully framed to provide the correct size openings for firestop spacers and support boxes. Fig. 19-27 shows the installation of a firestop spacer in an attic area.

When installed on an exterior wall, the chimney and fireplace unit can be located in a special projection called a chase. This is a box-like structure that is built as a part of the floor, wall, or roof frame, Fig. 19-28 and Fig. 19-29. The outer walls of the chase are usually finished to match the outer walls of the house as shown in Fig. 19-30. Insulate the walls of the chase and seal them against infiltration. Be sure insulation does not touch the flue pipe.

Fig. 19-24. This free-standing fireplace is made of heavy gauge steel with porcelain finish. It requires a 36 in. clearance from combustible wall. (Malm Fireplaces Inc.)

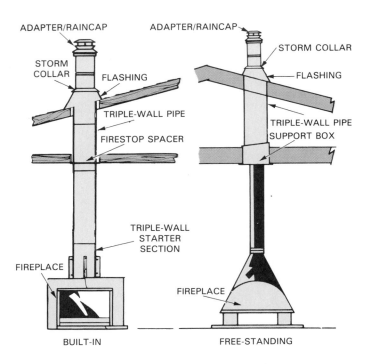

Fig. 19-26. Chimney system for prefabricated fireplaces. (Preway Inc.)

Fig. 19-28. Typical framing for a chimney chase. The roof framework will hold the chase steady on three sides. (American Plywood Assoc.)

Fig. 19-27. Installing firestop spacer after cutting and framing ceiling opening. Triple-walled pipe is supported by fireplace in room below. It is a good practice to close off vertical channels of any kind in a building frame. (Majestic Co.)

GLASS ENCLOSURES

Glass enclosures improve the efficiency of fireplaces and can be installed on either masonry structures or prefabricated units. They reduce the amount of heated room air that escapes up the chimney, even when there is no fire burning. Effi-

ciency is improved because the rate of burning can be controlled by adjusting draft vents and dampers. Enclosures consist of tempered plate glass doors hinged to a metal frame. Fig. 19-31 shows a pair of bi-fold doors that open from the center. For narrow openings, two swinging doors are commonly used. Manufacturers provide complete instructions for installing and operating their units.

TEST YOUR KNOWLEDGE — UNIT 19

1. The overall size of a chimney will depend on the materials from which it is constructed and the number and size of the _____.
2. Wood framing should be located at least _____ away from a masonry chimney.
3. The offset angle in a chimney should be limited to _____ degrees.
4. To prevent heat loss from a room when the fireplace is not in operation, the fireplace must be equipped with a _____.
5. The length of the throat along the face should be equal to the _____ of the fireplace opening.
6. All surfaces of the smoke chamber should be

Fig. 19-29. Another type of chase construction. To make use of standard length studs, the chase is built in two sections. Uprights are joined at the midpoint with overlapping horizontal strips. (American Plywood Assoc.)

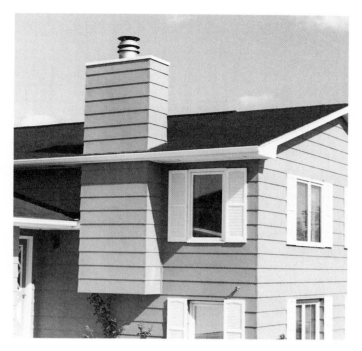

Fig. 19-30. View of a completed chase. Very little weight is supported at the base of the framework since the fireplace is a metal shell.

finished with a 1/2 in. thickness of _____.

7. As a general rule, the cross-sectional flue area should equal about _____ of the total area of the fireplace opening.

8. Prefabricated chimneys consist of pipe sections with double or _____ walls.

9. When a prefabricated metal chimney is used for an incinerator, the inside pipe should be coated with porcelain or made of _____.

10. An outside air intake on a prefabricated fireplace provides air for _____ (burning, circulating).

11. A chase is made rigid with decorative sheathing and with support from wall and _____ framing.

Fig. 19-31. Tempered glass doors on this prefabricated fireplace save energy. (Majestic Co.)

OUTSIDE ASSIGNMENTS

1. Write a report on the historical development of the fireplace. Start with the simple designs used by primitive humans and highlight developments through the years. Place special emphasis on the materials used for building fireplaces and chimneys. Is it possible to determine:
   a. When metal parts became popular?
   b. When safe designs with wood structures became common?
   Ask your school librarian to suggest books that would be helpful.

2. Get a copy of the building code that applies to your locality and study the sections that deal with chimneys, fireplaces, and venting systems. Make a list of the specific requirements concerning residential structures. Include information about masonry and prefabricated metal chimneys for heating plants, incinerators, and fireplaces. Also include requirements and restrictions for the use and installation of prefabricated fireplaces. Make a report to your class.

Interior view of low slope post-and-beam structure. Laminated ceiling beams (about 8 x 18 in.) are spaced 8 ft. O.C. Deck is 2 x 12 pine planking, tongue and grooved. Note V-groove in center of each plank. (Boise-Cascade Corp.)

Traditional design in post-and-beam construction follows design of barns. Joints are mortise and tenon secured with wooden pegs. Note strength of frame and heavy braces at the corners.   (Timberpeg)

Unit 20

POST-AND-BEAM CONSTRUCTION

Post-and-beam construction consists of large framing members—posts, beams, and planks. These are spaced farther apart than conventional framing members. Frames of this type are similar to "mill construction" once used for barns and heavy timber buildings. It is often used today for residential work since it permits greater flexibility in contemporary and traditional designs than conventional framing methods. See Fig. 20-1 and Fig. 20-2 for a comparison.

Post-and-beam construction is also known as "plank-and-beam" construction. It is often combined with conventional framing. For example, the walls might be built conventionally and the roof

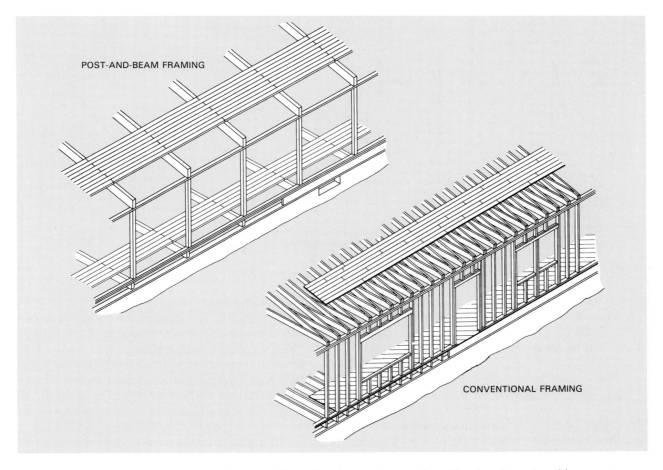

POST-AND-BEAM FRAMING

CONVENTIONAL FRAMING

Fig. 20-1. Comparison of typical post-and-beam framing and conventional framing. In post-and-beam construction framing around large windows and doors is simplified since headers can be eliminated. (National Forest Products Assoc.)

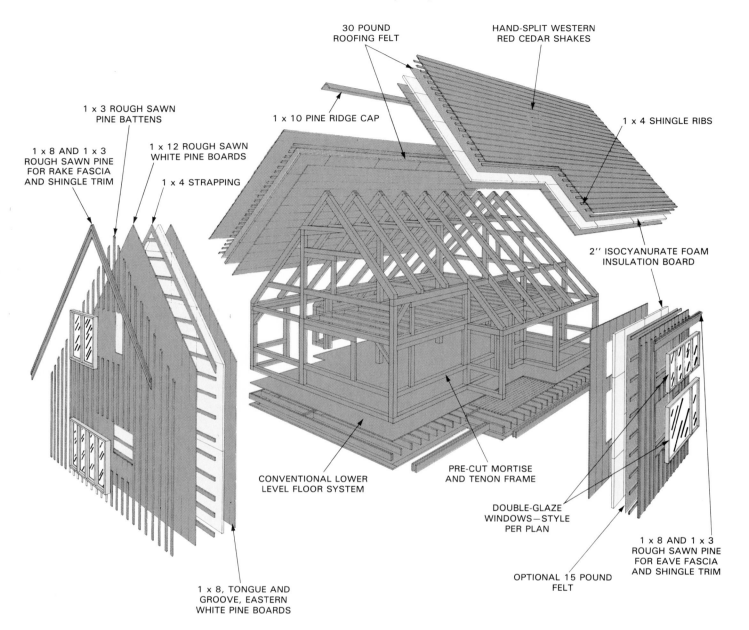

30 POUND
ROOFING FELT

HAND-SPLIT WESTERN
RED CEDAR SHAKES

1 x 10 PINE RIDGE CAP

1 x 3 ROUGH SAWN
PINE BATTENS

1 x 8 AND 1 x 3
ROUGH SAWN PINE
FOR RAKE FASCIA
AND SHINGLE TRIM

1 x 12 ROUGH SAWN
WHITE PINE BOARDS

1 x 4 STRAPPING

1 x 4 SHINGLE RIBS

2'' ISOCYANURATE FOAM
INSULATION BOARD

CONVENTIONAL LOWER
LEVEL FLOOR SYSTEM

PRE-CUT MORTISE
AND TENON FRAME

DOUBLE-GLAZE
WINDOWS—STYLE
PER PLAN

1 x 8 AND 1 x 3
ROUGH SAWN PINE
FOR EAVE FASCIA
AND SHINGLE TRIM

OPTIONAL 15 POUND
FELT

1 x 8, TONGUE AND
GROOVE, EASTERN
WHITE PINE BOARDS

Fig. 20-2. This rustic variation of the post-and-beam house follows the traditional design of colonial barn framing. Conventional flooring system is used on the first level; heavy beams are used on the second floor.   (Timberpeg)

framed with beams and planks. In such a case the term ''plank-and-beam'' could be applied to the roof structure. Similarly, it would be correct to refer to a floor structure as a ''plank-and-beam'' system.

ADVANTAGES

Advantages of this construction are the distinctive architectural effect created by the exposed beams in the ceiling and the added height. See Fig. 20-3. The underside of the roof planks may serve as the ceiling surface thus providing a saving in material.

Post-and-beam framing may also provide some saving in labor. The pieces are larger and fewer in number and can usually be assembled more rapidly than conventional framing.

One of the chief structural advantages is the simplicity of framing around door and window openings. Loads are carried by posts spaced at wide intervals in the walls. Large openings can be framed without the need for headers. Window walls, characteristic of contemporary architectural styling, can be formed by merely inserting window frames between the posts. Another advantage: wide overhangs can be built by simply extending the heavy roof beams, Fig. 20-4.

Fig. 20-3. Exposed ceiling beams provide an attractive architectural feature while adding to ceiling height. (Georgia-Pacific Corp.)

Fig. 20-4. A wide roof overhang is easy to construct when posts, beams, and planks are used.

In addition to its flexibility in design, post-and-beam construction also provides high resistance to fire. Wood beams do not transmit heat and collapse like unprotected metal beams. Exposure of wood beams to flame results in a slow loss of strength.

Most limitations of post-and-beam construction can be resolved through careful planning. The plank floors, for example, are designed to carry moderate, uniform loads. Therefore, extra framing must be provided under bearing partitions, bath tubs,

refrigerators, and other places where heavy loads are likely. Good lateral stability must be given the framework, especially the walls. This might be provided with various types of bracing. It is more common, however, to enclose some of the wall area with large panels and use conventional stud constructions as shown in Fig. 20-5. Absence of concealed spaces in outside walls and ceilings, makes installation of electrical wiring, plumbing, and heating somewhat more difficult.

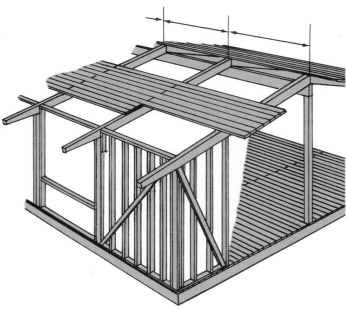

Fig. 20-5. Conventional braced-stud sections can be used to provide lateral bracing. Refer to Fig. 20-2 for another type of bracing sometimes used.

FOUNDATIONS AND POSTS

Foundations for post-and-beam framing may consist of continuous walls or simple piers located under each post. Either type of foundation must rest on footings that meet the requirements of local building codes. See Unit 6.

Posts must be strong enough to support the load and also large enough to provide full bearing surfaces for the ends of the beams. In general, posts should not be less than 4 x 4 in. nominal size. They may be made of solid stock or built up from 2 in. pieces. Where the ends of beams are joined over a post, the bearing surface should be increased with bearing blocks as shown in Fig. 20-6.

When posts extend any great height without lateral bracing, a larger cross-sectional area is required to prevent buckling. Requirements are usually

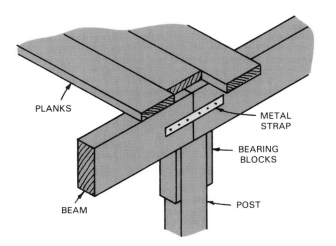

Fig. 20-6. Provide adequate bearing surface where beams are joined over a post. Heavy steel plate may replace the bearing blocks.

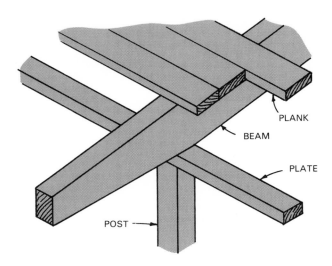

Fig. 20-7. Roof beams must rest directly over supporting posts.

listed in local building codes through a l/d ratio (also called slenderness ratio). The l represents the length in inches and the d stands for the smallest cross-sectional dimension (actual size). For example, a 4 x 4 piece 8 ft. long would have a ratio of about 27. This is within the limits usually prescribed.

Distance between posts will be determined by the basic design of the structure. This spacing must be carefully engineered. Usually posts are spaced evenly along the length of the building and within the allowable free span of the floor or roof planks. With today's emphasis on modular dimensions, construction costs will be cheaper if post-and-beam positions occur at standard increments (increases) of 16, 24, and 48 in.

In single-story construction, a plate is attached to the top of the posts in about the same way as conventional framing. The roof beams are then positioned directly over the posts as shown in Fig. 20-7.

FLOOR BEAMS

Beams for floor structures may be solid, glue-laminated, or built-up, Fig. 20-8. Sometimes the built-up beams are formed with spacer blocks be-

tween the main members. Box-beams, described in another section of this Unit, may also be used.

For one-story structures, where under-the-floor appearance is not critical, standard dimension lumber can be nailed together to form any size of beam. Fig. 20-9 illustrates how 2 x 8s have been nailed together and then installed 4 ft. on center. The decking being applied is 1 1/8 in. plywood with a special tongue-and-groove edge.

Design of sills for a plank-and-beam floor system can be similar to regular platform construction, Fig. 20-10. When it is desirable to keep the silhouette of the structure low (floor level near grade level) the beams can be supported in pockets in the foundation wall.

BEAM DESCRIPTIONS

In general, it is best to use solid timbers when beam sizes are small and when a rustic architectural appearance is desired. Where high stress factors demand large sizes and a finished appearance is required, it is usually more economical to use laminated beams. They are manufactured in a wide range of sizes and finishes.

    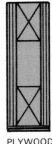

| SOLID | VERTICAL LAMINATED | HORIZONTAL LAMINATED | SPACED WITH WOOD BLOCKING | REINFORCED WITH STEEL PLATE | PLYWOOD BOX BEAM |

Fig. 20-8. Beams can be solid or built up from smaller dimension lumber.

492

Fig. 20-9. Floor system with beams spaced 4 ft. O.C. Decking is 1 1/8 in. plywood with edges tongue and grooved. (American Plywood Assoc.)

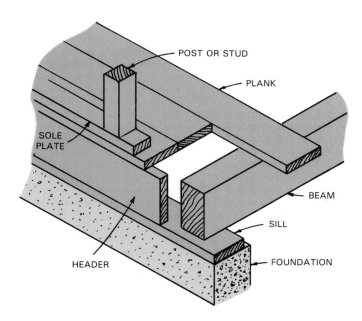

Fig. 20-10. Typical sill construction for a post-and-beam frame. Add blocking under the post when it is not located over the beam.

Solid timbers are available in:
1. A range of standard cross sections.
2. Lengths of 6 ft. and longer.
3. Longer lengths in multiples of 1 or 2 feet.
   Surface finish is either rough sawn or planed.

When beams are exposed, appearance becomes an important consideration. Fig. 20-11 illustrates on-the-job treatment that may be applied to exposed beams.

Beam sizes must be based on the span (spacing between supports), deflection permitted, and the load they must carry. Design tables are available from lumber manufacturers which may be used to determine sizes for simple buildings. See Fig. 20-12. Refer to Unit 7 for additional information on calculating beam loads.

ROOF BEAMS

Beam supported roof systems are of two basic types:
1. Transverse beams that are similar to exposed rafters on wide spacings.
2. Longitudinal beams that run parallel to the supporting side walls and ridge beam. Fig. 20-13 shows both types.

Longitudinal beams, also called purlin beams, are usually larger in cross section than transverse beams because they have greater spans and carry heavier loads. The longitudinal beam permits many variations in end-wall design. Extensive use of glass and extended roof overhangs are special features.

Either type of beam must be adequately supported either on posts or stud walls that incorporate a heavy top plate.

493

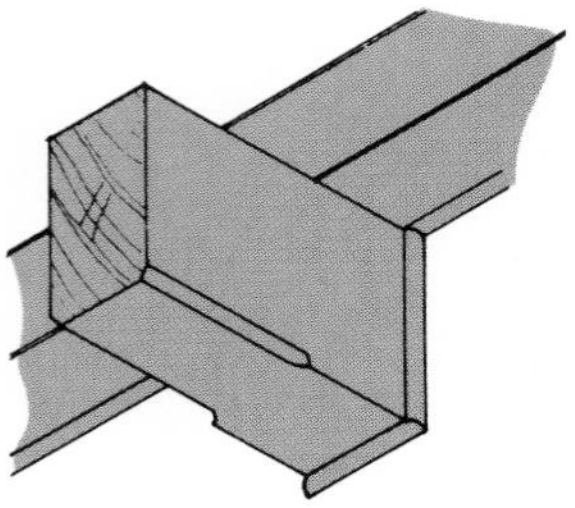

SOLID BEAM

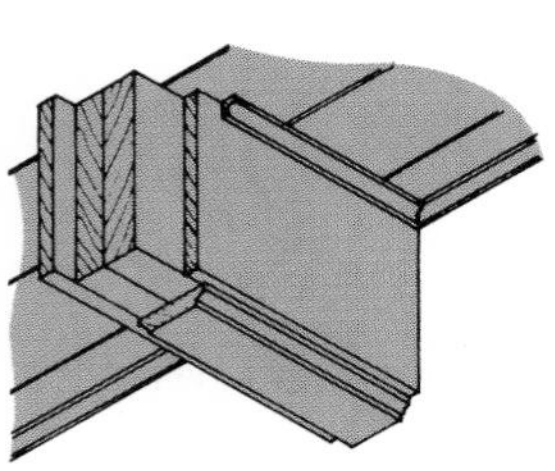

CASED BEAM

Fig. 20-11. Either solid or built-up beams can be surface treated for better appearance when they are to be left exposed.

When supported on posts, the connection can be reinforced with a wide panel frame that extends to the top of the beam, Fig. 20-14. A similar method of supporting a ridge beam is shown in Fig. 20-15.

Transverse beams can be joined to the sides of the ridge beam or supported on top as illustrated in Fig. 20-16. Metal tie plates, hangers, and straps are required to absorb the horizontal thrust.

Flat roof designs often consist of a plank-and-beam system. Details of construction are about the same as illustrated for low, sloping roofs.

Size (Actual)	Wt. per Lineal Ft.	SIMPLE SPAN IN FEET													
		10	12	14	16	18	20	22	24	26	28	30	32	34	36
		Load Bearing Capacity—Lbs. per Lineal Ft.													
3''x5¼''	3.7 lbs.	151	85	—	—	—	—	—	—	—	—	—	—	—	—
3''x7¼''	4.9 lbs.	362	206	128	84	—	—	—	—	—	—	—	—	—	—
3''x9¼''	6.7 lbs.	566	448	300	199	137	99	—	—	—	—	—	—	—	—
3''x11¼''	8.0 lbs.	680	566	483	363	252	182	135	102	—	—	—	—	—	—
4½''x9¼''	9.8 lbs.	850	673	451	299	207	148	109	—	—	—	—	—	—	—
4½''x11¼''	12.0 lbs.	1,036	860	731	544	378	273	202	153	—	—	—	—	—	—
3¼''x13½''*	10.4 lbs.	1,100	916	784	685	479	347	258	197	152	120	—	—	—	—
3¼''x15''*	11.5 lbs.	1,145	1,015	870	759	650	473	352	267	206	163	128	104	—	—
5¼''x13½''*	16.7 lbs.	1,778	1,478	1,266	1,105	773	559	415	316	245	193	154	124	101	—
5¼''x15''*	18.6 lbs.	1,976	1,647	1,406	1,229	1,064	771	574	438	342	269	215	174	142	116
5¼''x16½''*	20.5 lbs.	2,180	1,810	1,550	1,352	1,155	933	768	586	457	362	290	236	183	160
5¼''x18''*	22.3 lbs.	2,378	1,978	1,688	1,478	1,308	1,113	918	766	598	474	382	311	254	204

*Horizontally Laminated Beams

TABLE 1

example: Clear Span = 18′ 0″
Beam Spacing = 8′ 0″
Dead Load = 8 lbs./sq. ft. (decking + roofing)
TABLE I Live Load = 20 lbs./sq. ft. (snow)
ROOF BEAM Total Load = (20 + 8) (8) = 224 lbs./lineal ft.
From Table I—Select 3″x11¼″ beam with capacity of 252 lbs./lin. ft.

Size (Actual)	Wt. per Lineal Ft.	SIMPLE SPAN IN FEET													
		10	12	14	16	18	20	22	24	26	28	30	32	34	36
		Load Bearing Capacity—Lbs. per Lineal Ft.													
3''x5¼''	3.7 lbs.	114	64	—	—	—	—	—	—	—	—	—	—	—	—
3''x7¼''	4.9 lbs.	275	156	84	55	—	—	—	—	—	—	—	—	—	—
3''x9¼''	6.7 lbs.	492	319	198	130	89	—	—	—	—	—	—	—	—	—
3''x11¼''	8.0 lbs.	590	491	361	239	165	119	—	—	—	—	—	—	—	—
4½''x9¼''	9.8 lbs.	738	479	298	196	134	96	—	—	—	—	—	—	—	—
4½''x11¼''	12.0 lbs.	900	748	541	359	248	178	131	92	—	—	—	—	—	—
3¼''x13½''*	10.4 lbs.	956	795	683	454	316	228	169	128	98	—	—	—	—	—
3¼''x15''*	11.5 lbs.	997	884	756	626	436	315	234	178	137	108	—	—	—	—
5¼''x13½''*	16.7 lbs.	1,541	1,283	1,095	732	509	367	271	205	158	123	96	—	—	—
5¼''x15''*	18.6 lbs.	1,713	1,423	1,219	1,009	703	508	376	286	221	173	137	109	—	—
5¼''x16½''*	20.5 lbs.	1,885	1,568	1,340	1,170	939	678	505	384	298	235	187	151	—	—
5¼''x18''*	22.3 lbs.	2,058	1,710	1,464	1,278	1,133	886	660	503	391	309	247	200	—	—

*Horizontally Laminated Beams

TABLE 2

example: Clear Span = 18′ 0″
Beam Spacing = 5′ 0″
Dead Load = 7 lbs./sq. ft. (decking + covering)
TABLE II Live Load = 40 lbs./sq. ft. (furniture, occupants, etc.)
FLOOR BEAM Total Load = (40 + 7) (5) = 235 lbs./lineal ft.
From Table II—Select 4½″x11¼″ beam with capacity of 248 lbs./lin. ft.

Fig. 20-12. Check tables to determine allowable spans for vertical glue-laminated beams. Table 1 is intended for roof beams where a deflection of 1/240 of span is permitted. Table 2 is for floor beams and is based on an allowable deflection of 1/360 or less. (Weyerhaeuser Co.)

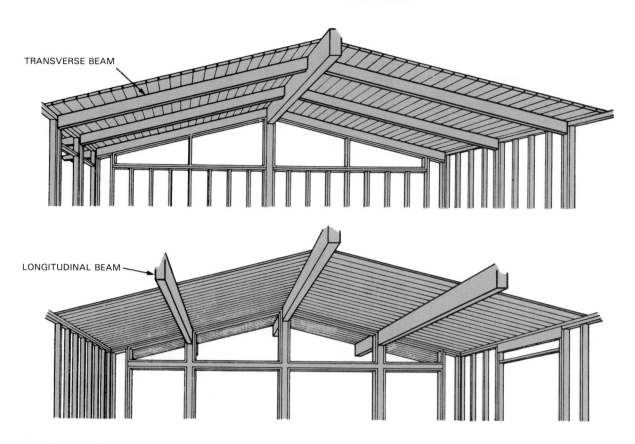

TRANSVERSE BEAM

LONGITUDINAL BEAM

Fig. 20-13. Plank-and-beam roof construction methods. Top. Transverse beams. Bottom. Longitudinal beams.

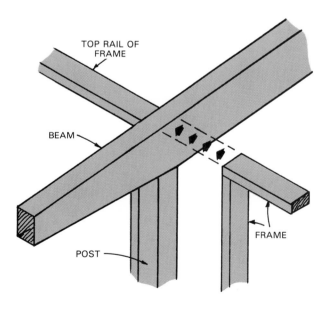

TOP RAIL OF FRAME

BEAM

POST

FRAME

Fig. 20-14. Transverse beam which bears on a post will need support against lateral (side-to-side) movement. Filler panel frame at right provides needed reinforcement.

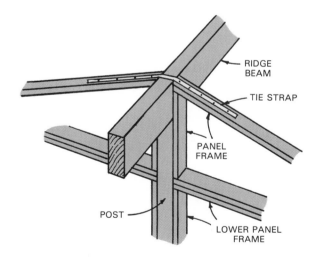

RIDGE BEAM

TIE STRAP

PANEL FRAME

POST

LOWER PANEL FRAME

Fig. 20-15. Ridge beam is held in place by panel frame and metal tie strip. Same detail of construction may be applied to transverse beams.

FASTENERS

A post-and-beam frame consists of a limited number of joints. Therefore the loads and forces ex-erted upon the structure are concentrated at these points. Regular nailing patterns used in conventional framing will usually not provide a satisfactory connection and the joints will need to be reinforced with special metal connectors. See Fig. 20-17. To increase the holding power of metal connectors, they

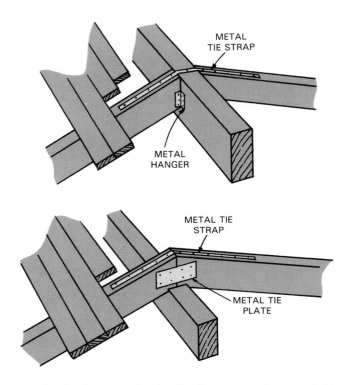

Fig. 20-16. Construction details of transverse beam and ridge beam. Top. Beam attached to side of ridge. Below. Beam supported on top of ridge.

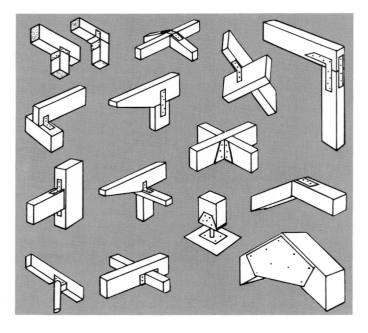

Fig. 20-17. Many different kinds of metal fasteners are made for post-and-beam construction. (Western Wood Products Assoc. and Timber Engineering Co.)

should be attached with lag screws or bolts.

Since beam structures are usually exposed, some connectors will likely detract from the appearance. Concealed devices will need to be substituted. Steel or wood dowel pins of appropriate size can be used.

Notches and gains cut in the structural members may provide an interlocking effect or a recess for metal connectors. Fig. 20-18 shows a heavy beam-and-truss system. Note how the metal plates and fasteners blend with the rough surface of the structural members to provide a special architectural effect.

Use extra care when assembling exposed posts and beams. Tool and hammer marks will detract from the final appearance.

Fig. 20-18. Heavy beam-and-truss system supports roof beams at midpoint and ridge. All members are laminated and finished with a rough sawn surface that masks the laminate joints. Note the use of metal fasteners.
(Boise-Cascade Corp.)

PARTITIONS

Interior partitions are more difficult to construct under an exposed beam ceiling. Except for a load-bearing partition under a main ridge beam, it is usually best to make the installation after beams and planks are in place. Partitions running perpendicular to a sloping ceiling should have regular top plates with filler sections installed between the beams.

Partitions parallel to transverse beams will have a sloping top plate. Sometimes it is best to construct these partitions in two sections. First build a conventional lower section the same height as the side walls. Then add a triangular section above.

When nonbearing partitions run at a right angle

to a plank floor, no special framing is necessary. However, when nonbearing partitions run parallel, Fig. 20-19, additional support must be provided. Replace the sole plate with a small beam or add the beam below the plank flooring.

PLANKS

Planks for floor and roof decking can be anywhere from 2 to 4 in. thick depending on the span. Edges may be tongue and grooved or they may be grooved for a spline joint that can be assembled into a tight, strong surface. Fig. 20-20 illustrates standard designs. Identification numbers are those listed by the Western Wood Products Association. When planks are end-matched, Fig. 20-21, the joints need not meet over beams.

Planks are stronger and stiffer if they continue over more than one span. This rule can also be applied to beams and other supports and is illustrated in Fig. 20-22.

Roof planks should be selected carefully, especially when the faces will be exposed. Fig. 20-23 shows the application of planking with a V-joint along the edge. Solid materials should have a moisture content that will correspond closely to the

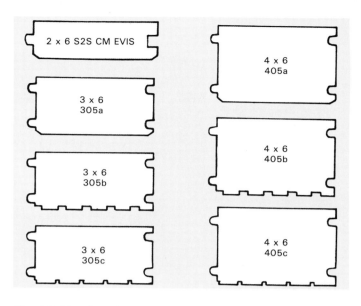

Fig. 20-20. Standard plank patterns. Edges are tongue and groove. Faces are machined.

E.M.C. of the interior of the structure when it has been placed in service. Because of the large cross-sectional size of posts, beams, and planks, special precautions should be observed in selecting material with proper M.C. levels. Otherwise, difficulties due

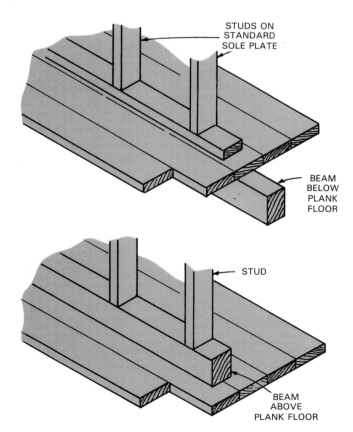

Fig. 20-19. Two methods of supporting nonbearing partitions which run parallel to flooring planks.

Fig. 20-21. End-matched planks are being installed over floor beams.   (Weyerhaeuser Co.)

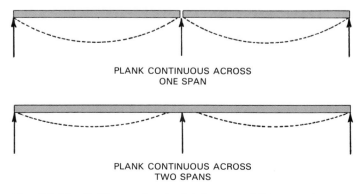

Fig. 20-22. Stiffness of a plank is increased if it extends continuously over two or more spans.

Fig. 20-23. Four inch double tongue-and-groove planks are being laid on glue-laminated beams. V-groove facing makes an attractive, durable ceiling.

to excessive swelling or shrinkage may be encountered.

In cold climates, plank roof structures directly over heated areas require insulation and a vapor barrier. Thickness of the insulation will depend on the climate. Refer to Unit 13. The insulation should be a rigid type that will support the finished roof surface and workers. An approved vapor barrier should be installed between the planks and the insulation as shown in Fig. 20-24.

Several types of heavy structural composition board, 2 to 4 in. thick, are available for roof decks. The panel sizes are large, and the material is lightweight. See Fig. 20-25. Edges usually have some type of interlocking joint that provides a tight, smooth deck. When the underside (ceiling side) is prefinished, no further decoration is usually necessary. Always follow the manufacturer's recommendations when selecting and installing these materials.

STRESSED SKIN PANELS

Roof and floor decking, and also wall sections can be formed with plywood stressed skin panels. These can be designed to carry structural loads over wide spans as illustrated in Fig. 20-26.

Stressed skin panels are made by gluing sheets of plywood (skins) to longitudinal framing members or other core materials. They form a structural unit with a supporting action similar to a series of built-

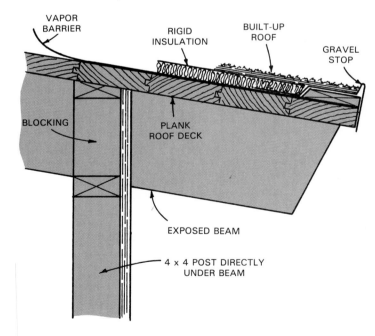

Fig. 20-24. Cross section shows application of vapor barrier and insulation to a plank decking. This type of covering is necessary when the deck is located directly over heated space.

up wooden I-beams. See Fig. 20-27.

Panels are usually produced in factories where rigid specifications in design and construction can be maintained. Fig. 20-28 provides general

Fig. 20-25. Application of special composition board decking. It must be strong enough to carry workers.

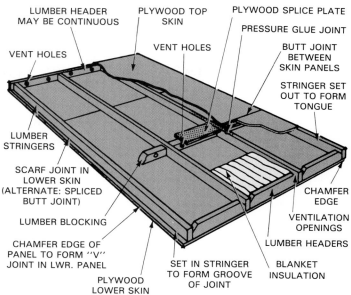

Fig. 20-28. General construction details for stressed skin panels.

LUMBER HEADER MAY BE CONTINUOUS
PLYWOOD TOP SKIN
PLYWOOD SPLICE PLATE
PRESSURE GLUE JOINT
VENT HOLES
VENT HOLES
BUTT JOINT BETWEEN SKIN PANELS
STRINGER SET OUT TO FORM TONGUE
LUMBER STRINGERS
SCARF JOINT IN LOWER SKIN (ALTERNATE: SPLICED BUTT JOINT)
CHAMFER EDGE
LUMBER BLOCKING
VENTILATION OPENINGS
LUMBER HEADERS
CHAMFER EDGE OF PANEL TO FORM "V" JOINT IN LWR. PANEL
SET IN STRINGER TO FORM GROOVE OF JOINT
BLANKET INSULATION
PLYWOOD LOWER SKIN

Fig. 20-26. Stressed skin panels form the roof deck. Note the length of the span. (Georgia-Pacific Corp.)

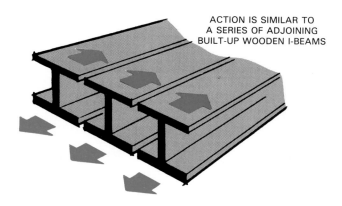

ACTION IS SIMILAR TO A SERIES OF ADJOINING BUILT-UP WOODEN I-BEAMS

Fig. 20-27. How a stressed skin panel provides support.

guidelines and constructional details. Note that insulation can be installed easily as a part of the manufacturing process.

Sandwich panels with plywood skins and cores of such material as foamed polystyrene or paper honeycomb are similar to the stressed skin panel. They do not provide as much rigidity and strength, however. These are used for curtain walls and various installations where the major support is carried by other components. Skins can be made from a wide variety of sheet materials including plywood, hardboard, plastic laminates, and aluminum.

BOX BEAMS

Modern box beams made of plywood webs offer a structural unit that can span distances up to 120 ft. The high strength-to-weight ratio offers a tremendous advantage in commercial structures where wide, unobstructed areas are required.

Basic design features of plywood box beams, illustrated in Fig. 20-29, consist of one or more vertical plywood webs which are laminated to seasoned lumber flanges. The flanges are separated at regular intervals by vertical spacers (stiffeners) which help distribute the load between the upper and lower flange. Spacers also prevent buckling of the plywood webs. The strength of the unit depends, to a large extent, on the quality of the glue bond between the various members. Plywood box beams must be carefully designed and fabricated under controlled conditions.

LAMINATED BEAMS AND ARCHES

Laminated wood beams and arches are available in many shapes and sizes. They have opened new

499

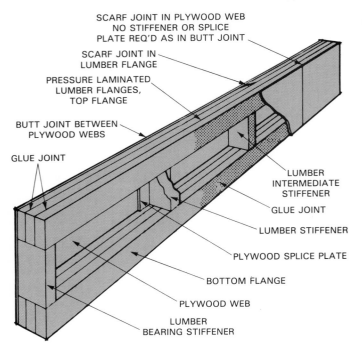

SCARF JOINT IN PLYWOOD WEB
NO STIFFENER OR SPLICE
PLATE REQ'D AS IN BUTT JOINT

SCARF JOINT IN
LUMBER FLANGE

PRESSURE LAMINATED
LUMBER FLANGES,
TOP FLANGE

BUTT JOINT BETWEEN
PLYWOOD WEBS

GLUE JOINT

LUMBER
INTERMEDIATE
STIFFENER

GLUE JOINT

LUMBER STIFFENER

PLYWOOD SPLICE PLATE

BOTTOM FLANGE

PLYWOOD WEB

LUMBER
BEARING STIFFENER

TYPICAL CROSS SECTIONS SHOWING BEAM TYPES

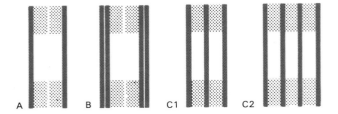

A   B   C1   C2

Fig. 20-29. Box beams have great strength for their weight. These are the basic construction details.

Fig. 20-30. Gracefully soaring parabolic arches of laminated wood are often used in church construction.

dimensions of design in modern construction. Beside its natural beauty, laminated wood offers strength, safety, economy, and permanence. See the graceful arches in Fig. 20-30.

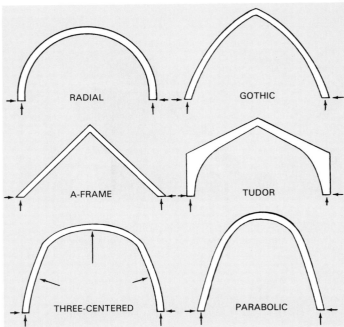

RADIAL   GOTHIC

A-FRAME   TUDOR

THREE-CENTERED   PARABOLIC

Fig. 20-31. Types of laminated wood arches. Arrows indicate the support and lateral thrust that must be provided.

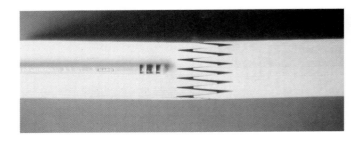

Fig. 20-32. Finger joint used to join ends of laminations. Joint is formed with a special cutter head. Individual laminations should not exceed 2 in. in net thickness.
(American Institute of Timber Construction)

Fig. 20-33. Laminated beams provide a clear span of 48 ft. Purlins, spaced at 8 ft. will support 4 x 8 ft. prefabricated roof panels. Note the metal connectors used to fasten the purlins to the beams. (Boise-Cascade Corp.)

Two views of a post-and-beam house under construction. Conventional construction is used for walls and partitions. Roof uses transverse beams and planking for exposed beam ceiling.

Most laminated structural members are made of softwoods. They are manufactured in industrial plants specializing in such production. When they arrive at the building site they are prefinished.

In residential work, beams are usually straight or tapered; however, in institutional and commercial buildings they are often formed into curves, arches, and other complicated shapes. Some of the standard forms are illustrated in Fig. 20-31.

In the fabrication of beams and arches, lumber is carefully selected and machined to size. To get the required length, pieces must often be end-joined. Since end-grain is hard to join, a special finger joint may be used. A number of these joints, Fig. 20-32, may be required in each ply. The joints are staggered at least 2 ft. from a similar joint in an adjacent layer. Fig. 20-33 shows an installation view of laminated beams that will support the roof of a large commercial building.

TEST YOUR KNOWLEDGE - UNIT 20

1. The slenderness ratio of a post compares the total height in inches with the _____ (largest, smallest) dimension of its cross section.
2. Two common types of roof beams include longitudinal beams and _____ beams.
3. This live load on a residential floor is usually figured at _____ lb. per sq. ft. and represents the weight of furniture, occupants, and equipment.
4. Planks for floor and roof decking are available in thicknesses of 2 to _____ in.
5. A fabricated building component, similar to a stressed skin unit with a core of foamed plastic or paper is called a _____ panel.
6. In plywood box beam construction, the flanges are separated at regular intervals with spacer blocks which distribute the load and prevent buckling of the _____.

OUTSIDE ASSIGNMENTS

1. Visit a building supply center and secure descriptive literature about floor and roof decking that can be used in post-and-beam construction. Include both solid and laminated planks and composition panels. Be sure to obtain prices. Study these materials thoroughly for qualities, characteristics, and installation procedures. Prepare a written or oral report.
2. Build a mock-up of a stress skin panel in which you can experiment with the design and size of the various parts. Use a scale of about 3″ = 1′-0″. Skins might be made of 1/8 in. plywood or 1/16 in. veneer.

Top. Sectionalized (modular) home being placed on treated wood foundation. Sections are factory built and completely finished inside and out. (American Plywood Assoc.) Bottom. Double-end sawing machine cuts bottom chord members for roof trusses. Each sawing unit (left and right) has three blades that can be set at various angles. (Speed Cut, Inc.)

502

Unit 21

PREFABRICATION

The term prefabrication means to cut and assemble parts and sections in factories. The term includes shipment to the building site for final assembly. Prefabrication also refers to buildings that are partially or fully erected before leaving the manufacturing plant.

In co-operation with industry, the modern builder has found:
1. Fast building methods.
2. Strong parts.

To make good parts, many steps must be done under controlled conditions with special tools. For example, stressed-skin panels and plywood box beams require accurate glue applications, special presses, and handling equipment. Often the simplest prefabrication process in modern plants is done on large production equipment, Fig. 21-1. The machine reduces the amount of human energy required and increases production. Speed is increased both in the plant and on the building site.

COMPONENTS

In house construction today, many factory-built components (parts) are used for finish work and for framing work. Some of these are windows, door units, soffit systems, stairs, and built-in cabinetwork.

Fig. 21-1. Roller coating machine applies finish to kitchen cabinet doors.
(National Homes Corp.)

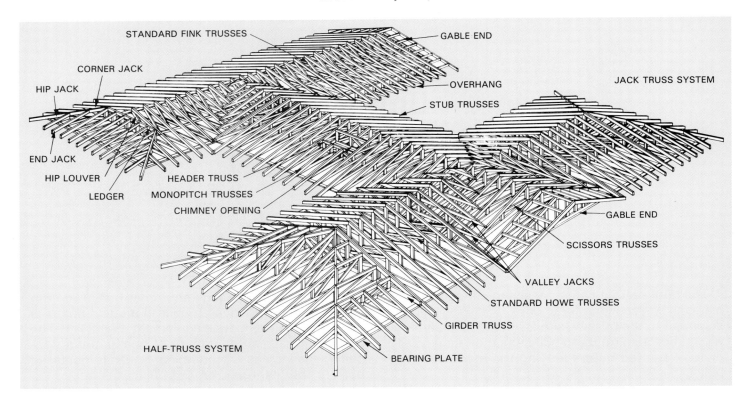

STANDARD FINK TRUSSES

GABLE END

CORNER JACK

HIP JACK

JACK TRUSS SYSTEM

OVERHANG

STUB TRUSSES

END JACK

HIP LOUVER

HEADER TRUSS

LEDGER

MONOPITCH TRUSSES

CHIMNEY OPENING

GABLE END

SCISSORS TRUSSES

VALLEY JACKS

STANDARD HOWE TRUSSES

GIRDER TRUSS

HALF-TRUSS SYSTEM

BEARING PLATE

Fig. 21-2. Prefabricated roof framing includes all the types of roof trusses used to form hips, valleys, overhangs, and gable ends. (Gang-Nail Systems, Inc.)

One framing component is the roof truss. Manufacturing plants can usually furnish either simple Fink trusses (described in Unit 9) or matched units for a complex roof as shown in Fig. 21-2. Roof trusses must be carefully designed by structural engineers and built according to exact specifications. High production double-end saws, Fig. 21-3, cut truss members to length. Special presses and jig tables fasten the assembly with "gang-nail" connectors, Fig. 21-4. Completed roof trusses are shown being transported to the construction site in Fig. 21-5.

The floor truss, Fig. 21-6, is another example of a prefabricated structure. It permits a wide unsupported span with a minimum of material. It also has open spaces for heating and air conditioning ducts, electrical conduit, and plumbing lines. Fig. 21-7 shows a floor truss being built.

Wall and roof panels are made in various shapes and sizes. Panels provide structural strength in addition to forming inside and outside surfaces. Fig. 21-8 shows curved stressed-skin panels constructed of lumber ribs and plywood surfaces.

Fig. 21-3. Double-end saw with four cutting heads speeds production of truss members.

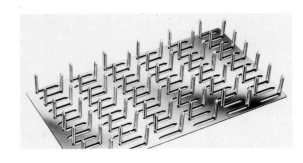

Fig. 21-4. "Gang-Nail" connector is available in several sizes and gauges of steel. It must be applied with approved presses.

Fig. 21-5. Trussed roof system being transported to construction site. (Gang-Nail Systems, Inc.)

Fig. 21-6. Floor truss system provides ample space for installation of plumbing lines. (Gang-Nail Systems, Inc.)

Fig. 21-7. Building a metal web truss. Press (not shown) forces toothed plates into chords, top and bottom.

Two other types of panels are flat stressed-skin panels and flat sandwich panels. Both are described in Unit 20

Sandwich panels made with a foamed plastic core provide enough strength for walls and partitions in

Fig. 21-8. Installing curved stressed-skin panels which copy some effects of vaulted ceilings.

some structures. See Fig. 21-9. These lightweight panels may have an aluminum outer skin and a hardboard or plywood inside surface. Special metal channels and angles are available which reudce labor at the building site.

PREFABRICATING BUILDINGS

A building in various stages of completion can be prepared in factories and then shipped as a ''package'' to the building site. Single-family homes are built most often. However, small commercial buildings, farm structures, and multi-family homes are also produced.

There are four basic types of prefabricated houses and other buildings: the precut, the panelized, the sectionalized, and the mobile-home. Combinations of these are used.

Fig. 21-9. Vacation home made of stressed-skin floor panels and sandwich wall panels. Such components are basic to all prefabricated construction.

PRECUT

For a modern precut house, lumber is cut, shaped, and labeled to reduce labor and save time on the building site. Manufacturers of this type of house include materials needed to form the outside and inside surfaces. Also shipped are such millwork items as windows, doors, stairs, and cabinets. Optional items include electrical, plumbing, and heating equipment.

PANELIZED

In panelized prefabrication, flat sections of the structure are fabricated on assembly lines. Large woodworking machines cut framing members to length and required angles, Fig. 21-10. Parts are stored and delivered to the assembly stations as needed. Wall and floor frames are formed by placing the various members in positioning jigs on the production line. The parts are fastened with pneumatic nailers, Fig. 21-11. Any electrical wiring

or other facility is installed soon after the frame is built.

As the completed frames move along conveyor lines, wall surface materials are placed in position and nailed. Nailing is done with powered gang-nailers and/or staplers. See Fig. 21-12. A special type of staple is being used to attach high R-value foamed plastic sheathing in Fig. 21-13. Farther along the line, insulation is put in, Fig. 21-14. Some of the wall sheathing steps may be repeated to close the panel. Then siding is attached, Fig. 21-15.

In other areas of the plant, roof units are prepared. For post-and-beam structures, panels that form both the roof and ceiling surfaces are common. To avoid

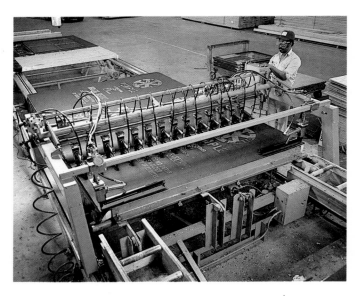

Fig. 21-12. A gang of pneumatic-powered nailers secure fiberboard sheathing to a wall frame in less than two minutes. (Duo-Fast Corp.)

Fig. 21-10. Control end of a double-end saw used to cut framing stock in a plant that makes panels. Guards are removed to show blades.

Fig. 21-11. Automatic machine, called a panel extruder, assembles 60 floor frames a day. Unit is fully computerized. (Cardinal Industries, Inc.)

Fig. 21-13. Stapling foamed plastic sheathing to wall frame. Special steel strap is used for bracing. (Citation Homes)

Fig. 21-14. Installing blanket insulation in wall panel. Insulation is arranged around any electrical equipment. (Wausau Homes Inc.)

Fig. 21-16. Painting the ceiling side of roof panels used in a post-and-beam type of prefabricated building. (Wausau Homes Inc.)

Fig. 21-15. Applying aluminum siding to the outside of a wall panel. Look closely at the left side; this is not regular corrugated aluminum. (National Homes Corp.)

Fig. 21-17. Prefabricating stressed-skin floor panels. Full length joists and scarf-glued plywood sheets (foreground) are assembled with glue to form units as large as 8 x 24 ft. (Wausau Homes Inc.)

painting in high places after erection, the ceiling side is painted at waist level, Fig. 21-16.

Although most prefabricated houses of the panelized type use a first floor deck built by conventional methods, some manufacturers design and build floor panels. Full length joists are assembled with headers, Fig. 21-17. Then long plywood sheets, formed with scarf joints, are glued in place. The resulting stressed-skin construction is rigid and strong. The panels prevent nail pops and squeaks. For an example of a stressed-skin floor panel being installed, refer again to Fig. 21-9.

As the panels near the end of the production lines, they receive a final inspection. After inspection, each panel is numbered for easy assembly and is then ready for storage or loading into a trailer, Fig. 21-18. When all materials and millwork to complete the "package" are loaded, the truck is sent to the building site.

Fig. 21-18. Finished wall panel being loaded on trailer to the right. Panel includes electrical wiring with outside door installed. Drywall has been painted. (Wausau Homes Inc.)

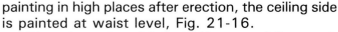

Kitchen cabinets and other built-in units are usually made in plants specializing in these items. The units may be shipped either to the home fabricator or directly to the building site.

The panel type of prefabrication is not limited to one residential style or size. Multilevel, two story, and split-level types are available. Although most manufacturing plants produce houses, some firms also supply panels and other parts for school buildings, apartments, and small commercial structures.

SECTIONALIZED OR MODULAR

In this type of prefabrication, entire sections (commonly called modules) of the structure are built and finished in manufacturing plants, Fig. 21-19. The sections or modules are then hauled to the site where they are assembled. Widths of a module seldom exceed 14 feet. Trucks and roads cannot handle wider units.

An advantage of sectionalized construction is that nearly all of the detail finish work can be done at the factory. Kitchen cabinets can be attached to the walls and other built-in features can be installed. Also wall, floor, and ceiling surfaces can be applied and finished.

Electrical wiring, heating and air conditioning ducts, plumbing lines, and even plumbing fixtures can be installed under close control in manufacturing plants. A section which has a group of plumbing and heating facilities is often included with other sections that consist mainly of panels. Sections of this nature which include most of the utility hook-ups are called mechanical cores. Core sizes match the Unicom system. See Unit 5, page 109.

Mechanical cores group the kitchen, bath, and utilities in one unit that requires only three connections at the site. A mechanical core with a bathroom and one kitchen wall is a common unit. Some units are designed to include heating and air conditioning equipment as well as electrical and plumbing equipment, Fig. 21-20.

Disadvantages of the sectionalized type of prefabrication are the problems involved with storage and transportation, and the need for large power cranes to handle the units at the construction site.

Fig. 21-20. Crane places 5 ton module on masonry foundation. Units are 95 percent complete when they arrive on site. (Cardinal Industries, Inc.)

MOBILE HOMES

Mobile homes are mounted on a chassis, Fig. 21-21, and do not need a permanent foundation. They are defined by manufacturers as a trailer longer than 28 ft. and heavier than 4500 lb.

The most common width is 12 to 14 ft. Lengths may be as great as 68 ft.

Designs sometimes include an ''expansible'' feature that permits wider living rooms. A second section is carried inside while the mobile home is being hauled. It is attached to the side after the home is on the site. The second section can be like a bay window or can be more square.

Production methods used in building mobile homes are like those used for standard prefabricated housing. Some of the structures, however, are different. For example, the floor must be one rigid unit and usually consists of a wood frame attached to a welded steel chassis, Fig. 21-22.

A mobile home is finished and fully equipped at the factory. It is then pulled to the dealer or directly to a trailer park. Utility hook-ups are ready to be

Fig. 21-19. Final assembly line. Kitchen and living room modules are put together.

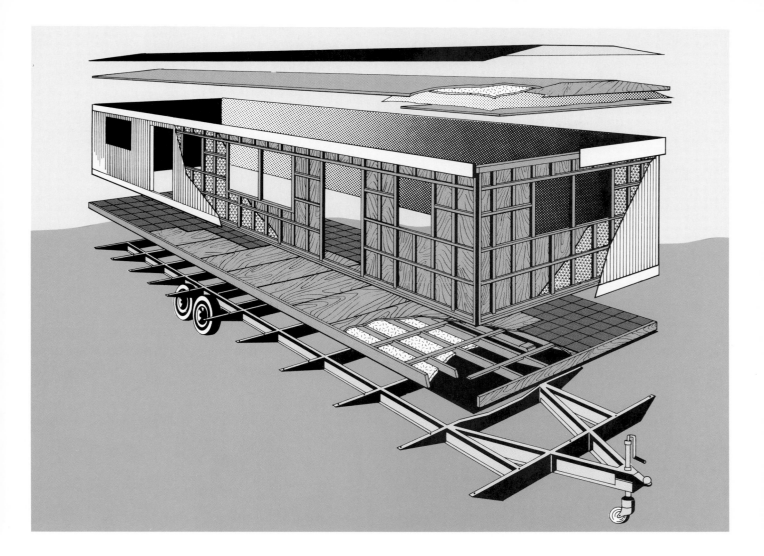

Fig. 21-21. Exploded view of a mobile home. All structural members above the steel chassis are wood. Plywood deck is securely screwed to floor frame which is assembled with glue. Outside walls consist of 2 x 3 studs with 1 x 2 horizontal rails glued in place. Inside surface is 3/16 in. prefinished plywood. Prepainted aluminum panels cover outside. (Redman Industries Inc.)

made at the site. Interior furnishings include all major appliances, carpeting, drapes, furniture, and lamps. Fig. 21-23 shows an inside view of a large mobile home.

Fig. 21-22. Wooden floor frame is attached to steel chassis. Floor must be fully insulated. Note plumbing and electrical lines. (Redman Industries Inc.)

ON-SITE CONSTRUCTION

At the building site, the foundation should have been built. The various prefabricated units were loaded on the trailer so they can be removed in the proper order for matching the edges. The floor deck is built in the standard way or is assembled from panels. Then:
1. Mechanical core units are set in place.
2. Walls and partitions are joined.
3. Ceiling-roof units are installed.
4. Roof panels close the structure.
Fig. 21-24 shows in four views the construction of one modern prefabricated home. The shell of a home can be finished in one day.

When prefabricated panels are finished on both the inside and outside surface, they are called closed panels. Today, the terms "manufactured" and "industrialized" are used when referring to prefabricated housing.

The purpose of prefabrication is to save time and

Fig. 21-23. This double-width mobile home has wood framing under the floor to reduce heat flow and side stresses. Sectionalized construction was used. The sections can be taken apart to move the home to a new site. (Marshfield Homes, Inc.)

FLOOR PANELS HAVE BEEN LAID ONTO THE FOUNDATION AND A MECHANICAL CORE FOR THE KITCHEN IS BEING LOWERED INTO PLACE. NOTE BATH MODULE IN PLACE. ALSO NOTE SUN ANGLE AS WORK CONTINUES.

FRONT WALL PANEL FOR LIVING ROOM IS BEING INSTALLED. OUTSIDE SURFACE HAS NOT BEEN APPLIED SINCE THIS SECTION OF THE HOUSE WILL BE FINISHED WITH BRICK VENEER. TRAILER IN FIRST PHOTOGRAPH HELD PANELS IN PROPER ORDER.

CRANE LOWERS CEILING-ROOF PANELS INTO PLACE OVER BEDROOM AREA. NOTE HOW THE PANELS ARE HINGED TOGETHER. THE COMBINED UNIT WAS CLOSED FOR SHIPMENT. THE UNIT IS OPENED FOR FITTING AT THE BUILDING SITE.

INSTALLING FINAL ROOF PANEL ON GARAGE. NO CEILING PANELS WERE REQUIRED. NOTE THAT CARPENTERS HAVE STARTED SHINGLE WORK OVER BEDROOM AREA. THIS PREFABRICATED HOME WAS ENCLOSED IN LESS THAN A DAY.

Fig. 21-24. Sequence shows on-site erection of a modern prefabricated home. (Wausau Homes Inc.)

Fig. 21-25. When components are made alike, they are easy to handle. Here five prefabricated Howe trusses are being installed at one time. Also, modules simplify every phase of construction. Here the plumbing and electrical systems are ready to be connected between modules. (National Homes Corp.)

labor. One type of manufactured housing is built with panels. Fig. 21-25 shows another way to build a home. Bare roof trusses are often chosen for their low cost compared to finished roof panels. Sometimes the initial cost saving for bare trusses offsets the labor cost to sheathe them on the building site. Half as much labor and sheathing will be needed as for roof panels finished on both sides.

One disadvantage of prefabrication is soon discovered at the building site. Weather prevents work much of the time. The builder must choose a day without wind or rain. Rain will damage inside wall materials applied at the factory. Wind makes it hard to unload large panels.

There is also the problem of soft ground. It limits the size of the crane that can be used. Unless the reach of the crane is great, there is no way to safely bring large structures into wet areas.

In some ways, the weather is less important for prefabricated than for conventional construction. A skilled crew can erect and enclose a manufactured single-family home in 5 to 20 hours. A few weeks are usually required to complete the wiring, heating, plumbing, and decorating. With prefabrication, the inside work begins on dry structures. Often the inside gets wet before conventional construction is closed in. Therefore, the finish work goes faster on a manufactured home. For this reason and for other reasons, prefabrication will compete well with other building methods.

TEST YOUR KNOWLEDGE — UNIT 21

1. Prefabricated components are mass produced using conveyor lines and assembly _____.
2. _____-skin panels make floors, walls, and roofs rigid and strong.
3. A prefabricated section that includes a concentration of plumbing, heating, and other utility hook-ups is called a _____.
4. The chief advantage of the _____ type of prefabrication is that most of the inside and outside finish work can be completed at the factory.
5. Mobile homes are constructed on a metal chassis and do not require a permanent _____.
6. The outer structure of a panelized home can be finished in one _____.
7. Due to speed of construction, a prefinished home will probably not be:
   a. Well-built.
   b. Wet inside.
   c. Checked by a Building Inspector.

OUTSIDE ASSIGNMENTS

1. Visit a local builder or the manager of a building supply center that represents a ''prefab'' home manufacturer in your region. Get descriptive literature and information concerning sizes, design, fabrication features, material quality, and erection procedures. Also check prices for standard models and time required for delivery. Organize the information carefully and make an oral presentation to your class. Supplement your oral descriptions with appropriate visual aids.
2. Prepare a written report on the development of prefabrication as applied to building construction in this country. Highlight early experiments and devote most of your study to progress since World War II. Find information in reference books, encyclopedias, and trade magazines.
3. Visit a mobile home sales center and get information and descriptive literature concerning size, features, and price. Inspect models that are on display and give special attention to the quality of materials, work, and finish. Also note the quality of equipment and furnishings. Compare prices with costs of prefabricated and conventionally built houses. Prepare carefully organized notes and make an oral report to your class on your findings.

Passive solar structures are designed to capture the heat of the sun and usually have some provision for storage of the heat. Top left. New Jersey solar home features three-story solar glazing. A Trombe wall behind the glazing stores the heat of the sun's rays. (Robert Perron)   Top right. Massive brick wall and fireplace below act as solar heat collectors. (Pittsburgh Plate Glass Industries)   Left. Prefabricated sun space can be added as a remodeling project. Both walls and roof are glazed for maximum solar collection. ''Slim shades'' are installed between double glazing to shut out sun in warm weather.
(Rolscreen Co.)

Unit 22

PASSIVE SOLAR CONSTRUCTION

Because sunshine is free it makes sense to capture it and use it for heating buildings. Constructing buildings to take advantage of the sun requires some planning and an understanding of how to make the sun provide heat when it is most needed.

HOW RADIATION AND HEAT ACT

Solar radiation can travel through glass almost as well as it travels through the atmosphere. However, once transformed into heat energy (by striking a surface inside a glassed area) it cannot readily pass back through the glazing. This is known as the greenhouse effect. This effect is important to all solar construction. See Fig. 22-1.

HEAT TRANSFER PRINCIPLES

The very basis of solar construction is found in the way that heat moves from one place to another. Fortunately for us, heat *always travels from a higher temperature to a lower temperature.* It can travel by three different methods:
1. Conduction.
2. Convection.
3. Radiation.

CONDUCTION

In conduction, heat travels point-by-point through solid matter. One part of the solid body must be in touch with a heat source. Gradually the whole body becomes heated. For example, if the sun heats a block of concrete the entire block will eventually become hot. This is explained by the KINETIC THEORY OF MATTER. The molecules making up the concrete become more active as they become warmer, moving farther and faster and bumping into other molecules. This contact moves the heat from one molecule to another. See Fig. 22-2.

CONVECTION

Convection occurs only in fluids or gases. Heat causes them to expand and become lighter or less dense. Lighter elements always rise; cooler ele-

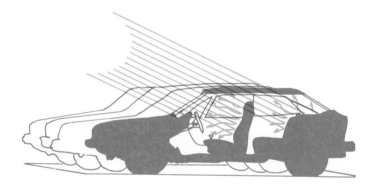

Fig. 22-1. Solar radiation can pass through glass much easier than heat. This is known as the "greenhouse effect." It explains why an automobile with windows closed will get so hot inside in direct sunlight. (Iowa Energy Policy Council)

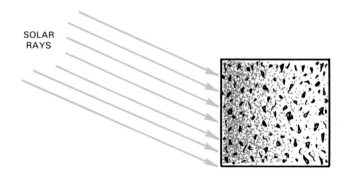

SOLAR RAYS

Fig. 22-2. In conduction, heat travels when warm molecules of a solid bump into cooler ones. Eventually, all of them are heated.

ments descend. Through this constant motion, heat moves through water or air, Fig. 22-3. Older furnaces used this principle. Heat rose through natural convection and spread through the living space. No mechanical means are needed to make the air move.

RADIATION

Radiant energy moves through space in waves. When waves strike a solid object, Fig. 22-4, the energy is absorbed by the solid matter. The internal energy of molecules in the body rises and its temperature rises. An electric heater works this way. So do the sun's rays.

THERMOSIPHONING

Thermosiphoning is simply the result of a liquid or gas expanding and rising. This principle is put to work in both active and passive solar heating. Sunlight is captured in a closed space where a liquid such as water or air is heated. The system operates by the action of the expanded water or air rising. Cooler air or water comes in to the space to take its place, Fig. 22-5.

TYPES OF SOLAR CONSTRUCTION

Adapting building to use of solar energy can be approached in two different ways: active solar and/or passive solar construction.

Active solar construction is actually a system of

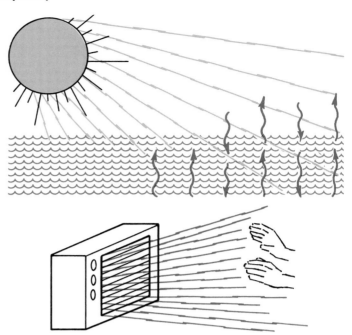

Fig. 22-4. Two examples of radiant heating. In radiation, energy moves in waves from source to solid object. Top. Sun warms body of water as radiation strikes the surface. Bottom. Electric heater works the same way.

collecting solar energy apart from the structure of the house. It is a separate system that may be added to a dwelling during or after it is constructed. It has little, if any, effect on the way the building is built.

Fig. 22-3. Convection is caused by heat contacting and entering liquids or gases. Heated gases and liquids expand, become lighter, and rise. Cooler gases and liquids are heavier so they settle. Left. Liquid action. Right. Convection in air.

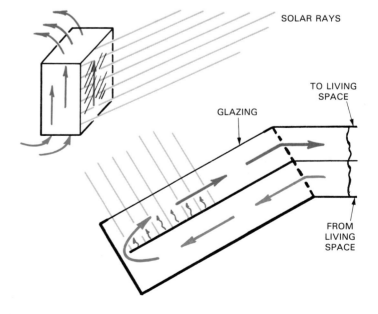

Fig. 22-5. Top. Simple example of thermosiphoning. Sun shines into open ended box. Heated air rises out the top, cooler air is pulled in the bottom. Bottom. When baffle is added, cold air moves in from lower side of box. Warm air rises along upper half.

In some cases the solar collectors, Fig. 22-6, are made a part of the roof. In other cases, the collectors are attached to the existing roof by a metal framework.

Active systems are so called because electrically operated blowers or pumps are required to carry heated air or fluid to where it will be used or stored. Fig. 22-7 is a simple sketch of an active solar system.

Passive solar construction, on the other hand, by its very nature, requires a change in the structural makeup of the building. It is called passive because it has few if any parts which must move or require power to make it work. Collection, storage, and transporting of heat is done naturally by the materials used in construction.

TYPES OF PASSIVE SOLAR ENERGY

Passive solar systems are named for the way in which they operate. The word "gain" refers to how the heat is picked up from solar radiation. There are three basic types of passive solar construction.
1. Direct gain.
2. Indirect gain.
3. Isolated gain or sun space.

Depending upon the individual house design, more than one system may be used, Fig. 22-8. Several approaches may be combined in the same house for greater efficiency, affordability, and beauty. No one floor plan is required and passive solar systems may be incorporated into many architectural styles. Passive designs work well with Cape Cods, colonials, contemporary, or ranch styles.

DIRECT GAIN SYSTEM

Direct gain means that the sun shines directly into the living space and heats it up. In the simplest of direct gain systems there may be no massive

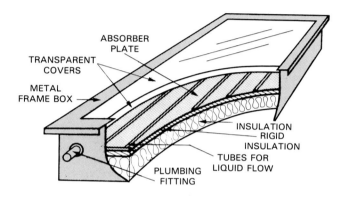

Fig. 22-6. Cutaway of a flat plate solar collector designed to heat liquid. It is used in an active solar system. (National Solar Heating and Cooling Center)

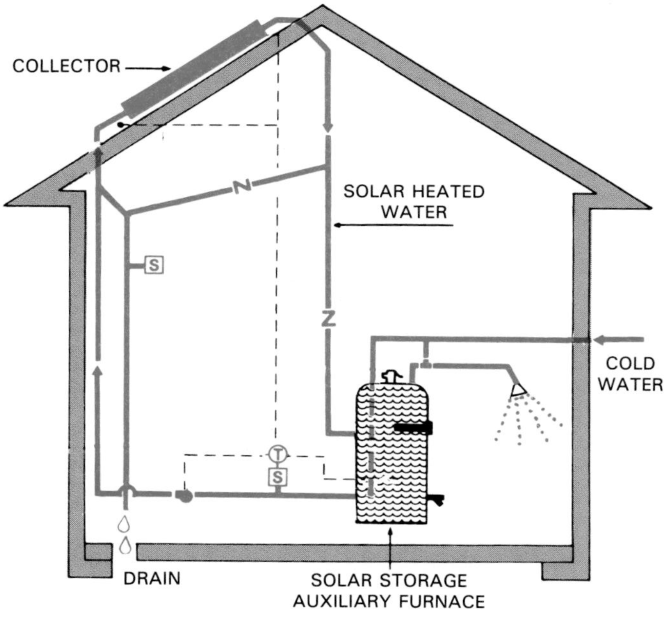

IN A PUMPED DRAINDOWN UNIT, SOLAR HEATED WATER FLOWS TO THE STORAGE TANK FOR DIRECT USE BY THE HOUSEHOLD. WHEN THE PUMP SHUTS OFF, WHATEVER WATER REMAINS IN THE COLLECTOR DRAINS AWAY BY GRAVITY FLOW.

Fig. 22-7. Schematic of an active solar system using roof collector illustrated in Fig. 22-6.

Fig. 22-8. Home using both active and passive solar energy. Note solar panels on roof. Solar room at right is an example of isolated sun space illustrated in Fig. 22-13. (Timberpeg)

structures to store the heat produced. The furniture soaks up some heat and so does the air in the room. Usually, though, there is some means of storing the heat for use at night or on cloudy days. Thick masonry walls or floors, a masonry fireplace, a heavy stone planter, or containers of water are examples of storage systems that may be used. Suitable materials include stone, adobe, brick, or concrete.

At night, stored heat is distributed to the living space by radiation and convection. If the storage is adequate, it should give off heat throughout the night.

One major disadvantage of direct gain systems is the wide range of heat fluctuation (changes). During the hours of strongest sunlight, heat can build up to uncomfortable levels. Fig. 22-9 is a simple drawing of a direct gain structure.

INDIRECT GAIN SYSTEM

In this system, the rays of the sun enter through glazing and heat up thermal mass rather than the room itself. Often the mass is a thick masonry or concrete wall located directly behind large windows or glide-by glass doors. Sometimes the wall is vented top and bottom. This allows hot air to flow upward and enter the room quickly by way of convection currents, Fig. 22-10.

Often called a Trombe wall for the French scientist, Felix Trombe, who designed it, the thermal storage wall is the most popular storage structure for indirect solar heating. The outer face is dark colored for greater absorption of heat. By conduction,

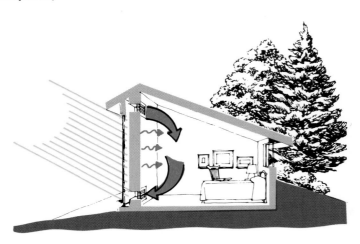

Fig. 22-10. The Trombe wall is the most popular method of storing and distributing solar heat by the indirect gain method. (Iowa Energy Policy Council)

the heat travels through the wall to its inner face. Then radiation and convection, together, distribute the heat throughout the space.

Thermal walls still perform their function even with windows cut in them. Inner masonry surfaces can be made to fit the interior design by covering them with stucco, gypsum plaster, ceramic tile, or other heat conducting materials. Pictures may be hung without materially affecting the wall's performance.

WATER STORAGE WALL

Water is an effective, inexpensive material for storage of solar heat. In fact, it can absorb about twice as much heat as rock. However, the materials for containing it are not usually cheap. Still, barrels and other empty containers that are rescued from salvage, can be used.

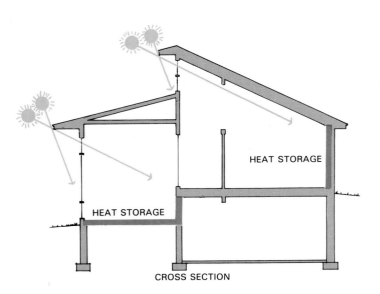

CROSS SECTION

Fig. 22-9. Basic design for a direct gain passive solar house. Floor and wall store direct sunlight. (District 1 Technical Institute, Eau Claire, Wisconsin)

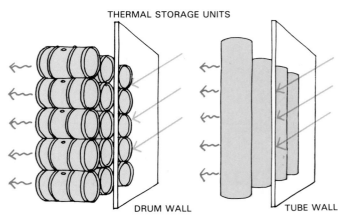

THERMAL STORAGE UNITS

DRUM WALL   TUBE WALL

Fig. 22-11. Water and phase change materials can also be used to store the sun's heat. (Pittsburgh Plate Glass Co.)

The water wall is placed directly behind a south-facing window. As with masonry storage, the water absorbs the solar heat and transfers it by radiation and convection.

A more sophisticated water wall may use phase change materials. These are materials which, like butter, change from a solid to a liquid state as they absorb heat. They are able to absorb much more heat than materials which do not change their form. Usually, a type of salt solution is used. Common glauber's salt is one of them. See Fig. 22-11.

It is generally easier and more trouble free to use masonry with conventional structural features such as walls, floors, and interior fireplaces. Such

materials and structures rarely need attention once they are in place.

ISOLATED GAIN SYSTEM

In an isolated gain system, the solar heat is collected and stored in an area remote (apart) from the living or working space. Its advantages are several:
1. The living/working area is not directly exposed to the sun.
2. The heat it collects is more easily controlled.

Fig. 22-12 shows a simple sketch of a thermosiphon isolated passive system. This is one type. Another is called a sun space.

The sun space, Fig. 22-13, is a common method of providing solar heat. It is a separate room with large areas of its south wall being glazing. Sometimes the room has thermal mass built into the inside wall and/or the floor. Additional thermal storage can be provided in water containers or bins of rock.

PASSIVE SOLAR ADVANTAGES

Passive designs have advantages over active solar heating and conventional heating:
1. Common building materials, glass, concrete, and masonry can be used. Other manufactured materials are not needed although they are available.
2. Conventional carpentry and masonry skills are sufficient. No additional skills need be learned.
3. Passive components do not wear out and need little maintenance. Life expectancy is greater than active systems.
4. Properly designed passive features can supply upwards of 50 percent of heat required.

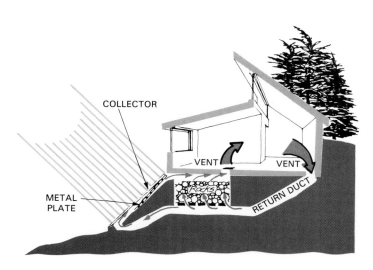

Fig. 22-12. Thermosiphoning isolated passive system. Collector at ground level is the thermosiphon. Heated air flows on its own power to storage. Cool air is pulled into the collector from below.

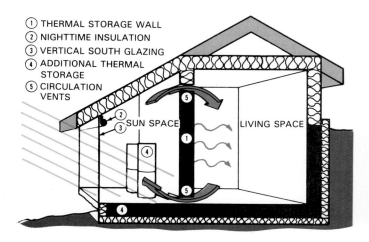

Fig. 22-13. Another type of isolated gain heating is the sun space. Left. This is a room with plenty of storage mass to save solar generated heat. Note how insulation protects floor and north wall from colder surroundings. (Iowa Energy Policy Council) Right. Solar sun room is an "add on" sun space. (Rolscreen Co.)

5. Passive solar is nonpolluting.
6. Passive systems are comfortable in cold weather.
7. Since heat is generated close to or in the space being heated, little heat is lost through transfer.

PASSIVE SOLAR DISADVANTAGES

Disadvantages of solar systems include:
1. Control of the heat is not as responsive as either conventional heat systems or active solar systems. Building occupants are accustomed to temperature swings of 3° to 5°F. Many passive systems have swings of 10° to 15°F.
2. It is not a simple matter to control heat and heat distribution in a passive system. Careful planning is required.

SOLAR HEAT CONTROL

Adequate control methods must be arranged when passive solar heating is being planned. There are times when solar heat is not wanted. In the summer, heat is not required. Then the same structural features which allow sunlight to enter must be sheltered from the sun. Shade can be provided either naturally or artificially.

Deciduous trees (those which lose their leaves in winter) should be located to protect south-facing windows in summer. It may be possible to locate a new building to take advantage of existing trees on the property. Fast-growing trees can also be planted to eventually provide this shade. They should be located to provide shade for west-facing walls as well.

Constructed protection includes outdoor roof structures including overhangs and shutters, Figs. 22-14 and 22-15.

OVERHANGS

The width of the protective overhang must be determined by three variable factors. See Fig. 22-16.
1. The height of the window or collector.
2. The height of the header above the window or collector.
3. The latitude of the construction site.

As a rule of thumb, the overhang in southern states (roughly latitude 36 degrees north) should be 25 percent of the combined height of header and window. In northern states (around 48 degrees latitude) the percentage should be upped to 50.

A simple mathematical formula, used with the chart in Fig. 22-17 will also provide a quick and quite accurate determination of how far the overhang should project:

$$\text{Projection} = \frac{(\text{window height} + \text{header height})}{F\ (\text{factor from Fig. 22-17})}$$

MOVABLE INSULATION

While outdoor structures and natural shade are more effective protection against the hot summer sun, movable insulation is sometimes the only practical protection possible. It is also effectively used to reduce nighttime heat losses through glass.

Movable insulation is usually sheet or blanket materials that can be put over glazing temporarily. It is placed over the glass only when you do not want radiation or heat passing through.

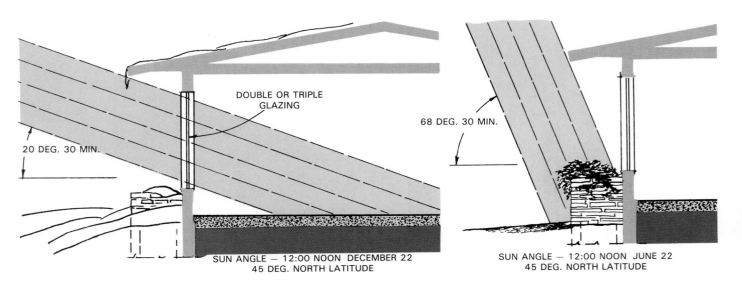

DOUBLE OR TRIPLE GLAZING

20 DEG. 30 MIN.

SUN ANGLE — 12:00 NOON DECEMBER 22
45 DEG. NORTH LATITUDE

68 DEG. 30 MIN.

SUN ANGLE — 12:00 NOON JUNE 22
45 DEG. NORTH LATITUDE

Fig. 22-14. Roof extensions can be used to control solar rays in passive heating design. Latitude of 45 deg. north is the general location of Portland, OR; Minneapolis, MN; and Montreal, Canada.

Fig. 22-15. Exterior shading includes deciduous tree, overhang, and hinged shutters. (HUD)

Movable insulation solves a major problem of glazing: its poor insulating quality. The insulation cuts nighttime heat losses and will help keep the building cool during summer days.

Triple glazing of window area will also cut down heat losses. However, it also reduces the solar radiation transmission from 74 percent (for double glazing) to 64 percent. Removable insulation is preferred because it provides better heat retention and does not cut down on solar efficiency of the window.

Movable insulation is available in a variety of types including:
1. Sheets of rigid insulation.
2. Framed and hinged insulation panels, Fig. 22-18.
3. Exterior mounted plastic or metal roller shades, Fig. 22-19.
4. Padded roller-mounted flexible cloth panels, Fig. 22-20.

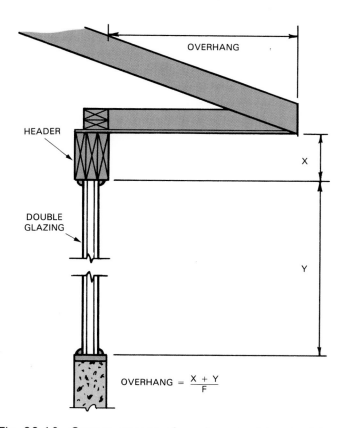

$$\text{OVERHANG} = \frac{X + Y}{F}$$

Fig. 22-16. Correct amount of overhang is determined by dividing height of window and header by a factor based on latitude.

LATITUDE (in Degrees)	APPROXIMATE LOCATION (State)	FACTOR* (F)
28	Central Florida	5.6 to 11.1
32	Central Texas	4.0 to 6.3
36	Northern Oklahoma	3.0 to 4.5
40	Northern Missouri	2.5 to 3.4
44	Iowa-Minnesota Border	2.0 to 2.7
48	Northern Minnesota	1.7 to 2.2
52	Southern Canada	1.5 to 1.8
56	Central Canada	1.3 to 1.5

*Higher of two factors provides 100% shading at noon on June 21; lower factor until August 1.

Fig. 22-17. Factors for finding width of overhang from south Texas to central Canada.

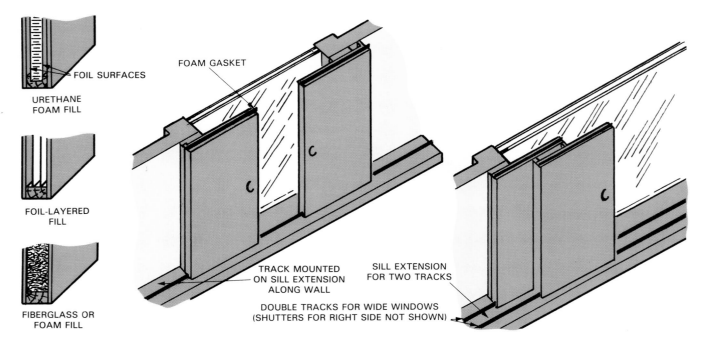

Fig. 22-18. Sliding shutters are designed to prevent loss of heat through solar glazing at night.

FOIL SURFACES

URETHANE FOAM FILL

FOIL-LAYERED FILL

FIBERGLASS OR FOAM FILL

FOAM GASKET

TRACK MOUNTED ON SILL EXTENSION ALONG WALL

SILL EXTENSION FOR TWO TRACKS

DOUBLE TRACKS FOR WIDE WINDOWS (SHUTTERS FOR RIGHT SIDE NOT SHOWN)

Fig. 22-19. Power operated plastic or metal shutters shut out unwanted sunlight, provide security, and shut out cold at night. Left. Opened. Right. Closed to block out sunlight in warm weather. (Pease Industries, Inc.)

5. Powered systems such as ''Beadwall'' which depend upon blowers to fill air cavities between layers of double glazing with insulating beads.

VENTING

Venting has an important role in controlling solar heat both in indirect and isolated gain systems. Trombe walls sometimes have them to provide daytime heating for living space. Refer once more to Fig. 22-10. Opening the vents allows convection to carry heated air into the room through the top vents. Meanwhile, cool air is drawn into the air space between the glazing and the Trombe wall. Closing the vents, top and bottom, will prevent heat from entering the living space.

Isolated sun space is often vented to the outdoors to exhaust unwanted solar heat during hot weather. The vent acts like a chimney. The hot air rises through the roof. If there are vents at floor level, cooler air will be pulled into the sun space. This provides additional cooling.

Fig. 22-20. Roll-up fabric insulating shade.

Fig. 22-21. Two methods for using venting for warm weather cooling. Top. Vent in glazing draws air from cool, shaded side of dwelling across room, through Trombe wall, and out. Bottom. Thermal chimney effect. Warm air rises and exhausts through vents at top of house. Cool air enters from open window in shaded wall. (Pittsburgh Plate Glass Co.)

Natural venting of buildings is also possible using the chimney effect. This is accomplished with a little sunlight and a vent or vents high up in the building. A window or skylight will let in sunlight to heat a "pocket" of inside air. The air rises and passes through the vent. Cooler air can then be brought into the building from another area of the living space. Fig. 22-21 shows two methods of using venting for cooling.

ORIENTATION

The first step in solar construction is to locate the building so one wall catches the sun's rays all day long. This is called orientation, Fig. 22-22.

Most of the windows should be placed in the south wall so that the solar radiation can be collected. Few windows will be placed on the north side. Garages should be placed to the northwest, if possible, to block cold winter winds.

ENERGY BALANCE

Homes experience heat gains from three sources on a typical winter day:
1. From the conventional heat source, such as a stove or furnace.
2. From solar radiation.
3. From appliances and occupants. This is called INTERNAL HEAT.

Energy use is reduced when the heating system can operate at the lowest level that will provide comfort. Solar and internal heat gains can permit a cutback in operation of the heating plant if the house design can balance internal heat gain against solar gain. This means locating heat producing

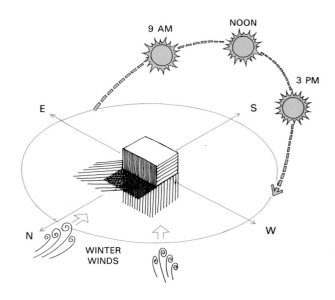

Fig. 22-22. Home is properly placed with respect to the sun. South facing wall will catch all the sunlight it possibly can. Walls can be 10 to 20 degrees off true south without losing much of their efficiency. (Pittsburgh Plate Glass Co.)

appliances on the opposite side of the house from space receiving solar gain. See Fig. 22-23 for best use of internal and solar gains in the northern hemisphere.

BUILDING PASSIVE SOLAR STRUCTURES

Thermal storage walls used in passive solar construction need to store enough heat to even out the wild fluctuations of temperature which could make a building uncomfortable. A number of materials have the ability to store heat. Among them are:

1. Water. Though cheap, it might be expensive to construct or purchase containers. It has greater heat capacity than solid materials.
2. Concrete and concrete block. Both have good heat storage capacity and will provide structural support for the building, as well. Cores of hollow block can be used as warm air ducts. However, the wall will hold more heat if they are filled with sand-mix concrete. Walls can be plastered or painted to conceal mass.
3. Brick. Its properties are similar to concrete or block. Attractive in appearance, it is easier to blend with decor. Though more expensive than concrete, its appearance often makes it the preferred material.

These materials are good choices for thermal storage, because of their cost, appearance, and suitability. Fig. 22-24 compares heat carrying capacity of these materials with other building materials.

Substance	Specific Heat (BTU/lb-°F)	Density (lbs/cu ft)	Heat Capacity (BTU/cu ft-°F)
*Water	1.00	62.4	62.4
Wood, oak	0.57	47.0	26.8
Fir, pine, and similar softwoods	0.33	32.0	10.6
Expanded polyurethane	0.38	1.5	0.57
Wool, fabric	0.32	6.9	2.2
Air	0.24	0.075	0.018
*Brick	0.20	123.0	25.0
*Concrete	0.156	144.0	22.0
Steel	0.12	489.0	59.0
Gypsum or plaster board	0.26	50.0	13.0
Plywood (Douglas Fir)	0.29	34.0	9.9

*Preferred materials.

Fig. 22-24. Heat carrying capacities of water, concrete, and brick make them good choices for heat storage. (Iowa Energy Policy Council)

SIZING THERMAL STORAGE SYSTEMS

Sizes of thermal storage systems are determined by the square footage of the living space. For example, in climates where the average winter temperature is 20° to 30°F, between 0.43 and 1 sq. ft. of masonry per sq. ft. of living area is adequate. Fig. 22-25 lists adequate mass for different areas by degree days. Consult the chart, Fig. 13-8 in Unit 13, for degree-day information on your locality.

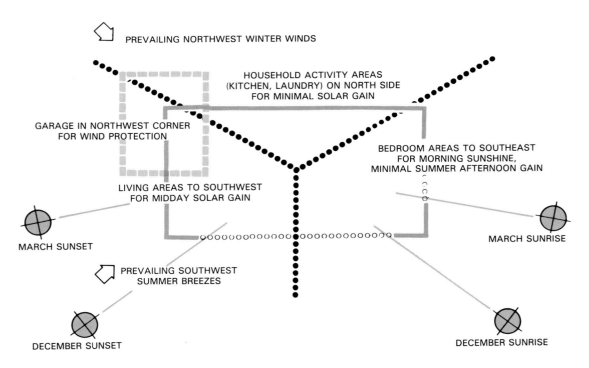

Fig. 22-23. Locating living areas for best balance and use of solar and internal heat gains. Appliances represent internal gains. (Forest Products Laboratory)

CLIMATE	WINTER TEMPERATURE For Coldest Months in Degree Days/Mo	SQUARE FOOTAGE FOR EACH SQUARE FOOT AREA	
		Masonry Wall	Water Wall
Cold	1500	0.72 to 1	0.55 to 1
	1350	0.62 to 1	0.45 to 0.85
	1200	0.51 to 0.93	0.38 to 0.70
	1050	0.43 to 0.78	0.31 to 0.55
Temperate	900	0.35 to 0.60	0.25 to 0.43
	750	0.28 to 0.46	0.20 to 0.34
	600	0.22 to 0.35	0.16 to 0.25

Fig. 22-25. Solar storage mass must be based on average winter temperature and the size of the area to be heated.

WALL THICKNESS

Storage wall thickness is important. A space will overheat if more energy is transmitted through the wall than is needed. This will happen if the wall is too thin. Further, a wall which is too thin will not be able to transmit heat throughout the night. It will have cooled before morning. Fig. 22-26 suggests thicknesses for storage walls of various materials.

The more rapidly the material conducts heat the thicker it must be. Otherwise, it may deliver its heat load too soon. This makes the living space uncomfortably warm.

SIZING DIRECT GAIN STORAGE

A rule of thumb for walls, slabs-on-grade, etc. used for storage in direct gain systems is to provide 150 lb. of concrete or masonry for every square foot of glazing. This assumes that the storage is subjected to direct sunlight.

At the same time, there should be at least 150 lb. of concrete or masonry for every 2 sq. ft. of area to be heated. It should be about 4 to 6 in. thick. If thinner than 4, it will not be able to store enough heat per square foot. Floors used for thermal storage should not be carpeted wall to wall. However, scatter rugs, furniture, and other obstructions will have little effect on the performance of the floor.

MATERIAL	THICKNESS (IN.)
Water	6 or more
Dense concrete and block	12 to 18
Brick	10 to 14
Adobe	8 to 12

Fig. 22-26. General recommendations for thickness of storage walls when constructed of various materials. Thicker walls will store more heat as well as delay the time when the wall will conduct heat to the living space.

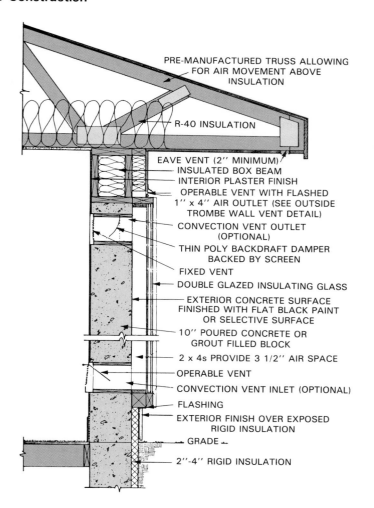

PRE-MANUFACTURED TRUSS ALLOWING FOR AIR MOVEMENT ABOVE INSULATION

R-40 INSULATION

EAVE VENT (2" MINIMUM)
INSULATED BOX BEAM
INTERIOR PLASTER FINISH
OPERABLE VENT WITH FLASHED 1" x 4" AIR OUTLET (SEE OUTSIDE TROMBE WALL VENT DETAIL)
CONVECTION VENT OUTLET (OPTIONAL)
THIN POLY BACKDRAFT DAMPER BACKED BY SCREEN
FIXED VENT
DOUBLE GLAZED INSULATING GLASS
EXTERIOR CONCRETE SURFACE FINISHED WITH FLAT BLACK PAINT OR SELECTIVE SURFACE
10" POURED CONCRETE OR GROUT FILLED BLOCK
2 x 4s PROVIDE 3 1/2" AIR SPACE
OPERABLE VENT
CONVECTION VENT INLET (OPTIONAL)
FLASHING
EXTERIOR FINISH OVER EXPOSED RIGID INSULATION
GRADE
2"-4" RIGID INSULATION

Fig. 22-27. Typical section of a Trombe wall. Note insulation in hollow header. Air space between wall and glazing can vary from 2 to 5 in. (Iowa Energy Policy Council)

EFFECT OF COLOR ON COLLECTING SURFACE

Since dark colors absorb heat more readily, the wall surface facing the glazing should be dark. Though black is most efficient as an absorber, other dark colors may be used with nearly the same efficiency. Dark blue, for example, works almost as well. Inside wall surfaces may be any color.

WALL CONSTRUCTION

Solid Trombe walls will store more heat than vented ones. The outside surface of the wall may reach 150°F or more on a sunny day. The north-facing wall, however, will maintain a fairly uniform temperature throughout a 24 hour period. Vents will be necessary if the living space requires extra warmth during the day. Then the air space between the wall and the glazing sets up a thermosiphon action (natural convection).

Since the Trombe wall is a bearing wall, construct

it based on the principles given in Unit 6, Footings and Foundations. Footings should be twice as wide as the wall's cross section and equal in thickness. Fig. 22-27 shows a vertical cross section through a typical Trombe wall. Refer also to Fig. 5-14, page 102, for a horizontal cross section.

Some provision for venting of unwanted heat in warm weather is advisable. Fig. 22-28 is a detail drawing for venting the Trombe wall at the header. For aesthetic purposes, it is alright to include windows in a Trombe wall. Adjustments should be made to the dimensions to assure sufficient area for proper heat collection and storage. Fig. 22-29 shows several designs which incorporate windows.

SPECIAL CONCERNS

It is desirable to construct thermal glazing so that the glazing can be removed occasionally for cleaning the glass or to repaint the Trombe wall. Thus, it would be helpful if the glass panels are supported in their own sash. Panels should not be so large that the windows are difficult to handle.

Single panes of glazing are not as efficient as double glazing, and triple glazing is not generally recommended. It cuts down on the effectiveness of the solar radiation and does not appreciably reduce the heat loss over double glazing.

DESIGNING THE ISOLATED GAIN SYSTEM

Basically, the isolated gain system is a small room attached to the main structure of the building. Fig. 22-30 shows, in simple sketches, a variety of designs for bringing the heat from the sun space to the living space. A greenhouse attached to a south exposure of a house is a good example of this type of structure.

The sun space is sometimes "embedded" in the house. That is, it is within the walls of the house and is separated from the house by only its interior partitions. One of these partitions can be a storage wall of concrete, masonry, or water.

PASSIVE THERMOSIPHON SYSTEM

Thermosiphoning space heating systems are not well researched or understood. Passive designers do suggest that the rock bed needed for storage have a cross-sectional area that is about 50 to 75 percent of the area of the collectors. For example, if the area of the collector is 10 x 20 ft. = 200 sq. ft., the cross section of the bed should be 100 or 150 sq. ft. Four-inch rock are recommended for easy circulation of the air around them. Depth of the bed should be 20 times the rock diameter, about 6 1/2 ft.

To find the proper volume of storage, multiply the collector area by a factor of 3.25 to 5. The 200 sq. ft. collector would require 650 to 1000 cu. ft. of storage space.

Rock bins need to be insulated against heat loss on their sides and bottoms. Dampers should be located in the ducts leading to the collector. These are closed every evening to prevent the drawing off of heat to the outside at night.

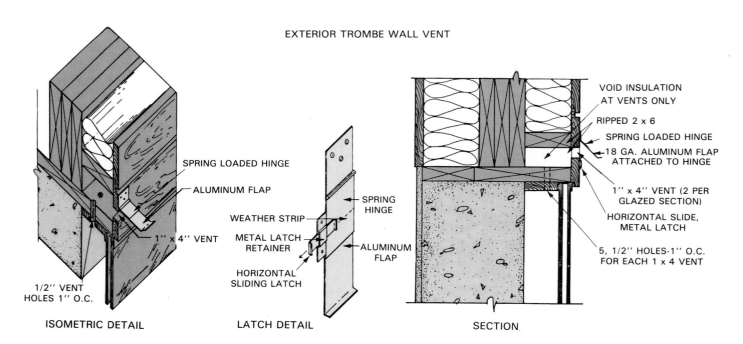

EXTERIOR TROMBE WALL VENT

SPRING LOADED HINGE
ALUMINUM FLAP
1" x 4" VENT
1/2" VENT HOLES 1" O.C.
ISOMETRIC DETAIL

SPRING HINGE
WEATHER STRIP
METAL LATCH RETAINER
HORIZONTAL SLIDING LATCH
ALUMINUM FLAP
LATCH DETAIL

VOID INSULATION AT VENTS ONLY
RIPPED 2 x 6
SPRING LOADED HINGE
18 GA. ALUMINUM FLAP ATTACHED TO HINGE
1" x 4" VENT (2 PER GLAZED SECTION)
HORIZONTAL SLIDE, METAL LATCH
5, 1/2" HOLES-1" O.C. FOR EACH 1 x 4 VENT
SECTION

Fig. 22-28. Method of venting Trombe wall to outside to disperse warm weather heat.

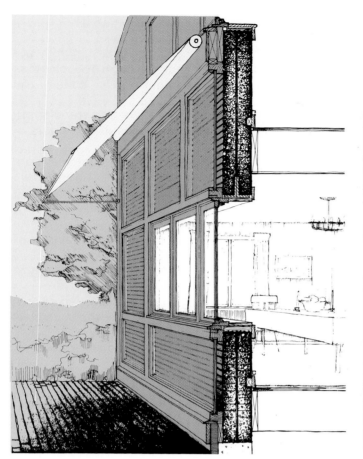

Fig. 22-29. Two different designs for windows in Trombe walls. Left. Section of brick wall with casement windows. Storage serves two levels. Roll-up awning takes place of overhang. Right. Inside and outside view of design which uses small windows in free-standing storage wall. Clerestory window, above, provides additional lighting. (Pittsburgh Plate Glass Co.)

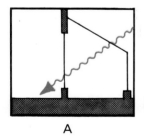
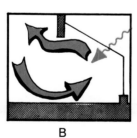
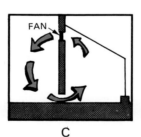
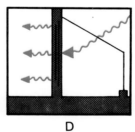
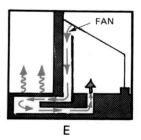

A   B   C   D   E

Fig. 22-30. There are many methods of bringing heat from sun space into living area. A—Solar rays are transmitted through sun space into living space. B—Natural air exchange takes place. C—Fan pulls in heated air. D—Storage wall delivers heat through radiation. E—Gravel storage bin stores heat which radiates to building above.

INSULATING PASSIVE SOLAR BUILDINGS

To take advantage of solar heating, buildings should be well insulated. Refer to Unit 13, Thermal and Sound Insulation. Fig. 22-31 is a cross section of a two-story house showing good insulating practice for passive solar construction.

When insulating, it pays to provide thermal barriers anywhere that heat could leak out. There should be no place that air can flow between inside and outside without going through insulation. Seal up all areas where air could possibly leak through cracks. Refer to Fig. 22-32 and Unit 5 for more information on passive solar structures.

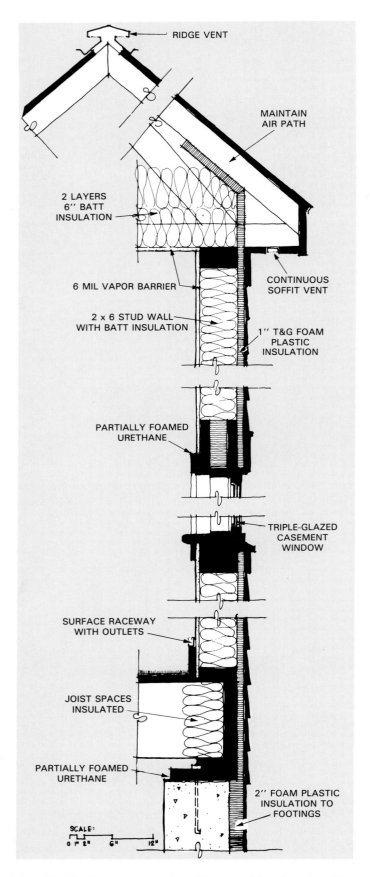

Fig. 22-31. Cross section of well-insulated dwelling. Good insulation goes hand-in-hand with solar heating. (Iowa Energy Policy Council)

TEST YOUR KNOWLEDGE — UNIT 22

1. Solar radiation passes through glass readily but heat cannot pass back through the glass as readily. This phenomenon is called the _____ _____.
2. List the three different ways heat is transferred.
3. _____ is simply the result of a liquid or gas expanding and rising.
4. The main characteristics of passive solar construction are (Select all correct statements):
   a. It has few, if any, moving parts that require mechanical or electrical energy for operation.
   b. It is usually a stationary, structural part of the house.
   c. Maintenance of the system is more expensive.
   d. Collection, storage, and transporting of heat is done usually naturally by the materials used in construction.
   e. There is always ductwork to carry the heat to living space.
5. List two disadvantages of passive solar energy.
6. What factors govern the size of the overhang above solar glazing?
7. What is orientation?
8. Internal heat is the heat gained from _____ and _____.
9. As a general rule how much heat storage (in pounds of concrete) should be provided for every square foot of glazing?
10. Triple glazing is more effective than double glazing in passive solar construction. True or False?
11. A storage bin must be constructed to house the rock needed for a thermosiphon system. The glazed area of the collector is 8 ft. by 20 ft. What amount of rock must the bin house?
    a. 160 to 180 cu. ft.
    b. 520 to 800 cu. ft.
    c. 1600 to 1800 cu. ft.

OUTSIDE ASSIGNMENTS

1. From your local libraries, select books on designing passive solar systems for housing. Report to your class on your findings.
2. Construct a small thermosiphoning unit and test it. Report to your class on its operation.
3. Write to companies offering passive solar structures. Study the literature and report to the class on items that seem to have practical advantages.
4. Using the charts and information in this chapter design an indirect gain system (using a Trombe wall). Size the glazing, storage wall, and overhang according to the conventional principles.

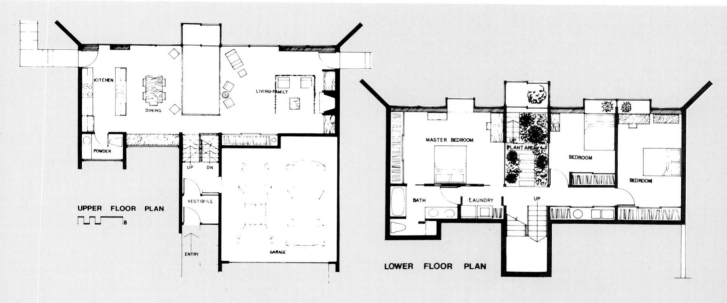

UPPER FLOOR PLAN

KITCHEN

DINING

LIVING-FAMILY

POWDER

UP   DN

VESTIBULE

ENTRY

GARAGE

LOWER FLOOR PLAN

MASTER BEDROOM

UP
PLANT AREA

BEDROOM

BEDROOM

BATH

LAUNDRY

UP

SUMMER EQUINOX
WINTER SUN ANGLE

SECTION THRU ENTRY

TABLE 1—THERMAL PERFORMANCE PREDICTION

Month	Heat Load GJ	Solar Heat GJ	SHF	Aux. Heat. Req. GJ
Sept.	.48	.50	100	0
Oct.	2.88	2.90	100	0
Nov.	7.04	5.92	84	1.11
Dec.	10.60	6.83	64	3.78
Jan.	11.91	7.83	66	4.07
Feb.	9.82	7.45	76	2.37
Mar.	7.83	6.56	84	1.27
Apr.	3.56	3.44	97	.12
May	1.24	1.26	100	0
June	0	0	0	0
Annual	55.37	42.67	77	12.72

Fig. 22-32. Iowa passive solar project. Its heating load is half that of comparable homes. (Iowa Energy Policy Council)

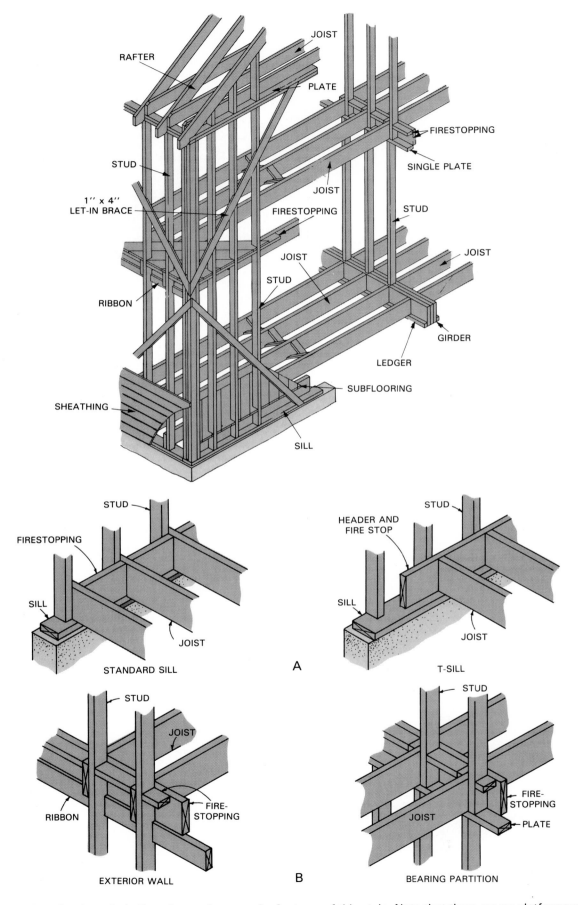

Fig. 23-1. Top. Section of a balloon frame shows major features of this style. Note that there are no platforms with plates and band joists between floors. Bottom. Details of a balloon frame. A—First floor level, two methods of building the sill. B—Second floor level. Left, method of attaching second floor joists to the studs. Right, framing for an interior bearing partition.

Unit 23

REMODELING

In some ways, remodeling work is more painstaking than new construction. Often, sections or components of the old building must be dismantled without destroying what remains or what will be reused. Often the carpenter must be able to visualize how a structure was built so that no damage is done to the structure or any of its systems during demolition or removal.

DESIGN OF OLD STRUCTURES

Many older homes were built using balloon framing. This method was popular in the United States from about 1850 to the 1930s. Its main feature is the long studs which run from the sill on the first floor all the way to the plate on which the rafters rest. These are called "building height" studs. They may be spaced anywhere from 12 to 24 in. O.C.

Second floor joists rest on a horizontal member called a ribbon or ledger. This member, which is let into the stud, is usually a 1 x 6. Fire stops, short pieces of blocking 2 in. thick, are installed between joists at each floor level. Their purpose is to prevent spread of fires from one part of the dwelling to another.

Fig. 23-1 shows details of balloon type construction. Additional information on housing designs will be found in Units 7, 8, 9, and 20. Fig. 23-2 shows details of post and beam construction.

HIDDEN STRUCTURAL DETAILS

Before demolition of any kind is attempted, study the building to determine what type of construction was used. This will make removal of structural elements much easier and safer.

If the house has a basement, examine the area. Try to determine if there are gas or fuel lines, electric circuits, or plumbing service running through the walls or floors where you will be working. Locate and close valves to stop gas and fuel flow in service lines. As a precaution, cut power to the electrical circuit by pulling the fuse or tripping a circuit breaker. If there are other services on the same circuit which you do not want interrupted, an electrician should be called to do the work.

Go outside and study the roof. Note where vents and chimneys are located. Greater care must be exercised in removing wall coverings where exhaust vents and plumbing stacks are located. See Fig. 23-3 and Fig. 23-4.

If additions are to be built that require excavating be sure to locate any underground utility services. Pipes and cables could accidentally be cut or damaged by digging. Sometimes, inspection of sills and joists in a basement will reveal markings indicating where underground services are located. Checking with local utility companies and city departments for buried lines is also advisable.

TEARING OUT OLD WORK

The first step in opening up a wall is to remove wall trim such as baseboards, cove strips at the ceiling, and trim around windows and doors. If these materials are to be reused, use care in removing them.

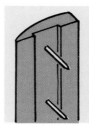

To salvage interior trim for reuse do not attempt to remove the old nails from the front. Use a small wrecking bar or a nippers and remove them from the back. This avoids splintering which would ruin the face.

Next, remove the wall covering to expose the wall frame. Most construction in the past 40 years has used drywall. In homes older then that, wood lath

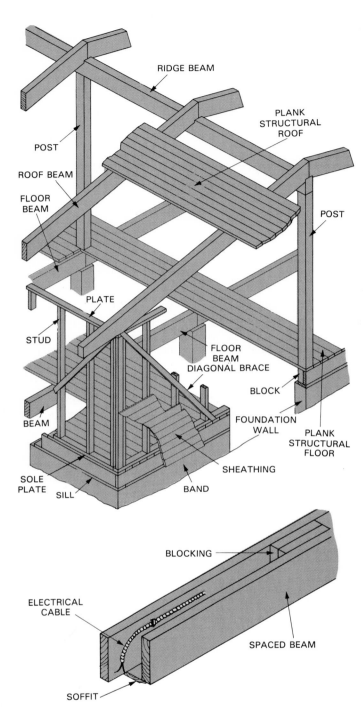

Fig. 23-2. Basic post and beam construction. Top. General construction detail. Bottom. Often pipes and electrical wiring are concealed in beams of this type construction.

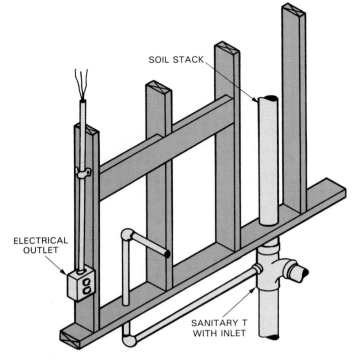

Fig. 23-3. Wiring and plumbing may be concealed behind wall coverings even when there are no telltale fixtures coming out of the wall.

Fig. 23-4. Check the roof for other features such as vents which are not visible from the basement or inside of the house. (Certainteed)

and plaster were used. Be sure to wear a mask to avoid breathing in dust and particles of construction materials.

Useful tools for demolition are a sledge hammer for removing heavy parts of a frame, a ripping or wrecking bar, a hammer or hatchet, and a rip chisel. A reciprocating saw is also useful for cutting framing. Most of these tools are shown in Unit 2. The

rip chisel is essential if any of the trim is to be saved and reused.

Many carpenters use a flat bladed garden spade to remove lath and drywall. This tool will remove drywall and lathing nails as it is run over the studs.

If the wall is a partition, removal of the covering will be all that is necessary to strip the framing. In outside walls, insulation and siding will need to be removed as well.

For good housekeeping, it is often useful to rent a large trash container like the one shown in Fig. 23-5. Then, debris can be cleared away immediately so that it does not clutter up the job site and present a safety hazard.

FINDING BEARING WALLS

Before removing a wall or a portion of it you will need to determine whether it is a bearing wall. Such walls will need shoring (temporary support) while the old wall is being removed. Permanent supports of some kind must be in place before the remodeling is completed.

Determining which are bearing walls is quite simple with outside walls. Interior partitions are another matter.

All outside walls support some weight of the structure. End walls of gabled, single story houses are the sole exception. They carry little weight except that of the wall section from the peak of the roof to the ceiling level. Outside walls running parallel to the ridge of the roof carry the most load since they provide most of the support for the roof.

Identifying which partitions (interior walls) support weight from structures above is not always simple. The following conditions are usually an indication of a bearing wall:
1. The wall runs down the middle of the length of the house.
2. Overhead joists are spliced over the wall indicating they depend on the wall for support.
3. The wall runs at right angles to overhead joists and breaks up a long span. (The joist may not be spliced over the wall.) Check span tables for

loadbearing ability of the joist.
4. The wall is directly below a parallel wall on the upper level. (If the walls are parallel to the run of the joists, the joists may have been doubled or tripled to carry the load of the upper wall. The only way to be certain is to break through the ceiling for a visual inspection.

Familiarize yourself with the framing principles used for the type of structure you are remodeling. Study the framing units in this book for details. Also see Fig. 23-6.

PROVIDING SHORING

Shoring must be provided to prevent sagging or collapse of the structure by removing all or sections of a bearing wall. Fig. 23-7, shows one type. It is simply a short wall of 2 x 4 plates and studs.

The wall can be assembled on the floor and lifted into place. To strengthen it use double plates top and bottom.

Measure distance from floor to ceiling and then allow 1/4 to 1/2 in. clearance. Nail stud to one top and one bottom plate. Attach the second top plate. Leave one of the bottom (or sole) plates off. Square up the wall and attach a 1 x 4 diagonal brace.

Lift the wall in place. If necessary, get a helper to do some of the lifting. (Keep it far enough from the partition so you will have room to work.) Slide the second plate under the bottom plate. Then, use wood shingle shims at regular intervals along the bottom of the wall to bring the temporary wall to full height.

A second shoring method is used by many carpenters. Larger dimension lumber — 4 x 6s, for example — are used for plates. Studs are replaced by several adjustable steel posts, Fig. 23-8.

Be sure that the shoring is resting on adequate

Fig. 23-5. Keep the job site clean. If there will be considerable debris, it is wise to rent a trash container like this that will be hauled away following the remodeling job. Left. Container on job site. Right. Typical scavenger service. (Homewood Scavenger Services, Inc.)

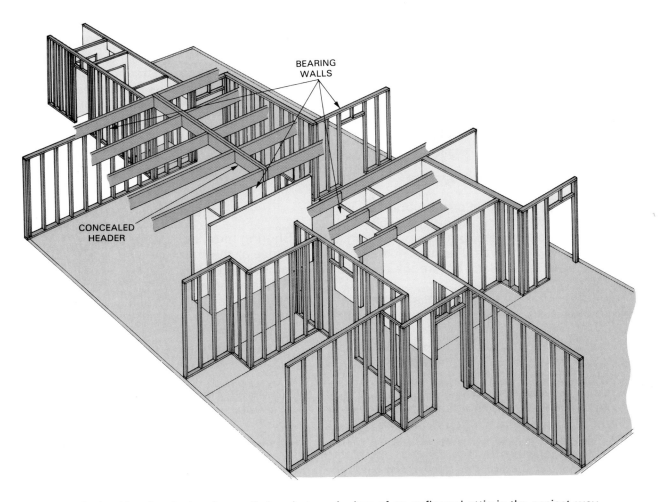

Fig. 23-6. Looking for the bearing walls in a house. A view of an unfloored attic is the easiest way.

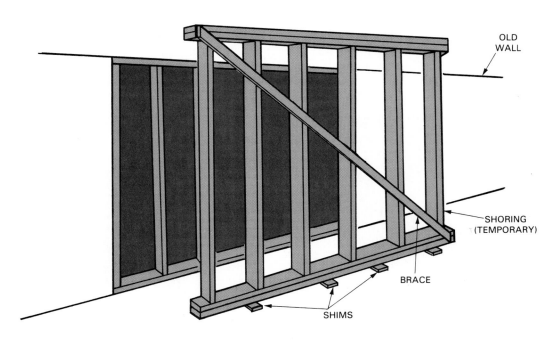

Fig. 23-7. Temporary support is provided by the shoring in the foreground. Two such supports are needed when a bearing partition must be opened up. Place one on each side of the partition.

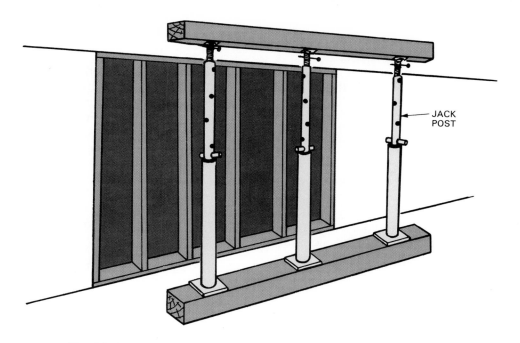

Fig. 23-8. Alternate method of shoring up to support ceiling structure uses jack posts. These are adjustable steel pillars.

support. That is, it should rest across several joists. If running parallel to floor joists lay down planking to distribute weight across several joists. Do the same for adequate support at the ceiling.

Above an interior bearing wall place shoring *on both sides of the wall. This is important!* If not done, the ceiling on the unsupported side will likely sag or collapse. Joists are generally lap joined over an interior bearing wall.

When shoring is in place, you can remove studs in the old bearing wall. Leave the shoring in place until a permanent support has been framed in to take the place of the bearing wall.

FRAMING OPENINGS IN A BEARING WALL

When studs are removed to open up a wall for a passageway or room addition, a header (also called a lintel) must be installed to provide support. The header can be built up from 2 in. lumber. Shim between them with thin plywood to build out the width to that of the framed wall, Fig. 23-9. On very

long spans, the carpenter may prefer to construct or purchase a box beam. Refer to Unit 20, Fig. 20-29, for information on its construction.

SUPPORTING HEADERS

Headers are supported by resting them on the top of short studs called trimmers or trimmer studs. Additional support is provided by nailing a full length stud alongside the trimmer at each end of the header. These are sometimes called "king studs." Stagger nail the trimmers to the king studs. Toenail the header to the trimmer and end nail it to the king studs, Fig. 23-10. On long spans, the header may

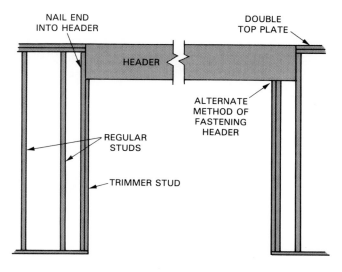

Fig. 23-10. Method of installing a header and supporting it. Alternate method of fastening at right is used on longer spans. It gives the wall more rigidity and doubled 2 x 4 post provides more support for the load.

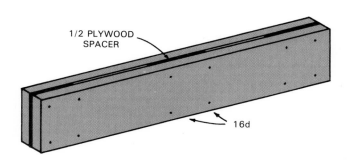

Fig. 23-9. Basic header construction. Use plywood spacer to bring it to same width as the wall.

be extended beyond the rough opening to the next full-length stud. This will help make the framed opening more rigid.

SIZING HEADERS

Headers which replace sections of bearing walls should be sized according to recommended standards for load-bearing ability. The size will be specified in the architectural drawings for the remodeling job. The carpenter should also be aware of what the standards are and how the sizes are

determined. Essential information is given in Fig. 23-11 and 23-12. A careful builder will also check local building codes before going ahead with the job.

To use the table in Fig. 23-11, find the column which describes the load. Move down the column until you find the span corresponding with the opening you need. Then read to the left for the depth needed for the header.

The tables in Fig. 23-12 will require some arithmetic. First compute the load using the table of average weights for different areas of the building.

	OUTSIDE WALLS			INSIDE WALLS			
Nominal depth of header (in inches)	Roof, with or without attic storage	Roof, with or without attic storage, + one floor	Roof, with or without attic storage, + two floors	Little attic storage	Full attic storage, or roof load, or little attic storage + one floor	Full attic storage + one floor, or roof load + one floor, or little attic storage + two floors	Full attic storage + two floors, or roof load + two floors
4	4'	2'	2'	4'	2'	No	No
6	6'	5'	4'	6'	3'	2' 6''	2'
8	8'	7'	6'	8'	4'	3'	3'
10	10'	8'	7'	10'	5'	4'	3' 6''
12	12'	9'	8'	12' 6''	6'	5'	4'

Fig. 23-11. Table of allowable spans for headers under different load conditions. This is a guide for sizing headers. Check local codes. Components of the structure, rather than load figures, are used.

Average weight of house by area	
Unit	Load/ft.²
Roof	40 lb.
Attic (low)	20 lb.
Attic (full)	30 lb.
Second floor	30 lb.
First floor	40 lb.
Wall	12 lb.

	Built-up wood header (double or triple on two 4'' x 4'' posts)							
Span (in feet)	Weight (in pounds) safely supported by:							
	2-2x6	2-2x8	2-2x10	2-2x12	3-2x6	3-2x8	3-2x10	3-2x12
4	2250	4688	5000	5980	3780	5850	7410	8970
6	1680	3126	5000	5980	2520	4689	7410	8970
8		2657	3761	5511		3985	5641	8266
10		2125	3008	4409		3187	4512	6613
12			2507	3674			3760	5511
14				3149				4723

	Steel plate header (on two 4'' x 4'' posts)								
Span (in feet)	Weight (in pounds) safely supported by wood sides and plate								
	2-2x8 + 7 1/2'' by			2-2x10 + 9 1/2'' by			2-2x12 + 11 1/2'' by		
	3/8''	7/16''	1/2''	3/8''	7/16''	1/2''	3/8''	7/16''	1/2''
10	6754	7538	8242	10,973	12,199	13,418	15,933	17,729	19,604
12	5585	6216	6827	9095	10,131	11,106	13,224	14,517	16,265
14	4756	5293	5811	7751	8623	9463	11,295	12,561	13,876
16		4481	5036	6746	7494	8221	9815	10,953	12,086
18			5942	6606	7158	8675	9652	10,647
20				6466	7746	8618	9408

Fig. 23-12. Another method of finding header sizes calculates the load. Top left. Basic load calculations for different parts of the house. Top right, Bottom. Recommendations for safely supporting different loads at different spans. Check local codes.

Once you know the total load, locate the load in the column for the correct span in Fig. 23-12. The size of the header needed will be at the top of the column.

If headroom is a concern, a steel reinforced header will allow greater strength with less depth. A steel plate is sandwiched between lengths of 2 in. lumber. The plate is called a flitch and can be made up by a metal fabricator.

COMPUTING THE LOAD

To compute the load on a header, refer to Fig. 23-13. First, you need to find the number of square feet. Multiply the length of the span you need by half the distance between load-bearing walls. (You can always assume that the header must support the load halfway to the next bearing wall.)

Thus, if the distance to the next bearing wall is

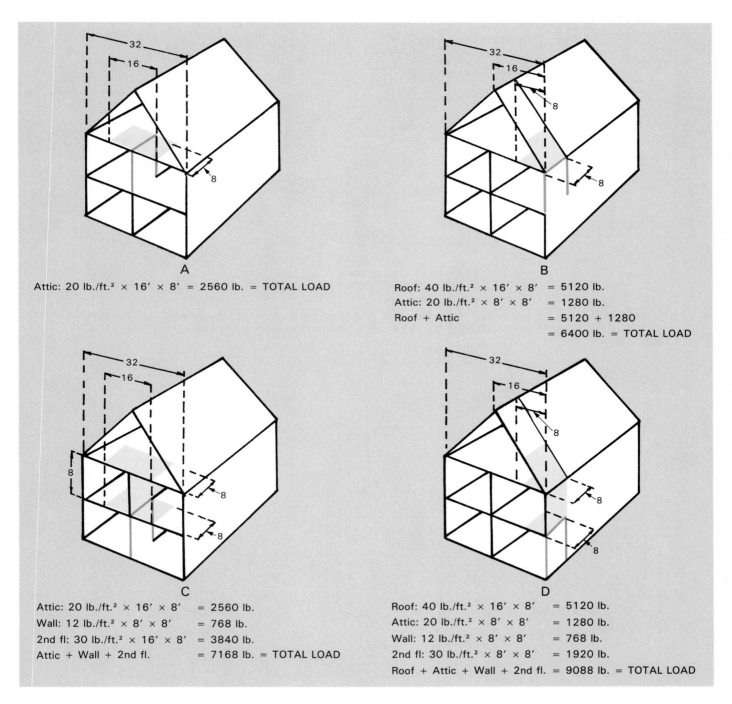

A

Attic: 20 lb./ft.² × 16′ × 8′ = 2560 lb. = TOTAL LOAD

B

Roof: 40 lb./ft.² × 16′ × 8′ = 5120 lb.
Attic: 20 lb./ft.² × 8′ × 8′ = 1280 lb.
Roof + Attic = 5120 + 1280
= 6400 lb. = TOTAL LOAD

C

Attic: 20 lb./ft.² × 16′ × 8′ = 2560 lb.
Wall: 12 lb./ft.² × 8′ × 8′ = 768 lb.
2nd fl: 30 lb./ft.² × 16′ × 8′ = 3840 lb.
Attic + Wall + 2nd fl. = 7168 lb. = TOTAL LOAD

D

Roof: 40 lb./ft.² × 16′ × 8′ = 5120 lb.
Attic: 20 lb./ft.² × 8′ × 8′ = 1280 lb.
Wall: 12 lb./ft.² × 8′ × 8′ = 768 lb.
2nd fl: 30 lb./ft.² × 8′ × 8′ = 1920 lb.
Roof + Attic + Wall + 2nd fl. = 9088 lb. = TOTAL LOAD

Fig. 23-13. How to compute loads. Find horizontal distances between bearing walls. Determine what parts of the house the header will be supporting. Multiply load per sq. ft. by the length and width being supported. Add loads together.

12 ft., the square footage is 6 ft. × the span. Add on any other loads which the header must support to find total load. On roof sections, load is figured the same as for the floor—use the horizontal distance covered by the roof.

CONCEALED HEADERS AND SADDLE BEAMS

There are times when it is not desirable to have a header extending below the ceiling level. In such cases, the header must be at the same level or above the joists they support.

A concealed header is butted against the ends of the joists it is supporting. A saddle beam is positioned above the joists it supports. See Fig. 23-14.

A concealed header is simple to add to an exterior wall in platform type constructions, Fig. 23-15. Simply attach a second piece of 2 in. lumber to the band joist. Add studding at each end to provide support for the header.

Because they stick above the joists, saddle beams can only be used in attic space. They should rest

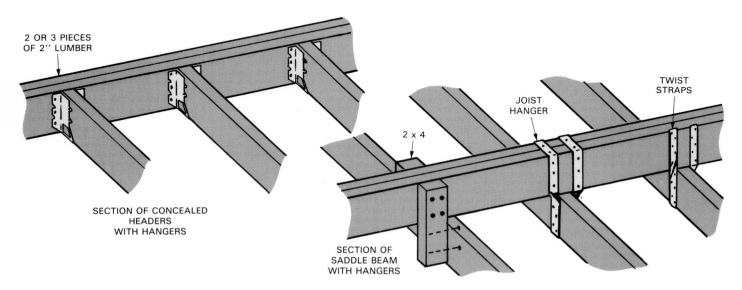

2 OR 3 PIECES OF 2" LUMBER

SECTION OF CONCEALED HEADERS WITH HANGERS

TWIST STRAPS

JOIST HANGER

2 x 4

SECTION OF SADDLE BEAM WITH HANGERS

Fig. 23-14. Concealed headers and saddle beams are used when support must be at ceiling height. Neither support member hangs lower than the ceiling joists. Note that joist hangers are used for these installations.

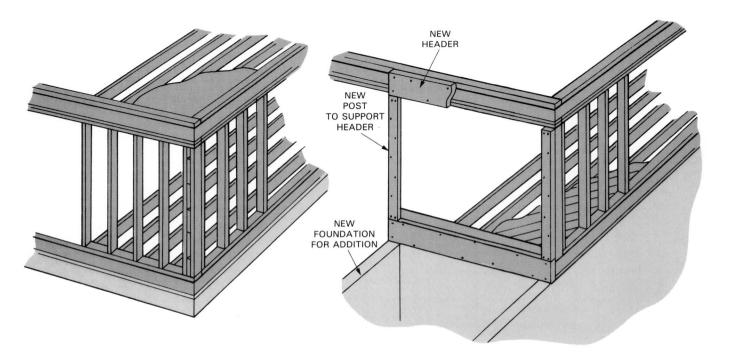

NEW HEADER

NEW POST TO SUPPORT HEADER

NEW FOUNDATION FOR ADDITION

Fig. 23-15. In remodeling a wall in platform construction, header can often be built by adding to a band joist on a second floor.

1. Carefully lay out the exact location of the partition. Snap a chalk line that will be used to locate one side of the sole plate. Lay out the top plate and sole plate.

2. Cut plates to length. Nail sole plate to the floor as shown. Space nails about 2 ft. O.C. and use a nail size that will extend through the subfloor.

3. Nail top plate to the two end studs and then install this assembly on the sole plate. Use a level as shown. If top plate and studs cannot be nailed to structural members (studs and ceiling joists), use toggle bolts.

4. Install studs on 16 or 24 in. centers. Toenail them to the top plate and sole plate. Use two 8d box nails on each side of the stud.

5. Because floors and walls are seldom perfectly level, it is usually best to measure for each stud. This means cutting each stud so that it fits snugly between the sole and top plate.

6. Study wall for best arrangement of standard drywall sheets. Fasten sheets with 1 1/2 in. drywall nails. Cut panels as needed using utility knife and square or straightedge. Space nails 7 to 8 in. apart.

7. After the gypsum panels are installed, flat joints and inside corners are reinforced with tape bedded in joint compound. Outside corners are reinforced with a metal corner bead.

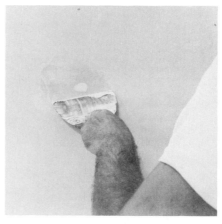

8. Cover nail heads and imperfections with compound as shown. Allow about 24 hours before applying a second coat.

Fig. 23-16. Steps for constructing a nonbearing partition. (Georgia-Pacific)

on top of joists supported by the remaining portions of the bearing wall.

It is recommended that loads be checked by an architect or engineer. He or she should also specify the header sizes.

BUILDING ADDITIONS ONTO OLDER HOMES

When building additions to older homes you need to check all dimensions of the old construction carefully. In particular, check the following:

1. Foundations. If you are using modern masonry units they may not match sizes of older masonry units. You will need to adjust level of footings or alter mortar joint thickness.
2. Ceiling heights. In balloon type framing, ceiling heights may vary from one house to another. Study the local codes to see if variances you plan are acceptable. Also, new framing methods may require architectural changes. These should be satisfied before construction begins.
3. Dimensions of framing members. Standards of dimension lumber (2 x 4s, 2 x 6s, etc.) were different then. If you are trying to match a roof-line, adjustments must be made. Remember, if the match cannot be made exactly, it is better to alter the plan.

SMALL REMODELING JOBS

Most of the units in this book deal with new construction. In general these same construction steps can be used with some modification for small remodeling jobs. Fig. 23-16 gives step-by-step instructions for installing a new nonbearing partition.

REPLACING AN OUTSIDE DOOR

An old, ill-fitting exterior door, in addition to being an eyesore, wastes energy. It permits excessive air infiltration and heat loss. New designs, especially the prehung units, make replacement easier. Standard replacement units are made with an insulated steel or steel clad door mounted in a steel or wood frame. Fig. 23-17 illustrates a unit with a steel frame and shows how it fits into the old door frame.

To install this unit, remove the old door, hinges, threshold, and trim. Then slide the new unit into the opening. Fig. 23-18 shows the procedure step-by-step.

When making such an installation follow the directions provided by the manufacturer of the unit. Replacement doors are generally available to replace standard sizes: 2'-6'' x 6'-8'' and 3'-0'' x 6'-8''.

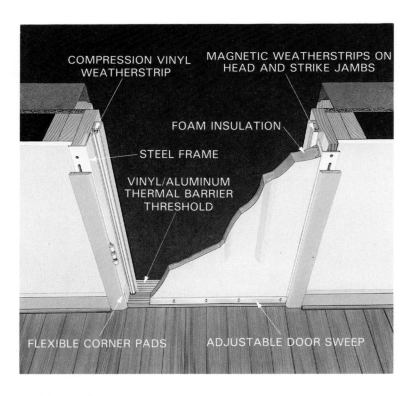

Fig. 23-17. Cutaway showing general details of modern replacement door unit. After alignment, frame is secured with long screws that extend into the wall studs. (General Products Co., Inc.)

1. Remove old door, hinges, strike plate, and threshold. Carefully pry off the interior trim. To secure a smooth separation between casing and paper or paint, pull a sharp knife or razor blade along the joint.

2. Nail wood shims in rabbet of old door frame to make a snug fit and plumb the opening. Be sure shims are located at each hinge and strike plate area. Apply a double-bead of caulking along base of opening. Set the new unit in place as shown. Check fit and nail through predrilled holes in front edge of frame.

3. Remove retainer bands and brackets, and open door carefully. Adjust and install screws on hinge side first. Use long screws that will go through the hinge, frame, and into the studs. Coating the screw threads with soap will make them easier to drive.

4. Adjust lock-side jamb and insert screws as specified. Install lockset according to manufacturer's recommendations. Steel replacement doors and frames will have openings that fit standard locksets. Chisel out wood behind strike plate as required.

5. Install weatherstrip. Use magnetic type for top and latch side, compression type for hinge side. Apply a 1/8 in. bead of caulk along the wood strips. With the door closed, position each strip against the door face and tack them into place. Check door operation and then complete the nailing as shown.

6. Adjust crown of threshold so there is smooth contact with door bottom weatherstrip. Apply a small bead of caulk along the outside edge of the threshold and bottom end of wood stops. Replace old interior trim or install new trim.

Fig. 23-18. Basic procedures to follow when installing a modern replacement door unit.   (Pease Industries Inc.)

INTERIOR DOORS

Remodeling may include replacing, moving, or adding interior doors and door jambs. The framing of doorways is explained in Unit 8. Installation of jambs and hanging of doors is explained in Unit 17.

Use of a special patented jamb clip presents another method of hanging an interior door, Fig. 23-19. The clips, which eliminate the need to use wedging, are made of 19 gauge cold-rolled steel. One size (4 1/2 in.) fits the standard rough opening consisting of 2 x 4 plus 1/2 in. of drywall on each side. For other jamb sizes the clip snaps in two at a breakaway point. Eight clips should be used with hollow core doors, 10 for solid core doors. Fig. 23-20 is a cutaway of a jamb clip installed. Follow the installation instructions in Fig. 23-21.

SOLAR RETROFIT

Older homes, with a long wall to the south where the sunlight is not blocked by other structures, are

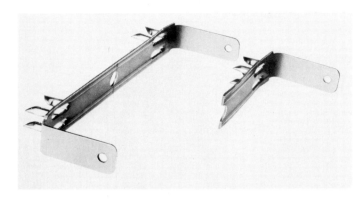

Fig. 23-19. Patented jamb clips allow jamb installation without shims. Half clip at right, made by snapping full unit in two, is for jambs wider than standard. (Panel Clip Co.)

Fig. 23-20. Cutaway of jamb and rough frame showing jamb clip attachment.

1. INSTALL CLIP ONTO DOOR JAMB THEN STRAIGHTEN OUT ALL OF THE EARS ON ONE SIDE OF THE JAMB.

2. POSITION INTO ROUGH OPENING.

3. BEND EARS BACK AND NAIL.

Fig. 23-21. Steps for attaching jamb clips and installing door.

likely candidates for passive solar retrofit. A first consideration, however, is to investigate the insulation levels in the house. For insulation standards and methods, refer to Unit 13, Thermal and Sound Insulation.

BASIC SOLAR DESIGNS

The three basic passive solar designs are:
1. Direct gain. Here south facing double glazed windows allow the sun direct access to living space. A masonry wall or floor is needed to act as storage for the solar heat.
2. Indirect gain. In this design, a storage wall is located a few inches from the glazing. The wall soaks up the solar heat and radiates it to the living space.
3. Isolated gain (sun space). These are separate spaces—greenhouse or thermosiphon, for example—that have solar storage systems for collecting and distributing the heat to other parts of the house.

Additional information on these systems and how to construct them are to be found in Unit 22, Passive Solar Construction. Probably, direct gain and isolated gain systems are more practical as part of a remodeling program.

Adding a direct gain system involves removing a section of a south-facing wall and replacing it with windows. If glassed area is small, no storage is required. However, for large amounts of glass, a storage wall or floor may be needed. One way of getting storage is to install a fireplace on the opposite wall where sunlight could strike it. A brick planter can also serve as heat storage. These are not always practical because of weight.

The amount of thermal mass is important. For a large expanse of glass, about 150 lb. of masonry is needed for every square foot of south-facing glass. This could present serious structural problems for many older homes when you consider that 200 sq. ft. of glazing requires 15 ton of masonry.

In most situations, solar retrofit—through adding sun space or a thermosiphon seems most practical.

THERMOSIPHON

Sometimes called a SOLAR FURNACE, the thermosiphon is basically a glazed box which captures heat from sunlight. Natural convection (rising of heated air) moves the air into living space. See Fig. 23-22. These units can be attached to window sills to provide solar heat for individual rooms. One ex-

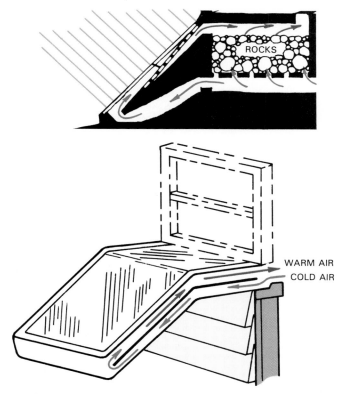

Fig. 23-22. Two thermosiphon units. Top. Large unit designed to heat stored rock. Bottom. Small window unit.

Fig. 23-23. Thermosiphon system that occupies part of a south-facing wall. Series of boxes are formed by attaching 1 or 2 in. lumber on edge to the side of the house. Glazing covers it to trap heat. Vents allow circulation of the heated air into the home.

perimental design involved glazing the entire south wall. At regular intervals, 6 in. wide boards were run edgewise vertically up the wall. Blocking was added to top and bottom enclosing the individual boxes. At first floor ceiling level, more blocking was added. Then vents were cut through the wall at floor and ceiling level on both floors. Finally, glazing was attached to the outer edge of the vertical boards.

As the sun shone through the glass, the air in the enclosed spaces heated up and set up a convection. Warm air rose entering the house at ceiling level. Cold air moved from the bottom. Fig. 23-23 is a simple diagram of the system.

TEST YOUR KNOWLEDGE - Unit 23

1. Before beginning a remodeling job, why is it wise to find out the type of construction used?
2. Before starting excavation for an addition you should determine what _____ are _____ beneath the ground.
3. List four indications that a wall is a bearing wall.
4. _____ is a temporary wall installed to prevent the collapse or sagging of a structure when part of a bearing wall is removed.
5. How deep should a built-up wooden header be which must span 8 ft. in a bearing partition and carry the load of a shallow attic above it?
6. When a header must be flush with the ceiling, one of two types of beam or header are used. Name them.
7. A home using direct gain solar heating must always include a method of storing heat gain. True or False?
8. A thermosiphon is a type of passive design called _____ gain.

OUTSIDE ASSIGNMENTS

1. Write a report on the steps you would follow to add a 12 x 16 ft. addition to the side of a ranch home built by the platform method. Include sketches for shoring and for final framing.
2. Compute the depth of the header needed to span an 8 ft. opening in a bearing partition of a one story house. The partition is in the center of a 28 ft. span. Use the load tables in Fig. 23-12 and Fig. 23-13.

Fig. 24-1. Scaffold for dormer work consists of steel "pump-jack" units attached to double 2 x 4 uprights. Units provide brackets for both platform and guardrail and can be raised along the uprights by food-pedal action. Note bracing at top and lower section of uprights. This bracing is very important. (Richard Mehmen)

Unit 24

SCAFFOLDS AND LADDERS

Carpenters use scaffolding (also called staging) to reach work areas which are too high to reach while standing on the ground or floor deck. Scaffolding provides rigid, raised platforms to support workers, tools, and materials with a high degree of safety. The height must help the worker avoid stooping and reaching.

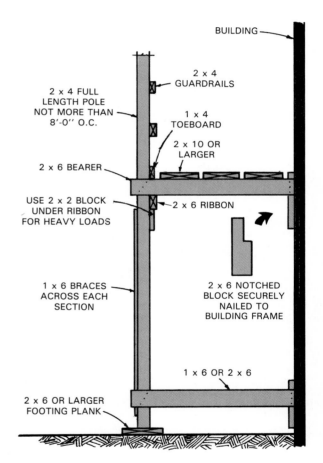

Fig. 24-3. A typical design for a single-pole scaffold. Horizontal distance between bearer sections should never exceed 8 ft.

The type of scaffold required depends on how many workers will be on it at one time, the distance above the ground, and whether it must support building materials.

TYPES OF SCAFFOLDING

Typical scaffolds are shown in Figs. 24-1, 24-2, and 24-3. In constructing wooden scaffolds, the

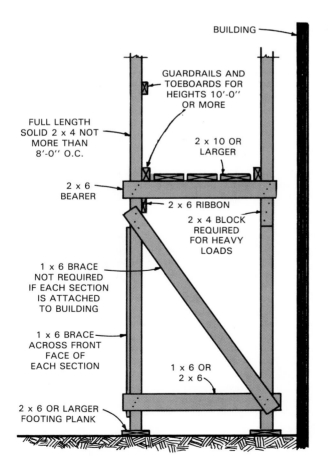

Fig. 24-2. A typical design for a double-pole scaffold. Structure can be extended upward to form several platforms. Maximum height should be limited to 18 ft.

543

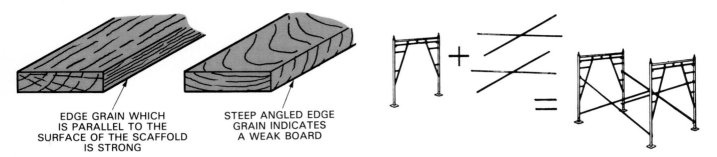

EDGE GRAIN WHICH
IS PARALLEL TO THE
SURFACE OF THE SCAFFOLD
IS STRONG

STEEP ANGLED EDGE
GRAIN INDICATES
A WEAK BOARD

Fig. 24-4. To build a safe platform, choose straight-grain lumber. Pick lumber for uprights the same way.

Fig. 24-5. Metal scaffold sections can be rapidly assembled from prefabricated trussed frames and diagonal braces.

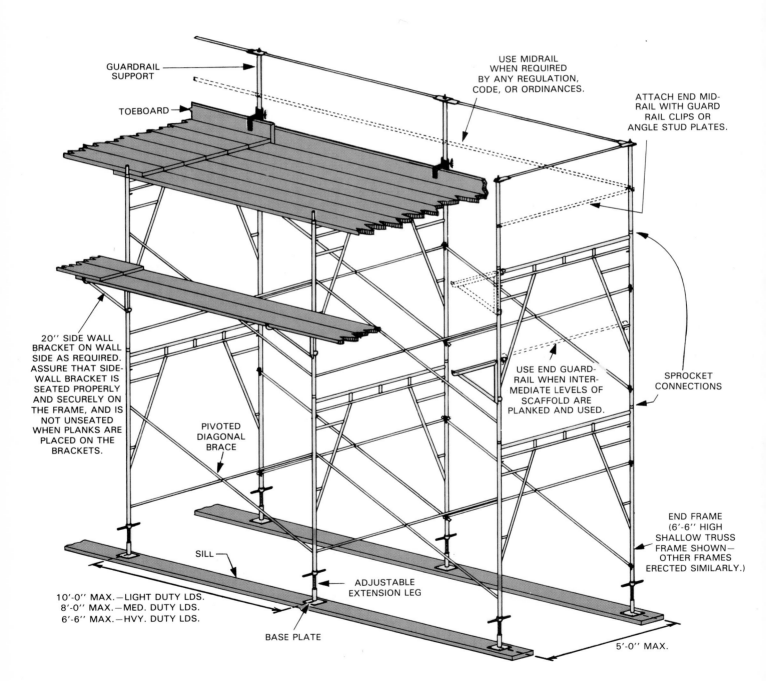

GUARDRAIL
SUPPORT

TOEBOARD

USE MIDRAIL
WHEN REQUIRED
BY ANY REGULATION,
CODE, OR ORDINANCES.

ATTACH END MID-
RAIL WITH GUARD
RAIL CLIPS OR
ANGLE STUD PLATES.

20'' SIDE WALL
BRACKET ON WALL
SIDE AS REQUIRED.
ASSURE THAT SIDE-
WALL BRACKET IS
SEATED PROPERLY
AND SECURELY ON
THE FRAME, AND IS
NOT UNSEATED
WHEN PLANKS ARE
PLACED ON THE
BRACKETS.

PIVOTED
DIAGONAL
BRACE

USE END GUARD-
RAIL WHEN INTER-
MEDIATE LEVELS OF
SCAFFOLD ARE
PLANKED AND USED.

SPROCKET
CONNECTIONS

END FRAME
(6'-6'' HIGH
SHALLOW TRUSS
FRAME SHOWN—
OTHER FRAMES
ERECTED SIMILARLY.)

SILL

10'-0'' MAX.—LIGHT DUTY LDS.
8'-0'' MAX.—MED. DUTY LDS.
6'-6'' MAX.—HVY. DUTY LDS.

ADJUSTABLE
EXTENSION LEG

BASE PLATE

5'-0'' MAX.

Fig. 24-6. A scaffold assembly made from sections joined vertically and horizontally.　(Patent Scaffolding Co.)

uprights should be made of clear, straight-grain 2 x 4s. The lower ends should be placed on planks to prevent settling into the ground.

Bearers (sometimes called cross ledgers) consist of 2 x 6s about 4 ft. long. Use at least three 16d nails at each end of the bearers to fasten them to the uprights. For the single-pole scaffold, one end of the bearer should be fastened to a 2 x 6 block securely nailed to the wall. Braces may be made of 1 x 6 lumber, fastened to uprights with 10d nails, and with 8d nails where they cross. For the platform, 2 x 10 planks free from large knots should be used. It is good practice to spike the planks to bearers to prevent slipping. Whether lumber is for planks or uprights, select lengths with parallel grain, Fig. 24-4.

Another scaffold type is the swinging scaffold. Swinging scaffolds are suspended from the roof or other overhead structures. These are used mainly by painters and others if only light equipment and materials are required.

MANUFACTURED SCAFFOLDING

In modern construction, many builders use sectional steel or aluminum scaffolding. The units are quickly and easily assembled from prefabricated frames. See Fig. 24-5. These box-shaped units can be stacked and/or assembled horizontally to build a scaffold of any size.

There are many styles of sectional steel scaffolding. Some types have adjustable legs. Frame sizes range from 2 to 5 ft. wide and from 3 to 10 ft. high. Various size bracing provides frame spacings of 5 to 10 ft. The basic units are set up and joined vertically and horizontally. Fig. 24-6 illustrates the construction details of a sectional scaffold. The frames can be equipped with casters when a rolling scaffold is desired.

BRACKETS, JACKS, AND TRESTLES

Metal wall brackets, Fig. 24-7, are used in residential construction because they can be:
1. Quickly attached to a wall.
2. Easily moved from one construction site to another.
3. Installed above overhangs or ledges.

Great care is needed when fastening the brackets to a wall. For light work at low levels, the type that is attached with nails may provide enough safety. Use at least four 16d or 20d nails. Be sure they penetrate sound framing lumber. Most carpenters prefer the greater safety provided by brackets that hook around studding.

Some metal wall brackets have posts for holding guardrails and toeboards, Fig. 24-8. A guardrail pro-

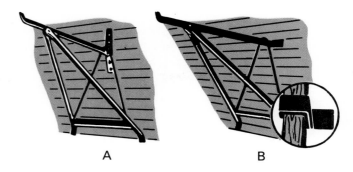

Fig. 24-7. Kinds of metal wall brackets. A—Attached with nails securely set in building frame. B—Hooked directly to studding.

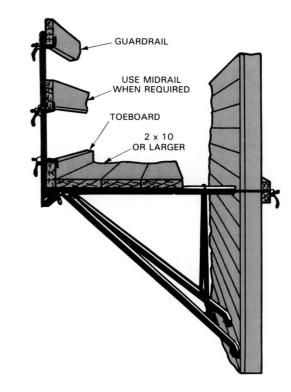

GUARDRAIL

USE MIDRAIL WHEN REQUIRED

TOEBOARD

2 x 10 OR LARGER

Fig. 24-8. This metal wall bracket is approved for work at heights of 10 ft. or more. A high scaffold must have a guardrail and toeboard. However, rails and other parts of the scaffold can be detached if not needed.

tects a worker on the scaffold and a toeboard protects people beneath from falling tools. A wire mesh screen on open sides is sometimes added.

The scaffold shown in Fig. 24-8 is fastened to the wall with a bolt. Fig. 24-9 is a more detailed view of the fastening system.

Wall brackets which support a wood platform should not be spaced more than 8 ft. apart. Metal or reinforced platforms can extend more than 8 ft. between supports. Fig. 24-10 shows a metal platform on metal wall brackets.

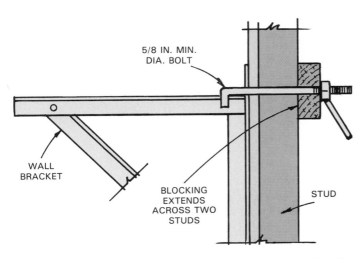

Fig. 24-9. A strong metal wall bracket is secured to wall with bolt and nut assembly. Bolt passes through wall alongside the stud. Bracket is offset to bear against the sheathing directly over the stud.

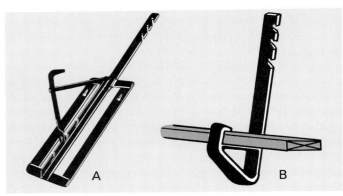

Fig. 24-11. Roofing brackets. A—Slotted end will hook on nails driven into the roof deck and frame. Arm that carries planking is adjustable. Overall length is 30 in. B—Bracket is used on steep roofs. All brackets require 16d common nails driven into rafter.

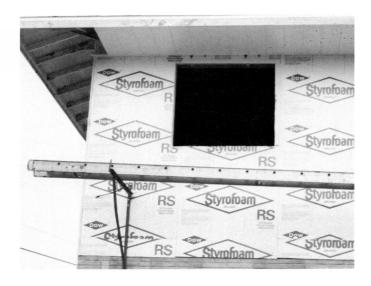

Fig. 24-10. Scaffolding formed with metal brackets and a patented stage section called a ladder plank. A structure of aluminum alloy supports a plywood strip. The unit provides nearly twice the maximum span specified for wood planking. Use below 10 ft. Lifeline recommended.

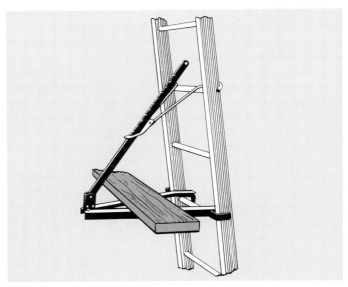

Fig. 24-12. A ladder jack. This type is held in place with hooks that fit around side rails.   (Patent Scaffolding Co.)

Roofing brackets, Fig. 24-11, are easy to use and provide safety when working on steep slopes. One type has an adjustable arm to support the plank. A level platform can be set up on nearly any angle of roof surface. Another type of roofing bracket is set at one angle.

Ladder jacks are used to support simple scaffolds for repair jobs. Setup requires two sturdy ladders of the same size and a strong plank. Fig. 24-12 shows a jack that hangs below the ladder and hooks

Fig. 24-13. Ladder jacks that project on the front side. Left. Method of raising jack while plank is supported on arm and shoulder. Use lifeline when performing this operation and do not raise the scaffold more than one rung at a time. Use below 10 ft. Right. Lifeline is recommended when seated.

to the side rails. Another type of jack, Fig. 24-13, is easily adjusted upward along the ladder.

Trestle jacks support low platforms for interior work. They are assembled as shown in Fig. 24-14. Follow the manufacturer's guideline for the proper size and weight of jack. Be sure the material used for the ledger (horizontal board held by the jacks) is sound and large enough to support the load.

SAFETY RULES FOR SCAFFOLDING

1. All scaffolds should be built under the direction of an experienced craftsperson.
2. Follow design specifications as listed in local and state codes. Inspect scaffolds daily before use.
3. Provide adequate pads or sills under scaffold posts.
4. Plumb and level scaffold members as each is set up.
5. Equip planked areas with proper guardrails, toeboards, and screens when required.
6. Power lines near scaffolds are dangerous. Consult power company for advice and procedure.
7. Do not use ladders or makeshift devices on top of scaffold platforms to increase the height.
8. Be certain the planking is heavy enough to carry the load with a safe span length.
9. Planking should be lapped at least 12 in. and extend 6 in. beyond all supports.
10. Do not permit planking to extend an unsafe distance beyond supports.
11. Remove all materials and equipment from the platform of rolling scaffolds before moving.
12. The height of the platform on rolling scaffolds should not exceed four times the smallest base dimension.

LADDERS

Types of wood and aluminum ladders commonly used by the carpenter are illustrated in Fig. 24-15. Stepladders range in size from 4 to 20 ft. A one-piece ladder, the single straight ladder, is available in sizes of 8 to 26 ft. Extension ladders provide lengths up to 60 ft.

Quality wood ladders are made from clear, straight-grained stock that is carefully seasoned. Ladders should be given a clear finish which permits easy visual inspection. When reconditioning a wood ladder, never use paint.

Basic care and handling of ladders is illustrated in Fig. 24-16. Always keep ladders clean. Do not let grease, oil, or paint collect on the rails or rungs. On extension ladders, keep all fittings tight. Lubricate the locks and pulleys. Replace any frayed or worn rope.

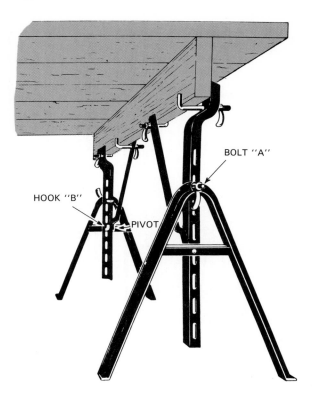

Fig. 24-14. Trestle jacks. To adjust height first loosen bolt A and then change hook B to another slot.
(Patent Scaffolding Co.)

To erect a ladder, place the lower end against a solid base so it cannot slide. Raise the top end to get under it. Walk toward the bottom end, grasping and raising the ladder rung by rung as you proceed. When vertical, lean it against the structure at the proper angle. (See Safety Rule 2 below.) Make sure bottom ends of both rails rest on a firm base.

SAFETY RULES FOR LADDERS

1. Always inspect a ladder before using it.
2. Place ladder so the horizontal distance from its lower end to the vertical wall is one-fourth the length of the ladder.
3. Before climbing the ladder, be sure both rails rest on solid footing.
4. Equip the rails with safety shoes, Fig. 24-17, when the ladder is used on surfaces where the bottom might slip.
5. Do not place ladder in front of doorway where the door can be opened toward the ladder.
6. Never place ladders on boxes or any unstable base to get more height.
7. Never splice together two short wood ladders to make a longer ladder.
8. Always face a ladder when climbing up or down.

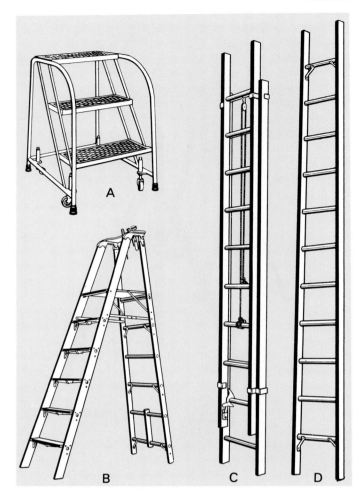

Fig. 24-15. Ladder types used by the carpenter. A—Safety rolling ladder. Drops firmly to floor when weight applied. Used for inside trim work. B—Stepladder with platform and toolholder. C—Extension ladder. D—Single straight ladder. (Patent Scaffolding Co.)

INSPECTION — LADDERS SHOULD BE INSPECTED FREQUENTLY. THOSE WHICH HAVE DEVELOPED DEFECTS SHOULD BE EITHER REPAIRED OR DESTROYED.

CARRYING — ALWAYS CARRY A LADDER OVER YOUR SHOULDER WITH FRONT END ELEVATED. BE SURE NOT TO DROP OR LET FALL. SUCH IMPACT WEAKENS A LADDER.

STORAGE — STORE HORIZONTALLY ON SUPPORTS TO PREVENT SAGGING. DO NOT STORE NEAR HEAT OR OUT IN WEATHER.

Fig. 24-16. Ladder care and handling.

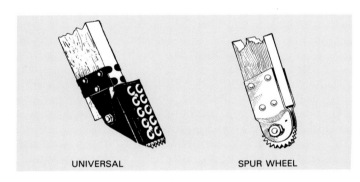

UNIVERSAL   SPUR WHEEL

Fig. 24-17. Safety shoes are required for all ladders used on smooth surfaces. (Tilley Ladder Co.)

9. Place ladder so work can be done without leaning beyond either side rail.
10. Be sure extension ladders have sufficient lap between sections. A 36 ft. length should lap at least 3 ft.; a 48 ft., at least 4 ft.
11. When a ladder is used to get onto a roof, it should extend above the roof at least 3 ft.
12. Both hands should be free to grasp when climbing a ladder. Use a hand line to raise or lower tools and materials.
13. Before mounting a stepladder, be sure it is fully open and locked, and all four legs are firmly supported.
14. Do not stand on either of the two top steps of a stepladder.
15. Do not leave tools on the top of a stepladder unless the ladder has a special holder.
16. Never use metal ladders where contact with electric current is possible.

TEST YOUR KNOWLEDGE — UNIT 24

1. Bearers that support the planking of wood scaffolds should be made of material with a nominal cross section of _____x_____ inches.
2. In wooden scaffolds, horizontal spacing of support sections should not exceed _____ feet.
3. Metal wall brackets which support a wood platform should not be spaced more than _____ ft. apart.
4. The height of the platform of a rolling scaffold unit should not exceed _____ times the smallest dimension of the base.
5. Single straight ladders are available in sizes _____ ft. to _____ ft.
6. The sides of a ladder are called _____.

7. When a ladder is used to climb onto a roof, it should extend at least _____ feet above the roof edge.

OUTSIDE ASSIGNMENTS

1. Get descriptive literature from a company that manufactures steel scaffolding. Check with your local building supply dealer or write directly to the company. From a study of this information, learn about the kinds and sizes of tubing that are used. Also learn how the various connecting devices operate. Try to find out approximate costs of this type of scaffolding. Prepare carefully organized notes and make an oral report to your class.

2. Make a study of the types of ladders used in construction and maintenance work and prepare a written report. Include information about the kind and quality of material used and how the ladders are assembled and finished. Also include commonly available sizes and list prices. Finally, develop a list of reasons a homeowner may have for buying or renting a ladder. Are these reasons the same for the builder?

Installing decorative sheathing with the aid of scaffolding. Scaffolds can be used on both straight and curved walls. Safety can be improved for the metal tube scaffold shown here.   (American Plywood Assoc.)

This public building could not be put up without the work of many carpenters. Carpenters are also needed to make the prefabricated parts used for construction.   (American Plywood Assoc.)

Work in a prefabricated housing plant requires a variety of carpentry skills. Skilled workers are completing modules of a house as the modules move along a final assembly line. (Cardinal Industries, Inc.)

Unit 25

CARPENTRY—A CAREER

Carpentry is a rewarding career. It is ideal for the person who has an interest in and aptitude for working with tools and materials. The trade requires the development of manual skills. These skills involve both thinking and doing. Carpentry also requires a thorough knowledge of materials and methods used in construction work.

OPPORTUNITIES

The carpenter belongs to the largest group of building trade workers. Jobs in carpentry are expected to grow about as fast as the average for all building trades. In the next decade, probably 80,000 more carpenters will be employed in the United States.

Job opportunities for you as a carpenter depend somewhat on your choice of work. You can choose to:
1. Construct new buildings.
2. Do remodeling, maintenance, and repair work.
3. Produce prefabricated buildings and parts.
Each type of work has good and bad features. Consider the following when choosing any type of carpentry work:
1. Winter layoffs.
2. Stress from working in the heat or cold.
3. Depression cycles in the building industry.
Also note that the demand for work on existing buildings is higher during a low cycle for new construction and prefabrication.

Nearly one out of three carpenters is self-employed. Self-employment has an advantage. Often profits are high due to low overhead.

The income earned by the carpenter is close to that received by other trades, and provides a good living. In addition, the carpenter has pride and satisfaction from doing quality work.

After gaining experience, the carpenter may want to undertake a small construction contract. This may be a first step toward the general contracting business. The experienced carpenter is usually able to handle the work of the estimator who figures labor and material costs.

The carpenter who prefers the sales and service aspects of construction work can often find a position with a lumber yard or building supply center as is shown in Fig. 25-1.

The carpenter can also join the customer service department of a company making prefabricated structures. The customer may need advice on installation of or choice of prefabricated units. Perhaps the carpenter will win repeat business for such a company.

Fig. 25-1. Work at a building supply center requires many carpentry skills. The carpenter has knowledge and experience which assure that the customer receives quality work.

TRAINING

A high school education is basic to a career in carpentry and other areas of building construction. This is as true for the apprentice-to-be as for the technician or engineer. Take as many woodworking and building construction courses as possible. Other courses are also valuable. Drafting is especially useful.

Students interested in the building trades sometimes avoid science and math. This is unfortunate because these studies are essential if you are to understand the technical aspects of modern methods and materials used in construction work. Include social studies and English since everyone should:

1. Be prepared to improve our society.
2. Be competent in reading, writing, and speaking.

After graduation from high school, you will probably have the skills to enter directly into an apprenticeship training program, Fig. 25-2. If circumstances permit, you may wish to enroll in a vocational-technical school in your area. There you can take advanced courses in carpentry and related areas. Some high schools offer these vocational courses as a part of the normal programs before graduation. If possible, take classes in concrete work, bricklaying, plumbing, sheet-metal, and electric wiring. The carpenter usually works closely with

tradespeople in these areas, Fig. 25-3, and a basic understanding of the methods and procedures will be very valuable. This is especially true if you want to become a supervisor.

APPRENTICESHIP

Our modern apprenticeship training program had its beginning in early times. It started as an arrangement between teenage children and their parents. This system passed the knowledge and skill of a trade on to each generation. As society and the economic structure became more complex, the trainee often left home and was placed under the guidance and direction of another master of the trade in the community. The student was called an apprentice and learned the trade of this master.

During the period of apprenticeship (sometimes it was as long as seven years) the apprentice often

Fig. 25-2. An apprenticeship program provides training on the job. Here a carpenter explains fastening techniques to an apprentice.
(United Brotherhood of Carpenters and Joiners of America)

Fig. 25-3. To produce a strong concrete wall, workers must understand trades other than their own. Concrete workers and carpenters combine their knowledge both when putting up forms and when pouring the concrete.

lived in the master's house. There were no wages, but board, room, and clothing were provided.

When the training was complete, the apprentice was granted the status of journeyman and could then work for wages. The word, journeyman, was derived from the fact that the apprentice who had completed the training period was then free to "journey" to other places in search of employment. The term journeyman is still used to denote a tradesperson who is fully qualified.

To gain more experience, the journeyman worked with other masters as an equal. Then this mature worker would often start a separate business.

This form of apprenticeship declined rapidly with the advent of the industrial revolution. A new kind of system developed. In it the apprentice lived at home and received wages for work done. Under this system, the apprentice was often exploited. Some became only low paid workers who received little training. Such practices continued in varying degrees until federal legislation was enacted which established standards and specific requirements for apprenticeship training programs.

Today, apprenticeship programs are carefully set up and supervised. Local committees consisting of labor and management provide direct control. There is help from schools, state and federal organizations and agencies, Fig. 25-4.

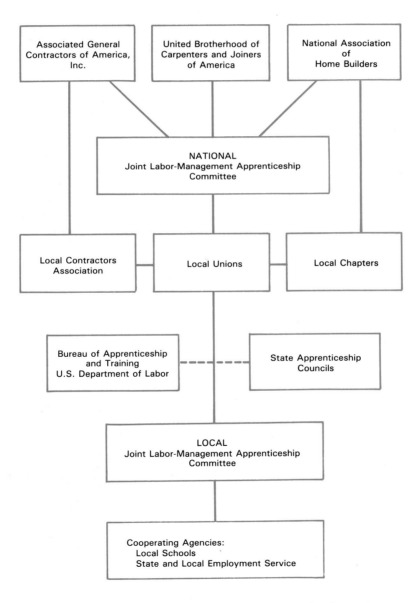

Fig. 25-4. The apprenticeship and training system for the carpentry trade is well organized. Additional information may be obtained from the bulletin National Carpentry Apprenticeship and Training Standards, prepared by the U.S. Department of Labor.

APPRENTICESHIP STAGES

The apprentice works under a signed agreement with an employer. The agreement includes the approval of local and state committees on apprenticeship training.

Applicants must be at least 17 years of age and shall satisfy the local committee that they have the ability or aptitude to master the trade. Then they are placed on a waiting list. The waiting period can last from 1 to 5 years depending on local demand.

The term of apprenticeship for the field of carpentry is normally four years. This may be adjusted for applicants with significant experience or those who may have completed certain advanced courses in vocational-technical schools.

In addition to the instruction and skills learned through regular work on the job an apprentice attends school classes in subjects related to the trade. These classes are usually held in the evening and total at least 144 hours per year. They cover technical information about tools, machines, methods, and processes. They provide practice in mathematical calculations, blueprint reading, sketching, layout work, estimating, and similar activities. A great deal of study is required to master the technical knowledge needed in carpentry work. See Fig. 25-5.

The apprentice works and is paid while learning. The wage scale is determined by the local committee. It usually starts at about 50 percent of the amount received by a journeyman carpenter. This scale is advanced regularly and may approach 90 percent of a journeyman's pay during the last year.

When the training period is complete and the apprentice has passed a final examination, he or she becomes a journeyman carpenter. A certificate which affirms this status is issued. This document is recognized throughout the country.

PERSONAL QUALIFICATIONS

To become a successful carpenter you must:
1. Be physically able to do the work.
2. Have a talent for working with the tools and materials of the trade.

You also need sincere interest and enthusiasm that will intensify your efforts as you study and practice the skills and "know-how" required.

In addition, you must develop certain desirable character traits. Honesty in all your dealings is very important, especially in the quality and quantity of work performed. You must show courtesy, respect, and loyalty to those with whom you work. Punctuality and reliability reflect your general attitude and are important not only during your training program but later when you enter regular employment.

Fig. 25-5. Before any work is done, the carpenter or another person will consult the building plan. Some training in blueprint reading is required for all carpenters.   (Orem Research)

The ability to cooperate and get along with others—students, co-workers, supervisors, and employers—is essential to success. Most people who fail in the carpentry trade do so because of a deficiency in personal characteristics, not because of their skill level.

The hazards associated with carpentry require that you develop a good attitude toward safety. This means that you must be willing to spend time learning the safest way to do your work. You must be willing to follow safety rules and regulations at all times.

Even after you complete your training program you must continue to perfect your skills and adjust to new methods and techniques, Fig. 25-6. Each day brings new materials and improved procedures to the construction industry. This presents a special challenge to those in the carpentry trade. You should read and study new books and manufacturer's literature, along with trade journals and

Fig. 25-6. Worker installs acoustical ceiling tile in a new shopping mall. System provides easy access to mechanical systems. (USG Acoustical Products Co.)

magazines in the building construction field. Much information can be obtained at association meetings and conventions where new products are exhibited. To be a successful carpenter, you will also need to keep informed on code changes, new zoning ordinances, safety regulations, and other aspects of construction work that apply to the local community in which you work.

During the next few years, the growth in building construction will result in a greatly increased demand for carpenters. Carpenters will be needed for the construction of homes, schools, churches, and stores. They will be needed to put up public, institutional, and other types of buildings.

For those who have ability, and are willing to work, the field of carpentry is unlimited.

OUTSIDE ASSIGNMENTS

Find out how to enter a carpentry apprenticeship program in your locality. Contact a local carpentry contractor, a local of a union, and a state apprenticeship agency. What skills do you have now? What skills you must learn? Report your findings to your class.

Carpentry apprentices compete in a statewide contest sponsored by a carpenters' union. Judging is based on the methods used as well as on the speed and accuracy of the work. (Master Builders of Iowa)

TECHNICAL INFORMATION

Acoustical Tile	AT	Dimension	Dim.	On Center	OC
Aggregate	Aggr.	Ditto	Do.	Opening	Opng.
Air Conditioning	Air Cond.	Double Strength Glass	DSG	Plaster	Plas.
Air dried	AD	Drawing	Dwg.	Plate	Pl.
Alternate	Alt.	Dressed and Matched	D & M	Plate Glass	Pl. Gl.
Aluminum	AL	Edge	Edg.	Plumbing	Plbg.
American Institute of Architects	A.I.A.	Edge Grain	EG	Precast	Prcst.
		Entrance	Ent.	Prefabricated	Prefab.
American Institute of Electrical Engineers	A.I.E.E.	Excavate	Exc.	Quart	Qt.
		Exterior	Ext.	Random	Rdm.
American Society for Testing Materials	A.S.T.M.	Federal Housing Authority	FHA	Refrigerator	Ref.
		Finish	Fin.	Reinforcing	Reinf.
American Standards Association, Inc.	A.S.A.	Fixture	Fix.	Revision	Rev.
		Flashing	Fl.	Rough	Rgh.
Apartment	Apt.	Flat Grain	FG	Rough Opening	Rgh. Opng.
Approximate	Approx.	Flooring	Flg.	Schedule	Sch.
Architectural	Arch.	Fluorescent	Fluor.	Screen	Scr.
Asbestos	Asb.	Foot or Feet	Ft.	Select	Sel.
Basement	Bsmt.	Footing	Ftg.	Service	Serv.
Bathroom	B	Foundation	Fdn.	Sheathing	Shthg.
Beam	Bm.	Furring	Fur.	Shelving	Shelv.
Better	Btr.	Gallon	Gal.	Shiplap	S/lap
Beveled	Bev.	Galvanized Iron	GI	Siding	Sdg.
Blocking	Blkg.	Glass	Gl.	Specifications	Spec.
Board Foot	Bd. Ft.	Hardwood	Hdwd.	Square	Sq.
Brick	Brk.	Horsepower	HP	Square Feet	Sq. Ft.
British thermal unit	Btu	Hose Bib	HB	Stairway	Stwy.
Building	Bldg.	Hot Water	HW	Standard	Std.
Bundle	Bdl.	Hundred	C	Steel	St. or Stl.
Cabinet	Cab.	Insulation	Ins.	Structural	Str.
Casing	Csg.	Interior	Int.	Surfaced One Side	S1S
Cement	Cem.	Kiln-dried	KD	Surfaced Two Sides	S2S
Cement Floor	Cem. Fl.	Kitchen	K	Surfaced Four Sides	S4S
Cement Mortar	Cem. Mort.	Lavatory	Lav.	Surfaced One Side and Two Edges	S1S2E
Center Line	CL	Length	Lgth.		
Center Matched	CM	Light	Lt.	Switch	Sw. or S
Closet	Cl or Clo.	Linen Closet	L Cl.	Temperature	Temp.
Column	Col.	Linoleum	Lino.	Thermostat	Thermo.
Common	Com.	Living Room	LR	Thousand	M
Concrete	Conc.	Masonry Opening	MO	Tongue and Groove	T & G
Concrete Block	Conc. B	Material	Matl.	Typical	Typ.
Conduit	Cnd.	Maximum	Max.	Unexcavated	Unexc.
Construction	Const.	Medicine Cabinet	MC	Ventilation	Vent.
Counter	Ctr.	Minimum	Min.	Water Closet	WC
Cubic Foot	Cu. Ft.	Miscellaneous	Misc.	Water Heater	WH
Cubic Yard	Cu. Yd.	Modular	Mod.	Weight	Wt.
Diagram	Diag.	Molding	Mldg.	Wood	Wd.
Diameter	Dia. or Diam.	Nosing	Nos.		

Abbreviation commonly used in building construction.

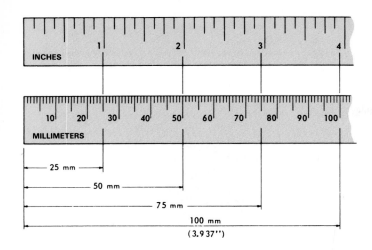

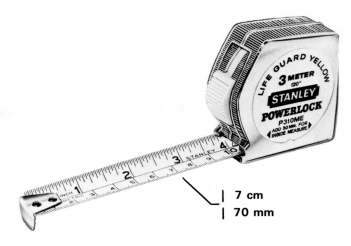

25 mm
50 mm
75 mm
100 mm
(3.937'')

Typical dual dimensioned rule. Numbers on the metric side read in centimeters (cm). Multiply these numbers by 10 (add 0) to convert them to millimeters (mm).

INCH TO MILLIMETERS CONVERSIONS

| EXACT | | ┌──── ROUND OUT ────┐ |
		0.5 mm	1 mm
1/32 IN.	0.794	1 mm	1 mm
1/16 IN.	1.588	1.5 mm	2 mm
1/8 IN.	3.175	3 mm	3 mm
1/4 IN.	6.350	6.5 mm	6 mm
3/8 IN.	9.525	9.5 mm	10 mm
1/2 IN.	12.700	12.5 mm	13 mm
5/8 IN.	15.875	16 mm	16 mm
3/4 IN.	19.050	19 mm	19 mm
7/8 IN.	22.225	22 mm	22 mm
1 IN.	25.400	25.5 mm	25 mm

CONVERSION ACCURACY

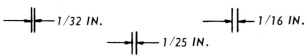

The size of a millimeter (1/25 in.) is about halfway between 1/32 in. and 1/16 in. as shown above. For general woodworking, rounding a converted figure to the nearest millimeter is acceptable practice. When greater accuracy is required, round to the nearest 1/2 millimeter. See table at left. For sizes not listed, add combinations of figures (exact) and then round to accuracy desired.

METRIC CONVERSIONS FOR LENGTH, AREA, AND VOLUME

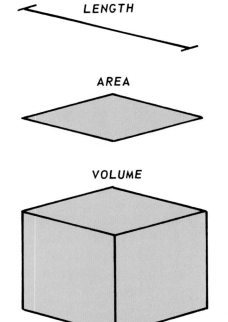

LENGTH

AREA

VOLUME

1 METER (m) = 1 000 MILLIMETERS (mm) = 100 CENTIMETERS (cm)

1 m = 3.28 FT.	1 FT. = .305 m
1 cm = 0.39 IN.	1 IN. = 2.54 cm
1 mm = 0.0394 IN.	1 IN. = 25.4 mm

1 SQUARE METER (m^2) = 10 000 SQUARE CENTIMETERS (cm^2)

$1\ m^2 = 1.2\ YD.^2$	$1\ YD.^2 = .836\ m^2$
$1\ m^2 = 10.8\ FT.^2$	$1\ FT.^2 = .093\ m^2$
$1\ cm^2 = .155\ IN.^2$	$1\ IN.^2 = 6.45\ cm^2$

1 CUBIC METER (m^3) = 1 000 CUBIC DECIMETER (cm^2)
$1\ dm^3 = 1$ LITER

$1\ m^3 = 1.31\ YD.^3$	$1\ YD.^3 = 0.765\ m^3$
$1\ m^3 = 35.3\ FT.^3$	$1\ FT.^3 = 0.028\ m^3$
$1\ cm^3 = .061\ IN.^3$	$1\ IN.^3 = 16.39\ cm^3$

Metric conversions for length, area, and volume.

METRICS IN CONSTRUCTION

Efficient building construction begins with standardized sizes for building parts. Designs geared for mass production are based on standard modules. Layouts (horizontal and vertical) use one specific size. Whole multiples of the size make up all larger measurements.

The U.S. customary system uses a basic module of 4 in. The layout grid is further divided into spaces of 16, 24, and 48 in. Modular design saves materials and time.

As the United States converts to the metric system, it has been recommended that the 4 in. module be replaced with a 100 millimeter (mm) module. Smaller sizes (submultiples) would include 25, 50, and 75 mm. Large multiples would measure 400, 600, and 1200 mm as shown in the drawing below.

The 100 mm module is slightly smaller than 4 in. and the conversion can hardly be detected in small measurements. In 48 in. however, the difference would be about 3/4 in. and a standard 4 ft. by 8 ft. plywood panel would be about 1 1/2 in. longer than the similar metric size (1200 mm by 2400 mm).

Manufacturers will need to change the size of many basic building materials. Also, some changes will be required in building components such as windows, doors, and cabinetwork. Although the conversion to the metric system will cause some difficulties, there will be many advantages. International trade of building materials and products will be easier, especially for such items as plywood and imported hardwoods. Leaders in building construction believe the change will provide an unusual opportunity to discard obsolete sizing and grading patterns and develop new and more efficient systems.

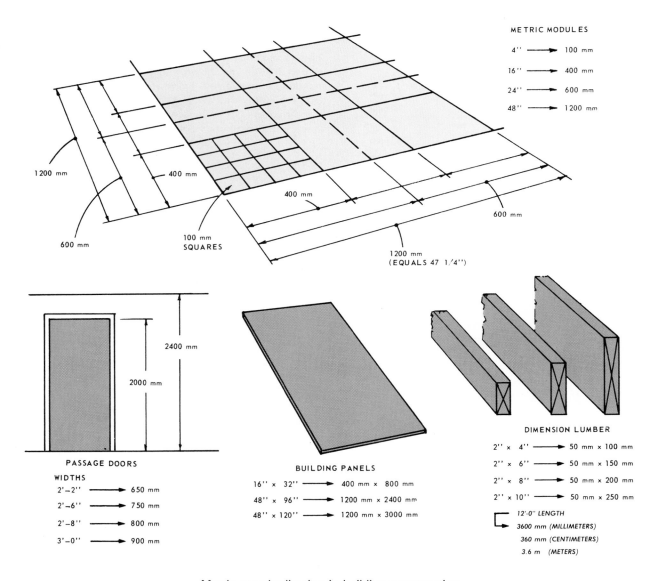

Metric standardization in building construction.

CEILING JOISTS

Species	Grade	2 x 4 16" oc	2 x 4 24" oc	2 x 6 16" oc	2 x 6 24" oc	2 x 8 16" oc	2 x 8 24" oc
DOUGLAS FIR-LARCH	2 & Better	11-6	10-0	18-1	15-7	23-10	20-7
	3	9-9	7-11	14-8	11-11	19-4	15-9
DOUGLAS FIR SOUTH	2 & Better	10-6	9-2	16-6	14-5	21-9	19-0
	3	9-5	7-9	14-2	11-7	18-9	15-3
HEM-FIR	2 & Better	10-9	9-5	16-11	13-11	22-4	18-4
	3	8-7	7-0	13-1	10-8	17-2	14-0
MOUNTAIN HEMLOCK	2 & Better	9-11	8-8	15-7	13-8	20-7	18-0
	3	8-11	7-3	13-3	10-10	17-6	14-4
MOUNTAIN HEMLOCK-HEM-FIR	2 & Better	9-11	8-8	15-7	13-8	20-7	18-0
	3	8-7	7-0	13-1	10-8	17-2	14-0
WESTERN HEMLOCK	2 & Better	10-9	9-5	16-11	14-6	22-4	19-1
	3	9-0	7-5	13-9	11-3	18-1	14-10
ENGELMANN SPRUCE ALPINE FIR (Engelmann Spruce-Lodgepole Pine)	2 & Better	9-11	8-8	15-6	12-8	20-7	16-8
	3	7-10	6-5	11-9	9-7	15-6	12-8
LODGEPOLE PINE	2 & Better	10-3	8-11	16-1	13-3	21-2	17-6
	3	8-4	6-9	12-7	10-3	16-7	13-6
PONDEROSA PINE-SUGAR PINE (Ponderosa Pine-Lodgepole Pine)	2 & Better	9-11	8-8	15-7	12-10	20-7	16-10
	3	8-0	6-6	12-0	9-10	15-10	12-11
WHITE WOODS (Western Woods)	2 & Better	9-7	8-5	15-1	12-6	20-2	16-5
	3	7-10	6-5	11-9	9-8	15-6	12-8
IDAHO WHITE PINE	2 & Better	10-3	8-5	15-3	12-6	20-0	16-5
	3	7-10	6-5	11-9	9-8	15-6	12-8
WESTERN CEDARS	2 & Better	9-7	8-5	15-1	13-2	20-0	17-6
	3	8-3	6-9	12-7	10-3	16-7	13-7

Design Criteria:
Strength—5 lbs. per sq. ft. dead load plus 10 lbs. per sq. ft. live load.
Deflection—Limited to span in inches divided by 240 for live load only.

FLOOR JOISTS

Species	Grade	2 x 8 16" oc	2 x 8 24" oc	2 x 10 16" oc	2 x 10 24" oc	2 x 12 16" oc	2 x 12 24" oc
DOUGLAS FIR-LARCH	2 & Better	13-1	11-3	16-9	14-5	20-4	17-6
	3	10-7	8-8	13-6	11-0	16-5	13-5
DOUGLAS FIR SOUTH	2 & Better	12-0	10-6	15-3	13-4	18-7	16-3
	3	10-3	8-4	13-1	10-8	15-11	13-0
HEM-FIR	2 & Better	12-3	10-0	15-8	12-10	19-1	15-7
	3	9-5	7-8	12-0	9-10	14-7	11-11
MOUNTAIN HEMLOCK	2 & Better	11-4	9-11	14-6	12-8	17-7	15-4
	3	9-7	7-10	12-3	10-0	14-11	12-2
MOUNTAIN HEMLOCK-HEM-FIR	2 & Better	11-4	9-11	14-6	12-8	17-7	15-4
	3	9-5	7-8	12-0	9-10	14-7	11-11
WESTERN HEMLOCK	2 & Better	12-3	10-6	15-8	13-4	19-1	16-3
	3	9-11	8-1	12-8	10-4	15-5	12-7
ENGELMANN SPRUCE ALPINE FIR (Engelmann Spruce-Lodgepole Pine)	2 & Better	11-2	9-1	14-3	11-7	17-3	14-2
	3	8-6	6-11	10-10	8-10	13-2	10-9
LODGEPOLE PINE	2 & Better	11-8	9-7	14-11	12-3	18-1	14-11
	3	9-1	7-5	11-7	9-5	14-1	11-6
PONDEROSA PINE-SUGAR PINE (Ponderosa Pine-Lodgepole Pine)	2 & Better	11-4	9-3	14-5	11-9	17-7	14-4
	3	8-8	7-1	11-1	9-1	13-6	11-0
WHITE WOODS (Western Woods)	2 & Better	11-0	9-0	14-0	11-6	17-0	14-0
	3	8-6	6-11	10-10	8-10	13-2	10-9
IDAHO WHITE PINE	2 & Better	11-0	9-0	14-0	11-6	17-1	14-0
	3	8-6	6-11	10-10	8-10	13-2	10-9
WESTERN CEDARS	2 & Better	11-0	9-7	14-0	12-3	17-0	14-11
	3	9-1	7-5	11-6	9-5	14-0	11-6

Design Criteria:
Strength—10 lbs. per sq. ft. dead load plus 40 lbs. per sq. ft. live load.
Deflection—Limited to span in inches divided by 360 for live load only.

Allowable clear spans for various sizes and species of joists.   (Western Wood Products Assoc.)

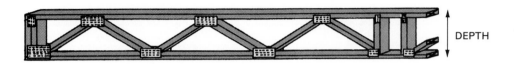

DEPTH

DOUGLAS FIR #1 DENSE MC 15

Span (Feet)	Depth (Inches)										Maximum load*
	12.0	14.0	16.0	18.0	20.0	22.0	24.0	26.0	28.0	30.0	
16.0	171.2	196.6	216.9	235.3	252.0	267.3	281.4	294.4	306.4	317.5	
18.0	124.7	166.6	185.2	202.2	218.0	232.6	246.1	258.8	270.5	281.5	
20.0	95.0	128.4	159.1	174.8	189.4	203.1	215.9	227.9	239.3	249.9	
22.0	75.1	100.2	129.6	150.4	165.5	178.1	190.1	201.4	212.2	222.3	
24.0	61.3	80.6	103.3	126.9	141.9	156.8	168.1	178.7	188.7	198.4	
26.0	51.4	66.6	84.4	104.9	121.4	134.2	146.9	159.1	168.5	177.5	
28.0	44.2	56.3	70.6	87.0	105.0	116.1	127.2	138.2	149.2	159.5	
30.0	38.7	48.6	60.2	73.5	88.6	101.4	111.1	120.8	130.4	140.0	
32.0	34.5	42.7	52.2	63.2	75.6	89.3	97.9	106.4	114.9	123:4	
34.0	31.3	38.1	46.0	55.2	65.5	77.1	86.9	94.5	102.1	109.6	

SOUTHERN PINE #1 DENSE KD

Span (Feet)	Depth (Inches)										Maximum load*
	12.0	14.0	16.0	18.0	20.0	22.0	24.0	26.0	28.0	30.0	
16.0	164.2	193.3	212.8	230.5	246.5	261.1	274.4	286.7	298.0	308.5	
18.0	119.8	163.5	182.2	198.7	213.9	227.9	240.9	252.9	264.1	274.6	
20.0	91.4	123.2	153.9	172.2	186.4	199.6	211.9	223.4	234.3	244.5	
22.0	72.4	96.3	124.4	145.0	162.0	175.5	187.0	198.0	208.3	218.1	
24.0	59.2	77.6	99.3	122.4	136.8	151.2	165.5	176.0	185.7	195.0	
26.0	49.8	64.3	81.3	100.8	117.0	129.4	141.6	153.9	166.0	174.9	
28.0	42.8	54.4	68.1	83.7	101.2	111.9	122.6	133.2	143.8	154.3	
30.0	37.6	47.1	58.2	70.9	85.2	97.8	107.1	116.4	125.7	134.9	
32.0	33.6	41.4	50.6	61.0	72.9	86.1	94.4	102.6	110.8	119.0	
34.0	30.5	37.0	44.6	53.4	63.3	74.3	83.8	91.1	98.4	105.7	

MACHINE RATED LUMBER 2100F-1.8E

Span (Feet)	Depth (Inches)										Maximum load*
	12.0	14.0	16.0	18.0	20.0	22.0	24.0	26.0	28.0	30.0	
16.0	156.3	192.0	211.2	228.6	244.3	258.6	271.7	283.8	294.9	305.1	
18.0	114.2	155.7	181.1	197.3	212.3	226.0	238.8	250.6	261.6	271.9	
20.0	87.4	117.5	153.0	171.2	185.1	198.1	210.3	221.7	232.3	242.3	
22.0	69.4	92.0	118.7	149.2	162.2	174.3	185.8	196.5	206.7	216.4	
24.0	56.9	74.3	94.8	118.4	142.8	154.1	164.7	174.9	184.5	193.6	
26.0	47.9	61.7	77.8	96.3	117.2	136.8	146.7	156.1	165.2	173.8	
28.0	41.4	52.4	65.3	80.1	96.9	115.5	131.2	139.9	148.4	156.5	
30.0	36.4	45.4	55.9	67.9	81.5	96.7	113.4	125.9	133.8	141.3	
32.0	32.7	40.0	48.7	58.6	69.8	82.3	96.1	111.2	121.0	128.1	
34.0	29.7	35.9	43.1	51.4	60.7	71.1	82.6	95.2	108.8	116.5	

*1. Loads are shown in pounds per lineal foot.
2. Deflection may govern the span limits.
3. Tables are consistent with manufacturer's load deflection policy.
4. These tables are limited to a simple span condition uniformly loaded.

5. Nail values are species specific.
6. Consult the manufacturer for special conditions.
7. Design is based on 0 percent stress increase on lumber and fasteners.

Loading tables for floor/ceiling trusses built with 2 x 4 chords.   (Gang-Nail Systems, Inc.)

LENGTHS OF COMMON RAFTERS

FEET OF RUN	2 in 12	2½ in 12	3 in 12	3½ in 12	4 in 12	4½ in 12
	Inclination (set saw at) 9°-28'	Inclination (set saw at) 11°-46'	Inclination (set saw at) 14°-2'	Inclination (set saw at) 16°-16'	Inclination (set saw at) 18°-26'	Inclination (set saw at) 20°-33'
	12.17 in. per ft. of run	12.26 in. per ft. of run	12.37 in. per ft. of run	12.5 in. per ft. of run	12.65 in. per ft. of run	12.82 in. per ft. of run
4'	4' 0-11/16"	4' 1-1/32"	4' 1-15/32"	4' 2"	4'-2-19/32"	4' 3-9/32"
5	5' 0-27/32"	5' 1-5/16"	5' 1-27/32"	5' 2-1/2"	5' 3-1/4"	5' 4-3/32"
6	6' 1-1/32"	6' 1-9/16"	6' 2-7/32"	6' 3"	6' 3-29/32"	6' 4-15/16"
7	7' 1-3/16"	7' 1-13/16"	7' 2-19/32"	7' 3-1/2"	7' 4-9/16"	7' 5-3/4"
8	8' 1-3/8"	8' 2-3/32"	8' 2-31/32"	8' 4"	8' 5-3/32"	8' 6-9/16"
9	9' 1-17/32"	9' 2-7/16"	9' 3-11/32"	9' 4-1/2"	9' 5-27/32"	9' 7-3/8"
10	10' 1-23/32"	10' 2-19/32"	10' 3-23/32"	10' 5"	10' 6-1/2"	10' 8-7/32"
11	11' 1-7/8"	11' 2-7/8"	11' 4-1/16"	11' 5-1/2"	11' 7-5/32"	11' 9-1/32"
12	12' 2-1/32"	12' 3-1/8"	12' 4-7/16"	12' 6"	12' 7-13/16"	12' 9-27/32"
13	13' 2-7/32"	13' 3-3/8"	13' 4-13/16"	13' 6-1/2"	13' 8-15/32"	13' 10-21/32"
14	14' 2-3/8"	14' 3-21/32"	14' 5-3/16"	14' 7"	14' 9-3/32"	14' 11-1/2"
15	15' 2-9/16"	15' 3-29/32"	15' 5-9/16"	15' 7-1/2	15' 9-3/4"	16' 0-5/16"
16	16' 2-23/32"	16' 4-5/32"	16' 5-15/16"	16' 8"	16' 10-13/32"	17' 1-1/8"
INCHES OF RUN	These lengths are to be added to those shown above when run involves inches					
1/4"	1/4"	1/4"	1/4"	1/4"	1/4"	9/32"
1/2"	1/2"	1/2"	1/2"	17/32"	17/32"	17/32"
1"	1"	1"	1-1/32"	1-1/32"	1-1/16"	1-1/16"
2"	2-1/32"	2-1/32"	2-1/16"	2-3/32"	2-3/32"	2-1/8"
3"	3-1/16"	3-1/16"	3-3/32"	3-1/8"	3-5/32"	3-7/32"
4"	4-1/16"	4-3/32"	4-1/8"	4-5/32"	4-7/32"	4-9/32"
5"	5-1/16"	5-1/8"	5-5/32"	5-7/32"	5-9/32"	5-11/32"
6"	6-3/32"	6-1/8"	6-3/16"	6-1/4"	6-5/16"	6-13/32"
7"	7-3/32"	7-5/32"	7-7/32"	7-9/32"	7-3/8"	7-15/32"
8"	8-1/8"	8-3/16"	8-1/4"	8-11/32"	8-7/16"	8-17/32"
9"	9-1/8"	9-3/16"	9-1/4"	9-3/8"	9-15/32"	9-5/8"
10"	10-5/32"	10-7/32"	10-5/16"	10-7/16"	10-17/32"	10-11/16"
11"	11-5/32"	11-1/4"	11-11/32"	11-15/32"	11-19/32"	11-3/4"

FEET OF RUN	5 in 12	5½ in 12	6 in 12	6½ in 12	7 in 12	7½ in 12	8 in 12
	Inclination (set saw at) 22°-37'	Inclination (set saw at) 24°-37'	Inclination (set saw at) 26°-34'	Inclination (set saw at) 28°-27'	Inclination (set saw at) 30°-15'	Inclination (set saw at) 32°-0'	Inclination (set saw at) 33°-41'
	13.00 in. per ft. of run	13.20 in. per ft. of run	13.42 in. per ft. of run	13.65 in. per ft. of run	13.89 in. per ft. of run	14.15 in. per ft. of run	14.42 in. per ft. of run
4	4' 4"	4' 4-13/16"	4' 5-11/16"	4' 6-19/32"	4' 7-9/16"	4' 8-19/32"	4' 9-11/16"
5	5' 5"	5' 6"	5' 7-1/8"	5' 8-1/4"	5' 9-15/32"	5' 10-3/4"	6' 0-1/8"
6	6' 6"	6' 7-3/32"	6' 8-1/2"	6' 9-29/32"	6' 11-11/32"	7' 0-29/32"	7' 2-17/32"
7	7' 7"	7' 8-13/32"	7' 9-15/16"	7' 11-9/16"	8' 1-7/32"	8' 3-1/16"	8' 4-15/16"
8	8' 8"	8' 9-19/32"	8' 11-3/8"	9' 1-7/32"	9' 3-1/8"	9' 5-7/32"	9' 7-3/8"
9	9' 9"	9' 10-13/16"	10' 0-25/32"	10' 2-27/32"	10' 5"	10' 7-11/32"	10' 9-25/32"
10	10' 10"	11' 0"	11' 2-7/32"	11' 4-1/2"	11' 6-29/32"	11' 9-1/2"	12' 0-7/32"
11	11' 11"	12' 1-3/32"	12' 3-5/8"	12' 6-5/32"	12' 8-13/16"	12' 11-21/32"	13' 2-5/8"
12	13' 0"	13' 2-13/32"	13' 5-1/32"	13' 7-13/16"	13' 10-11/16"	14' 1-13/16"	14' 5-1/32"
13	14' 1"	14' 3-19/32"	14' 6-15/32"	14' 9-15/32"	15' 0-9/16"	15' 3-31/32"	15' 7-15/32"
14	15' 2"	15' 4-13/16"	15' 7-7/8"	15' 11-1/8"	16' 2-15/32"	16' 6-1/8"	16' 9-7/8"
15	16' 3"	16' 6"	16' 9-5/16"	17' 0-3/4"	17' 4-11/32"	17' 8-1/4"	18' 0-5/16"
16	17' 4"	17' 7-7/32"	17' 10-23/32"	18' 2-13/32"	18' 6-1/4"	18' 10-13/32"	19' 2-23/32"
INCHES OF RUN	These lengths are to be added to those shown above when run involves inches						
1/4"	9/32"	9/32"	9/32"	9/32"	9/32"	5/16"	5/16"
1/2"	17/32"	9/16"	9/16"	9/16"	9/16"	1-9/32"	5/8"
1"	1-3/32"	1-3/32"	1-1/8"	1-1/8"	1-5/32"	1-3/16"	1-7/32"
2"	2-5/32"	2-7/32"	2-1/4"	2-9/32"	2-5/16"	2-11/32"	2-13/32"
3"	3-1/4"	3-5/16"	3-3/8"	3-13/32"	3-15/32"	3-17/32"	3-19/32"
4"	4-11/32"	4-13/32"	4-15/32"	4-9/16"	4-5/8"	4-23/32"	4-13/16"
5"	5-13/32"	5-1/2"	5-19/32"	5-11/16"	5-25/32"	5-29/32"	6"
6"	6-1/2"	6-19/32"	6-23/32"	6-13/16"	6-15/16"	7-1/32"	7-7/32"
7"	7-9/16"	7-23/32"	7-13/16"	7-31/32"	8-1/8"	8-1/4"	8-13/32"
8"	8-21/32"	8-13/16"	8-15/16"	9-1/8"	9-1/4"	9-7/16"	9-5/8"
9"	9-3/4"	9-29/32"	10-5/32"	10-1/4"	10-13/32"	10-19/32"	10-13/16"
10"	10-27/32"	11"	11-3/16"	11-3/8"	11-9/16"	11-25/32"	1'0"
11"	11-29/32"	12-3/32"	1' 0-5/16"	1' 0-1/2"	1' 0-23/32"	1' 0-31/32"	1' 1-7/32"

Common rafter table. Subtract one-half ridge board thickness. (Building Supply News)

LUMBER--Lumber shall be of a good grade of sufficient quality to permit the following allowable unit stresses:
c = 900#/□" Compression parallel to grain.
f = 900#/□" Extreme fiber in bending.
E = 1,600,000#/□" Modulus of elasticity.

CONNECTORS--Timber connectors shall be 2-1/2" diameter split rings and Trip-L-Grip framing anchors.

BOLTS--Bolts shall be 1/2" diameter machine bolts with 2" x 2" x 1/8" plate washers, 2-1/8" diameter cast or malleable iron washers, or ordinary cut washers.

DIMENSIONS--Dimensions shown will provide approximately 1/2" camber at bottom chord panel points. Utilize full uncut length of bottom chord pieces by increasing the spacing of connectors in the splice.

SPAN L	DIMENSIONS A	B	C	DESIGN STRESSES U₁	U₂	L₁	L₂	V₁	D₁
20'-0"	5'-5"	5'-3⅝"	2'-7⅝"	1756	1450	1614	1472	430	430
22'-0"	5'-11½"	5'-10¹/16"	2'-10¹³/16"	1932	1595	1775	1619	473	473
24'-0"	6'-6"	6'-4⁷/16"	3'-2"	2108	1740	1936	1766	516	516
26'-0"	7'-0½"	6'-10⅞"	3'-5¼"	2282	1885	2097	1913	559	559
28'-0"	7'-7"	7'-5¼"	3'-8⁷/16"	2459	2030	2258	2060	602	602
30'-0"	8'-1½"	7'-11¹¹/16"	3'-11⅝"	2634	2175	2420	2207	645	645
32'-0"	8'-8"	8'-6¹/16"	4'-2¹³/16"	2810	2320	2581	2354	688	688

LUMBER

Span	2" x 6" No.	Length	2" x 4" No.	Length	Total F.B.M.
20'-0"	2	12'-0"	2	12'-0"	53
22'-0"	2	14'-0"	2	10'-0"	60
24'-0"	2	14'-0"	2	10'-0"	63
26'-0"	2	16'-0"	2	12'-0"	70
28'-0"	2	16'-0"	2	12'-0"	73
30'-0"	2	18'-0"	2	14'-0"	79
32'-0"	2	20'-0"	2	14'-0"	83

HARDWARE

No.	Item	Size
11	Split Rings	2½" Diam.
2	Trip-L-Grip	Type A
1	Bolt	½" x 7½"
4	Bolts	½" x 6"
14	Washers	½"

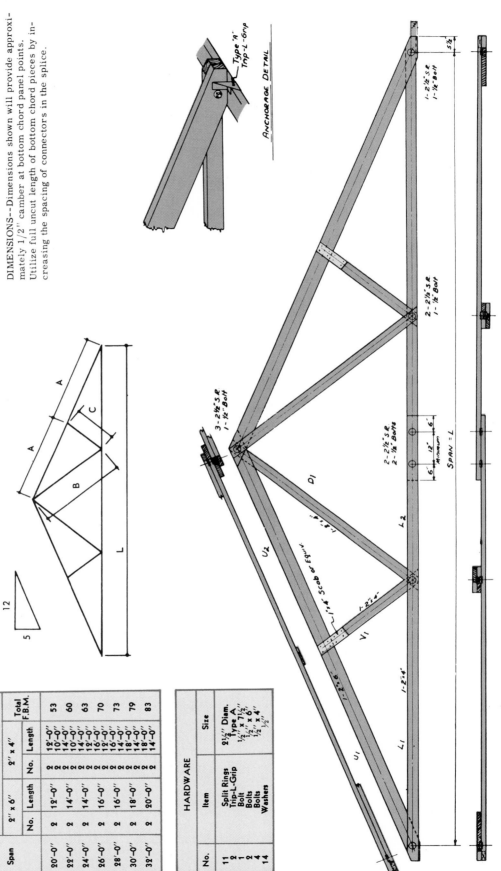

Detail sheet of a Fink truss designed for a roof slope of 5 in 12. (Timber Engineering Co.)

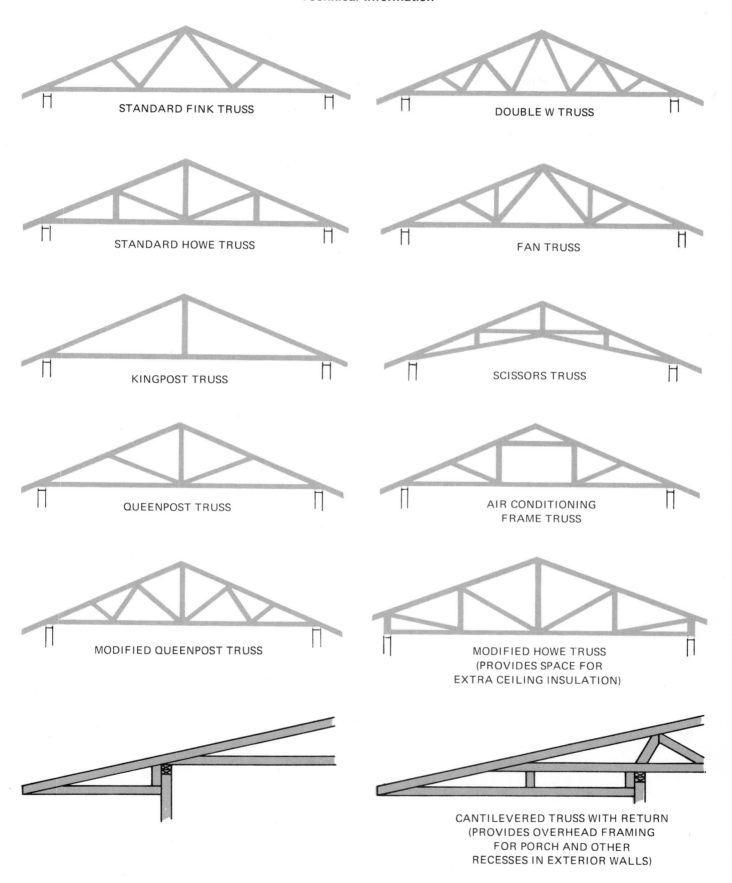

STANDARD FINK TRUSS

DOUBLE W TRUSS

STANDARD HOWE TRUSS

FAN TRUSS

KINGPOST TRUSS

SCISSORS TRUSS

QUEENPOST TRUSS

AIR CONDITIONING
FRAME TRUSS

MODIFIED QUEENPOST TRUSS

MODIFIED HOWE TRUSS
(PROVIDES SPACE FOR
EXTRA CEILING INSULATION)

CANTILEVERED TRUSS WITH RETURN
(PROVIDES OVERHEAD FRAMING
FOR PORCH AND OTHER
RECESSES IN EXTERIOR WALLS)

Types of roof trusses and overhangs. The Fink truss is also called a W truss. Any type of truss should always be constructed according to designs developed from engineering data.

EXTERIOR MATERIALS

Wood bevel siding, ½ x 8, lapped........ R-0.81
Wood bevel siding, ¾ x 10, lapped....... R-1.05
Wood siding shingles, 16," 7½" exposure ... R-0.87
Aluminum or Steel, over sheathing,
   hollow-backed...................... R-0.61
Stucco, per inch...................... R-0.20
Building paper........................ R-0.06
½" nail-base insulating board sheathing.... R-1.14
½" insulating board sheathing, regular
   density............................. R-1.32
²⁵⁄₃₂" insul. board sheathing, regular density . R-2.04
Insulating-board backed nominal ⅜"....... R-1.82
Insulating-board backed nominal ⅜"
   foil backed......................... R-2.96
Plywood ¼"........................... R-0.31
Plywood ⅜"........................... R-0.47
Plywood ½"........................... R-0.62
Plywood ⅝"........................... R-0.78
Hardboard ¼"......................... R-0.18
Hardboard, medium density siding ⁷⁄₁₆".... R-0.67
Softwood board, fir pine and similar softwoods
   ¾"................................. R-0.94
   1½"................................ R-1.89
   2½"................................ R-3.12
   3½"................................ R-4.35
Gypsumboard ½"....................... R-0.45
Gypsumboard ⅝"....................... R-0.56

MASONRY MATERIALS

Concrete blocks, three oval cores
   Cinder aggregate, 4" thick............. R-1.11
   Cinder aggregate, 12" thick............ R-1.89
   Cinder aggregate, 8" thick............. R-1.72
   Sand and gravel aggregate, 8" thick..... R-1.11
   Sand and gravel aggregate, 12" thick ... R-1.28
   Lightweight aggregate (expanded clay,
     shale, slag, pumice, etc.), 8" thick..... R-2.00
Concrete blocks, two rectangular cores
   Sand and gravel aggregate, 8" thick..... R-1.04
   Lightweight aggregate, 8" thick......... R-2.18
Common brick, per inch.................. R-0.20
Face brick, per inch.................... R-0.11
Sand-and-gravel concrete, per inch....... R-0.08

GLASS

U-VALUES

	Glass Only (Winter)
Single-pane glass	1.16
Double-pane ⅝" insulating glass (¼" air space)	.58
Double-pane xı insulating glass	.55
Double-pane 1" insulating glass (½" air space)	.49
Double-pane xı insulating glass with combination (2" air space)	.35

Glass U-Values obtained from PPG and Cardinal Glass Company.

INSULATION

Fiberglass 2" thick..................... R-7.00
Fiberglass 3½" thick.................... R-11.00
Fiberglass 6" thick..................... R-19.00
Fiberglass 12" thick.................... R-38.00
Styrofoam Board ¾" thick................ R-4.05
Styrofoam Board 1" tongue & groove...... R-5.40

ROOFING

Asphalt shingles....................... R-0.44
Wood shingles, plain & plastic film faced ... R-0.94

SURFACE AIR FILMS

Inside, still air
Heat flow UP (through horizontal surface)
   Non-reflective...................... R-0.61
   Reflective.......................... R-1.32
Heat flow DOWN (through horizontal
   surface)
   Non-reflective...................... R-0.92
   Reflective.......................... R-4.55
Heat flow HORIZONTAL (through vertical
   surface)
   Non-reflective...................... R-0.68

Outside
Heat flow any direction, surface any position
   15 mph wind (winter)................. R-0.17
   7.5 mph wind (summer)............... R-0.25

EXAMPLE CALCULATIONS
(to determine the U value of an exterior wall)

Wall Construction	Insulated Wall Resistance
Outside surface (film), 15 mph wind.....	0.17
Wood bevel siding, lapped............	0.81
½" ins. bd. sheathing, reg. density......	1.32
3½" air space.....................	
R-11 insulation....................	11.00
½" gypsumboard...................	0.45
Inside surface (film)................	0.68
Totals.........................	14.43

For insulated wall, $U = \dfrac{1}{R} = \dfrac{1}{14.3} = 0.07$

TEMPERATURE CORRECTION FACTOR

Correction Factor is an ASHRAE standard to be applied for varying outdoor design temperatures. As follows:

If design temperature is:	−20	−10	0	+10	+20
Then correction factor is:	0.778	0.875	1.0	1.167	1.40

*Additional resistance values can be obtained from ASHRAE Handbook of Fundamentals published by the American Society of Heating, Refrigerating and Air-Conditioning Engineers.

Insulation values for common building materials. The method used to calculate U and R values for a wall structure can also be used for ceilings and floors. Temperature correction factors are used by designers of heating and cooling systems.
(Andersen Corp.)

GUIDE TO HARDWOOD FLOORING GRADES

Flooring is bundled by averaging the lengths. A bundle may include pieces from 6 inches under to 6 inches over the nominal length of the bundle. No piece is shorter than 9 inches. Quantity with length under 4 ft. held to stated percentage of total footage in any one shipment of item. 3/4 inch added to face length when measuring length of each piece.

UNFINISHED OAK FLOORING (Red & White Separated)

SELECT AND BETTER

CLEAR (Plain or Quarter Sawn)**
Best Appearance
Best grade, most uniform color, limited small character marks. Bundles 1¼ ft. and up. Average length 3¾ ft.

SELECT & BETTER (Special Order)
A Combination of Clear and Select grades

SELECT (Plain or Quarter Sawn)**
Excellent Appearance
Limited character marks, unlimited sound sap. Bundles 1¼ ft. and up. Average length 3¼ ft.

NO. 1 COMMON
Variegated Appearance
Light and dark colors; knots, flags, worm holes and other character marks allowed to provide a variegated appearance, after imperfections are filled and finished. Bundles 1¼ ft. and up. Average length 2¾ ft.

NO. 2 COMMON (Red & White may be mixed)
Rustic Appearance
All wood characteristics of species. A serviceable, economical floor after knot holes, worm holes, checks and other imperfections are filled and finished. Bundles 1¼ ft. and up. Average length 2¼ ft.

BEECH, BIRCH & HARD MAPLE

SECOND AND BETTER — THIRD AND BETTER

***FIRST GRADE WHITE HARD MAPLE** (Spec. Order)
Same as FIRST GRADE except face all bright sapwood.

***FIRST GRADE RED BEECH & BIRCH** (Spec. Order)
Same as FIRST GRADE except face all red heartwood.

FIRST GRADE
Best Appearance. Natural color variation, limited character marks, unlimited sap. Bdles. 2 ft. & up. 2 & 3 ft. bdles. up to 33% footage.

SECOND & BETTER GRADE
Excellent Appearance. A combination of FIRST and SECOND GRADES. Bdles. 2 ft. & up. 2 & 3 ft. Bdles. up to 40% footage. (NOTE: 5% 1¼ ft. bdles. allowed in SECOND & BETTER jointed flg. only.)

SECOND GRADE
Variegated Appearance. Varying sound wood characteristics of species. Bdles. 2 ft. & up. 2 & 3 ft. Bdles. up to 45% footage.

THIRD & BETTER GRADE
A combination of FIRST, SECOND and THIRD GRADES. Bundles 1¼ ft. and up. 1¼ ft. to 3 ft. bundles as produced up to 50% footage.

THIRD GRADE
Rustic Appearance
All wood characteristics of species. Serviceable, economical floor after filling. Bundles 1¼ ft. and up. 1¼ ft. to 3 ft. bundles as produced up to 65% footage.

PECAN FLOORING

***FIRST GRADE RED** (Spec. Order)
Same as FIRST GRADE except face all heartwood.

***FIRST GRADE WHITE** (Spec. Order)
Same as FIRST GRADE except face all bright sapwood.

FIRST GRADE
Excellent Appearance. Natural color variation, limited character marks, unlimited sap Bdles. 2 ft. & up. 2 & 3 ft. Bdles. up to 25% footage.

***SECOND GRADE RED**
(Special Order Only)
Same as SECOND GRADE except face all heartwood.

SECOND GRADE
Variegated Appearance
Varying sound wood characteristics of species. Bundles 1¼ ft. and up. 1¼ ft. to 3 ft. bundles as produced up to 40% footage.

THIRD GRADE
Rustic Appearance
All wood characteristics of species. A serviceable, economical floor after filling. Bundles 1¼ ft. and up. 1¼ ft. to 3 ft. bundles as produced up to 60% footage.

PREFINISHED OAK FLOORING (Red & White Separated-graded after finishing)

STANDARD AND BETTER — TAVERN AND BETTER

***PRIME**
(Special Order Only)
Excellent Appearance
Natural color variation, limited character marks, unlimited sap. Bundles 1¼ ft. and up. Average length 3½ ft.

STANDARD & BETTER GRADE
Combination of STANDARD and PRIME. Bundles 1¼ ft. and up. Average length 3 ft.

STANDARD GRADE
Variegated Appearance
Varying sound wood characteristics of species. A sound floor. Bundles 1¼ ft. and up. Average length 2¾ ft.

***TAVERN & BETTER GRADE**
(Special Order Only)
Combination of PRIME, STANDARD and TAVERN. All wood characteristics of species. Bundles 1¼ ft. and up. Average length 3 ft.

TAVERN GRADE
Rustic Appearance
All wood characteristics of species. A serviceable, economical floor. Bundles 1¼ ft. and up. Average length 2¼ ft.

*1¼ FT. SHORTS (Red & White may be mixed)
Unique Variegated Appearance. Lengths 9 inches to 18 inches. Bundles average nominal 1¼ ft. Production limited.

*NO. 1 COMMON & BETTER SHORTS
A combination grade, CLEAR, SELECT, & NO. 1 COMMON 9 inches to 18 inches.

*NO. 2 COMMON SHORTS
Same as No. 2 COMMON, except length 9 inches to 18 inches.

** Quarter Sawn — Special Order Only
*Check with supplier for grade and species available.
†NESTED FLOORING: Random length tongued and grooved, end-matched flooring is bundled end to end continuously to form 8 ft. long (nominal) bundles. Regular grade requirements apply.
‡NESTED FLOORING: If put up in 8 ft. nested bundles, 9 to 18 inch pieces will be admitted in ¾" x 2¼" as follows in the species of Beech, Birch & Hard Maple: FIRST GRADE, 4 pcs. per bundle; SECOND GRADE, 8 pcs.; THIRD GRADE, as develops. Average lengths: FIRST GRADE, 42 inches; SECOND GRADE, 33 inches; THIRD GRADE, 30 inches.

PREFINISHED BEECH & PECAN FLOORING
*TAVERN & BETTER GRADE (Special Order Only)
Combination of PRIME, STANDARD and TAVERN. All wood characteristics of species. Bundles 1¼ ft. and up. Average length 3 ft.

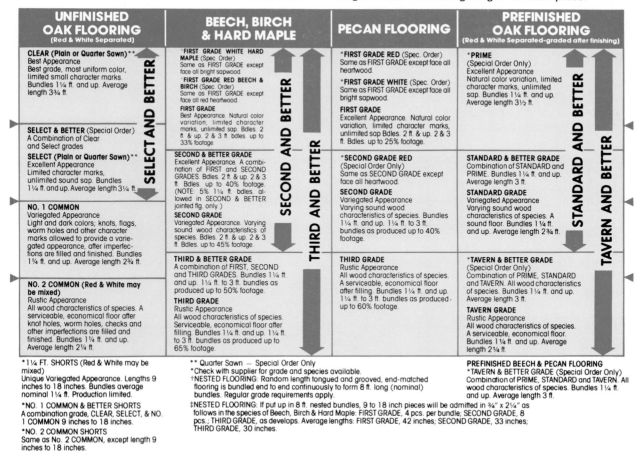

STANDARD SIZES, COUNTS, & WEIGHTS

''Nominal'' is the size designation used by the *trade*, but it is not always the actual size. Sometimes the actual thickness of hardwood flooring is 1/32 inch less than the so-called nominal size.

''Actual'' is the *mill* size for thickness and face width, excluding tongue width. ''Counted'' size determines the board feet in a shipment. Pieces less than 1 inch in thickness are considered to be 1 in.

Nominal	Actual	Counted	Weights M Ft.
TONGUE AND GROOVE-END MATCHED			
** ¾x3¼ in.	¾x3¼ in.	1x4 in.	2210 lbs.
¾x2¼ in.	¾x2¼ in.	1x3 in.	2020 lbs.
¾x2 in.	¾x2 in.	1x2¾ in.	1920 lbs.
¾x1½ in.	¾x1½ in.	1x2¼ in.	1820 lbs.
** ⅜x2 in.	11/32x2 in.	1x2½ in.	1000 lbs.
** ⅜x1½ in.	11/32x1½ in.	1x2 in.	1000 lbs.
** ½x2 in.	15/32x2 in.	1x2½ in.	1350 lbs.
** ½x1½ in.	15/32x1½ in.	1x2 in.	1300 lbs.
SQUARE EDGE			
** 5/16x2 in.	5/16x2 in.	face count	1200 lbs.
** 5/16x1½ in.	5/16x1½ in.	face count	1200 lbs.

Nominal	Actual	Counted	Weights M Ft.
SPECIAL THICKNESSES (T and G, End Matched)			
** 33/32x3¼ in.	33/32x3¼ in.	5/4x4 in.	2400 lbs.
** 33/32x2¼ in.	33/32x2¼ in.	5/4x3 in.	2250 lbs.
** 33/32x2 in.	33/32x2 in.	5/4x2¾ in.	2250 lbs.
JOINTED FLOORING — i.e., SQUARE EDGE			
** ¾x2½ in.	¾x2½ in.	1x3¼ in.	2160 lbs.
** ¾x3¼ in.	¾x3¼ in.	1x4 in.	2300 lbs.
** ¾x3½ in.	¾x3½ in.	1x4¼ in.	2400 lbs.
** 33/32x2½ in.	33/32x2½ in.	5/4x3¼ in.	2500 lbs.
** 33/32x3½ in.	33/32x3½ in.	5/4x4¼ in.	2600 lbs.

**Special Order Only

NAIL SCHEDULE

Tongue And Groove Flooring Must Be Blind Nailed		
¾x1½, 2¼ & 3¼ in.	2 in. machine driven fasteners, 7d or 8d screw or cut nail.	10-12 in. apart*
¾x3 in. to 8 in.** Plank	2 in. machine driven fasteners, 7d or 8d screw or cut nail.	8" apart into and between joists.

*If subfloor is ½ inch plywood, fasten into each joist, with additional fastening between.
**Plank Flooring over 4" wide must be installed over a subfloor.

Following Flooring Must Be Laid On A Subfloor		
½x1½ & 2 in.	1½ in. machine driven fastener, 5d screw, cut steel or wire casing nail.	10 in. apart
⅜x1½ & 2 in.	1¼ in. machine driven fastener, or 4d bright wire casing nail.	8 in. apart
Square-Edge Flooring As Follows, Face-Nailed—Through Top Face		
5/16x1½ & 2 in.	1 inch 15 gauge fully barbed flooring brad. 2 nails every 7 inches.	
5/16x1⅓ in.	1 inch 15 gauge fully barbed flooring brad. 1 nail every 5 inches on alternate sides of strip.	

Tables showing hardwood flooring grades, sizes, counts, and weights make flooring selection easier. Follow the recommended nailing schedule for best results. (National Oak Flooring Manufacturer's Assoc.)

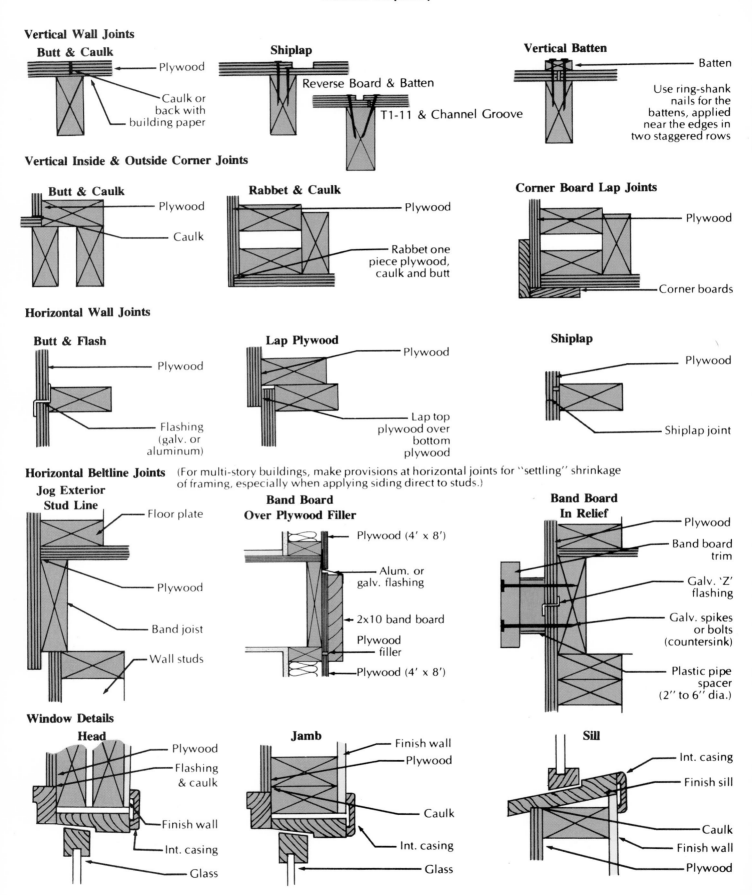

Approved joint details for plywood siding. (American Plywood Assoc.)

Technical Information

— One-hour assembly —
resilient channel ceiling system

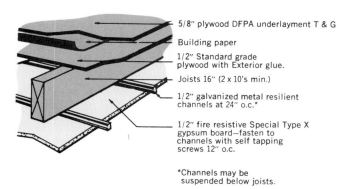

5/8" plywood DFPA underlayment T & G

Building paper

1/2" Standard grade plywood with Exterior glue.

Joists 16" (2 x 10's min.)

1/2" galvanized metal resilient channels at 24" o.c.*

1/2" fire resistive Special Type X gypsum board—fasten to channels with self tapping screws 12" o.c.

*Channels may be suspended below joists.

— One-hour assembly —
T-bar grid ceiling system

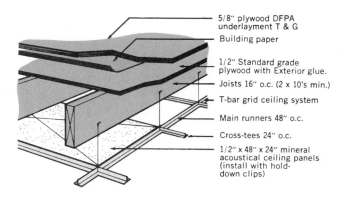

5/8" plywood DFPA underlayment T & G

Building paper

1/2" Standard grade plywood with Exterior glue.

Joists 16" o.c. (2 x 10's min.)

T-bar grid ceiling system

Main runners 48" o.c.

Cross-tees 24" o.c.

1/2" x 48" x 24" mineral acoustical ceiling panels (install with hold-down clips)

— One-hour interior shear wall construction

1/2" fire resistive special Type X gypsum board*

2 x 4 studs @ 16" o.c.

3/8" plywood shear panels

*Regular 1/2" gypsum board may be used when mineral wool or glass fiber batts are used in wall cavity.

Insulation batts in wall cavity also used for sound transmission control

— One-hour exterior wall construction

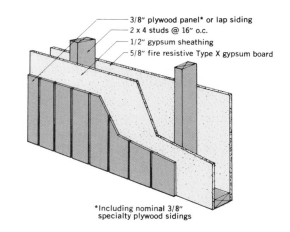

3/8" plywood panel* or lap siding

2 x 4 studs @ 16" o.c.

1/2" gypsum sheathing

5/8" fire resistive Type X gypsum board

*Including nominal 3/8" specialty plywood sidings

— Treated stressed skin panel construction

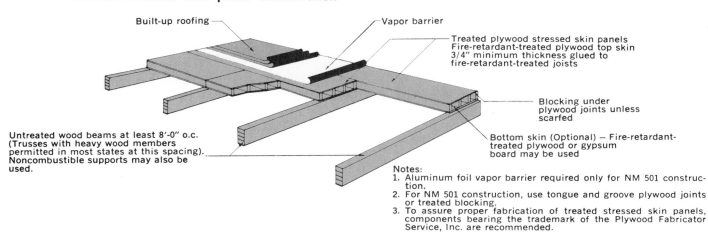

Built-up roofing

Vapor barrier

Treated plywood stressed skin panels Fire-retardant-treated plywood top skin 3/4" minimum thickness glued to fire-retardant-treated joists

Blocking under plywood joints unless scarfed

Untreated wood beams at least 8'-0" o.c. (Trusses with heavy wood members permitted in most states at this spacing). Noncombustible supports may also be used.

Bottom skin (Optional) — Fire-retardant-treated plywood or gypsum board may be used

Notes:
1. Aluminum foil vapor barrier required only for NM 501 construction.
2. For NM 501 construction, use tongue and groove plywood joints or treated blocking.
3. To assure proper fabrication of treated stressed skin panels, components bearing the trademark of the Plywood Fabricator Service, Inc. are recommended.

Fire resistant construction. All assemblies shown provide a one-hour rating.   (American Plywood Assoc.)

City and State	Yearly Total	City and State	Yearly Total
Birmingham, Alabama	2551	Omaha, Nebraska	6612
Anchorage, Alaska	10864	Las Vegas, Nevada	2709
Phoenix, Arizona	1765	Reno, Nevada	6332
Tucson, Arizona	1800	Concord, New Hampshire	7383
Little Rock, Arkansas	3219	Trenton, New Jersey	4980
Los Angeles, California	2061	Albuquerque, New Mexico	4348
Sacramento, California	2502	Albany, New York	6875
San Diego, California	1458	Buffalo, New York	7062
San Francisco, California	3015	New York, New York	4871
Denver, Colorado	6283	Raleigh, North Carolina	3393
Pueblo, Colorado	5462	Bismarck, North Dakota	8851
Hartford, Connecticut	6235	Cincinnati, Ohio	4410
Wilmington, Delaware	4930	Cleveland, Ohio	6351
Washington, D.C.	4224	Oklahoma City, Oklahoma	3725
Jacksonville, Florida	1239	Portland, Oregon	4635
Miami, Florida	214	Philadelphia, Pennsylvania	5144
Orlando, Florida	766	Pittsburgh, Pennsylvania	5987
Atlanta, Georgia	2961	Providence, Rhode Island	5954
Savannah, Georgia	1819	Charleston, South Carolina	2033
Honolulu, Hawaii	0	Columbia, South Carolina	2484
Boise, Idaho	5809	Sioux Falls, South Dakota	7839
Chicago, Illinois	5882	Knoxville, Tennessee	3494
Springfield, Illinois	5429	Memphis, Tennessee	3232
Indianapolis, Indiana	5699	Nashville, Tennessee	3578
Des Moines, Iowa	6588	Dallas, Texas	2363
Topeka, Kansas	5182	El Paso, Texas	2700
Louisville, Kentucky	4660	Houston, Texas	1396
Baton Rouge, Louisiana	1560	Salt Lake City, Utah	6052
New Orleans, Louisiana	1385	Burlington, Vermont	8269
Portland, Maine	7511	Richmond, Virginia	3865
Baltimore, Maryland	4654	Seattle-Tacoma, Washington	5145
Boston, Massachusetts	5634	Charleston, West Virginia	4476
Detroit, Michigan	6232	Milwaukee, Wisconsin	7635
Minneapolis, Minnesota	8382	Cheyenne, Wyoming	7381
Jackson, Mississippi	2239	Montreal, Canada	7899
Kansas City, Missouri	4711	Quebec, Canada	8937
St. Louis, Missouri	5000	Toronto, Canada	6827
Great Falls, Montana	7750	Vancouver, Canada	5515

Yearly Fahrenheit heating degree days are based on the temperature difference from 65 deg. F. Data is taken from a publication of the United States Weather Bureau. Check figures for your locale since the degree days may vary 20% within a state. (Masonry Industry Committee)

Technical Information

Better construction methods and materials used by builders today produce houses that are tighter, more draft-free than those built a half century ago.

In such snug shelters, moisture created inside the house often saturates insulation and causes paint to peel off exterior walls.

The use of a vapor barrier on the warm side of walls and ceilings will minimize this condensation, but adequate ventilation is needed to remove the moisture from the house.

FHA requires that attics have a total net free ventilating area not less than 1/150 of the square foot area, except that a ratio of 1/300 may be provided if:

(a) a vapor barrier having a transmission rate not exceeding one perm is installed on the warm side of the ceiling, or

(b) at least 50% of the required vent area is provided by ventilators located in the upper portion of the space to be ventilated, with the balance of the required ventilation provided by eave or cornice vents.

There are many types of screened ventilators available, including the handy miniature vents illustrated below, which can be used for problem areas or to supplement larger vents.

INSTALLING MINIATURE VENTS

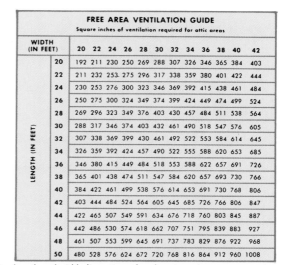

FREE AREA VENTILATION GUIDE
Square inches of ventilation required for attic areas

LENGTH (IN FEET) \ WIDTH (IN FEET)	20	22	24	26	28	30	32	34	36	38	40	42
20	192	211	230	250	269	288	307	326	346	365	384	403
22	211	232	253	275	296	317	338	359	380	401	422	444
24	230	253	276	300	323	346	369	392	415	438	461	484
26	250	275	300	324	349	374	399	424	449	474	499	524
28	269	296	323	349	376	403	430	457	484	511	538	564
30	288	317	346	374	403	432	461	490	518	547	576	605
32	307	338	369	399	430	461	492	522	553	584	614	645
34	326	359	392	424	457	490	522	555	588	620	653	685
36	346	380	415	449	484	518	553	588	622	657	691	726
38	365	401	438	474	511	547	584	620	657	693	730	766
40	384	422	461	499	538	576	614	653	691	730	768	806
42	403	444	484	524	564	605	645	685	726	766	806	847
44	422	465	507	549	591	634	676	718	760	803	845	887
46	442	486	530	574	618	662	707	751	795	839	883	927
48	461	507	553	599	645	691	737	783	829	876	922	968
50	480	528	576	624	672	720	768	816	864	912	960	1008

Using length and width dimensions of each rectangular or square attic space, find one dimension on vertical column, the other dimension on horizontal column. These will intersect at the number of square inches of ventilation required to provide 1/300th.

1-INCH
Install in paneling used in basement rooms. Can be painted over to match panel finish.

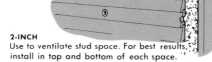

2-INCH
Use to ventilate stud space. For best results, install in top and bottom of each space.

2½-INCH
Made especially to plug the 2½-in. hole cut to blow insulation between studs.

3-INCH
For ventilating rafter space in flat-roof buildings, or other jobs requiring fairly large free area.

4-INCH
For hard-to-reach spots needing large ventilating area. Large enough for venting soffits.

1. Drill or cut a hole the same size as the ventilator.

2. Insert the ventilator. Tension ridges hold ventilator in place . . . no nails or screws are needed.

3. Tap into place with a hammer, using a wood block to protect the margin. The louvers are recessed so there's no danger of damage during installation.

RIDGE VENT provides 18 sq. in. of net free area per lineal foot. Installed quickly over a 1½" gap in the sheathing at the ridge.

ROOF VENTS fit over openings cut between rafters to pull hot air out of attic.

TRIANGLE VENT fits snugly under the roof gable to provide large vent areas at the highest point of the gable end.

ATTIC VENTS are installed in the gable end, usually above the level of the probable level of a future ceiling should the attic be finished later.

SOFFIT VENT replaces a portion of the soffit material to provide continuous ventilation along its entire length.

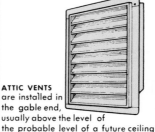
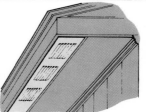

BRICK VENTS are exactly the size of a brick, can be laid in any brick wall. Screened to meet FHA specs.

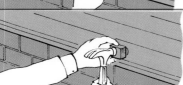

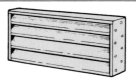

CEMENT BLOCK VENTS are designed to be mortared into the same space as an 8x16" cement block. At least four should be used to vent crawl space.

Modern ventilators and vent applications.

NAILS

COMMON APPLICATIONS

Joining	Size & Type	Placement
Wall Framing		
Top plate	8d common 16d common	
Header	8d common 16d common	
Header to joist	16d common	
Studs	8d common 16d common	
Wall Sheathing		
Boards	8d common	6" o.c.
Plywood 5/16", ⅜", ½"	6d common	6" o.c.
Plywood (⅝", ¾")	8d common	6" o.c.
Fiberboard	1¾" galv. roofing nail 8d galv. common nail	6" o.c. 6" o.c.
Foamboard	Cap nail, length sufficient for penetration of ½" into framing	12" o.c.
Gypsum	1¾" galv. roofing nail 8d galv. common nail	6" o.c. 6" o.c.
Subflooring	8d common	10"-12" o.c.
Underlayment	(1¼" x 14 ga. annular underlayment nail)	6" o.c. edges 12" o.c. face
Roof Framing		
Rafters, beveled or notched	12d common	
Rafter to joist	16d common	
Joist to rafter and stud	10d common	
Ridge beam	8d & 16d common	
Roof Sheathing		
Boards	8d common	
Plywood (5/16", ⅜", ½")	6d common	12" o.c. and 6" o.c. edges
Plywood (⅝", ¾")	8d common	12" o.c. and 6" o.c. edges

*Aluminum nails are recommended for maximum protection from staining.

Joining	Size & Type	Placement
Roofing, Asphalt		
New construction shingles and felt	⅞" through 1½" galv. roofing	**4 per shingle**
Re-roofing application shingles and felt	1¾" or 2" galv. roofing	**4 per shingle**
Roof deck/ Insulation	Thickness of insulation plus 1" insulation roof deck nail	
Roofing, Wood Shingles		
New construction	3d-4d galv. shingle	**2-3 per shingle**
Re-roofing application	5d-6d galv. shingle	**2-3 per shingle**
Soffit	6d-8d galv. common	**12" o.c. max.**
Siding*		
Bevel and lap	Aluminum nails are recommended for optimum performance	**Consult siding manufacturer's application instructions**
Drop and shiplap		
Plywood		
Hardboard	Galvanized hardboard siding nail Galvanized box nail	**Consult siding manufacturer's application instructions**
Doors, Windows, Mouldings, Furring		
Wood strip to masonry	Nail length is determined by thickness of siding and sheathing. Nails should penetrate at least 1½" into solid wood framing.	
Wood strip to stud or joist		
Paneling		
Wood	4d-8d casing-finishing	**24" o.c.**
Hardboard	2" x 16 ga. annular	**8" o.c.**
Plywood	3d casing-finishing	**8" o.c.**
Gypsum	1¼" annular drywall	**6" o.c.**
Lathing	4d common blued	**4" o.c.**
Exterior Projects:		
Decks, patios, etc.	8d-16d hot dipped galvanized common	

NOTE: Usage may vary somewhat due to regional differences and preferences.

DRYWALL SCREWS

Description	No.	Length	Applications
Bugle Phillips 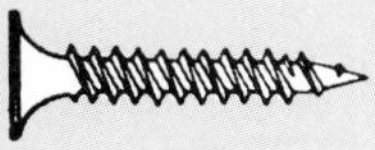	1E 2E 3E 4E 5R 6R 7R 8R	6x1 6x1⅛ 6x1¼ 6x1⅝ 6x2 6x2¼ 8x2½ 8x3	For attaching drywall to metal studs from 25 ga. through 20 ga.
Coarse Thread 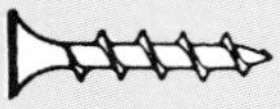	1C 2C 3C 4C 5C 6C	6x1 6x1⅛ 6x1¼ 6x1⅝ 6x2 6x2¼	For attaching drywall to 25 ga. metal studs, and attaching drywall to wood studs
Pan Framing 	19	6x7/16	For attaching stud to track up to 20 ga.

Description	No.	Length	Applications
HWH Framing 	21 22 35	6x7/16 8x9/16 10x¾	For attaching stud to track up to 20 ga. where hex head is desired
K-Lath 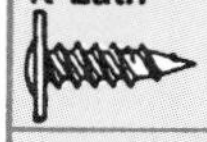	28	8x9/16	For attaching wire lath, K-lath to 20 ga. studs
Laminating 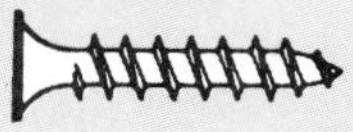	8	10x1½	Type G laminating screw for attaching gypsum to gypsum, a temporary fastener
Trim Head 	9 10	6x1⅝ 6x2¼	Trim head screw for attaching wood trim and base to 25 ga. studs

Nail chart shows common applications, size, type, and placement. (Georgia-Pacific) Drywall screw chart gives description, number, length, and applications. (Compass International)

Technical Information

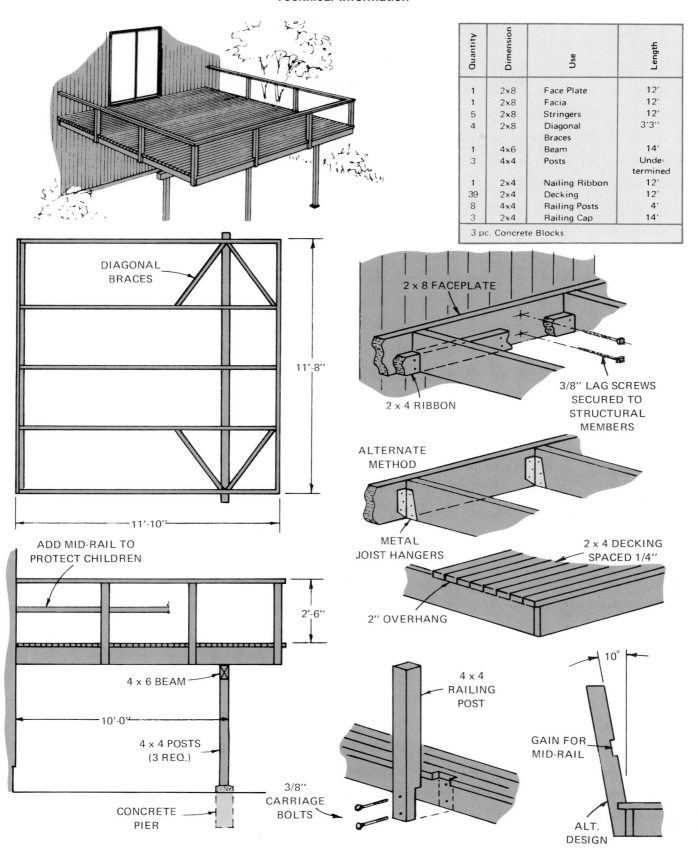

Quantity	Dimension	Use	Length
1	2x8	Face Plate	12'
1	2x8	Facia	12'
5	2x8	Stringers	12'
4	2x8	Diagonal Braces	3'3"
1	4x6	Beam	14'
3	4x4	Posts	Undetermined
1	2x4	Nailing Ribbon	12'
39	2x4	Decking	12'
8	4x4	Railing Posts	4'
3	2x4	Railing Cap	14'
3 pc. Concrete Blocks			

DIAGONAL BRACES

11'-8"

11'-10"

2 x 8 FACEPLATE

2 x 4 RIBBON

3/8" LAG SCREWS SECURED TO STRUCTURAL MEMBERS

ALTERNATE METHOD

METAL JOIST HANGERS

2 x 4 DECKING SPACED 1/4"

2" OVERHANG

ADD MID-RAIL TO PROTECT CHILDREN

2'-6"

4 x 6 BEAM

10'-0"

4 x 4 POSTS (3 REQ.)

CONCRETE PIER

3/8" CARRIAGE BOLTS

4 x 4 RAILING POST

10°

GAIN FOR MID-RAIL

ALT. DESIGN

General construction details for sun deck. Use weatherproof metal fasteners and pressure treated wood. The American Wood Preservers Bureau (AWPB) provides standards and technical requirements for pressure treating.

PARTITIONS

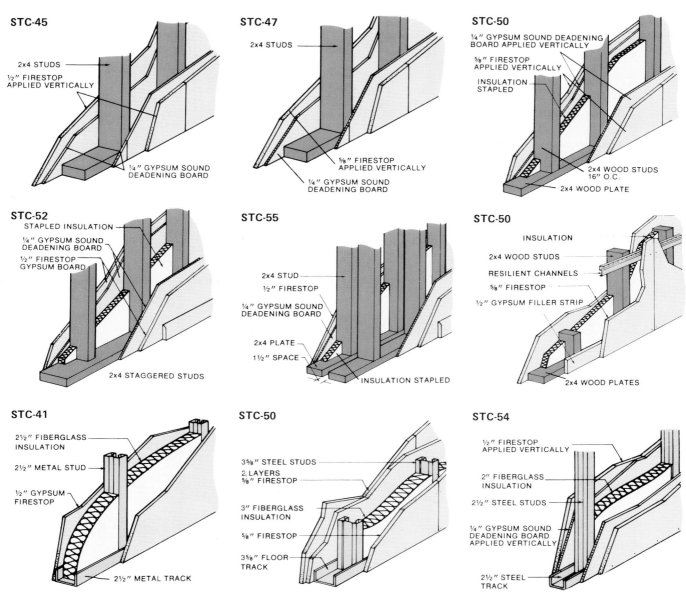

STC-45
- 2x4 STUDS
- ½″ FIRESTOP APPLIED VERTICALLY
- ¼″ GYPSUM SOUND DEADENING BOARD

STC-47
- 2x4 STUDS
- ⅝″ FIRESTOP APPLIED VERTICALLY
- ¼″ GYPSUM SOUND DEADENING BOARD

STC-50
- ¼″ GYPSUM SOUND DEADENING BOARD APPLIED VERTICALLY
- ⅝″ FIRESTOP APPLIED VERTICALLY
- INSULATION STAPLED
- 2x4 WOOD STUDS 16″ O.C.
- 2x4 WOOD PLATE

STC-52
- STAPLED INSULATION
- ¼″ GYPSUM SOUND DEADENING BOARD
- ½″ FIRESTOP GYPSUM BOARD
- 2x4 STAGGERED STUDS

STC-55
- 2x4 STUD
- ½″ FIRESTOP
- ¼″ GYPSUM SOUND DEADENING BOARD
- 2x4 PLATE
- 1½″ SPACE
- INSULATION STAPLED

STC-50
- INSULATION
- 2x4 WOOD STUDS
- RESILIENT CHANNELS
- ⅝″ FIRESTOP
- ½″ GYPSUM FILLER STRIP
- 2x4 WOOD PLATES

STC-41
- 2½″ FIBERGLASS INSULATION
- 2½″ METAL STUD
- ½″ GYPSUM FIRESTOP
- 2½″ METAL TRACK

STC-50
- 3⅝″ STEEL STUDS
- 2 LAYERS ⅝″ FIRESTOP
- 3″ FIBERGLASS INSULATION
- ⅝″ FIRESTOP
- 3⅝″ FLOOR TRACK

STC-54
- ½″ FIRESTOP APPLIED VERTICALLY
- 2″ FIBERGLASS INSULATION
- 2½″ STEEL STUDS
- ¼″ GYPSUM SOUND DEADENING BOARD APPLIED VERTICALLY
- 2½″ STEEL TRACK

FLOORS

STC-41
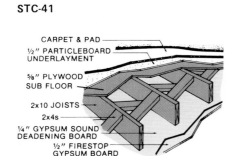
- CARPET & PAD
- ½″ PARTICLEBOARD UNDERLAYMENT
- ⅝″ PLYWOOD SUB FLOOR
- 2x10 JOISTS
- 2x4s
- ¼″ GYPSUM SOUND DEADENING BOARD
- ½″ FIRESTOP GYPSUM BOARD

STC-47
- 1″x4″ TONGUE & GROOVE FINISHED FLOORING
- RESIN BUILDING PAPER
- 1″x6″ TONGUE & GROOVE SUBFLOORING
- ½″ FIRESTOP GYPSUM BOARD
- 2″x10″ WOOD JOISTS
- RESILIENT METAL FURRING CHANNELS

STC-53
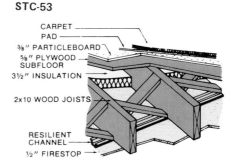
- CARPET
- PAD
- ⅜″ PARTICLEBOARD
- ⅝″ PLYWOOD SUBFLOOR
- 3½″ INSULATION
- 2x10 WOOD JOISTS
- RESILIENT CHANNEL
- ½″ FIRESTOP

How to build partition and floor structures with high STC (sound transmission class) ratings. Ratings are based on sound tests conducted according to ASTM-E90 (Georgia-Pacific)

ACKNOWLEDGMENTS

The author wishes to thank the individuals and organizations listed below for the valuable information, photographs, and other illustrations they so willingly provided.

Abitibi-Price Corporation, Troy, MI
Acoustical and Board Products Association, Palatine, IL
Agricultural Extension Service, University of Minnesota, Minneapolis, MN
Alcoa Building Products, Inc., Pittsburg, PA
Aluminum Siding Association, Chicago, IL
American Institute of Timber Construction, Englewood, CO
American Plywood Association, Tacoma, WA
American Standard, Inc., Chicago, IL
American Wood Preservers Bureau, McLean, VA
Amerock Corporation, Rockford, IL
Andersen Corporation, Bayport, MN
Architectural Woodwork Institute, Arlington, VA
Armstrong World Industries, Incorporated, Lancaster, PA
Asphalt Roofing Manufacturers Association, New York, NY
Barnes Builders Supply, Cedar Falls, IA
Beecham Home Improvement Products, Dayton, OH
Berger Scientific Supplies, Boston, MA
Binks Mfg. Company, Franklin Park, IL
Bird and Son, East Walpole, MA
Black & Decker Mfg. Company, Towson, MD
Blandin Wood Products Company, Grand Rapids, MI
Boise-Cascade Corporation, Boise, ID
Borden, Incorporated, New York, NY
Robert Bosch Power Tool Corporation, New Bern, NC
Bostitch, Incorporated, East Greenwich, RI
Brammer Mfg. Company, Chicago, IL
Building Dept., City of Cedar Falls, Cedar Falls, IA
The Burke Company, San Mateo, CA
Cardinal Industries Incorporated, Columbus, OH
C-E Morgan Building Products, Oshkosh, WI
Century Wood Products, Kansas City, MO
CertainTeed Corporation, Valley Forge, PA
Dennis C. Christensen, Contractor, Cedar Falls, IA
Citation Homes, Spirit Lake, IA
Colonial Stair and Woodwork Company, Jeffersonville, OH
Columns, Incorporated, Pearland, TX
Conestoga Wood Specialties, Incorporated, East Earl, PA
Construction Craftsman Magazine, Washington, DC
Crown Aluminum Industries Corporation, Pittsburg, PA
Dexter Industries, Incorporated, Deer Park, NY
Dexter Lock, Grand Rapids, MI
Dickinson Homes, Incorporated, Kingsford, MI
District 1 Technical Institute, Eau Claire, WI
Duo Fast Corporation, Franklin Park, IL
Eaton Yale & Towne Incorporated, White Plains, NY
Ekco Building Products Company, Canton, OH
Evans Products Company, Portland, OR
Fine Hardwoods, American Walnut Association, Chicago, IL
Fir and Hemlock Door Association, Portland, OR
Flintkote Company, New York, NY
Foley Manufacturing Company, Minneapolis, MN
Follansbee Steel Corporation, Follansbee, WV
Forest Products Laboratory, Madison, WI
Forestry Suppliers, Incorporated, Jackson, MS
Formica Corporation, Cincinnati, OH
Gamble Brothers. Incorporated, Louisville, KY
Gang-Nail Systems, Incorporated, Miami, FL
Garlinghouse Company, Incorporated, Topeka, KS
Gee Lumber Company, South Holland, IL
General Products Company, Incorporated, Fredericksburg, VA
Georgia-Pacific Corporation, Portland, OR
Goldblatt Tool Company, Kansas City, KS
Gold Bond Building Products, Charlotte, NC
Gory Associated Industries, Incorporated, North Miami, FL
Greenlee Brothers & Company, Rockford, IL
Grosse Steel Company, Incorporated, Cedar Falls, IA
Gypsum Association, Evanston, IL
Haas Cabinet Company, Incorporated, Sellersburg, IN
Handy Andy Home Improvement Center, Chicago Heights, IL
Home Planners, Incorporated, Detroit, MI
Homewood Scavenger Services, Inc., East Hazelcrest, IL
HUD, Washington, DC
Ideal Company, Waco, TX
Independent Nail and Packing Company, Bridgewater, MA
International Conference of Building Officials, Whittier, CA
International Paper Company, Longview, WA
Iowa Energy Policy Council, Des Moines, IA
I-XL Furniture Company, Goshen, IN
Jordan Millwork Company, Sioux Falls, SD
Kentron Div., North American Reiss, Belle Mead, NJ
Kitchen Kompact, Incorporated, Jeffersonville, IN
Libbey-Owens, Ford Glass Company, Toledo, OH
Louisiana-Pacific Corporation, Portland, OR
Majestic Company, Incorporated, Huntington, IN
Malm Fireplaces, Incorporated, Santa Rosa, CA
Manville Building Materials Corporation, Denver, CO
Maple Flooring Manufacturers Association, Oshkosh, WI

Marshfield Homes, Incorporated, Marshfield, WI
Masonite Corporation, Chicago, IL
Master Builders of Iowa, Des Moines, IA
Memphis Hardwood Flooring Company, Memphis, TN
Millers Falls Company, Greenfield, MA
Mineral Fiber Products Bureau, New York, NY
Mohawk Flush Doors, Incorporated, South Bend, IN
National Association of Home Builders, Washington, DC
National Forest Products Association, Washington, DC
National Homes Corporation, Lafayette, IN
National Lock Company, Rockford, IL
National Oak Flooring Manufacturers Association, Memphis, TN
National Solar Heating and Cooling Center, Rockville, MD
National Wood, Window, and Door Association, Des Plaines, IL
Nicholson File Company, Providence, RI
Nichols Wire and Aluminum Company, Davenport, IA
Oak Flooring Institute, Memphis, TN
Omni Products, Addison, IL
Orem Research, Hinsdale, IL
Osmose, Buffalo, NY
Overhead Door Corporation, Dallas, TX
Owens-Corning Fiberglas Corporation, Toledo, OH
The Panel Clip Company, Farmington, MI
Paslode Corporation, A Signode Company, Lincolnshire, IL
Pass & Seymour, Incorporated, Syracuse, NY
Patent Scaffolding Company, Long Island City, NY
Frank Paxton Lumber Company, Des Moines, IA
Pease Industries, Incorporated, Hamilton, OH
Pemko Manufacturing Company, Emeryville, CA
Perlite Institute, Incorporated, New York, NY
Pittsburgh Plate Glass Company, Pittsburgh, PA
Porter-Cable Corporation, Jackson, TN
H. K. Porter Company, Incorporated, Pittsburgh, PA
Portland Cement Association, Skokie, IL
PPG Industries, Incorporated, Pittsburgh, PA
Preway Incorporated, Wisconsin Rapids, WI
Proctor Products Company, Kirkland, WA
Red Cedar Shingle and Handsplit Shake Bureau, Seattle, WA
Redman Industries Incorporated, Dallas, TX
Reinke Shakes, Hebron, NB
Richard Mehmen, Contractor, Cedar Falls, IA
Richmond Screw Anchor Company, Brooklyn, NY
Riviera Cabinets, Incorporated, Chesapeake, VA
Robbins Incorporated, Cincinnati, OH
Rock Island Millwork, Waterloo, IA
Rockwool Industries, Incorporated, Denver, CO
Rolscreen Company, Pella, IA
Ronthor, Plastics Div., U.S. Mfg. Corporation, New York, NY
Senco Products, Incorporation, Cincinnati, OH
Shakertown Corporation, Cleveland, OH
Simpson Timber Company, Seattle, WA
Skil Corporation, Chicago, IL
Skyline Corporation, Elkhart, IN
Southern Forest Products Association, New Orleans, LA
Speed Cut, Incorporated, Corvallis, OR
Spotnails, Incorporated, Rolling Meadows, IL
Stanley Door Systems, Farmington, CT
Stanley Works, New Britain, CT
Stenson-Warm-Grimes-Port Architects, Incorporated, Waterloo, IA.
Superior Fireplace Company, Fullerton, CA
Supradur Mfg. Corporation, New York, NY
Symons Corporation, Des Plaines, IL
TECO, Chevy Chase, MD
Therma-Tru, Toledo, OH
John S. Tilley Ladders Company, Incorporated, Davenport, IA
Timber Engineering Company, Washington, DC
Timberpeg, Claremont, NH
Harold Truax, Mason, Waterloo, IA
TrusWal Systems, Incorporated, Troy, MI
United Brotherhood of Carpenters and Joiners of America, Washington, DC
United States Gypsum Company, Chicago, IL
United States Steel Corporation, Pittsburgh, PA
Universal Form Clamp Company, Chicago, IL
Upson Company, Lockport, NY
Vermiculite Institute, Minneapolis, MN
Village of Flossmoor, Flossmoor, IL
Visador Company, Jasper, TX
Wallace Murray Corporation, Nampa, ID
Waterloo Iowa Daily Courier, Waterloo, IA
Wausau Homes Incorporated, Wausau, WI
Western Wood Mldg. & Mlwk. Producers Association, Portland, OR
Western Wood Products Association, Portland, OR
Weyerhaeuser Company, Tacoma, WA
Whirlpool Corporation, Benton Harbor, MI
David White Instruments, Div. of Realist, Inc., Menomonee Falls, WI
Wood Conversion Company, St. Paul, MN
Daniel Woodhead Company, Northbrook, IL

GLOSSARY

ACOUSTICAL MATERIALS: Types of tile, plaster, and other materials which absorb sound waves. Generally applied to interior wall surfaces to reduce reverberation or reflection of the waves.

ADHESIVE: A substance capable of holding material together by surface attachment. A general term that includes glue, cement, mastic, and paste.

AGGREGATE: Materials such as sand, rock, and gravel used to make concrete.

AIR CONDITIONING: Control of temperature, humidity, movement, and purity of air in buildings.

AIR DRIED: Wood seasoned by exposure to the atmosphere, in the open or under cover, without artificial heat.

AIR-ENTRAINING AGENTS: Chemical additives which improve the workability and freeze-thaw durability of mortar as it ages.

ALTERATION: Any change in the facilities, structural parts, or mechanical equipment of a building which does not increase the cubic content.

ANCHOR BOLTS: Bolts embedded in concrete used to hold structural members in place.

ANCHOR STRAPS: Strap fastener which is embedded in concrete or masonry walls to hold sills in place.

ANNUAL RINGS: Rings or layers of wood which represent one growth period of a tree. In cross section the rings may indicate the age of the tree.

APRON: A piece of horizontal trim applied against the wall immediately below the stool. Conceals rough edge of plaster.

AREAWAY: An open space around a basement window or doorway. Provides light, ventilation, and access.

ASPHALT: A residue from evaporated petroleum. It is insoluble in water but is soluble in gasoline and melts when heated. Used for waterproofing roof coverings, exterior wall coverings, and flooring tile.

ASTRAGAL: An interior molding attached to one of a pair of doors or window sash in order to prevent swinging through; also used with sliding doors to insure tighter fitting where doors meet.

BACKBAND: A narrow rabbeted molding applied to the outside corner and edge of interior window and door casing to create a "heavy trim" appearance.

BACKFILL: The replacement of earth around foundations after excavating.

BACKING BOARD (also, backer board): In a two-layer drywall system, the base panel of gypsum drywall. It uses gray liner paper as facing and is not suitable as a top surface.

BALLOON FRAMING: A type of building construction with upright studs which extend from the foundation sill to the rafter plate. Its use is decreasing in favor of platform framing and other construction styles.

BALUSTER: Turned and/or square spindle-like vertical stair member which supports the stair rail.

BALUSTRADE: A railing consisting of a series of balusters resting on a base, usually the treads, which supports a continuous stair or hand rail.

BASEMENT: The part of a house that is partly or wholly in the ground.

BASE SHOE: Small narrow molding used around the perimeter of a room where the baseboard meets the finish floor.

BATTEN: A strip of wood placed across a surface to cover joints.

BATTER: The slope, or inclination from the vertical, of a wall or other structure or portion of a structure.

BATTER BOARD: A temporary framework used to assist in locating corners when laying out a foundation.

BAY: One of the intervals or spaces into which a building plan is divided by columns, piers, or division walls.

BAY WINDOW: A rectangular, curved, or polygonal window, or group of windows usually supported on a foundation extending beyond the main wall of a building.

BEAM: A principal structural member used between posts, columns, or walls.

BEARING PARTITION: A partition which supports a vertical load in addition to its own weight.

BEARING WALL: A wall which supports a vertical load in addition to its own weight.

BEDDING: A filling of mortar, putty, or other substance used to secure a firm bearing.

BED MOLDING: A molding applied where two surfaces come together at an angle. Commonly used in cornice trim especially between the plancier and frieze.

BENCH MARK: A mark on a permanent object fixed to the ground from which land measurements and elevations are taken.

BEVEL: To cut to an angle other than a right angle, such as the edge of a board or door.

BEVEL SIDING: Used as finish covering on the exterior of a structure. It is usually manufactured by ''resawing'' dry, square surfaced boards diagonally to produce two wedge-shaped pieces.

BID: An offer to supply, at a specified price: materials, supplies, and equipment, or the entire structure or sections of the structure.

BIRD'S-MOUTH: A notch cut on the underside of a rafter to fit it to the top plate. Not a full notch if rafter ends flush with top plate instead of overhanging.

BLEMISH: Any defect, scar, or mark that tends to detract from the appearance of wood.

BLIND NAILING: Refers to tongue-and-groove flooring. Nails are placed at the root of the tongue where they will be hidden. The nails pierce the subfloor at a 45 degree angle.

BLIND STOP: A member applied to the exterior edge of the side and head jamb of a window to serve as a stop for the top sash and to form a rabbet for storm sash, screens, blinds, and shutters.

BLUE STAIN: A stain caused by a fungus growth in unseasoned lumber—especially pine. It does not affect the strength of the wood.

BOARD: Lumber less than 2 in. thick.

BOARD FOOT: The equivalent of a board 1 ft. square and 1 in. thick.

BRACKET: A projecting support for a shelf or other structure.

BRICK CONSTRUCTION: A type of construction in which the exterior walls are bearing walls made of brick.

BRICK MOLDING: A molding for window and exterior door frames. Serves as the boundary molding for brick or other siding material and forms a rabbet for the screens and/or storm sash or combination door.

BRICK VENEER CONSTRUCTION: A type of construction in which a wood-frame construction has an exterior surface of single brick.

BRIDGING: Pieces fitted in pairs from the bottom of one floor joist to the top of adjacent joists, and crossed to distribute the floor load. Sometimes pieces of solid stock of a width equal to the joist are used.

BUILT-UP ROOF: A roofing composed of several layers of rag felt or jute saturated with coal tar, pitch, or asphalt. The top is finished with crushed slag or gravel. Generally used on flat or low-pitched roofs.

BUTT: Type of door hinge. One leaf is fitted into space routed into the door frame jamb and the other into the edge of the door.

CABINET: Case or box-like assembly consisting of shelves, doors, and drawers, used primarily for storage.

CABINET DRAWER GUIDE: A wood strip used to guide the drawer as it slides in and out of its opening.

CABINET DRAWER KICKER: Wood cabinet member placed immediately above and generally at the center of a drawer to prevent tilting down when pulled out.

CAMBER: A slight arch in a beam or other horizontal member which prevents it from bending into a downward or concave shape due to its weight or load it is to carry.

CANT STRIP: A triangular shaped strip of wood used under shingles at gable ends or under the edges of roofing on flat decks.

CASED OPENING: An interior opening without a door that is finished with jambs or trim.

CASEIN GLUE: An adhesive of casein and hydrated lime suitable for gluing oily woods and for laminating wood which has a high moisture content.

CASEMENT: A window in which the sash swings on its vertical edge, so it may be swung in or out.

CASING: The trimming around a door or window, either outside or inside, or the finished lumber around a post or beam.

CAULK: To seal and waterproof cracks and joints, especially around window and exterior door frames. (Also calk.)

CHAIR RAIL: An interior molding applied along the wall of a room to prevent the chair from marring the wall.

CHAMFER: Corner of a board beveled at a 45 deg. angle. Two boards butt-jointed and with chamfered edges form a V-joint.

CHASE: A wood frame jutting from an outside wall which supports a prefabricated chimney. A prefabricated fireplace is often enclosed.

CHECK RAILS: Meeting rails of a double-hung window which are made thicker to fill the opening between the top and bottom sash. They are usually beveled.

CLEAT: A strip of wood fastened across a door to add strength. Also a strip fastened to a wall to support a shelf, fixture, or other objects.

CLOSET POLE: A round molding installed in clothes closets to accommodate clothes hangers.

COLLAR BEAM: A tie beam connecting rafters considerably above the wall plate. It is also called a rafter tie.

COLUMN: Upright supporting member circular or rectangular in shape.

COMMERICAL STANDARD: A voluntary standard that establishes quality, methods of testing, certification, rating, and labeling of manufactured items. It provides a uniform base for fair competition.

CONDUCTORS: Pipes for conducting water from a roof to the ground or to a receptacle or drain; downspout.

CONDUIT, ELECTRICAL: A pipe or tube, usually metal, in which wiring is installed.

CONTACT CEMENT: Neoprene rubber-based adhesive

which bonds instantly upon contact of parts being fastened.

CONVENIENCE OUTLET: Electrical outlet into which may be plugged portable equipment such as lamps.

COPE: To cut or shape the end of a molded wood member so it will cover and fit the contour of an adjoining piece of molding.

CORBEL OUT: To extend outward from the surface of a masonry wall one or more courses to form a supporting ledge.

CORNER BEAD: Molding used to protect corners. Also a metal reinforcement placed on corners before plastering.

CORNER BRACES: Diagonal braces let into studs to reinforce corners of frame structures.

CORNICE: Exterior trim of a structure at the meeting of the roof and wall; usually consists of panels, boards, and moldings.

COUNTERFLASHING: Flashing used on chimneys at the roof-line to cover shingle flashing and prevent moisture entry.

COVE MOLDING: Molding with a concave profile used primarily where two members meet at a right angle.

CRIPPLE JACK (also, cripple rafter): A rafter which intersects neither the plate nor the ridge and is terminated at each end by hip and valley rafters.

CRIPPLE STUD: A stud used above a wall opening. Extends from the header above the opening to the top plate. Also used beneath a wall opening between sole plate and rough sill.

CUPOLA: Small vented four-sided structure installed on a roof. Adds decoration to the building and provides ventilation for the attic.

CURTAIN WALL: A wall, usually nonbearing, between piers or columns.

DADO: A rectangular groove cut in wood across the grain.

DEAD BOLT (also called a dead lock): Special door security consisting of a hardened steel bolt and a lock. Lock is operated by a key on the outside and by either a key or handle on the inside.

DEAD LOAD: The weight of permanent, stationary construction included in a building.

DECAY: Disintegration of wood substance due to action of wood destroying fungi.

DEGREE DAY: Method of measuring the harshness of climate for insulation and heating purposes. A degree day is the product of one day and the number of degrees the mean temperature is below 65 °F.

DIMENSION LUMBER: Lumber 2 to 5 in. thick, and up to 12 in. wide.

DIMENSIONAL STABILITY: The ability of a material to resist changes in its dimensions due to temperature, moisture, and physical stress.

DIRECT GAIN SYSTEM: Passive solar construction in which the sun shines directly into living space to heat it.

DOOR FRAME: An assembly of wood parts that form an enclosure and support for a door. Door frames are classified as exterior and interior.

DOOR STOP: A molding nailed to the faces of the door frame jambs to prevent the door from swinging through.

DORMER: A projecting structure built out from a sloping roof. Usually includes one or more windows.

DRIP CAP: A molding which directs water away from a structure to prevent seepage under the exterior facing material. Applied mainly over window and exterior door frames.

DRIP GROOVE: Semicircular groove on the underside of a drip cap or the lip of a window sill which prevents water from running back under the member.

DROP SIDING: Siding, usually 3/4 in. thick and machined into various patterns. Drop siding has tongue and groove or shiplap joints.

DRY ROT: A term loosely applied to many types of decay but especially to that which, when in an advanced stage, permits the wood to be easily crushed to a dry powder.

DRYWALL: Sheet materials used for wall covering which do not need to be mixed with water before application. See Gypsum Wallboard.

EASED EDGE: Corner slightly rounded or shaped to a slight radius.

EAVES: The lower part of a roof that projects over an exterior wall. Also called the overhang.

ELECTRIC MOISTURE METER: Meter used to determine the moisture content of wood. Action is based on electrical resistance or capacitance which varies with change in moisture content.

ELEVATION: The height of an object above grade level. Also means a type of drawing which shows the front, rear, and sides of a building.

EQUILIBRIUM MOISTURE CONTENT: The moisture content at which wood neither gains nor loses moisture when surrounded by air at a given relative humidity and temperature.

ESCUTCHEON: In builders' hardware, a protective plate or shield containing a key hole.

EXPANSION JOINT: A bituminous fiber strip used to separate blocks or units of concrete to prevent cracking due to dimensional change caused by shrinkage and variation in temperature.

FACADE: Main or front elevation of a building.

FACE NAIL: A nail driven perpendicular to the surface of a piece.

FACTORY AND SHOP LUMBER: Lumber intended to be cut up for use in further manufacture. It is graded on the basis of the percentage of the area which will produce a limited number of cuttings of a specified, or a given minimum, size, and quality.

FASCIA: A wood member used for the outer face of a box cornice where it is nailed to the ends of the rafters and lookouts.

FENESTRATION: The placement or arrangement and sizes of the windows and exterior doors of a building.

FIBER BOARD: A broad term used to describe sheet

material of widely varying densities; manufactured from wood, cane, or other vegetable fibers.

FIBER SATURATION POINT: The stage in the drying or wetting of wood at which the cell walls are saturated and the cell cavities are free from water. It is assumed to be 30 percent moisture content, based on oven dry weight and is the point below which shrinkage occurs.

FIRE STOP: A block or stop used in wall of building between studs to prevent the spread of fire and smoke through air space.

FIRE WALL: A wall which subdivides a building to restrict the spread of fire.

FLASHING: Sheet metal or other material used in roof and wall construction (especially around chimneys and vents) to prevent rain or other water from entering.

FLAT ROOF: A roof which is level, or which is pitched only enough to provide for drainage.

FLUE: The space or passage in a chimney through which smoke, gas, or fumes rise. Each passage is called a flue, which with the surrounding masonry, makes up the chimney.

FLUSH: Adjacent surfaces even, or in same plane (with reference to two structural pieces).

FOOTING: The spreading course or courses at the base or bottom of a foundation wall, pier, or column.

FOUNDATION: The supporting portion of a structure below the first-floor construction, or grade, including the footings.

FRAMING: The timber structure of a building which gives it shape and strength; including interior and exterior walls, floor, roof, and ceilings.

FRIEZE: A boxed cornice wood trim member attached to the structure where the soffit (plancier) and wall meet.

FURRING: Narrow strips of wood spaced to form a nailing base for another surface. Furring is used to level, to form an air space between the two surfaces, and to give a thicker appearance to the base surface.

GABLE: That portion of a wall contained between the slopes of a double-sloped roof or that portion contained between the slope of a single-sloped roof and a line projected horizontally through the lowest elevation of the roof construction.

GAIN: Notch or mortise cut to receive the end of another structural member or a hinge and other hardware.

GAMBREL ROOF: A roof slope formed as if the top of a gable (triangular) roof were cut off and replaced with a less steeply sloped cap. This cap still has a peaked ridge in the center.

GIRDER: A large or principal beam used to support concentrated loads at particular points along its length.

GLAZING: The process of installing glass into sash and doors. Also refers to glass panes inserted in various types of frames.

GLAZING COMPOUND: A plastic substance of such consistency that it tends to remain soft and rubbery when used in glazing sash and doors.

GLUE BLOCK: A wood block, triangular or rectangular in shape, which is glued into place to reinforce a right-

angle butt joint. Sometimes used at the intersection of the tread and riser in a stairs.

GRADE BEAM: Thickened and reinforced section of a slab foundation designed to rest on supporting piling.

GROUND: A strip of wood assisting the plasterer in making a straight wall and in giving a place to which the finish of the room may be nailed.

GROUND FAULT INTERRUPTER: An electrical safety device which can be installed either in an electrical circuit or at an outlet. It is able to detect a short circuit and shut off power automatically. Used as a protection against electrical shock.

GROUNDING: A system used for electrical safety. An electrical wire runs from the exposed metal of a power tool to a third prong on the power plug. When used with a grounded receptacle, this wire directs harmful currents away from the operator.

GROUT: A thin mortar used in masonry work.

GUSSET: A panel or bracket of either wood or metal attached to the corners or intersections of a frame to add strength and stiffness.

GUTTER: Wood or metal trough attached to the edge of a roof to collect and conduct water from rain or melting snow.

GYPSUM WALLBOARD: Wall covering panels consisting of a gypsum core with facing and backing of paper.

HALF STORY: That part of a building situated wholly or partly within the roof frame, finished for occupancy.

HARDBOARD: A board material manufactured of wood fiber, formed into a panel having a density range of approximately 50 to 80 lb. per cu. ft.

HEADER: Horizontal structural member that supports the load over an opening, such as a window or door. Also called a lintel.

HEADROOM: The clear space between floor line and ceiling, as in a stairway.

HEARTWOOD: The wood extending from the pith or center of the tree to the sapwood, the cells of which no longer participate in the life processes of the tree.

HEAT TRANSMISSION COEFFICIENT: Hourly rate of heat transfer for one square foot of surface when there is a temperature difference of one deg. F. of the air on the two sides of the surface.

HIP ROOF: A roof which rises from all four sides of a building.

HOLLOW-BACK: Removal of a portion of the wood on the unexposed face of a wood member to more properly fit any irregularity in bearing surface.

HOLLOW CORE DOOR: Flush door with a core assembly of strips or other units which support the outer faces.

HORN: The extension of a stile, jamb, or sill.

HOSE BIB: A water faucet that is threaded so a hose connection can be attached; a sill cock.

I BEAM: A steel beam with a cross section that resembles the letter I.

INCINERATOR: A device which consumes household waste by burning.

INDIRECT GAIN SYSTEM: Passive solar construction in

which solar heat is stored in structures of masonry, water, or other medium and then passed along to living space by radiation, conduction, or convection.

INSULATION: (Thermal) Any material high in resistance to heat transmission that is placed in structures to reduce the rate of heat flow.

INTERIOR TRIM: General term for all the molding, casing, baseboard, and other trim items applied within the building by finish carpenters.

IN-THE-WHITE: Natural or unpainted; the natural unfinished surface of the wood.

ISOLATED GAIN SYSTEM: Passive solar construction in which solar generated heat is stored in a separate sun space. It is transported to living space by mechanical means.

JACK RAFTER: A short rafter framing between the wall plate and a hip rafter; or a hip or valley rafter and ridge board.

JALOUSIE: A series of small horizontal overlapping glass slats, held together by an end metal frame attached to the faces of window frame side jambs or door stiles and rails. The slats or louvers move simultaneously like a Venetian blind.

JAMB: The top and two sides of a door or window frame which contact the door or sash; top jamb and side jambs.

JIG: A device used to position material for accurate cutting or assembly. Most often used in factories making prefabricated building components.

JOINERY: A term used by woodworkers when referring to the various types of joints used in a structure.

JOIST: One of a series of parallel framing members used to support floor and ceiling loads, and supported in turn by larger beams, girders, or bearing walls.

KERFING: Longitudinal saw cuts or grooves of varying depths (dependent on the thickness of the wood member) made on the unexposed faces of millwork members to relieve stress and prevent warping; members are also kerfed to facilitate bending.

KILN DRIED: Wood seasoned in a kiln by means of artificial heat, controlled humidity, and air circulation.

KNOCKED DOWN: Unassembled; refers to structural units requiring assembly after being delivered to the job.

KNOT: Branch or limb embedded in the tree and cut through during lumber manufacture.

KRAFT PAPER: A brown building paper which resists puncturing. Kraft paper is used to face some blanket insulation materials.

LALLY COLUMN: A cylindrically shaped steel member used to support beams and girders. Sometimes filled with concrete.

LATH: A building material of wood, metal, gypsum, or insulating board, fastened to frame of building to act as a plaster base.

LAZY SUSAN: A circular revolving cabinet shelf used in corner kitchen cabinet unit.

LEADER: A vertical pipe that carries rainwater from the gutter to the ground or a drain. Also downspout.

LEDGER: A strip attached to vertical framing or structural members to support joists or other horizontal framing. Similar to a ribbon strip.

LET IN: Refers to any kind of notch in a stud, joist, block, or other piece which holds another piece. Somewhat like log cabin construction. The item which is supported in the notch is said to be "let in."

LEVEL-TRANSIT: A surveying instrument used to check the plumb of walls in new structures. The telescope tube can swing in a vertical arc for comparing forward and backward readings.

LIGHT CONSTRUCTION: Construction generally restricted to conventional wood stud walls, floor and ceiling joists, and rafters. Primarily residential in nature although it does include small commercial buildings.

LINEAL FOOT: Having length only, pertaining to a line one foot long as distinguished from a square foot or cubic foot.

LINTEL: A horizontal structural member which supports the load over an opening such as a door or window.

LIVE LOAD: The total of all moving and variable loads that may be placed upon a building.

LOCK BLOCK: A block of wood which is joined to the inside edge of the stile of a hollow core door and to which the lock is fitted. Flush doors have a lock block on each stile.

LOOKOUT: Structural member running between the lower end of a rafter and the outside wall. Used to carry the underside of the overhang; plancier or soffit.

LUMBER: The product of the saw and planing mill not further manufactured than by sawing, resawing, passing lengthwise through a standard planing machine, and crosscutting to length. Some matching of ends and edges may be included.

MAJOR MODULE: A unit of measure for modular construction. In the conventional system of units, 48 in. is the length of a major module. In the SI metric system, a major module is 1200 mm long.

MANSARD ROOF: A type of curb roof in which the pitch of the upper portion of a sloping side is slight and that of the lower portion steep. The lower portion is usually interrupted by dormer windows.

MASONRY: Stone, brick, hollow tile, concrete block, or tile, and sometimes poured concrete and gypsum block, or other similar materials, or a combination of same, bonded together with mortar to form a wall, pier, buttress, etc.

MATCHED LUMBER: Lumber that is edge dressed and shaped to make a close tongue-and-groove joint at the edges or ends. Also generally includes lumber with rabbeted edges.

MECHANICAL CORE: A prefabricated building module which contains one or more of the following utilities: electrical, plumbing, heating, ventilating, and air conditioning. Floor, ceiling, and wall framing are fully formed at the factory. Modules are joined at the building site.

MECHANICAL EQUIPMENT: In architectural and engineering practice: All equipment included under the general heading of plumbing, heating, air conditioning, gasfitting, and electrical work.

MEDALLION: A raised decorative piece, sometimes used on flush doors.

MEETING RAIL: The bottom rail of the upper sash, and the top rail of the lower sash of a double-hung window. Also called a check rail.

MILLWORK: The term used to describe products which are primarily manufactured from lumber in a planing mill or woodworking plant; including moldings, door frames and entrances, blinds and shutters, sash and window units, doors, stairwork, kitchen cabinets, mantels, cabinets, and porch work.

MINOR MODULE: A unit of measure for modular construction. In the conventional system of units, 24 in. is the length of a minor module. In the SI metric system, a minor module is 600 mm long.

MODULAR COORDINATION: The dimensioning of a structure and use of building materials based on a common unit of measurement called a module.

MOISTURE CONTENT: The amount of water contained in wood expressed as a percentage of the weight of oven-dry wood.

MOLDING: A relatively narrow strip of wood, usually shaped to a curved profile throughout its length. Used to accent and emphasize the ornamentation of a structure and to conceal surface or angle joints.

MONOLITHIC: Term used for concrete construction poured and cast in one unit without joints.

MOULDER: A woodworking machine designed to run moldings and other wood members with regular or irregular profiles. Also called a sticker.

MR (moisture resistant) WALLBOARD: A type of gypsum wallboard processed to resist the effects of moisture and high humidity. It is used as a base under ceramic tile and other nonabsorbent finishes used in showers and tub alcoves.

MULLION: A slender bar or pier forming a division between units of windows, screens, or similar frames generally nonstructural.

MUNTIN: Vertical member between two panels of the same piece of panel work. The vertical and horizontal sashbars separating the different panes of glass in a window.

NET FLOOR AREA: The gross floor area, less the area of the partitions, columns, and stairs and other floor openings.

NEWEL: The main post at the start of the stairs and the stiffening post at the landing; a stair newel.

NOMINAL SIZE: As applied to timber or lumber, the ordinary commercial size by which it is known and sold.

NONBEARING PARTITION: A partition extending from floor to ceiling which supports no load other than its own weight.

NOSING: The part of a stair tread which projects over the riser, or any similar projection; a term applied to the rounded edge of a board.

ON CENTER: A method of indicating the spacing of framing members by stating the measurement from the center of one member to the center of the succeeding one.

OPEN-GRAIN WOOD: Woods with large pores, such as oak, ash, chestnut, and walnut.

ORIEL WINDOW: A window that projects from the main line of an enclosing wall of a building and is carried on brackets, corbels, or a cantilever.

ORIENTED STRAND BOARD: A formed panel consisting of layers of compressed strand-like particles arranged at right angles to each other.

PARAPET: A low wall or railing along the edge of a roof, balcony, or bridge. The part of a wall that extends above the roof line.

PARGETING: Thin coat of plaster applied to stone or brick to form a smooth or decorative surface.

PARTICLEBOARD: A formed panel consisting of particles of wood flakes, shavings, slivers, etc., bonded together with a synthetic resin or other added binder.

PARTITION: A wall that subdivides space within any story of a building.

PARTY WALL: A wall used jointly by two parties under easement agreement and erected at or upon a line separating two parcels of land that may be held under different ownership.

PASSIVE SOLAR CONSTRUCTION: Structures of glass, wood, and masonry which collect, transport, and store heat from the energy of the sun.

PENNY: Term used to indicate nail length; abbreviated by the letter d. Applies to common, box, casing, and finishing nails.

PHASE-CHANGE MATERIALS: Material which changes from solid to liquid to gaseous state and which can store a great amount of heat energy.

PHOTOSYNTHESIS: The process plants use to store the sun's energy. Trees use sunlight to convert carbon dioxide and water into leaves and wood.

PIER: A column of masonry, usually rectangular in horizontal cross section. Used to support other structural members.

PILASTER: A part of a wall that projects not more than one-half of its own width beyond the outside or inside face of a wall. Chief purpose is to add strength but may also be decorative.

PILE: A heavy timber, or pillar of metal or concrete, forced into the earth or cast in place to form a foundation member.

PITCH: Inclination or slope, as of roofs or stairs. Rise divided by the span.

PLAN: A drawing representing any one of the floors or horizontal cross sections of a building, or the horizontal plane of any other object or area.

PLANCIER: The underside of an eave or cornice, usually horizontal.

PLASTER: A mixture of lime, cement, and sand, used to cover outside and inside wall surfaces.

PLAT: A map, plan, or chart of a city, town, section, or

subdivision indicating the location and boundaries of individual properties.

PLATE: A horizontal structural member placed on a wall or supported on posts, studs, or corbels to carry the trusses of a roof or to carry the rafters directly. Also a sole or base member of a partition or other frame.

PLATFORM FRAMING: A system of framing a building where the floor joists of each story rest on the top plates of the story below (or on the foundation wall for the first story) and the bearing walls and partitions rest on the subfloor of each story.

PLOT PLAN: A view from above a building site. The plan shows distances from a structure to property lines. Sometimes called a site plan.

PLUMB: Exactly perpendicular or vertical; at right angles to the horizon or floor.

PLUMBING: The work or business of installing pipes, fixtures, and other apparatus for bringing in the water supply and removing liquid and water-borne wastes. This term is used also to denote the installed fixtures and piping of a building.

PLUMBING STACK: A general term for the vertical main of a system of soil, waste, or vent piping.

PLUNGE CUTTING: A cutting method used to make a starting hole for a saber saw. The saw is held with the blade teeth almost flush with the wood surface. A cut is made completely through as the saw is tilted to a normal position.

POLYSTYRENE PANELS: Rigid insulation manufactured from expanded beads of plastic.

POLYVINYL RESIN EMULSION GLUE (also, white glue): Wood adhesive intended for interiors. Made from polyvinyl acetates which are thermoplastic and not suited for temperatures over 165 °F.

PORTICO: A porch or covered walk consisting of a roof supported by columns. A porch with a continuous row of columns.

PREFABRICATED CONSTRUCTION: Type of construction so designed as to involve a minimum of assembly at the site, usually comprising a series of large units manufactured in a plant.

PRESERVATIVE: Substance that will prevent the development and action of wood-destroying fungi, borers of various kinds, and other harmful insects that deteriorate wood.

PURLINS: Horizontal roof members used to support rafters between the plate and ridge board.

PUSH STICK: A pole or strip used to push a workpiece when cutting with power saws, jointers, and other power tools. Pushing a board by hand is usually unsafe with power equipment.

QUARTER ROUND: Small molding with a cross section of one-fourth of a cylinder.

QUARTER-SAWED: Lumber that has been cut at about a 90 deg. angle to the annular growth rings.

QUIT-CLAIM DEED: A deed whereby the grantor conveys without warranty, to the grantee, whatever interest he or she possesses in the property.

RABBET: A rectangular shape consisting of two surfaces cut along the edge or end of a board.

RAFTER: One of a series of structural members of a roof designed to support roof loads. The rafters of a flat roof are sometimes called roof joists.

RAIL: Cross or horizontal members of the framework of a sash, door, blind, or other assembly.

RAKE: The trim members that run parallel to the roof slope and form the finish between the roof and wall at a gable end.

RAMP: Inclined plane connecting separate levels.

REINFORCED CONCRETE CONSTRUCTION: A type of construction in which the principal structural members, such as floors, columns, and beams, are made of concrete poured around steel bars or steel meshwork in such a manner that the two materials act together to resist force.

RELATIVE HUMIDITY: Ratio of amount of water vapor in air in terms of percentage to total amount it could hold at the same temperature.

RESILIENT: The ability of a material to withstand temporary deformation, the original shape being assumed when the stresses are removed.

RETAINING WALL: Any wall subjected to lateral pressure other than wind pressures. Example: a wall built to support a bank of earth.

RIBBON: A narrow board attached to studding or other vertical members of a frame that adds support to joists or other horizontal members.

RISER: The vertical stair member between two consecutive stair treads.

ROLL ROOFING: Consists of mineral granules on asphalt saturated felt or fiberglass. Roll roofing is the uncut form of mineral surfaced shingle material.

ROOFING: The materials applied to the structural parts of a roof to make it waterproof.

ROOF RIDGE: The horizontal line at the junction of the top edges of two roof surfaces where an external angle greater than 180 deg. is formed.

ROTARY CUT VENEER: Veneer cut on a lathe which rotates the log against a broad cutting knife. The veneer is cut in a continuous sheet much the same as paper is unwound from a roll.

ROUGHING-IN: The work of installing all pipes in the drainage system and all water pipes to the point where connections are made with the plumbing fixtures. Also applies to partially completed electrical wiring and other mechanical aspects of the structure.

ROUGH LUMBER: Lumber that has been cut to rough size with saws but which has not been dressed or surfaced.

ROUGH OPENING: An opening formed by the framing members.

R VALUE: A number which specifies the efficiency of an insulating material like fiberglass batting, foam, or other similar material. A detailed definition is given in Unit 13.

SADDLE: A small gable type roof placed in back of a chimney on a sloping roof to shed water and debris.

SAPWOOD: The layers of wood next to the bark, usually

lighter in color than the heartwood, that are actively involved in the life processes of the tree. More susceptible to decay than heartwood. Sapwood is not essentially weaker or stronger than heartwood of the same species.

SASH: The framework which holds the glass in a window.

SCAFFOLD: A temporary structure or platform used to support workers and materials during building construction.

SCALE: A term which specifies the size of a reduced size drawing. For example, a plan is drawn to 1/4'' scale if every 1/4'' represents 1' on the real structure.

SCANTLING: Lumber with a cross section ranging from 2 in. x 4 in. to 4 in. x 4 in.

SCARFING: Joining the ends of stock together with a sloping lap-joint so they appear to be a single piece.

SCOTIA: A concave molding consisting of an irregular curve. Used under the nosing of stair treads and for cornice trim.

SCREED: A tool used in concrete work to level and smooth a horizontal surface. Consists of a 3 to 5 ft. wood or metal strip attached to a pole.

SCUTTLE: An opening in a ceiling which provides access to the attic.

SEASONING: Removing moisture from green wood to improve its serviceability.

SECOND GROWTH: Timber that has grown after the removal of a large portion of the previous stand.

SECTION DRAWING: A type of drawing which shows how a part of a structure looks when cut by a vertical plane. The face remaining after a real cut would look like the drawing.

SELVAGE: The part of the width of roll roofing which is smooth. For example, a 36 in. width has a granular surfaced area 17 in. wide and a 19 in. wide selvage area.

SEPTIC TANK: A sewage-settling tank intended to retain the sludge in immediate contact with the sewage flowing through the tank, for a sufficient period to secure satisfactory decomposition of organic sludge solids by bacterial action.

SETTING BLOCK: A wood block placed in the glass groove or rabbet of the bottom rail of an insulating glass sash to form a base or bed for the glass.

SHAKES: Handsplit shingles.

SHEATHING: The structural covering. Consists of boards or prefabricated panels that are attached to the exterior studding or rafters of a structure.

SHEATHING PAPER: A building material used in wall, floor, and roof construction to resist the passage of air.

SHIM: A thin strip of wood sometimes wedgeshaped, for plumbing or leveling wood members. Especially helpful when setting door and window frames.

SHIPLAP: Lumber with edges that have been rabbeted to form a lap joint between adjacent pieces.

SHORING: Lumber and timbers used to prevent the sliding of earth adjoining an excavation. Also the timbers used as temporary bracing or support.

SHUTTER: A wood assembly of stiles and rails to form a frame which encloses panels used in conjunction with door and window frames. Also may consist of vertical boards cleated together.

SIDE OF TRIM: Trim required to finish one side of a door or window opening.

SIDING: The finish covering of the outside wall of a frame building. Many different types are available.

SILL: The lowest member of the frame of a structure, usually horizontal, resting on the foundation and supporting the uprights of the frame. Also the lowest member of a window or outside door frame.

SKYLIGHT: Glazing framed into a roof.

SLEEPER: A timber laid on or near the ground to support floor joists and other structures above. Also wood strips laid over or embedded in a concrete floor to which finish flooring is attached.

SOFFIT: The underside of the members of a building, such as staircases, cornices, beams, and arches. Relatively minor in size as compared with ceilings. Also called drop ceiling and furred-down ceiling.

SOFTWOODS: The botanical group of trees that have needle or scalelike leaves and are evergreen for the most part, cypress, larch, and tamarack being exceptions. The term has no reference to the actual hardness of the wood. Softwoods are often referred to as conifers, and botanically they are called gymnosperms.

SOIL STACK: A general term for the vertical main of a system of soil, waste, or vent piping.

SOLAR FURNACE: Another name for a thermosiphon.

SOLAR ORIENTATION: Placement of a structure on a building site to get the most benefit of sunlight.

SOLE PLATE: The lowest horizontal strip on wall and partition framing. The sole plate for a partition is supported by a wood subfloor, concrete slab, or other closed surface.

SPAN: The distance between structural supports such as walls, columns, piers, beams, girders, and trusses.

SPECIFICATION: A written document stipulating the kind, quality, and, sometimes, the quantity of materials and work quality required for a construction job.

SPECIFIC GRAVITY: The ratio of the weight of a body to the weight of an equal volume of water at some standard temperature.

SPLASH BLOCK: A small masonry block laid with the top close to the ground surface to receive roof drainage and carry it away from the building.

SPLINE: A small strip of wood that fits into a groove or slot of both members to form a joint.

SQUARE: Unit of measure—100 square feet—usually applied to roofing material and to some types of siding.

STAIRWAY, STAIR, OR STAIRS: A series of steps, with or without landings, or platforms, usually between two or more floors of a building.

STAIRWELL: The framed opening which receives the stairs.

STATION MARK: The point where a level-transit is

located. It is a reference point like a stake or paint-mark directly below the center of the instrument.

STEEL-FRAME CONSTRUCTION: A type of construction in which the structural parts are of steel or are dependent on a steel frame for support.

STEPPED FOOTING: A footing that changes grade levels at intervals to accommodate a sloping site.

STICKERS: Strips of wood used to separate the layers in a pile of lumber so air can circulate.

STILE: The upright or vertical outside pieces of a sash, door, blind, or screen.

STOOL: A molded interior trim member serving as a sash or window frame sill cap. Stools may be beveled-rabbeted or rabbeted to receive the window frame sill.

STOOP: A small porch, veranda, or platform, or a stairway, outside an entrance to a building.

STORY: A single floor of any structure.

STORY POLE: A strip of wood used to lay out and transfer measurements for door and window openings, siding and shingle courses, and stairways. Also called a rod.

STRAIGHTEDGE: A straight strip of wood or metal used to lay out or check the accuracy of work.

STRESSED SKIN: Two facings, one glued to one side and the other to the opposite side of an inner structural framework to form a panel. Facings may be of plywood or other suitable material.

STRIKE PLATE: A metal piece mortised into or fastened to the face of a door frame side jamb to receive the latch or bolt when the door is closed.

STRONGBACK: L-shaped wooden support attached to tops of ceiling joists to strengthen them, maintain spacing, and bring them to same level.

STRUCTURAL WINDOW WALL PANEL: A window unit framed into a wall panel at the factory. Also called a factory-assembled structural wall panel.

STUD: One of a series of vertical wood or metal structural members in walls and partitions. Plural—studs or studding.

SUBFLOOR: Boards or panels laid directly on floor joists over which a finished floor will be laid.

SURFACED LUMBER: Lumber that is dressed or finished by running it through a planer.

TAIL BEAM: A relatively short beam or joist supported by a wall on one end and by a header on the other.

TERMITE SHIELD: A shield, usually made of sheet metal, placed in or on a foundation wall or around pipes to keep termites out of the structure.

TERRAZZO FLOORING: A floor produced by embedding small chips of marble or colored stone in concrete and then grinding and polishing the surface.

THERMOSIPHON: A solar collector consisting, in its simplest form, of a flat box. It is glazed on one side. A baffle, parallel to the glazing, divides the box in half. As sun heats the upper half of the box, the heated air moves out one end. Cool air moves in from the lower half of the box. The box will be vented either into living space or into ductwork leading to storage.

THERMOSTAT: An instrument that controls automatically the opperation of heating or cooling devices by responding to changes in temperature.

THREE-WAY SWITCH: A switch designed to operate in conjunction with a similar switch to control one outlet from two points.

THRESHOLD: A wood member, beveled or tapered on each side, used to close the space between the bottom of a door and the sill or floor underneath. Sometimes called a saddle.

TIE BEAM (collar beam): A beam so situated that it ties the principal rafters of a roof together and prevents them from thrusting the plate out of line.

TIMBERS: Lumber 5 inches or larger in least dimension.

TOEBOARD: A board fastened horizontally slightly above planking to keep tools and materials from falling on workers below. Can be used on scaffolds or at an access hole. Board must be at least 4 in. wide. Not needed for scaffolds under 10 ft.

TOENAILING: To drive a nail at a slant with the initial surface in order to permit it to penetrate into a second member.

TOE SPACE: A recessed space at the floor line of a base kitchen cabinet or other built-in units. Permits one to stand close without striking the vertical space with the toes.

TRANSFORMER: A device for transforming the voltage characteristics of an electric current.

TRANSOM: A small opening above a door separated by a horizontal member (transom bar). Usually contains a sash or a louver panel hinged to the transom bar.

TRAP: A plumbing fitting or device designed to provide a liquid trap seal which will prevent the sewer gases from passing through and entering a building.

TREAD: The horizontal part of a step on which the foot is placed.

TRIM: The finish materials in a building, such as moldings applied around openings (window trim, door trim) or at the floor and ceiling of rooms (baseboard, cornice, picture molding).

TRIMMER: The beam or floor joist into which a header is framed. Adds strength to the side of the opening.

TRIMMER STUD: A stud which supports the header for a wall opening. The stud extends from the sole plate to the bottom of the header. It is parallel to and in contact with a full length stud that extends from sole plate to top plate.

TRIPLE WALL: A type of chimney flue made with three metal pipes, each inside another. The concentric arrangement provides safety from fire while its light weight makes installation easy.

TRIPOD: A support for a builder's level consisting of three sloping legs. Friction of the legs with the ground keeps them in place.

TROMBE WALL: Thick wall of masonry placed next to exterior glazing to store solar energy in passive solar construction. Named for a French physicist, Felix Trombe.

TRUSS: A structural unit consisting of such members as beams, bars, and ties; usually arranged to form triangles. Provides rigid support over wide spans with a minimum amount of material.

UNICOM SYSTEM: A term taken from "uniform manufacture of components." The system uses modules with sizes that are multiples of a standard size. A complete definition is given in Unit 5.

UNPROTECTED-METAL CONSTRUCTION: A type of construction in which the structural parts are of metal unprotected by fireproofing.

UREA-FORMALDEHYDE RESIN GLUE: Moisture resistant glue which hardens through chemical action when water is added to the powdered resin.

VALLEY: The internal angle formed by the two slopes of a roof.

VALLEY RAFTER: A rafter which forms the intersection of an internal roof angle.

VAPOR BARRIER: A watertight material used to prevent the passage of moisture or water vapor into or through structural elements (floors, walls, ceilings).

VENEERED WALL: A frame building wall with a masonry facing (example—single brick). A veneered wall is nonloadbearing.

VENEER PLASTER: Interior wall covering consisting of a gypsum lath base and surfacing of 1/8 in. gypsum plaster.

VENT: A pipe installed to provide a flow of air to or from a drainage system or to provide a circulation of air within such system to protect trap seals from siphonage and back pressure.

VENTILATION: The process of supplying and removing air by natural or mechanical means. Such air may or may not have been conditioned.

VERMICULITE: Mineral closely related to mica, with the faculty of expanding on heating to form lightweight material with insulating qualities. Used as bulk insulation, also as aggregate in insulating and acoustical plaster, and in insulating concrete floors.

WAFERBOARD (also waferwood): Construction panels made up of long, thin chips of wood. Wafers are coated with a waterproof resin and wax and then bonded with heat and pressure.

WAINSCOT: A lower interior wall surface (usually 3 to 4 ft. above the floor) that contrasts with the wall surface above. May consist of solid wood or plywood.

WALE: A horizontal wood or metal strip used on the outside of forms for concrete. Wales are used to keep the form walls from bending outward under the weight of poured concrete.

WALLBOARD: Wood pulp, gypsum, or other materials made into large rigid sheets that may be fastened to the frame of a building to provide a surface finish.

WALL TIE: Metal strip or wire used to bind tiers of masonry in cavity wall construction, or to bind brick veneer to a wood frame wall.

WARP: Any variation from a true or plane surface. Warp includes bow, crook, cup, and twist, or any combination thereof.

WATER REPELLENT: A solution, primarily paraffin wax and resin in mineral spirits, which upon penetrating wood retards changes in its moisture content.

WATER TABLE: A ledge or slight projection at the bottom of a structure which carries the water away from the foundation.

WEATHERING: The mechanical or chemical disintegration and discoloration of the surface of wood. It can be caused by exposure to light, the action of dust and sand carried by winds, and the alternate shrinking and swelling of the surface fibers that come with the continual variation in moisture content brought by changes in the weather. Weathering does not include decay.

WEATHERSTRIP: Narrow strips of metal, vinyl plastic, or other material, so designed that when installed at doors or windows, they will retard the passage of air, water, moisture, or dust around the door or window sash.

WEEPHOLE: A small hole, as in a retaining wall, to drain water to the outside. Commonly used at the lower edges of masonry cavity walls.

WET WALL: An interior wall finish surface usually consisting of 3/8 in. gypsum plaster lath and 1/2 in. gypsum plaster applied to the lath surface.

WHALER (also waler): A horizontal member used in concrete form construction to stiffen and support the walls of the form.

WIND ("i" pronounced as in kind): A term used to describe the warp in a board when twisted (winding). It will rest upon two diagonally opposite corners, if laid upon a perfectly flat surface.

WINDOW UNIT: Consists of a combination of the frame, window, weatherstripping, and sash activation device. May also include screens and/or storm sash. All parts are assembled as a complete operating unit.

WING: A lateral extension of a building from the main portion; one of two or more coordinate portions of a building which extend from a common junction.

WIRE GLASS: Glass having a layer of meshed wire incorporated approximately in the center of the sheet.

WP SERIES MOLDING PATTERNS: A recent molding series of more than 500 profiles which was a joint venture of the former West Coast Lumbermen's Association and the Western Pine Association now merged as Western Wood Products Association.

INDEX

Index